Convergence Technologies for 3G Networks

Convergence Technologies for 3G Networks

IP, UMTS, EGPRS and ATM

Jeffrey Bannister, Paul Mather and Sebastian Coope

at Orbitage Consultants

John Wiley & Sons, Ltd

Other Wiley Editorial Offices

John Wiley & Sons Inc., 111 River Street, Hoboken, NJ 07030, USA

Jossey-Bass, 989 Market Street, San Francisco, CA 94103-1741, USA

Wiley-VCH Verlag GmbH, Boschstr. 12, D-69469 Weinheim, Germany

John Wiley & Sons Australia Ltd, 33 Park Road, Milton, Queensland 4064, Australia

John Wiley & Sons (Asia) Pte Ltd, 2 Clementi Loop #02-01, Jin Xing Distripark, Singapore 129809

John Wiley & Sons Canada Ltd, 22 Worcester Road, Etobicoke, Ontario, Canada M9W 1L1

Wiley also publishes its books in a variety of electronic formats. Some content that appears in print may not be available in electronic books.

British Library Cataloguing in Publication Data

A catalogue record for this book is available from the British Library

ISBN 0-470-86091-X

Typeset in 10/12pt Times by Laserwords Private Limited, Chennai, India
Printed and bound in Great Britain by TJ International, Padstow, Cornwall
This book is printed on acid-free paper responsibly manufactured from sustainable forestry in which at least two trees are planted for each one used for paper production.

Contents

About the Authors

The three authors form part of the senior management team of Orbitage, a high-technology consultancy firm offering specialised expertise in many aspects of the telecommunication and information technology fields. Originally founded in 1998 and based in Kuala Lumpur, Malaysia, it has expanded to primarily cover the Asia-Pacific region, and has regional offices in Cyberjaya and Petaling Jaya, Malaysia, Singapore and Hong Kong. Orbitage is numbered among those companies to be awarded the prestigious Malaysian MSC status. In addition, Orbitage has a development team in the UK, and representatives in Finland and Ireland, as well as a regional office in Europe. Orbitage is also a distributor of NetHawk 2G/3G analysis tools.

Orbitage is highly regarded in Asia, and has provided consultancy and training services to a number of major organizations, including Nokia, Ericsson, Motorola, Singapore Telecom, Mobile One Singapore, StarHub Singapore, Telekom Malaysia, Maxis Malaysia, Celcom Malaysia, Telstra Australia, KGT Taiwan, TCC Taiwan, AIS and DTAC in Thailand, Vodaphone Ireland, and NetHawk Finland. Orbitage has been providing services to Nokia in Asia for a number of years as well as projects in China and Europe. Orbitage specialises in cross-training of professionals between the IT and telecommunications fields to enable them to become proficient Convergence Engineers.

Further information can be found at www.orbitage.com or by email to convergence@orbitage.com

Dr. Jeffrey Bannister   B.A., B.A.I., Ph.D., THEC Cert., MIEE, C.Eng.

Jeffrey is a co-founder and Telecommunications Specialist at Orbitage. A native of Ireland, he received his Ph.D. in Telecommunications/High-speed electronics from Trinity College in Dublin. He has over 15 years of experience, and holds an internationally recognized teaching qualification. Jeffrey has also been a lecturer, research fellow and course developer with the Dublin Institute of Technology, Temasek Polytechnic, Singapore, and

Trinity College in Dublin, as well as providing consultation to a number of companies in Europe and Asia. He has been living in Malaysia for the past 5 years.

Mr. Paul Mather   *B.Eng.(Hons), M.Sc.BA&IT, MasterCNE, MIEE, Cert. Ed., C.Eng.*

Paul is a co-founder of Orbitage and has been located in the ASEAN region for the last seven years, during which time he has been involved in course development, training and consultancy for a number of companies. Prior to his relocation from Blackpool, UK, he worked for a British college, where he was engaged as both a lecturer in Information Engineering and as the computer network manager. As a certified internal verifier of vocational qualifications, he has comprehensive experience in delivery, assessment and development of a variety of IT and Communication programs. He is credited with establishing the first Novell Educational Academic Partnership in the ASEAN region. In an industrial context, he has worked in the IT and Communications fields for over 18 years, this work has taken him to many countries as well as various oil and gas platforms in the North Sea.

Mr. Sebastian Coope   *B.Sc., M.Sc., THEC Cert.*

Sebastian is an IP/Software Specialist at Orbitage. From a small village called Bollington near Manchester originally, he received his Masters in Data Communications and Networking from Leeds Metropolitan University. He has worked in a wide range of roles as software engineer development and project manager, as well as consultant in the fields of network security and management. He has also worked as lecturer and consultant at both Temasek Polytechnic Singapore and the University of Staffordshire. At Orbitage he has led the team responsible for the development of mobile application products. He is also co-author of Computer Systems (Coope, Cowley and Willis), a university text on computer architecture.

DEDICATION

To my family, Canjoe, Siobhán, Avril and Norman, for their love, support and encouragement-J.B.

To Vivian, my wife and best friend for her unbridled love and support and also to my dad for ensuring I was suitably equipped for this journey-P.M.

I would like to dedicate my contribution in this book to Carole and little Al-S.C.

ACKNOWLEDGEMENTS

The authors would like to thank the following individuals for their assistance, contributions and support with this book. We would like to thank our colleagues at Orbitage, Roger Percival, Ruairi Hann, Siva Nadarajah, Annie Ling, Karen Wong, Vivian Koh, John

Ting and in particular to Tuan Ismail bin Tuan Mohamed for his continued support and encouragement. At NetHawk, we would like to thank Hannu Loponen, Ari Niskanen and Wong YeHim.

We would particularly like to extend our thanks to Sally Mortimore and Birgit Gruber at John Wiley for their advice, support and encouragement in overseeing this project to completion.

We would also like to take this opportunity to thank Kim Johnston, Reimund Nienaber, Paolo Zanier, Jarkko Lohtaja, Dawn Ho, Pearly Ong, Lee Wing Kai, Kamaliah Aza-hari, Idahayati Md. Dahari, Lewis Lourdesamy, Neela Tharmakulasingam, Adzahar Md. Sharipin, Jennifer Huang and Mark Deegan. Jeffrey would further like to acknowledge Dr. Brian Foley of Trinity College in Dublin for nurturing in him the thirst for, and skills of, lifelong learning and critical analysis.

1

Introduction

1.1 BACKGROUND TO CONVERGENCE

The telecommunications industry, and particularly the cellular industry, is currently going through a state of enormous transition. Many of the major cellular operators are now deploying a network to support packet switched data services and lead them to third generation (3G). This step to 3G involves a major change in the network infrastructure with the introduction of complex technologies such as asynchronous transfer mode (ATM), code division multiple access (CDMA) and the Internet protocol (IP). For forward-looking operators, this transition also requires a clear, strategic transformation of their business model to grasp and maximize on the benefits of the next generation's lucrative revenue streams. An operator requires both highly motivated staff with a substantial skill set as well as comprehensive, dynamic information systems. Also crucial is a clear understanding of the role the operator will play in this new model on the continuum from mere provision of a bit-pipe, to an organization offering full Internet service provider (ISP) capabilities and value-added services. This revised business model needs to incorporate integrated solutions for charging and billing, and provide a clear understanding of the new revenue streams available. Smooth convergence of network and telecommunications technologies and a proactive business strategy are pivotal to the success of the future mobile operator.

Many telecoms engineers have little experience in the new packet and IP technologies. To remain competitive it is essential that they learn the new packet switched skills quickly. The circuit switched skills will be required for a long time as circuit switching is not expected to disappear overnight and will probably be around for decades. However, new network components for telecoms networks will be based around packet switched technology.

Second generation cellular systems have been implemented commercially since the late 1980s. Since then, the systems have evolved dramatically in both size and reliability to achieve the level of quality subscribers expect of current networks. Mobile

Convergence Technologies for 3G Networks: IP, UMTS, EGPRS and ATM   J. Bannister, P. Mather and S. Coope
© 2004 John Wiley & Sons, Ltd   ISBN: 0-470-86091-X

network operators have invested heavily in the technology and the infrastructure, and it is unreasonable to expect this to be simply discarded when a new 3G system is proposed.

As a term, *convergence* has been coined by both the telecoms and datacoms industries. From a telecoms perspective, it is the expansion of the public switched telephone network (PSTN) to offer many services on the one network infrastructure. For Internet advocates, it is the death of the PSTN as its role is largely replaced by technologies such as voice over IP (VOIP). In reality, the truth lies somewhere in the middle, and it is here that the cellular industry takes the best of both worlds to create an evolved network, where the goal is the delivery of effective services and applications to the end user, rather than focusing on a particular technology to drive them. That said, the economy of scale and widespread acceptance of IP as a means of service delivery sees it playing a central role in this process.

1.2 THIRD GENERATION (3G)

Third generation or 3G is now the generally accepted term used to describe the next wave of mobile networks and services. First generation (1G) is used to categorize the first analogue mobile systems to emerge in the 1980s, such as the advanced mobile phone system (AMPS) and nordic mobile telephony (NMT). These systems provided a limited mobile solution for voice, but have major limitations, particularly in terms of interworking, security and quality. The next wave, second generation (2G), arrived in the late 1980s and moved towards a digital solution which gave the added benefit of allowing the transfer of data and provision of other non-voice services. Of these, the global system for mobile communication (GSM) has been the most successful, with its global roaming model. 3G leverages on the developments in cellular to date, and combines them with complementary developments in both the fixed-line telecoms networks and from the world of the Internet. The result is the development of a more general purpose network, which offers the flexibility to provide and support access to any service, regardless of location. These services can be voice, video or data and combinations thereof, but, as already stated, the emphasis is on the service provision as opposed to the delivery technology. The motivation for this development has come from a number of main sources, as follows:

- subscriber demand for non-voice services, mobile extensions to fixed-line services and richer mobile content;
- operator requirements to develop new revenue sources as mobile voice services and mobile penetration levels reach market saturation;
- operators with successful portfolios of non-voice services now unable to sustain the volume of traffic within their current spectrum allocation;
- equipment vendor requirements to market new products as existing 2G networks become mature and robust enough to meet current consumer demand.

It is arguable which of these weigh most heavily on the big push for the introduction of 3G networks, and which of these are justifiable. Certainly in Japan and Korea, where operators are now generating more traffic and revenue from non-voice services, the business case for 3G is present. These operators are no longer able to meet the subscriber demand for such applications, and have been a major impetus in 3G development, particularly NTT DoCoMo, arguably the most successful, and a pioneer in non-voice services. However, the situation in Japan and Korea is somewhat different to the rest of the world. There are a number of key factors that led to the growth of data services there:

- low Internet penetration, due largely to language factors;
- high existing mobile penetration (in Japan, the high cost and low efficiency of fixed-line services has partially fuelled this);
- large urban conurbation with sizeable proportion of the working population commuting on public transport, often for a long duration;
- low relative cost of mobile services.

This is evident in Japan, where the first driving application of DoCoMo's iMode service was provision of email.

However, the current situation outside of these exceptions is that thus far, consumer demand for data services has been limited, even now when there is widespread availability of data-capable mobile devices. Cost of new services has been a significant factor in this poor uptake as bandwidth charges are unrealistically high when compared to fixed-line equivalents, particularly now with the widespread availability of economical consumer digital subscriber line (DSL) services.

1.3   WHY UMTS?

The 3G standard proposed by the European Telecommunications Standards Institute (ETSI) with much joint work with Japanese standardization bodies is referred to as the universal mobile telecommunications system (UMTS). UMTS is one of a number of standards ratified by the International Telecommunications Union–Telecommunication Standardization Sector (ITU-T) under the umbrella of International Mobile Telephony 2000 (IMT2000), as discussed in the next section. It is currently the dominant standard, with the US CDMA2000 standard gaining ground, particularly with operators that have deployed cdmaOne as their 2G technology. At the time of writing, Japan is the most advanced in terms of 3G network deployment. The three incumbent operators there have implemented three different technologies: J-Phone is using UMTS, KDDI has a CDMA2000 network, and the largest operator NTT DoCoMo is using a system branded as FOMA (Freedom of Multimedia Access). FOMA is based around the original UMTS proposal, prior to its harmonization and standardization.

The UMTS standard is specified as a migration from the 2G GSM standard to UMTS via the general packet radio service (GPRS) and enhanced data rates for global evolution (EDGE), as shown in Figure 1.1. This is a sound rationale since as of December 2002, there were over 780 million GSM subscribers worldwide,[1] accounting for 71% of the

---

[1]Source: GSM Association, www.gsmworld.com.

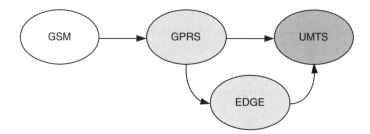

Figure 1.1   GSM evolution to UMTS

global cellular subscriber figures. The emphasis is on enabling as much of the GSM network as possible to continue to operate with the new system.

The goal of 3G is to provide a network infrastructure that can support a much broader range of services than existing systems so the changes to the network should reflect this. However, many of the mechanisms in the existing networks are equally applicable to supporting new service models, for example mobility management. For a successful migration, the manufacturers and suppliers of new 3G equipment understand that most licences granted for 3G network operation will be to existing 2G operators and thus the next step must be an evolution rather than a revolution. Operators in the main are expected to introduce GPRS functionality before taking the step to 3G. This will allow them to educate and develop the consumer market for these new services prior to major investment in new technology. This means that the Core Network will comprise the GSM circuit switched core and the GPRS packet switched core. The first release (Release 99) specification for UMTS networks is focused on changes to the Radio Access Network rather than the Core Network. This allows the Core Network to continue in functionality although changes will be made in areas of performance due to the higher data rates required by subscribers in the future networks. Maintaining this functionality allows the mobile network operators to continue using their existing infrastructure and progress to 3G in steps. The handover between UMTS and GSM offering worldwide coverage has been one of the main design criteria for the 3G system.

1.4   IMT2000 PROCESS

The IMT2000 is a global process, coordinated by the ITU-T to develop next generation mobile networks, and covers both the technical specifications and the frequency allocations. It was started in 1995 under the original heading of Future Plans for Land Mobile Telecommunications System (FPLMTS). IMT2000 is not a particular technology, but rather a system which should allow seamless, ubiquitous user access to services. The task is to develop a next generation network fulfilling criteria of ubiquitous support for broadband real-time and non-real-time services. The key criteria are

• high transmission rates for both indoor and outdoor operational environments;
• symmetric and asymmetric transmission of data;

- support for circuit and packet switched services;
- increased capacity and spectral efficiency;
- voice quality comparable to the fixed line network;
- global, providing roaming between different operational environments;
- support for multiple simultaneous services to end users.

The process is intended to integrate many technologies under one roof. Therefore, it should not be seen that wireless technologies from different regional standardization bodies, or supported by different manufacturers, are competing with each other, but rather that they can be included in the IMT2000 family. This is evident with the development of such interworking models as wireless LAN and 3G. A major enabler of the ITU-T vision is the emergence of software defined radio (SDR). With SDR, the air interface becomes an application, which enables a single mobile device to be able to operate with a variety of radio technologies, dynamically searching for the strongest signal, or the most appropriate network to connect to.

Thus far, the ITU-T has given the imprimatur of 3G to five different radio access technologies, as shown in Figure 1.2.

ITU-DS is the UMTS frequency division duplex (FDD) standard, ITU-MC is CDMA-2000, and ITU-TC covers both UMTS time division duplex (TDD) and time division synchronous CDMA. All these technologies are explained in Chapter 6. The IMT-SC system, UWC-136, is the EDGE standard (Chapter 4). The ITU-FT incorporates the European standard for cordless telephones, digital enhanced cordless telecommunications (DECT). DECT provides a local access solution which may be used, for example, in a home environment. The handset can automatically handover to a subscriber's domestic access point, providing dedicated resources. While the integration of DECT with GSM has been standardized, it has yet to see any exposure.

The development of these standards is under the control of two partnership organizations formed from a number of regional standardization bodies. The Third Generation Partnership Project (3GPP, www.3gpp.org) is responsible for UMTS and EDGE, while the Third Generation Partnership Project 2 (3GPP2, www.3gpp2.org) deals with CDMA2000 (Figure 1.3). DECT is the exception to this, with its standards developed solely by ETSI.

As can be seen, there is considerable overlap in terms of the bodies involved in the two organizations. The various bodies are described in Table 1.1.

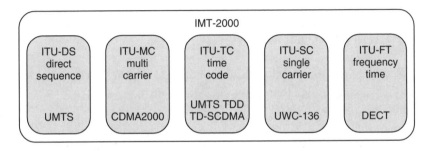

Figure 1.2   IMT2000 technologies

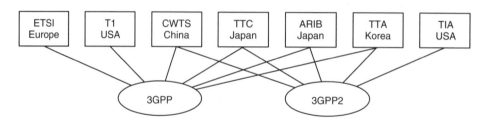

Figure 1.3 3G partnerships

Table 1.1 Standardization bodies

Body	Description
ETSI	The European Telecommunications Standards Institute is responsible for the production of standards for use principally throughout Europe, but standards may be used worldwide
T1	Committee T1 develops technical standards and reports in the US with regard to the interconnection and interoperability of telecommunications networks at interfaces with end user systems
CWTS	The China Wireless Telecommunication Standard group has the responsibility to define, produce and maintain wireless telecommunication standards in China
TTC	The Telecommunication Technology Committee is a Japanese organization whose role is to contribute to the standardization and dissemination of standards in the field of telecommunications
ARIB	The Association of Radio Industries and Businesses conducts investigations into new uses of the radio spectrum for telecommunications and broadcasting in Japan
TTA	The Telecommunications Technology Association is an IT standards organization that develops new standards and provides testing and certification for IT products in Korea
TIA	The Telecommunications Industry Association is the leading US trade association serving the communications and information technology industries

One of the tasks was to allocate a band of the frequency spectrum for this new system. Figure 1.4 shows the bands allocated, and compares this to the bands being used in both the US and Europe/Asia-Pacific regions, with the exception of Japan and Korea.

As can be seen, the allocated frequency is already extensively used in North America, and this presents deployment issues for 3G technologies. This is expanded in more depth in Chapter 6. In this frequency use chart, MSS is the mobile satellite system, which is the satellite component of 3G. Europe/Asia-Pacific has allocated all of the IMT2000 frequency to UMTS, with the exception of 15 MHz, which is already being used for DECT. The UMTS allocation is as follows:

- *UMTS FDD*: uplink: 1920–1980 MHz; downlink: 2110–2170 MHz
- *UMTS TDD*: uplink: 1900–1920 MHz; downlink: 2010–2025 MHz.

Most countries have now completed the licensing of these new bands for 3G services, many of them opting for an auction process. For UMTS, the basic carrier frequency is 5 MHz, and since it is a CDMA system it is possible to use only one frequency throughout

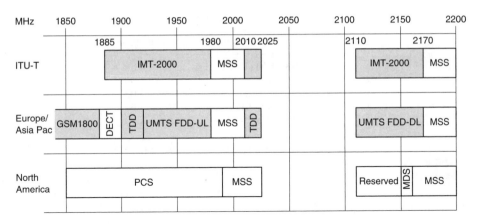

Figure 1.4 Cellular frequency usage

the system (see Chapter 2). For UMTS FDD, since there is 60 MHz of bandwidth available in UL/DL, this equates to 12 carriers. However, it is recommended that an operator be allocated three carrier frequencies. This is to tie in with the ITU-T principle of cell hierarchies, which provides for the following cell types:

- *Macro cell*: large area, outdoor general coverage
- *Micro cell*: small area, densely populated urban coverage
- *Pico cell*: indoor coverage.

Each cell type could be allocated a different carrier frequency, allowing for an overlay model. However, the decision of how to allocate frequencies is the remit of the national regulatory authority in a country. As an extreme example, consider the situation in the United Kingdom, which opted for the auction method. Five licences were allocated, as shown in Figure 1.5.

Licence A is allocated 15 MHz of FDD plus 5 MHz of TDD, and was reserved for a greenfield operator. Licence B consists of 15 MHz of FDD spectrum, and licences C–E 10 MHz of FDD plus 5 MHz of TDD each. After a controversial auction which concluded on 27 April 2000, the licences were sold as shown in Table 1.2.

Greenfield operator TIW UMTS (UK) Ltd is owned by the Canadian operator Telesystem International Wireless Inc. and is deploying UMTS in the UK as a joint venture with Hong Kong's Hutchison Whampoa, under the brand name 3. Commercial operation commenced at the end of December 2002. This rather cynical auctioning process worldwide has done little to aid the development of 3G, and has been widely criticized

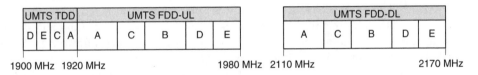

Figure 1.5 UK 3G spectrum licences

Table 1.2 UK 3G auction results

Licence	Operator	Fee (£ bn)
A	TIW UMTS (UK) Ltd	4.38
B	Vodaphone Ltd	5.96
C	BT (3G) Ltd	4.03
D	One2One Ltd	4.00
E	Orange 3G Ltd	4.10
Total		**22.47**

by many sources for the amount of capital it has taken out of the market. However, the various regulatory authorities have argued that the fees reflect the potential that the applicants expect from 3G in the long term.

1.5 ORGANIZATION OF THE BOOK

This book is intended to provide detailed and relevant information on the technologies related to the deployment of 3G systems, and focuses on UMTS. It is designed to cover the requisite knowledge to a reader coming from either a telecommunications or a computer networking background, examining how the different technologies are implemented in a UMTS context, and how the system evolves to deliver the service model. Throughout the text, examples of procedures are illustrated using trace files captured from UMTS networks to demonstrate their operation in practice.

Chapter 2 discusses the general principles on which packet-based networks are built, highlighting their use for the transport of real-time traffic. Added to this is the complication of a wireless interface to the network, and the mechanisms for providing multiple access are also explored. In particular, an overview of the principles and operation of the CDMA technique is presented, as this forms the central basis of the wireless physical layer of most 3G technologies.

Chapter 3 begins the description of cellular systems with a detailed explanation of the operation of the GSM. Aside from the access network, much of the existing GSM network is reused in UMTS, particularly at the higher layers such as connection and mobility management. In particular, the model for support of roaming within GSM and the basic security architecture are important components carried forward and expanded upon in UMTS. GSM is built around the signalling system 7 (SS7) protocol suite as used in the fixed-line PSTN, with extensions to support users accessing through a wireless interface.

Chapter 4 introduces the first major evolutionary step of GSM, the general packet radio service (GPRS). The GSM network has been designed and optimized for the delivery of one application, voice calls. Other services offered are considered supplementary. Chapter 2 explains why this type of network is not well suited for data transport, due to the vastly different requirements of the traffic. GPRS adds a network infrastructure based around the IP protocol which is designed with the needs of this data traffic is mind. It is also an essential building block of the UMTS network. Also described here is EDGE, which builds on GSM/GPRS to create a relatively high-speed network, without the major capital expenditure of UMTS. EDGE can be seen as a 3G solution in itself, or as a technology to complement a UMTS roll-out.

Chapter 5 describes the IP protocol suite and, in particular, its application to both the GPRS and UMTS Release 99 networks. Central to this is the ability of IP to provide mechanisms for quality of service (QoS), reliability and secure communication. The basic operation of IPv6 is discussed as it may be used as an application data bearer in UMTS. Also addressed are the related IP protocols for support of CDMA2000 networks.

Chapter 6 explores the architecture and operation of the UMTS network. It links what has gone before in GSM and GPRS with the new radio access network that forms the basis of the UMTS network. The chapter focuses on the operation of the first release of UMTS, Release 99, but explains the changes to this network as it evolves to an all-IP infrastructure. The operation of signalling protocols throughout the network is described in significant detail. A basic overview of the operation of the CDMA2000 system is also presented for reference.

Chapter 7 explains the application of ATM technology as a transport layer within the UMTS radio access network. At the time of development of UMTS, ATM was the only technology that could support the different types of traffic on the same infrastructure, while guaranteeing performance and meeting rigorous QoS requirements. In addition, ATM is a proven technology at integrating with both ISDN and IP networks, which is essentially the technologies around which the UMTS core network domains are based. A key feature of the application of ATM in a UMTS context is the extensive use of adaptation layer 2 (AAL2) as a transport for both real-time and non-real-time applications in the radio access network, a utilization not previously seen. Pivotal to the application of AAL2 is the ability to dynamically establish and release AAL2 connections using the AAL2 signalling protocol, and its operation is also explained.

Chapter 8 discusses the use of IP in UMTS as the network evolves to Release 4. In Release 4, the traditional circuit switched core network infrastructure of GSM is replaced with an IP-based softswitch architecture. This chapter explains the operation of new protocols to support this architecture, where the role of the mobile switching centre (MSC) is split into control using an MSC server and traffic transfer with a media gateway (MGW). The real-time extensions to IP for support of voice transport, the real-time transport protocol and the real-time transport control protocol (RTP/RTCP), are covered here. The MSC server uses a protocol to control its media gateways, and the operation of the media gateway control protocol (MEGACO), as specified for UMTS, is explained. For call control, the bearer-independent call control (BICC) protocol is specified between MSC servers, and the signalling transfer (sigtran) protocol stack is used for the transport of SS7 signalling over an IP network. Both are also explained.

Chapter 9 looks to UMTS Release 5, where IP use is extended through the UTRAN to the BTS. The various transport options for using IP in UTRAN are described. The session initiation protocol (SIP) is explained, as it is now the protocol specified for VoIP, mobility management and instant messaging in UMTS. This chapter also looks to other IP protocols and their possible use within UMTS, such as multi-protocol label switching (MPLS).

A list of the current versions of the specifications can be found at http://www.3gpp.org/specs/web-table_specs-with-titles-and-latest-versions.htm, and the 3GPP ftp site for the individual specification documents is http://www.3gpp.org/ftp/Specs/latest/

2

Principles of Communications

2.1 CIRCUIT- AND PACKET SWITCHED DATA

Many practical communication systems use a network which allows for full connectivity between devices without requiring a permanent physical link to exist between two devices. The dominant technology for voice communications is circuit switching. As the name implies, it creates a series of links between network nodes, with a channel on each physical link being allocated to the specific connection. In this manner a dedicated link is established between the two devices.

Circuit switching is generally considered inefficient since a channel is dedicated to the link even if no data is being transmitted. If the example of voice communications is considered, this does not come close to 100% channel efficiency. In fact, research has shown that it is somewhat less that 40%. For data which is particularly bursty this system is even more inefficient. Generally before a connection is established, there is a delay; however, once connected, the link is transparent to the user, allowing for seamless transmission at a fixed data rate. In essence, it appears like a direct connection between the two stations. Some permanent type circuits such as leased lines do not have a connection delay since the link is configured when it is initially set up. Circuit switching is used principally in the public switched telephone network (PSTN), and private networks such as a PBX or a private wide area network (WAN). Its fundamental driving force has been to handle voice traffic, i.e. minimize delay, but more significantly permit no variation in delay. The PSTN is not well suited to data transmission due to its inefficiencies; however, the disadvantages are somewhat overcome due to link transparency and worldwide availability.

The concept of packet switching evolved in the early 1970s to overcome the limitations of the circuit switched telecommunications network by implementing a system better suited to handling digital traffic. The data to be transferred is split into small packets, which have an upper size limit that is dependent on the particular type of network. For example, with asynchronous transfer mode (ATM) the cell size is fixed at 53 bytes whereas

Convergence Technologies for 3G Networks: IP, UMTS, EGPRS and ATM   J. Bannister, P. Mather and S. Coope
© 2004 John Wiley & Sons, Ltd   ISBN: 0-470-86091-X

an Ethernet network carries frames that can vary in size from 64 bytes up to 1500 bytes. A packet contains a section of the *data* plus some additional control information referred to as a *header*. This *data*, which has been segmented at the transmitter into packet sizes that the network can handle, will be rebuilt into the original data at the receiver. The additional *header* information is similar in concept to the address on an envelope and provides information on how to route the packet, and possibly where the correct final destination is. It may also include some error checking to ensure that the data has not been corrupted on the way. On a more complex network consisting of internetworking devices, packets that arrive at a network node are briefly stored before being passed on, once the next leg of the journey is available, until they arrive at their destination. This mechanism actually consists of two processes, which are referred to as *buffering* and *forwarding*. It allows for much greater line efficiency since a link between nodes can be shared by many packets from different users. It also allows for variable rates of transmission since each node retransmits the information at the correct rate for that link. In addition, priorities can be introduced where packets with a higher priority are transmitted first. The packet switched system is analogous to the postal system. There are two general approaches for transmission of packets on the network: datagrams and virtual circuits.

2.1.1   Datagram approach

With the datagram approach, each packet is treated independently, i.e. once on the network, a packet has no relation to any others. A network node makes a routing decision and picks the best path on which to send the packet, so different packets for the same destination do not necessarily follow the same route and may therefore arrive out of sequence, as illustrated in Figure 2.1. The headers in the figure for each of the packets will have some common information, such as the address of the receiver, and some information which is different, such as a sequence number. Reasons for packets arriving out of sequence may be that a route has become congested or has failed. Because packets can arrive out of order the destination node needs to reorder the packets before reassembly. Another possibility with datagrams is that a packet may be lost if there is a problem at a node; depending on the mechanism used the packet may be resent or just discarded. The Internet is an example of a datagram network; however, when a user dials in to an ISP via the PSTN (or ISDN), that link will be a serial link, most probably using the PPP protocol (see Chapter 5). This access link is a circuit switched connection in that the bandwidth is dedicated to the user.

2.1.2   Virtual circuits

Since the packets are treated independently across the network, datagram networks tend to have a high amount of overhead because the packet needs to carry the full address of the final destination. This overhead on an IP network, for example, will be a minimum of 20 bytes. This may not be of significance when transferring large data files of 1500 bytes or so but if voice over IP (VoIP) is transferred the data may be 32 bytes or less and here

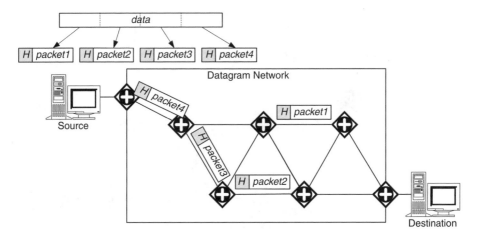

Figure 2.1   Datagram packet switched network

it is apparent that the overhead is significant. This approach establishes a virtual circuit through the nodes prior to sending packets and the same route is used for each packet. The system may not guarantee delivery but if packets are delivered they will be in the correct order. The information on the established virtual circuit is contained in the header of each packet, and the nodes are not required to make any routing decisions but forward the packets according to the information when the virtual circuit was established. This scheme differs from a circuit switched system as packets are still queued and retransmitted at each node and they do have a header which includes addressing information to identify the next leg of the journey. The header here may be much reduced since only localized addressing is required, such as 'send me out on virtual circuit 5' rather than a 4-byte address for the IP datagram system. There are two types of virtual circuit, permanent and switched:

- A permanent virtual circuit is comparable to a leased line and is set up once and then may last for years.
- A switched virtual circuit is set up as and when required in a similar fashion to a telephone call. This type of circuit introduces a setup phase each and every time prior to data transfer.

Figure 2.2 shows a network containing a virtual circuit. Packets traverse the virtual circuit in order and a single physical link, e.g. an STM-1 line, can have a number of virtual circuits associated with it.

The term *connectionless* data transfer is used on a packet switched network to describe communication where each packet header has sufficient information for it to reach its destination independently, such as a destination address. On the other hand, the term *connection-oriented* is used to denote that there is a logical connection established between two communicating hosts. These terms, *connection-oriented* and *connectionless,* are often incorrectly used as meaning the same as virtual circuit and datagram. Connection-oriented

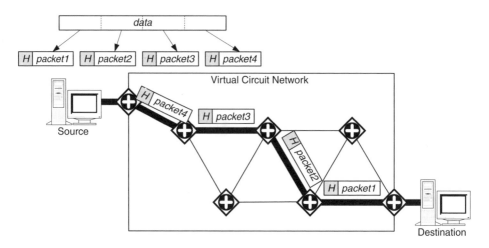

Figure 2.2   Virtual circuit

and connectionless are services offered by a network, whereas virtual circuits and datagrams are part of the underlying structure, thus a connection-oriented service may be offered on a datagram network, for example, TCP over IP.

2.2   ANALOGUE AND DIGITAL COMMUNICATIONS

In an analogue phone system, the original voice signal is directly transmitted on the physical medium. Any interference to this signal results in distortion of the original signal, which is particularly difficult to remove since it is awkward to distinguish between the signal and noise as the signal can be any value within the prescribed range. When the signals travel long distances and have to be amplified the amplifiers introduce yet further noise. Also, it is extremely easy to intercept and listen in to the transmitted signal. With digital transmission, the original analogue signal is now represented by a binary signal. Since the value of this signal can only be a 0 or a 1, it is much less susceptible to noise interference and when the signal travels long distances repeaters can be used to regenerate and thus clean the signal. A noise margin can be set in the centre of the signal, and any value above this is considered to be of value 1, and below of value 0, as illustrated in Figure 2.3. The carrier does not generally transport as much information in a given time when compared to an analogue system, but this disadvantage is far outweighed by its performance in the face of noise as well as the capability of compressing the data. Furthermore, an encryption scheme can be added on top of the data to prevent easy interception. For this reason, all modern cellular communication systems use digital encoding.

2.2.1   Representing analogue signals in digital format

Since the telephone exchange now works on a digital system in many countries, this necessitates the transmission of analogue signals in digital format. For example, consider

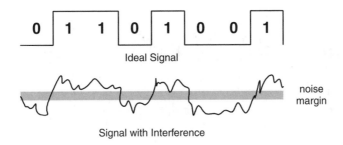

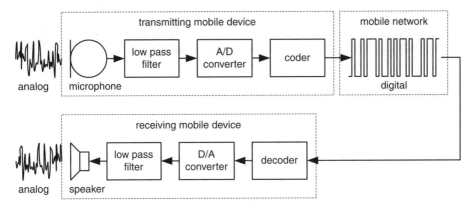

Figure 2.3 Digital transmission

Figure 2.4 Digital transmission of analogue signal

transmitting voice across the mobile telephone network. Figure 2.4 shows such a system. The analogue voice is filtered, digitized into a binary stream and coded for transmission. It will travel across the mobile network(s) in digital form until it reaches the destination mobile device. This will convert from digital back to analogue for output to the device's loudspeaker. Converting the analogue signal to digital and then back to analogue does introduce a certain amount of noise but this is minimal compared to leaving the signal in its original analogue state.

2.3 VOICE AND VIDEO TRANSMISSION

Before real-time analogue data can be transmitted on a digital packet switched network it must undergo a conversion process. The original analogue signal must be sampled (or measured), converted to a digital form (quantized), coded, optionally compressed and encrypted.

2.3.1 Sampling

Sampling is the process whereby the analogue signal is measured at regular intervals and its value recorded at each discrete time interval. It is very important that the signal is

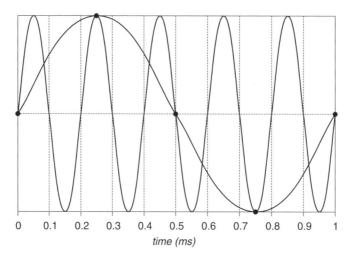

time (ms)

Figure 2.5   Aliasing

sampled at a rate higher than twice the highest frequency component of the original ana-logue signal otherwise a form of interference called *aliasing* may be introduced. Consider the problem highlighted in Figure 2.5. Here a 1 kHz signal is being sampled at 4000/sec-ond (4 kHz). However, there is a 5 kHz component also present, and the two produce the same result after sampling. For this reason the signal is filtered before sampling to remove any high-frequency components. For the PSTN, the signal is filtered such that the highest frequency is 3.4 kHz and sampling takes place at 8 kHz. Once the signal has been sampled it can then be generally compressed by encoding to reduce the overall amount of data to be sent. This encoded data is then bundled in packets or cells for transmission over the network. The exact amount of data that is carried in each packet is important. Packing a lot of data per packet causes a delay while the packet is being filled. This is referred to as *packetization delay*, and is described in Section 2.3.6. On the other hand, if the packets are not filled sufficiently this can lead to inefficiency as most of the packet can be taken up by protocol headers.

2.3.2 Coding and CODECs

When converting information from an audio or video stream into digital data, large amounts of information can be generated. Consider, for example, capturing a *single* frame on a 24-bit true colour graphics screen with a resolution of 1024 × 768 bits. Without com-pression this will generate 1024 × 768 × 3 (3 bytes = 24 bits of colour) = 2 359 296 or 2.25 megabytes of data. Sending 24 frames per second when capturing a video image will produce 54 megabytes of data every second, yielding a required data rate of 432 Mbps, which is unsustainable on the wireless network.

  To reduce the amount of data in the transmission the information is compressed before sending. Many techniques have been employed for both video and audio data but all compression algorithms use one of two basic types of method:

- **Lossless compression** removes redundancy from the information source and on decompression reproduces the original data exactly. This technique is used by graphics compression standards such as GIF and PNG. One technique used for PNG compression is the colour lookup table. Without compression the colour image on a screen requires each colour to be represented by 3 bytes (24 bits), even though there may be 256 or fewer different colours within a particular image. To compress the image each 3-byte code is replaced with a single byte and the actual 3-byte colour data stored in a separate table. This will produce a three-fold saving, less the small space to store the colour table of 768 bytes, and will involve little extra processing of the original image data.

- **Lossy compression**, on the other hand, relies on the fact that there is a lot of information within the image that the eye will not notice if removed. For example, the human eye is less sensitive to changes in colour than changes in intensity when looking at information in a picture. Consequently when images are compressed using the JPEG standard, the colour resolution can be reduced by half when scanning the original image. Lossy compression tends to produce higher compression rates than lossless compression but only really works well on real-world images, for example photographs. Lossless compression techniques such as PNG are more suitable for simple graphics images such as cartoons, figures or line drawings.

A CODEC is a term which refers to a coder/decoder and defines a given compression/decompression algorithm or technique. For audio compression the technique used for voice data is generally different to that used for music or other audio data. The reason for this is that voice CODECs exploit certain special human voice characteristics to reduce the bandwidth still further. These voice CODECs work well with a voice signal but will not reproduce music well since the CODEC will throw away parts of the original signal not expected to be there. Table 2.1 shows a summary of popular audio CODECs that are currently in use. Some of these are already used in wireless cellular networks such as GSM; others are recommended for use with UMTS and IP. Note that in the table, all the CODECs are optimized for voice apart from MP3, which is used predominantly on the Internet for music coding. The specific CODEC for voice used in UMTS is the adaptive multirate (AMR) CODEC, which is described in more detail in Chapter 6.

When choosing a voice CODEC, a number of characteristics have to be taken into consideration. Ideally a requirement is to use the least bandwidth possible but this generally comes at the expense of quality. The *mean opinion score* (MOS) defines the perceived quality of the reproduced sound: 5 means excellent, 4 good, 3 fair, 2 poor and 1 bad. The

Table 2.1   Audio CODECs

Standard	Bit rate (kbps)	Delay (ms)	MOS	Sample size
G.711	64	0.125	4.3	8
GSM-FR	13	20	3.7	260
G723.1	6.3	37.5	3.8	236
G723.1	5.3	37.5	3.8	200
UMTS AMR	12.2−4.75	Variable	Variable	Variable
MP3	Variable	Variable	Variable	Variable

MOS for a given CODEC is relatively subjective, as it is calculated by asking a number of volunteers to listen to speech and score each sample appropriately. G.711, which is the standard pulse code modulation (PCM) coding technique used for PSTN digital circuits, scores well. However, it uses a lot of bandwidth and is therefore not suitable for a wireless link. Generally as the data rate reduces, so does the MOS; however, surprisingly, G.723.1 scores better than standard GSM coding. The reason for this is that G.723.1 uses more complex techniques to squeeze additional important voice data into the limited bandwidth.

MP3 (MPEG layer 3 audio) is of interest in that it can provide a variable compression service based on either a *target data rate* or *target quality*. With target data rate the CODEC will try to compress the data down until the data rate is achieved. For music with a high dynamic range (for example classical music) higher rates may be required to achieve acceptable levels of reproduction quality. One report states that a data rate of 128 kbps with MP3 will reproduce sound which is very difficult to distinguish from the original. But, again, all reports are subjective and will very much depend on the source of the original signal.

It is also possible to set the MP3 CODEC to target a given quality. In this mode the data rate will go up if the complexity of the signal goes up. The problem with this type of mode of transmission is that it is difficult to budget for the correct amount of bandwidth on the transmission path.

When packing the voice data into packets it is important to be able to deliver the data to the voice decompressor fast enough so that delay is kept to a minimum. For example, if a transmitter packs one second's worth of speech into each packet this will introduce a packing/unpacking delay between the sender and receiver because one second's worth of data will have to be captured before each packet can be sent. For the higher-rate CODECs such as G.711, packing 10 milliseconds of data per packet would produce a data length of 80 bytes and a packing latency of only 0.01 seconds. For the CODECs which support lower date rates the minimum sample size is longer. With G.723.1 for example, the minimum sample size is 37.5 ms, which will introduce a longer fixed delay into the link. Also, if each packet contains one voice sample this will result in a packet length of only 30 bytes, which can result in inefficiencies due to the header overhead, For example, the header overhead for an IP packet is 20 bytes + higher-layer protocols (TCP, UDP, RTP, etc.) and for this reason header compression is generally used.

When looking at the video CODECs in Table 2.2 it can be seen that most of them do support a range of bit rates, which allows the encoder to pick a rate that suits the channel that it has available for the transmission. Standards such as MPEG-1 and MPEG-2 were designed for the storage, retrieval and distribution of video content. MPEG-1 is used in

Table 2.2 Video CODECs

Name	Bandwidth	Resolution
H.261	$N \times 64$ kbps	352×288 (CIF)
H.263	10 kbps–2 Mbps	180×144 (QCIF)
H26L	Variable	Variable
MPEG-1	1.5 Mbps	352×240
MPEG-2	3–100 Mbps	352×240 to 1920×1080
MPEG-4	Variable	Variable

the video CD standard at a fixed resolution of 352×240. MPEG-2, on the other hand, provides a wide range of resolutions from standard TV to high definition TV (HDTV) and for this reason is the predominant coding standard for DVD and digital TV transmission. The other CODECs are more suited and optimized for distribution over a network where bandwidth is at a premium, such as the cellular network. H.261 and H.263 were both designed to support video telephony and are specified as CODECs within the H.323 multimedia conferencing standard. Of particularly interest to UMTS service providers will be MPEG-4 and H26L. H26L is a low bit rate CODEC especially designed for wireless transmission. It has a variable bit rate and variable resolution. MPEG-4 was also designed to cope with narrow bandwidths and has a particularly complex set of tools to help code and improve the transmission of audio channels. This high-quality audio capability is of interest to content providers looking to deliver music and movies on demand over the radio network. Within MPEG-4 there are a number of different coding profiles defined, and the appropriate profile is chosen depending on the data rate available and the reliability of the channel. H26L has now been specified as one of the coding profiles within MPEG-4. It should be noted that while support for transport of video is a requirement of a 3G network, it is considered an application and, unlike voice, the coding scheme used is not included within the specification.

2.3.3   Pulse code modulation

Historically, the most popular method for performing this digitizing function on the analogue signal is known as pulse code modulation (PCM). The technique samples the analogue signal at regular intervals where the rate of sampling is twice the highest frequency present in the signal. This sampling rate is defined as the rate required to completely represent the analogue signal.

For the telephone network the assumption is made that the signals are below 4 kHz (actually 300 Hz to 3.4 kHz). Therefore the sample rate needed is 8000 samples per second. Each sample must be converted to digital. To do this, each analogue level is assigned a binary code. If 256 levels are required, then eight bits are used to split the amplitude up; an amplitude of zero is represented by binary 0000 0000, and a maximum amplitude by binary 1111 1111. Figure 2.6 shows a simple example of PCM, with 16 levels, i.e. 4 bits. For an 8-bit representation, with 8000 samples per second, a line of 64 kbps is required for the digital transmission of the voice signal. This is the standard coding scheme used for the fixed-line PSTN and ISDN telephone networks.

2.3.4   Compression

Compression involves the removal of redundancy from a data stream. This can be achieved through a number of techniques:

- **Run length encoding** replaces multiple occurrences of a symbol with one occurrence and a repetition count.

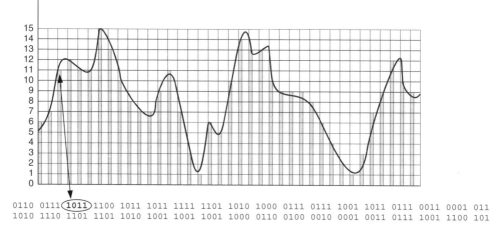

0110 0111 (1011) 1100 1011 1011 1111 1101 1010 1000 0111 0111 1001 1011 0111 0011 0001 011
1010 1110 1101 1101 1010 1001 1001 1001 1000 0110 0100 0010 0001 0011 0111 1001 1100 101

Figure 2.6   Pulse code modulation

- **Dictionary** replaces multiple symbols with single tokens which can be looked up in a dictionary.
- **Huffman** sends shorter codes for symbols which occur more often and longer codes for less frequent symbols.

Since the voice coding process removes most of the redundancy from the voice data itself, compression is largely used on the packet headers. Many schemes of header compression have been proposed and they are widely used on voice packets since these tend to be short and thus the header overhead (the percentage of data taken by the header) tends to be significant. Refer to Chapter 5 for more details.

2.3.5   Comfort noise generation and activity detection

To avoid transmitting unnecessarily, most systems use activity detection so that when a speaker is not talking, active background noise is not transmitted down the channel. In this case the CODEC usually encodes a special frame which informs the receiver to generate low-level noise (comfort noise), which reassures the listener they have not been cut off.

2.3.6   Packetization delay

Before the data is transmitted on the packet switched network is must be placed in the packets or cells. Some further explanation and examples of this packetization delay are presented in Chapter 7 in the context of the UMTS ATM transmission network. As discussed, the longer a packet is, the longer the delay suffered when forwarding the packet between switches or routes. In essence, the forwarding delay of a packet is just L/B where L is the length of the packet in bits and B the data rate of the link in bits per second. If the packet length is doubled, the forwarding delay is doubled. For very fast local area network (LAN) links this delay does not present a major problem. For example with a gigabit Ethernet link, a 1500 byte packet will take 12 μs to fill with voice data:

$$\text{Delay} = 1500 \times 8 \text{ bits}/1 \text{ Gbps} = 12 \text{ μs}$$

This delay does not include any component resulting from buffering or processing times, and therefore with a heavily loaded router or switch, actual packet delays may be somewhat larger.

There are essentially two basic forms of data transport available with IP networks, UDP and TCP. With UDP the service does not guarantee delivery of data. Since packets are never retransmitted the protocol will not add to the transit delay. With TCP the service is reliable but delays can be introduced when packets in error are retransmitted. For these reasons UDP and not TCP is used for VoIP data transport.

2.3.7 Erlang and network capacity

Voice networks use the *Erlang* as a standard measure of capacity. The Erlang is a measure of total voice traffic in one hour, usually classified as the busy hour (BH), which is the 60-minute interval during a 24-hour period in which the traffic load is at a peak. One Erlang is equivalent to one user talking for one hour on one telephone.

Consider that there are 45 calls in a one-hour period, and each call lasts for 3 minutes. This equates to 135 minutes of calls. In hours, this is $135/60 = 2.25$ Erlangs.

There are some variations in the Erlang model. The most common one is the Erlang B, which is used to calculate how many lines are required to meet a given minimum call blocking, usually 2–3%, during this BH. For cellular systems, it is used to estimate capacity per cell at base stations. The Erlang B formula assumes that all calls that are blocked are cleared immediately. This means that if a user attempts to connect and cannot, they will not try again. An extended form of Erlang B factors in that a certain percentage of users who are blocked will immediately try again. This is more applicable to the cellular environment, since if blocked, many users will immediately hit the redial button. The Erlang C model is the most complex since it assumes that a blocked call is placed in a queue until the system can handle it. This model is useful for call centres.

2.3.8 Voice over IP (VoIP)

The use of IP to transport voice traffic is one of the most remarkable developments in telecommunications in recent times. The development of the Internet as a global network means that through the use of VoIP, the Internet (and intranets) can be developed into a global telecommunications network. VoIP is a key enabler for the development of 3G networks as the infrastructure moves to use IP packet switching exclusively.

Since there is already a very prominent abundance of circuit switched telecommunications networks available, one might ask what benefits there are to be gained through the use of VoIP:

- *Lower transmission costs*: due to economies of scale, and open and widespread competition in the packet-switching market, the costs of transmission bandwidth have been pushed extremely low.
- *Data/voice integration*: many corporations already have an extensive data communications infrastructure. By using this to transmit voice, phone and fax, operating costs can be reduced. In particular, an organization which has a data communications network spanning international boundaries can avoid costly long-distance tariffs.
- *Flexible enhanced service*: data sent over IP can be encrypted for security, redirected to email voice mail services and routed via the Internet or PSTN. VoIP local area networks are ideal for building such solutions as customer call centre systems.
- *Bandwidth consolidation*: packet switching uses bandwidth considerably more efficiently than circuit switching. When there is no data to be sent, no bandwidth is used. This is distinct from circuit switched networks, where the circuit is allocated the full rate for the duration of the call.

There are also a number of problems associated with VoIP. The Internet itself is not well suited to the transport of real-time sensitive traffic since it offers poor performance in terms of delay and jitter. This is being addressed via a number of solutions to provide quality of service. Effective Internet telephony protocols have only recently been in place and equipment support is somewhat limited. With the development and widespread vendor support for session initiation protocol (SIP), this problem is largely solved. Finally there still remains a question mark over whether VoIP will still hold its cost/benefit advantage now that enhanced service provider (ESP) status has been removed from ISPs in the USA. This scheme had meant that ISPs were not required to pay access fees for telco local access facilities, giving ISPs advantages in competing for voice customers. The technical details of SIP are outlined in Chapter 9.

For VoIP, the delay must be minimal (telco standard minimum delay <100 ms) with no variation in delay. However, bandwidth requirements are modest, depending on the CODEC used, and unlike most data applications, some loss is acceptable but must be under a certain threshold for the call quality to be acceptable.

2.3.9   Quality of service

Quality of service (QoS) relates to providing performance guarantees to those applications that require it. Older packet switched protocols such as IP were originally intended to support transport of data traffic, for which the best-effort model is suitable, where of paramount importance is that data is delivered accurately and reliably, with delay and delay variation of little importance. However, when packet switched networks are required to transport real-time voice and video applications, the situation is much different and now these mechanisms are required to provide guarantees of performance. As an example, ATM technology builds this QoS mechanism in as a central aspect of the protocol. With IP, the QoS solutions must be incorporated into the protocol suite. There are two basic approaches to provide traffic with QoS: guaranteed QoS and class of service (CoS).

With guaranteed QoS, the network is expected to provide a minimum service defined by a set of quality parameters, including such things as minimum and average data rates, maximum delay and jitter as well as maximum packet loss rate. This type of service requires that the network has had some resources dedicated for the duration of the data transfer. This resource allocation can be done statically so the resources remain allocated even if the channel is not being used (as in the case of ATM permanent virtual circuits; PVCs) or dynamically before each call is made. Within IP the protocol defined which provides guaranteed QoS called the resource reservation protocol (RSVP).

CoS, on the other hand, splits the traffic into priority groups. The network simply guarantees to send high-priority traffic first, which works well provided the network has been scaled carefully to carry the total required volume of traffic. The protocol which provides CoS on IP networks is called DiffServ.

RSVP and DiffServ are presented in Chapter 5 while QoS in the context of ATM is explained in Chapter 7.

2.4   MULTIPLE ACCESS

In any communications system with many users, whether it be a fixed line or a wireless scheme, those users share some resource. Some mechanism must be employed to enable this resource sharing, and this is referred to as a multiple access scheme. In the wireless domain, the resource that is shared is frequency. For cellular communications, a change in generation has generally meant a change in the multiple access scheme that is implemented. The first generation of cellular systems used frequency division multiple access (FDMA); the majority of second generation systems use time division multiple access (TDMA) and most of the third generation schemes use code division multiple access (CDMA). In addition, a shift has been made from the original analogue system to a digital communications system.

2.5   FREQUENCY DIVISION MULTIPLE ACCESS (FDMA)

As previously stated, a wireless system has the resource of frequency to share among many users. The first approach to solving this problem is to split the available frequency into a number of channels, each with a narrow slice of the frequency. This concept is shown in Figure 2.7(a). Each user in the system that wishes to communicate is allocated a frequency channel, and each channel has a certain gap, known as a guard band, between it and the next channel so that the two do not interfere with each other. Once all the channels are in use, a new user to the system must wait for a channel to become free before communication can commence. Therefore, the system is limited in capacity as it can only support as many simultaneous users as there are channels. This is known as a

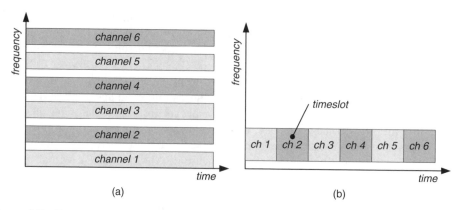

Figure 2.7 Frequency division multiple access scheme

hard capacity system. Another problem is that if there is any external interference at a particular frequency, then a whole channel may be blocked.

The concept of FDMA can be considered in the context of radio broadcasting. There is a certain allocation of frequency resources, for example 88 MHz to 108 MHz for FM, and each radio station in a particular region is given one channel within this on which it transmits.

2.6 TIME DIVISION MULTIPLE ACCESS (TDMA)

As wireless communications systems are expected to support more and more simultaneous users, there are clearly severe limitations with the FDMA scheme. A more efficient channel usage is required. With TDMA, a frequency channel is divided up into a number of slices of time, as shown in Figure 2.7(b). Here, a user is allocated a particular time slot, which repeats periodically. In the diagram, the frequency is split into six time slots; a user is allocated one slot in every six. Providing that the time slices are small enough and occur frequently enough, a user is oblivious to the fact that they are only being allocated a discrete, periodic amount of time. In this manner, the capacity can be dramatically increased and hence the efficiency of our system. Again, this is referred to as a *hard capacity* type network.

As an example, the global system for mobile communications (GSM) employs both a TDMA and FDMA approach. As with other mobile phone systems, an area to be covered is split up into a number of cells, each of which is operating at a particular frequency (frequency channel). Within a cell, the frequency being used is further split into time slots using the TDMA principle. If more capacity is required, either more cells, packed closer together, can be introduced, or another frequency channel can be deployed in a cell, increasing the number of available time slots, and hence, the number of simultaneous users. This does add some complication to the system, since the frequencies being used must be carefully planned so no two frequencies that are the same may border each other. This is the idea of *frequency reuse*; that is, a frequency can be used more than once in the system as long as there is a sufficient distance between the repeated usage locations. This idea is shown in Figure 2.8, where seven different frequencies, A, B, C, D, E, F and G, are being reused.

Typically in rural areas these cells are of the order of 10 km across but in areas of high usage (such as city centres) this may be reduced considerably to a few tens of metres.

Another advantage of the smaller cells is that less transmission power is required. This in turn means that the battery of the mobile devices can be smaller and lighter, thus reducing the overall weight of the devices. A single base station can control a number of cells, with each cell using a different frequency. More effective coverage of a highway, for example, can be attained through the use of sectored base stations as illustrated in Figure 2.9. A sectored site is typically used to cover a larger geographical area. Note

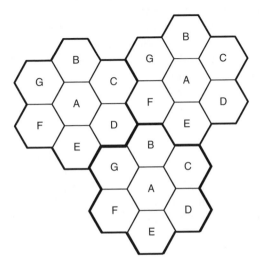

Figure 2.8   Cellular frequency reuse

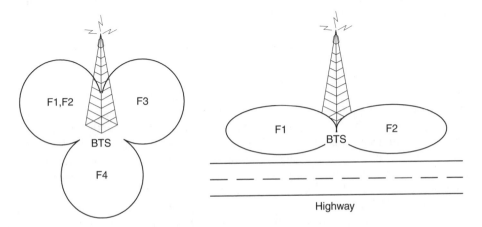

Figure 2.9   Sectoring a base station for efficient coverage

that in GSM the terms *cell* and *sector* are synonymous. A cell may also have more than a single frequency allocated to it, as illustrated in Figure 2.9. A transceiver unit (TRX) is the physical device located at the base station which controls each of these separate frequencies. A cell having a number of frequencies will therefore have a number of TRXs. In GSM, a TRX can handle at maximum eight full-rate simultaneous users.

2.7   CODE DIVISION MULTIPLE ACCESS (CDMA)

If the previous multiple access schemes are considered in terms of efficiency, each of them involves only one user transmitting on a particular channel at a particular time, which is clearly inefficient. For example, with GSM, in a given cell, only one user is transmitting at any time; all other active users are waiting for their time slot to come around. If a mechanism could allow more than one user to transmit at a time; then the resource usage could be dramatically improved. CDMA is such a scheme, where all users are transmitting at the same frequency at the same time. The effect of interference that users cause to each other is discussed under the heading of noise. Having a system that is limited by a noise target rather than specifically allocating resources for the sole use of a particular mobile device is known as a *soft capacity* system. Evidently, allowing multiple users to transmit simultaneously is not the central issue; providing some system to separate them out again is where the difficulty lies. This is the role of the codes. Much of the development work associated with CDMA was accomplished by the eminent mathematician Andrew J. Viterbi, who is also a cofounder of Qualcomm Inc. Thus, many of the patents associated with CDMA are held by Qualcomm, which has resulted in considerable debate with regard to the ownership of CDMA technology and much litigation against manufacturers of UMTS network equipment.

CDMA is part of a general field of communications known as spread spectrum. Spread spectrum describes any system in which a signal is modulated so that its energy is spread across a frequency range that is greater than that of the original signal. In CDMA, it is the codes that perform this spreading function, and also allow multiple users to be separated at the receiver. The two most common forms of CDMA are:

- *Frequency hopping (FH)*: with FH, the transmitted signal on a certain carrier frequency is changed after a certain time interval, known as the hopping rate. This has the effect of 'hopping' the signal around different frequencies across a certain wide frequency range. At a particular instant in time, the signal is transmitted on a certain frequency, and the code defines this frequency. This system is used for many communications systems, including the 802.11b wireless LAN standard and Bluetooth. These systems both use the unlicensed 2.4 GHz band, which is inherently subject to interference due to the large number of radio systems sharing that band, not to mention the effects of microwave ovens. By using a large number of frequencies, the effect of interference on the signal is substantially reduced, since the interference will tend to be concentrated in a particular narrow frequency range. FH is also employed in military communications, where the secrecy of the code and the rejection of interference in the form of a jamming signal make it extremely effective.

- *Direct sequence (DS)*: with DS, a binary modulated signal is 'directly' multiplied by a code. The code is a pseudo-random sequence of ±1, where the bit rate of the code is higher than the rate of the signal, usually considerably higher. This has the effect of spreading the signal to a wideband. At the receiver, the same code is used to extract the original signal from the incoming wideband signal. A bit of the code is referred to as a *chip*, and the defining parameter for such a system is the chip rate.

DS-CDMA is the form used for the air interface in UMTS, known as wideband CDMA (WCDMA), with a chip rate of 3.84 Mchip/s.

The origin of the spread spectrum and CDMA concept is generally accredited to the 1930s Austrian-born Hollywood actress Hedy Lamarr, and her pianist George Antheil, who filed a US patent for a 'Secret Communications System' in 1942 at the height of the Second World War. The system used a piano type system to perform frequency hopping on a signal. Neither of the two ever made any money from the patent, which subsequently expired.

2.7.1   DS-CDMA signal spreading

According to information theory, as the frequency spectrum a signal occupies is expanded, the overall power level decreases. In CDMA, the user signals are spread up to a wideband by multiplication by a code. Consider a narrowband signal, say, for example, a voice call. When viewed in the frequency spectrum, it occupies some frequency and has some power level, as illustrated in Figure 2.10(a). Once the frequency is spread across a wideband, the total power of this signal is substantially reduced.

Now consider that another user has the same procedure performed on it and is also spread to the same wideband. The total system power is increased by a small amount as the two users are transmitted at the same time. Therefore, each new user entering the system will cause the power of the wideband to increase. The idea is shown in Figure 2.10(b).

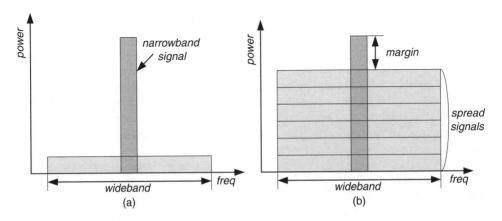

Figure 2.10   Signal spreading

At the receiver, the process of extracting one user is performed; the mechanism of how this can be implemented is described in the next section. The regenerated signal needs to be retrieved with enough power that it can be perceived above the level of the remaining spread signals. That is, it needs to be of a sufficient strength, or margin, above the rest of the signals so that the signal can be accurately interpreted. Considering this as a signal to interference ratio (SIR), or carrier to interference (C/I) ratio, the noise affecting one signal is the remaining spread signals that are transmitting at that frequency. This SIR is classified in CDMA as E_b/N_0. Literally, this means the energy per bit, E_b, divided by the noise spectral density, N_0. However, it is really a measure of the minimum required level the signal should be above the noise which is contributed by the other transmitting users. For mobile device measurements of the quality of the signals from the network, it uses a pilot channel, which is broadcast by each cell. The mobile device measures E_c/I_0, the energy level of this pilot channel, E_c, compared to the total energy received, I_0.

Another important characteristic is the rejection of unwanted narrowband noise signals. If a wideband signal is affected by a narrowband noise signal, then since the spreading function is commutative, the despreading operation while extracting the wanted signal will in turn spread the narrowband noise to the wideband, and reduce its power level. The rejection of the interference effects of wideband noise from other users is the role of convolution coding, which is described in Section 2.9.1.

This implies that the important factor that will affect how easily signals can be interpreted after they are despread is the power level in the system. The lower the power that the original signals are transmitted with, the lower the noise in the system. It is therefore essential that each user in the system transmits with an optimum power level to reach the receiver with its required power level. If the power level is too high, then that user will generate noise, which in turn affects the performance of all the other users. If there is too little power, then the signal which reaches the receiver is of too low quality, and it cannot be accurately 'heard'.

An analogy to this idea is a party at which all the guests are talking at the same time. At some point, with too many guests, the overall noise level rises to a point where none of the guests' individual conversations can be heard clearly.

There are two solutions to the problem of noise levels. First, an admission control policy is required that monitors the number of users and the noise level, and once it reaches some maximum tolerable level, refuses admission of further users. In a cellular system, such admission control needs to be considered not only for one cell, but also for the effects that noise levels within that cell have on neighbouring cells. In the party analogy, the effect on the neighbours should be considered. In conjunction with admission control, load control should also be implemented to try to encourage some users to leave a cell which has too many users, and consequently in which the noise level is too high.

The second solution is to implement power control. Each user needs to transmit with just enough power to provide a clear signal at the receiver above the noise floor. This should be maintained regardless of where the users are located with respect to the receiver, and how fast they are moving. Power control needs to be performed frequently to ensure that each user is transmitting at an optimum level. For more details, please refer to Section 2.8.2.

In direct sequence spread spectrum the signal is spread over a large frequency range. For example, a telephone speech conversation which has a bandwidth of 3.1 kHz would be spread over 5 MHz when transferred over the UMTS WCDMA system. The bandwidth has increased but the information transfer rate has remained constant. This is achieved by using a technique which introduces a code to represent a symbol of the transmitted message. A code is made up of a number of binary digits (bits), each one of which is referred to as a chip. The whole code consisting of all of the chips representing a symbol takes up the same time span as the original symbol. Thus if a single symbol is represented by a code of 8 chips, the chip rate must be 8 × the symbol rate. For example, if the symbol rate were 16 kbps then the chip rate (assuming 8 chips per symbol) would be 128 kbps. This higher data rate requires a larger frequency range (bandwidth). Figure 2.11(a) illustrates the data (symbols) to be spread (1001). Figure 2.11(b) indicates the 8-chip code '10010110'. Figure 2.11(c) combines parts (a) and (b) into a single waveform which represents the original data but which has been spread over a number of chips. This combining is achieved through the use of an exclusive-OR function.

The ratio of the original signal to the spread signal is referred to as the *spreading factor* and is defined as:

$$\text{Spreading factor (SF)} = \text{chip rate/symbol rate}$$

Thus in the above example, the SF is 8. Hence, variable data rates can be supported by using variable length codes and variable SF to spread the data to a common chip rate.

When considering CDMA systems, it is useful to define how the different signals interact with each other. Correlation is defined as the relationship or similarity between signals. For pulse-type waveforms, such as CDMA codes, the *cross-correlation* between two signals is defined as:

$$R_{12}(\tau) = \int v_1(t)v_2(t + \tau)\,dt$$

where R_{12} is the correlation between two signals v_1 and v_2, and τ is their relative time offset.

For the code to be effective, the receiver must know the specific code (in this case 10010110) which is being used for transmission and it must also be synchronized with this transmission. On reception the receiver can then simply reintroduce the correct code which is multiplied with the incoming signal and reproduce the actual symbol sent by the transmitter. The receiver also needs to know the actual number of chips that represent a

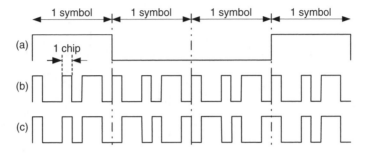

Figure 2.11   Spreading of data

symbol (spreading factor) so that the chips can be regenerated to the sent symbol through averaging the value of the chips over the symbol time. This is achieved through integration, where the chips are summed over the total time period of the symbol they represent.

The principle of correlation is used at the receiver to retrieve the original signal out of the noise generated by all the other users' wideband signal. Consider Figure 2.12. Notice that the logic levels of 0 and 1 have been replaced by the binary coded real values 1 and -1, respectively. The original data is coded and the resulting signal is transmitted. At the receiver, the received signal is multiplied by the code and the result is integrated over the period of each baseband bit to extract the original data. Since the receiver has four chips over which to integrate, the procedure yields a strong result at the output.

However, consider now that the receiver does not know the correct code. Then the integration process will result in a signal which averages to around zero (see Figure 2.13). For both of these, the relative strength of the desired signal and the rejection of other signals is proportionate to the number of chips over which the receiver has to integrate, which is the SF. Large SFs result in more processing gain and hence the original signals do not need so much transmission power to achieve a target quality level.

As can be seen, the longer the symbol time (i.e. lower data rate and higher chip rate), the longer the integration process, thus the higher the amplitude of the summed signal. This is referred to as *processing gain* (G_p) and is directly proportional to the SF used. For example, if the symbols were spread over 8 chips then the G_p will be 8; if spread over 16 chips, G_p would be 16. This means that the processing gain is higher for lower data rates than for higher data rates, i.e. lower data rates can be sent with reduced power since it is easier to detect them at the receiver. The processing gain can be used for link

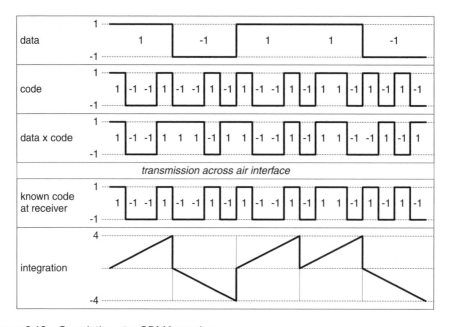

Figure 2.12   Correlation at a CDMA receiver

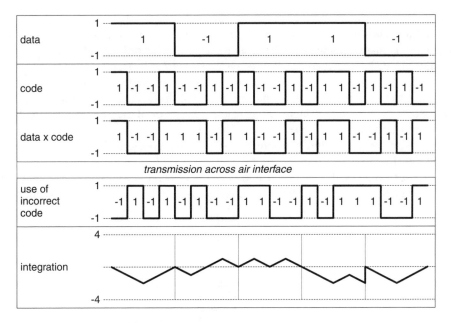

Figure 2.13 Correlation with incorrect code

budget calculations as follows:

$$G_p = 10 \log_{10} \text{chip rate/data rate}$$

Here, the data rate of the application can be used instead of the symbol rate, since it may be considered that what is lost in terms of bandwidth by the process of convolution coding and rate matching is gained again in terms of signal quality improvement.

As an example, consider that for voice, 12.2 kbps are required. The processing gain for this may be calculated as follows:

$$G_p = 10 \log_{10} 3.84 \text{ Mbps}/12.2 \text{ kbps} = 25 \text{ dB}$$

Thus higher data rates require more power and the limiting factor here is that the mobile devices can only supply of the order of 200–300 mW. Therefore to achieve higher data rates, the mobile device must be situated physically closer to the base station.

2.7.2 Orthogonal codes and signal separation

The signals that are all being transmitted at the same time and frequency must be separated out into those from individual users. This is the second role of the code. Returning to the party analogy, if this was a GSM party, then the problem is solved easily. All guests must be quiet and each is then allowed to speak for a certain time period; no two guests speak at the same time. At a CDMA party, all users are allowed to speak simultaneously, and they are separated by speaking in different languages, which are the CDMA codes.

All of the codes that are used must be unique and have ideally no relationship to each other. Mathematically speaking, this property is referred to as *orthogonality*. The system can support as many simultaneous users as it has unique or orthogonal codes.

Orthogonal codes are used in CDMA systems to provide signal separation. As long as the codes are perfectly synchronized, two users can be perfectly separated from each other. To generate a tree of orthogonal codes, a Walsh–Hadamard matrix is used. The matrix works on a simple principle, where the next level of the tree is generated from the previous as shown in Figure 2.14(a). The tree is then built up following this rule, with each new layer doubling the number of available codes, and the SF, as shown in Figures 2.14(b) and 2.15.

For perfect orthogonality between two codes, for example, it is said that they have a cross-correlation of zero when $\tau = 0$. Consider a simple example using the following two codes:

$$\text{Code } 1 = 1 - 1 \; 1 - 1$$

$$\text{Code } 2 = 1 - 1 - 1 \; 1$$

$$H_M = \begin{bmatrix} H_{M/2} & H_{M/2} \\ H_{M/2} & -H_{M/2} \end{bmatrix}$$

(a)   (b)

Figure 2.14   Orthogonal code matrix

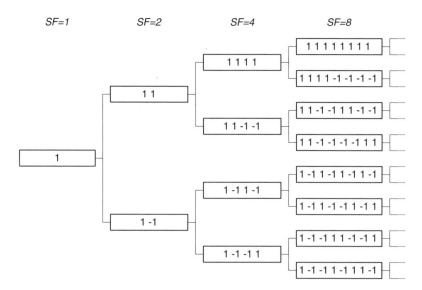

Figure 2.15   Channelization code tree

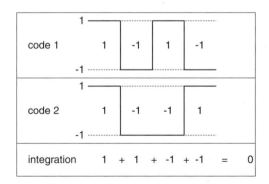

Figure 2.16   CDMA cross-correlation

To verify if these two have a zero cross-correlation, they are tested in the above equation, first multiplied together and then integrated, as shown in Figure 2.16. The result is zero, indicating that indeed they are orthogonal.

The number of chips which represent a symbol is known as the SF or the processing gain. To support different data rates within the system, codes are taken from an appropriate point in the tree. These types of orthogonal codes are known as orthogonal variable spreading factors (OVSF).

In the 3G WCDMA system the chip rate is constant at 3.84 Mchips/s. However, the number of chips that represent a symbol can vary. Within this system as laid down by the specifications, the minimum number of chips per symbol is 4 which would give a data rate of $3\,840\,000/4 = 960\,000$ symbols per second. The maximum SF or number of chips per symbol is 256,[1] which would give a data rate of $3\,840\,000/256 = 15\,000$ symbols per second. Thus it can be seen that the fewer chips used to represent a symbol, the higher the user data rate. The actual user data rate must be rate matched to align with one of these SF symbol rates. This process is described in more detail in Chapter 6.

Although orthogonal codes demonstrate perfect signal separation, they must be perfectly synchronized to achieve this. Another drawback of orthogonal codes is that they do not evenly spread signals across the wide frequency band, but rather concentrate the signal at certain discrete frequencies. As an example, consider that the code '1 1 1 1' will have no spreading effect on a symbol.

2.7.3   PN sequences

Another code type used in CDMA systems is the pseudo-random noise (PN) sequence. This is a binary sequence of ±1 that exhibits characteristics of a purely random sequence, but is deterministic. Like a random sequence, a PN sequence has an equal number of +1s and −1s, with only ever a difference of 1. PN sequences are extremely useful as they fulfil two key roles in data transmission:

---

[1] The specifications actually allow for 512; however, a number of restrictions apply when this is used.

1. Even spreading of data: when multiplied by a PN sequence, the resultant signal is spread evenly across the wideband. To other users who do not know the code, this appears as white noise.

2. Signal separation: while PN sequences do not display perfect orthogonality properties, nevertheless they can be used to separate signals. At the receiver, the desired signal will show strong correlation, with the other user signals exhibiting weak correlation.

Another property of PN sequences is that they exhibit what is known as autocorrelation. This is defined as the level of correlation between a signal and a time-shifted version of the same signal, measured for a given time shift, i.e. υ_1 and υ_2 in the previous correlation equation. For a PN sequence, the autocorrelation is at a maximum value, N, when perfectly time aligned, i.e. $\tau = 0$. N is the length in numbers of bits of the PN sequence. This single peak drops off quickly at $\pm T_c$, where T_c is the width of a chip of the code (see Figure 2.17).

This allows a receiver to focus in on where the signal is, without a requirement for the transmitter and receiver to be synchronized. In comparison, the autocorrelation of time-shifted orthogonal codes results in several peaks, which means that this signal locking is much more problematic.

PN sequences are generated using shift registers with a predefined set of feedback taps. The position of the taps is defined by what is known as a generator polynomial. A simple three-stage shift register arrangement is shown in Figure 2.18.

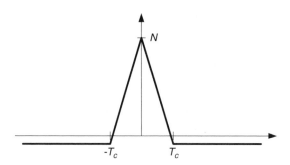

Figure 2.17 Autocorrelation of PN sequences

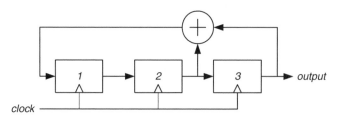

Figure 2.18 Three-stage shift register

clock	stage			output
	1	2	3	
1	0	1	0	0
2	1	0	1	1
3	1	1	0	0
4	1	1	1	1
5	0	1	1	1
6	0	0	1	1
7	1	0	0	0
8	0	1	0	0
9	...	...	...	...

Figure 2.19   Shift register states

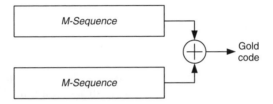

Figure 2.20   Gold code generator

From a certain starting configuration in the registers, the outputs of stages 2 and 3 are fed back to the input of the first stage via a modulo-2 adder. Any initial configuration is allowed except 000, since this results in a constant output of zero. Consider that the starting state is 010, then the stages for each clock cycle will be as shown in the state diagram (Figure 2.19).

At clock cycle 8, the sequence repeats, so the generated output sequence is 0101110. So for an M-stage shift register, a sequence of length $2^M - 1$ can be generated. These are referred to as *M-sequences*.

An improved form of PN sequences known as Gold codes are generated by using two such generators which are then combined (Figure 2.20). These Gold codes display better autocorrelation properties and allow much larger numbers of codes to be generated.

2.8   MULTIPATH PROPAGATION AND DIVERSITY

A transmission from a mobile device is more or less omnidirectional, and this is also the case for base stations which have only one cell. Base stations which are sectorized will have directional antennas, which will transmit only over a certain range. For example, a three-sectored site will have three antennas which each transmit over the range of 120 degrees. From the point of view of the mobile device, it would be ideal if a transmission were unidirectional; however, this is impractical since it would mean that the antenna of the mobile device would need to point towards the base station at all times. In this ideal situation the device could transmit with reduced power, thus causing less interference to other users and increasing the device's battery life. In the cellular environment, much of the power transmitted is actually in the wrong direction. In urban areas there is considerable reflection of the signal from surrounding buildings. This is actually a reason why cellular systems work, since the mobile device can thus be out of direct line of sight of the BTS and its signal will still be received. The reflected signals travel further distances than the direct line of sight transmission and therefore arrive slightly later, with greater attenuation and possible phase difference (see Figure 2.21).

It would be advantageous if these time-shifted versions in the multipath signal could be combined at the receiver with the effect that a much stronger signal is received. The autocorrelation property of the PN sequence is again used. Since the received signal resolves into a single peak around the chip width, then as long as the multipath profile is of a duration longer than the chip width, a number of peaks will be observed, each one representing a particular multipath signal. Figure 2.22 shows an example where three time-shifted paths have been resolved.

The number of paths that can be successfully resolved is related to the ratio of chip width to multipath profile. For WCDMA, the chip rate is constant at 3.84 Mchips/s, giving a constant chip time of approximately 0.26 μs. Typically for an urban area, a multipath profile is of the order of 1–2 μs over which there are signals arriving with sufficient power to be successfully resolved. Over this period, this means there is adequate time to resolve about three or four signals. In terms of distance, a time difference of 0.26 μs equates to 78 m, which means that to be resolved, a multipath must have a path length of at least 78 m greater than the direct signal.

CDMA systems harness this property through the use of a rake receiver. The rake receiver is so called since it has a number of fingers which resemble a garden rake. Figure 2.23 shows a simplified diagram of a rake receiver with three fingers. A rake receiver is a form of correlation receiver, so each finger is fed the same received signal,

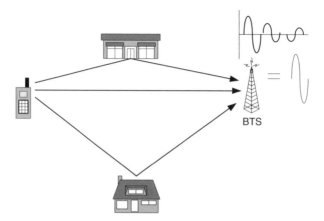

Figure 2.21   Multipath propagation

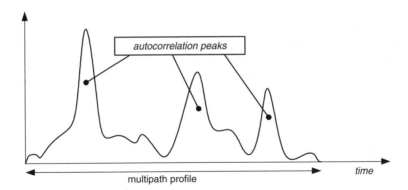

Figure 2.22   Multipath autocorrelation peaks

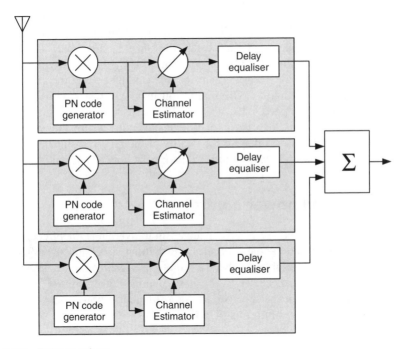

Figure 2.23   Rake receiver

which is correlated against the expected code to give an autocorrelated peak. This is fed into a channel estimator, which drives a phase adjuster to rectify the phase of the signal to be closer to that originally transmitted. This is needed since the phase of the different paths will have been altered, depending on the path they have taken, and the objects off which they have been reflected. Each finger has a delay equalizer so that the resolved peaks can be time aligned before passing to a summing unit where they are combined. This process is known as maximal ratio combining (MRC).

Because this combined signal is stronger, it is possible that the BTS may tell the mobile device to reduce its transmitting power. Any process of combining multiple versions of the same signal to provide a more powerful, better quality signal is known as *diversity*. In CDMA, this multipath diversity is referred to as *microdiversity*.

Further improvements may also be made at a base station by use of multiple antennas, separated in space, known as spatial diversity. Each antenna will receive the same signal, but with a small time shift compared to the other antennas, thus enabling combination of these signals. In WCDMA, up to four such antennas may be used to improve the signal quality.

2.8.1   Soft handover

A key advantage of CDMA systems is the principle of soft handover. Since each cell operates at the same frequency, it is possible for the mobile device to communicate simultaneously with more than one cell. Thus when a handover is required, the connection to the target cell can be established before the original connection is dropped. This is in contrast to traditional cellular TDMA/FDMA systems, where a handover requires that the connection is first dropped and then established at the target cell, since the cells are at different frequencies. In a CDMA device, during an active call, typically the rake receiver uses one finger to make measurements of surrounding cells at the same frequency as potential candidates for handover. Originally, soft handover was seen as advantageous since it resulted in fewer dropped calls, because the user is never disconnected from the network. However, now it is also used to provide diversity where the multiple active connections can be combined to improve the quality of the received signal. It is usual for a mobile device to be able to connect to up to three cells concurrently.

2.8.2   Fading and power control

The CDMA system needs a power control mechanism to overcome the effects of multiple users with different propagation characteristics transmitting simultaneously. This is often referred to as the *near–far* problem, where a remote user can easily be drowned out by a user that is physically much closer to the base station. Power control endeavours to ensure that signals arriving at the receiver are almost equal in power, and at a level that meets the quality requirements in terms of SIR.

The three main features here are:

- attenuation due to increase in distance from the receiver;
- fading variations due to specific features of the environment;
- fading variations due to the movement of the mobile device.

Radio waves propagating in free space are modelled by an inverse square law whereby as the distance between the transmitter and receiver doubles, the signal loses half of its power. Thus in the equation below a is generally regarded to be of value 2 and x indicates the distance in metres:

$$P_{\text{receive}} = \frac{P_{\text{transmit}}}{x^a}$$

This is not necessarily the case in a cellular system, where the terrain and buildings can have a major effect on the propagation model, and thus a is usually considered to be greater than 3. For example, in metropolitan areas, $a = 4$ for planning purposes. As a user moves around, the power level at the receiver will fluctuate. These fluctuations can be broken down into two general categories: slow and fast fading.

Slow fading or shadow fading is as a result of obstructions, which will result in changes in received power level. Multiple versions of the same signal will form constructive and destructive interference at the receiver as the relative time shifts vary due to different path lengths and reflection/refraction characteristics of the surrounding environment. It is more pronounced in urban areas, with significant changes in received signal strength occurring over tens of metres.

Fast fading, or Rayleigh fading, is due to the Doppler shift, where the apparent wavelength of the transmitted signal will increase as the mobile device moves towards the receiver and decrease as the device moves in the opposite direction of the receiver. This appears at the receiver as a change of phase of the transmitted signal. Generally a number of paths with different Doppler shifts will arrive at the receiver with changed phase shifts. As these *multipaths* are combined at the receiver, the signal will exhibit peaks and troughs of power corresponding to signals that are received in phase, and thus reinforce each other, and out of phase, where they cancel each other out. These variations are much faster than those occurring with environmental factors and can cause significant differences in power levels over relatively short distances. Consider the WCDMA system, where the transmit/receive frequency is in the 2 GHz range. The wavelength of this is 150 mm, and thus relatively small movements of the mobile device of the order of 75 mm will result in a different interference pattern, and consequently a different power level. This is why power control must be performed, and performed rapidly, in the system to attempt to maintain an ideal, even received power level. In the WCDMA system, as will be seen in Chapter 6 power control is performed 1500 times a second. In the IS-95 CDMA system, it is 800/second.

2.9   PROTECTING THE DATA

Despite the shift to data being transferred in digital format, there are still major problems in sending data across the air. In a fixed-line communications system, most of the problems of data transfer and 'data loss' are down to such issues as congestion, where data is stuck in a traffic jam, or buffer overflow, where a network device is being asked to process too much data. What is no longer considered to be a problem is the reliability of the medium over which the data is travelling. Consider a fibre optic cable, which can now be regarded as the standard for data transfer once out of the local loop. Fibre cables cite bit error rate figures of the order of 10^{-20}, and generally bit errors that do occur are bit inversions, that is, a 1 that should be a 0 and vice versa. When this order of error rate is achieved, one can assume that the medium is completely reliable. In fact, many high-speed communications systems use this to their advantage; for example, as will be seen later, ATM provides no error protection whatsoever on data, and does not require a destination to acknowledge receipt of data. In general, fixed-line schemes provide, at best, an error checking mechanism on data, usually in the form of a cyclic redundancy check (CRC). Should data arrive with errors, a rare occurrence, the sender is asked to retransmit, if that level of reliability is required. For example, Ethernet transmits frames of 1500 bytes of payload over which there is a 4-byte CRC, which introduces a relatively low overhead on the data.

However, a wireless communications system is notorious for corrupting data as it travels across the air. So far, cellular systems are focused on voice transmission, which is extremely tolerant of errors. Typically, a voice system can sustain about 1% of error before the errors become audible. With the introduction of mobile data solutions, more often the information being carried across the air is data, such as an IP packet. Unfortunately, data systems are very intolerant of errors, and generally require error-free delivery to an application. For that reason, cellular systems must now implement more rigorous error control mechanisms.

If a simple error checking scheme was introduced, there would be too much retransmission, and the system would spend the majority of the time retransmitting data, thus lowering the overall throughput. A better and more reliable scheme is required. The solution is to implement forward error correction (FEC). With this, a correction code is transmitted along with the data in the form of redundant bits distributed throughout the data, which allows the receiver to reconstruct the original data, removing as many errors as possible. For an efficient and robust wireless communications system, it is essential that a good FEC scheme is used to improve the quality of transmissions.

A problem common to all FEC schemes is the amount of overhead required to correct errors. If a very simple FEC scheme is considered, in which each bit is merely repeated to make the channel robust, then, as shown below, the amount of information to be transmitted is doubled. However, what is lost in bandwidth, by increasing the amount of information to be sent, is gained in the quality of the signal that is received.

Data: 1010101 1010100100100 1

Transmission: 1100110011001111001100110000110000110000 11

The standard terminology is that the data coming from a user application is quantified in bits per second. However, the actual transmission is quantified as *symbols per second*, since this transmission consists of data plus FEC bits. In the case above, one bit is represented by two symbols.

2.9.1   Convolution coding

A popular FEC scheme is convolution coding with Viterbi decoding. Convolution coding is referred to as a channel coding scheme since the code is implemented in a serial stream, one or a few bits at a time. The principle of convolution coding was also developed by Viterbi (1967).

Convolution coding is described by the code rate, k/n, where k is the number of bits presented to the encoder, and n the number of symbols output by the encoder. Typical code rates are $\frac{1}{2}$ rate and $\frac{1}{3}$ rate, which will double and triple the quantity of data respectively.

For example, to transmit a user application which generates a data rate of 144 kbps with $\frac{1}{2}$ rate convolution coding, the transmission channel will be operating at 288 ksps.

At the receiver, the data is restored by a Viterbi decoder. This has the advantage that it has a fixed decoding time and can be implemented in hardware, introducing minimal latency into the system. Current commercial Viterbi decoders can decode data at a rate in excess of 60 Mbps at the time of writing.

By implementing convolution coding, as already mentioned, there is a tradeoff in that the bandwidth is either doubled ($\frac{1}{2}$ rate) or tripled ($\frac{1}{3}$ rate). However, the upside is that a good convolution coding scheme will provide a 5 dB gain across the air interface for a binary or quadrature phase shift keying (BPSK or QPSK) modulation scheme. This means that a coded signal can be received with the same quality as an uncoded signal, but with 5 dB less transmit power.

Turbo coding is an advanced form of convolution coding which uses parallel concatenation of two turbo codes. Turbo coding, developed in 1993 at the research and development centre of France Telecom, provides better results than standard convolution coding. Turbo coding is recommended for error protection of higher data rates, where it will typically provide bit error rates of the order of 10^{-6}.

Both codes are designed to reduce the interference effects of random noise, or additive white Gaussian noise (AWGN). In CDMA systems, the source of most of this noise is other wideband user signals.

There are many other FEC schemes available, such as Hamming codes and Reed–Solomon codes.

2.9.2 Interleaving

Despite the dramatic improvements that a FEC scheme such as convolution coding introduces to the wireless system, it is not specifically designed to eliminate burst errors. Unfortunately, across the air interface errors usually occur in bursts where chunks of data are lost. Some additional protection is required to cope with the reality of the air interface.

To solve this, blocks of data are interleaved to protect against burst errors. Consider the transmission of the alphabet. To transmit, it is first split into blocks, as shown in Figure 2.24. These blocks are then transmitted column by column.

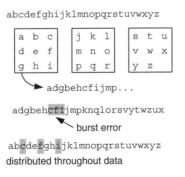

Figure 2.24 Principle of interleaving

If, subsequently, there is a burst error in the data, once the interleaving process is reversed, this error is distributed through the data, and can then be corrected by the convolution coding mechanism. This concept is illustrated in the lower part of Figure 2.24.

2.10 SUMMARY

This chapter addresses the basic concepts of both packet switched networks and cellular systems. Crucial to these is the transport of voice over a packet network, and the basic issues with regard to this are highlighted. For any cellular system, a multiple access mechanism must be present to allow many subscribers to share the resources of the network, and the main methods used in cellular are described. Arguably the most complex aspect of 3G is the use of CDMA as the air interface of choice, and the key principles of CDMA are described here, as well as the mechanisms to address problems of loss of data in radio transmission.

FURTHER READING

A. S. Tanenbaum (2003) *Computer Networks*, 4th edn. Prentice Hall, Upper Saddle River, NJ.

H. Taub, D. Schilling (1986) *Principles of Communication Systems*. 2nd edn. McGraw-Hill, New York.

A. J. Viterbi (1995) *CDMA: Principles of Spread Spectrum Communication*. Addison-Wesley, Reading, MA.

A. J. Viterbi (1967) 'Error bounds for convolutional codes and an asymptotically optimum decoding algorithm', *IEEE Transactions on Information Theory* **IT-13**, 260–269.

H. Holma, A. Toskala (2002) *WCDMA for UMTS*, 2nd edn. John Wiley&Sons, Chichester.

J. Laiho, A. Wacker, T. Novosad (2002) *Radio Network Planning and Optimisation for UMTS*, John Wiley&Sons, Chichester.

A list of the current versions of the specifications can be found at http://www.3gpp.org/specs/web-table_specs-with-titles-and-latest-versions.htm, and the 3GPP ftp site for the individual specification documents is http://www.3gpp.org/ftp/Specs/latest/

3

GSM Fundamentals

The roots of the development of the global system for mobile communications (GSM) began with a group formed by the European Conference of Postal and Telecommunications Administrations (CEPT) to investigate the development of a standard mobile telephone system to be used throughout Europe. This group was known as the Groupe Special Mobile, or GSM for short, and this is initially where the acronym GSM came from; however, it now is widely understood to stand for global system for mobile communications. A unified telephone system was desirable since Europe is made up of many separate countries each with their own government, language, culture and telecommunication infrastructure, much of which was still in the hands of state-run monopolies. As there is much trade between these countries, a mobile network which would free users to roam internationally from country to country was seen as a valuable asset.

The other major region to discuss in parallel is movements in mobile communications in the USA. Mobile technology was advancing there also, but the motivation to provide roaming capabilities was not such a fundamental requirement, since it is one country. There was and is considerable regionalization of communications in the USA and this was reflected in the proliferation of mobile devices, where operators only needed to cater for the domestic market.

GSM was eventually adopted as a European standard by the European Telecommunications Standards Institute (ETSI). It has been standardized to operate on three principal frequency regions, being 900 MHz, 1800 MHz and 1900 MHz.

GSM is by far the most successful of the second generation cellular systems, and has seen widespread adoption not only across Europe but also throughout the Asia-Pacific region, and more recently, the Americas. Some of the large mobile network operators in the USA are also introducing GSM, either as a migration step towards the UMTS flavour of 3G or simply in addition to the current offerings. Two such operators that are deploying GSM are Voicestream and AT&T wireless. Other existing systems include IS-136 (TDMA), IS-95 or cdmaOne (CDMA), and PDC. Japan tried to mimic the success of GSM with its home-grown PDC system, intending for this to spread throughout Asia.

Convergence Technologies for 3G Networks: IP, UMTS, EGPRS and ATM   J. Bannister, P. Mather and S. Coope
© 2004 John Wiley & Sons, Ltd   ISBN: 0-470-86091-X

However, the popularity of GSM brought to bear the economies of scale and thus PDC is only evident in Japan. The success of GSM in Asia is not surprising as many of the original ideas in the design of a network that would transcend political borders are also relevant in Asia.

Currently, at time of writing, GSM technology has over 70% global market share of second generation cellular systems. As networks evolve to 3G, GSM should not be seen as becoming redundant, but rather GSM is an integral part of the 3G UMTS network infrastructure as it also evolves to the GSM-EDGE radio access network (GERAN).

3.1 GENERAL ARCHITECTURE

Figure 3.1 shows the general architecture for a GSM network. The various functional blocks are explained in the following subsections.

Mobile station (MS)

The MS consists of the mobile equipment (ME; the actual device) and a smart card called the subscriber identity module (SIM). The SIM offers personal mobility since the user can remove the SIM card from one mobile device and place it in another device without informing the network operator. In contrast, most other 2G systems require a registration update to the operator. The SIM contains a globally unique identifier, the international mobile subscriber identity (IMSI), as well as a secret key used for authentication and other security procedures. The IMSI (or a variation of it for security purposes) is used throughout the network as the identifier for the subscriber. This system enables a subscriber to change the mobile equipment and still be able to make calls, receive calls and receive other subscriber information by simply transferring the SIM card to the new device. Any calls made will appear on a single user bill irrespective of changes in the mobile device. The mobile equipment is also uniquely identifiable by the international mobile equipment identity (IMEI). The IMEI and IMSI are independent, thus providing the user flexibility by separating the concept of subscriber from access device. Many operators still issue 'locked' mobile devices where the equipment is tied for use only on a particular operator's network. A mobile device not equipped with a SIM must also still be able to

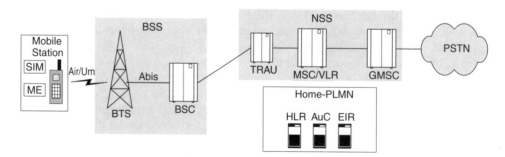

Figure 3.1 GSM general architecture

Figure 3.2 GSM IMEI and IMSI

make emergency calls. To protect the call from undesirable snooping or listening in, the IMSI will not always be transmitted over the cell to identify the subscriber. Instead a temporary IMSI (T-IMSI) identifier is used and changed at regular intervals. Note that for extra security the whole data stream is encrypted over the air interface. Figure 3.2 shows the 15-digit IMEI number on the left and the SIM card, which incorporates the 15-digit IMSI.

Base station subsystem (BSS)

The base station subsystem (BSS) is composed of three parts, the base transceiver station (BTS), the base station controller (BSC), which controls the BTSs, and the transcoding and rate adaption unit (TRAU).

Base transceiver station (BTS)

The BTS houses the radio transceivers (TRXs) that define a cell and handle the radio link with the mobile station. As was seen, each TRX can handle up to eight full-rate users simultaneously. If more than eight full-rate users request resources within the TRX then they will receive a busy tone, or a network busy message may be displayed on the mobile device. It is possible to increase the number of simultaneous users in a cell by increasing the number of TRXs, hence the number of frequencies used. When a mobile device moves from one cell to another the BTS may change. Within the GSM system a mobile device is connected to only one BTS at a given time. The first TRX in a cell can actually only handle a maximum of seven (possibly less) simultaneous users since one channel on the downlink is used for broadcasting general system information through what is known as the broadcast and control channel (BCCH). The BTS is also responsible for encrypting the radio link to the mobile device based on security information it receives from the core network.

Base station controller (BSC)

The BSC manages the radio resources for one or more BTSs. It handles the radio channel setup, frequency hopping and handover procedures when a user moves from one cell to another. When a handover occurs, the BSC may change; it is a design consideration that this will not change with the same regularity as a BTS change. A BSC communicates with the BTS through time division multiplex (TDM) channels over what is referred to as the Abis interface, generally implemented using E1 or T1 lines. If the numerous BTSs and the corresponding BSC are in close proximity then this link may be a fibre optic or copper cable connection. In some cases, there are a large number of BTSs in close proximity but quite some distance away from the controlling BSC. In such cases it may be more efficient to relay the calls from each of the BTSs to a single BTS via microwave links. This type of link may be very cost effective since generally the running costs of a point-to-point microwave link may be free. Of course this has to be weighed against the cost of the purchasing and deployment of the equipment. The collector BTS can then connect to the BSC via another microwave link or via a landline cable. A problem with the above system is that if the collector BTS fails then calls from the other BTSs may also fail. To overcome this problem it is possible to have two collector BTSs both sending the calls to the BSC. This forms a redundant link and if one collector BTS fails then this does not present such a large problem, as is illustrated in Figure 3.3(b).

Transcoding and rate adaption unit (TRAU)

The central role of the second generation systems is to transfer speech calls and the system has been designed and optimized for voice traffic. The human voice is converted to binary in a rather complex process. GSM is now quite an old system and as such the original encoding method used (LPC-RPE[1]) is not as efficient as some of the more recently developed coding systems such as those used in other cellular systems. There have been many developments in digital signal processing (DSP) which have enabled good voice quality to be transmitted at lower data rates. Although the TRAU is actually

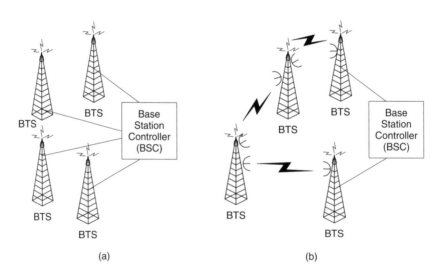

(a)   (b)

Figure 3.3   Base station connectivity

---

[1]Linear predictive coding with regular pulse excitation (LPC-RPE) provides a digital model of the vocal tract and vocal chords, excited by a signal which is air from the lungs.

seen as being logically part of the BSS, it usually resides close to the MSC since this has significant impact on reducing the transmission costs. The voice data is sent in a 16 kbps channel through to the TRAU from the mobile device via the BTS and BSC. The TRAU will convert this speech to the standard 64 kbps for transfer over the PSTN or ISDN network. This process is illustrated in Figure 3.4, where over the air interface, speech uses 13 kbps (full-rate) and data 9.6 or 14.4 kbps, with each of these requiring a 16 kbps link through the BSS.

As has been mentioned, digital voice data is robust in the face of errors, and can handle substantial bit error rates before the user begins to notice signal degradation. This is in stark contrast to data such as IP packets, which is extremely error intolerant and a checksum is generally used to drop a packet which contains an error. Table 3.1 lists the adaptive multirate (AMR) speech CODECS which are implemented in UMTS. Also indicated on the diagram are the enhanced full-rate (EFR) bit rates for the second generation GSM, TDMA and PDC systems for comparison. The GSM EFR uses the algebraic code excited linear prediction (ACELP) algorithm and gives better quality speech than full-rate (FR) using 12.2 kbps. A half-rate (HR) method of speech coding has also been introduced in to the standards, which is known as code excited linear prediction-vector sum excited linear prediction (CELP-VSELP). This method will enable two subscribers to share a single time slot.

Network switching subsystem (NSS)

The NSS comprises the circuit switched core network part of the GSM system. The main element is the mobile switching centre (MSC) switch and a number of databases referred

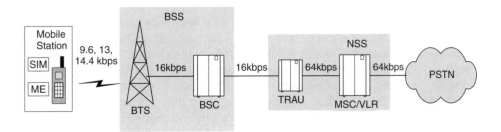

Figure 3.4 Transcoding

Table 3.1 CODEC bit rates

CODEC	Bit rate (kbps)
AMR_12.20	12.2 (GSM EFR)
AMR_10.20	10.2
AMR_7.95	7.95
AMR_7.40	7.4 (TDMA EFR)
AMR_6.70	6.7 (PDC EFR)
AMR_5.90	5.9
AMR_5.15	5.15
AMR_4.75	4.75

to as the visitor location register (VLR) and home location register (HLR). The HLR is always in the home network for roaming subscribers and thus any data exchange may have to cross international boundaries. The MSC and VLR are usually combined and are located in the visited network.

Mobile switching centre (MSC)

This acts like a normal switching node for a PSTN or ISDN network. It also takes care of all the additional functionality required to support a mobile subscriber. It therefore has the dual role of both switching and management. When a mobile device is switched on and requests a connection to a mobile network, it is principally the MSC that processes this request, with the BSS merely providing the access to facilitate this request. If the request is successful then the MSC registers the mobile device within its associated VLR (see below; most manufacturers tend to combine the VLR functionality with the MSC). The VLR will update the HLR with the location of this mobile device, and the HLR may be either in the same network, or a different network in the case of a roaming user. The MSC deals with registration, authentication (the MSC requests information from the authentication centre but it is the MSC which actually does the authentication), mobile device location updating and routing of calls to and from a mobile user. An MSC which provides the connectivity from the mobile network to the fixed network, e.g. ISDN or PSTN, is known as a gateway-MSC (G-MSC).

Home location register (HLR)

When a subscriber registers with an operator, they enter into what is known as a service level agreement (SLA). This operator's mobile network is known as the home network or home public land mobile network (H-PLMN). The HLR is a huge database located within this home network which stores administrative information about the mobile subscriber. The information stored for a user in the HLR will include their IMSI, service subscription information, service restrictions and supplementary services. The HLR is also expected to know the location of its mobile users. It actually knows their location only to the VLR with which the mobile device is registered. The HLR also only knows the location of a mobile device which is switched on and has registered with some mobile operator's network. This is the case even if the mobile is in a different country connected to another mobile operator's network, as long as a roaming agreement exists between the two mobile operators. The GSM system provides all the technical capabilities to support roaming; however, this roaming agreement is also required so that both operators can settle billing issues arising from calls made by visiting mobile subscribers.

Visitor location register (VLR)

The VLR is another database of users and is commonly integrated with an MSC. Unlike the HLR, where most information is of a permanent nature, the VLR only holds temporary information on subscribers currently registered within its vicinity. This vicinity covers the subscribers in the serving area of its associated MSC. When a mobile device enters a new area, the mobile device may wish to connect to this network and if so informs the MSC of its arrival. Once the MSC checks are complete, the MSC will update the VLR. A message is sent to the HLR informing it of the VLR which contains the location of the mobile. If the mobile device is making or has recently made a call, then the VLR will know the location of the mobile device down to a single cell. If the mobile device has requested and been granted attachment to a mobile network, but not made any calls recently, then the location of the mobile device will be known by the VLR to a location area, i.e. a group of cells and not a single cell. A mobile device that is attached to a mobile network where a roaming agreement is in force, i.e. is not in its H-PLMN, is said to be in a visited PLMN (V-PLMN).

Equipment identity register (EIR)

The EIR is a list of all valid mobiles on the network. If a terminal has been reported stolen or the equipment is not type approved then it may not be allowed to operate in the network. The terminals are identified by their unique IMEI identifier.

Authentication centre (AuC)

The AuC is a database containing a copy of the secret key present in each of the users' SIM cards. This is used to enable authentication and encryption over the radio link. The AuC uses a challenge–response mechanism, where it will send a random number to the mobile station; the mobile station encrypts this and returns it. The AuC will now decrypt the received number and if it is successfully decrypted to the number originally sent, then the mobile station is authenticated and admitted to the network.

3.2 MOBILITY MANAGEMENT

To make and receive calls, the location of the mobile device has to be known by the network. It would be extremely inefficient if a user needed to be paged across an entire network, and almost impossible to support roaming to other networks. Each cell broadcasts its globally unique identity on its broadcast channel, which is used by the mobile device for location purposes. Mobility management is the mechanism that the network uses for keeping a dynamic record of the location of all of the mobile devices currently active in the network. In this context, location does not refer specifically to the geographical location of the mobile device, but rather its location with respect to a cell in which it is currently located. However, for the development of cellular towards third generation, geographical location becomes important as an enabler for location-based services (LBS).

The major benefit of the cellular telephone over a fixed landline is the mobility that it presents to the subscriber. Initially, this mobility was merely allowing the user to move around and be tracked within a certain area; however, now mobility extends to cover the concept of roaming. Unfortunately, the provision of mobility makes the network much more complex to design and operate. As a subscriber moves from one location to another, the strength of the signal it receives from the base station to which it is currently listening will fluctuate, and, conversely, the signal received by the base station from the mobile device will also vary. Both the network and the mobile device must constantly monitor the strength of the signal, with the mobile device periodically reporting the information it has measured to the network. The mobile device also monitors the strength of other cells in the vicinity. When the signal strength gets too weak from a particular base station, a handover (also known as a handoff) to a base station in another cell may take place. The network must try to guarantee that in the event of a handover, the user call is not dropped and there is a smooth transition from cell to cell, even if the user is moving quite rapidly, as is the case for a motorist. Figure 3.5 indicates the information that is stored within the network when a mobile device is in idle and dedicated mode. The HLR, which is in the home network, knows which VLR has information regarding the particular subscriber. The information the VLR holds depends on the connection state of the mobile device: in idle mode only the location area (LA) is known whereas in dedicated mode the actual cell is known.

Much of the GSM mobile network is designed and implemented in a hierarchical manner and it can be seen from Figure 3.6 that as a subscriber's geographical location changes, there may be rather frequent movements from one cell to another. The change of a cell from one base station to another is relatively simple if the BTSs are controlled by the same BSC. The change of a BSC is more complex and hence will require more signalling but will occur less frequently since each BSC controls a number of BTSs. A change of the MSC is also possible but, again, this should be rather infrequent for most users. If a user is in a vehicle and moving at high speed, then a number of MSC handovers may take place during a prolonged voice call. However, this will probably occur rarely as the vehicle will likely have crashed or the driver been arrested before handover occurs! This system of handover enables a subscriber to continue with a call in progress while moving from one geographical area to another. This is illustrated in Figure 3.6.

- When User 1 changes from one cell to another, a cell update is required. As noted, this does not require much in the way of signalling.
- When User 2 changes cell, a cell update and a BSC update are required. This will require more signalling, with the MSC controlling the change in BSC.
- When User 3 changes cell, a cell update, a BSC update and an MSC update are required. This is a much more complex task, which will require a greater amount of signalling.

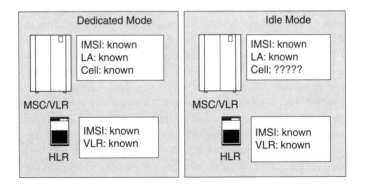

Figure 3.5   GSM idle and dedicated mode

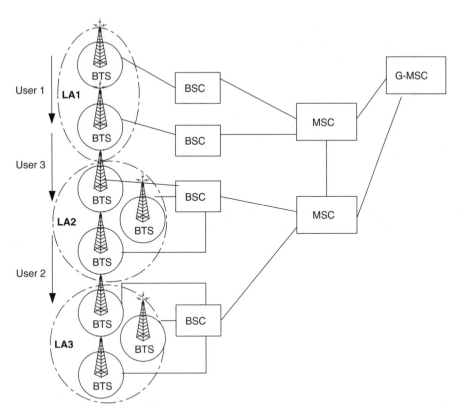

Figure 3.6 Location area updates

Note that these updates only take place when a mobile device has a call in progress, or in what is referred to as dedicated mode. Mobile devices which do not have a call in progress but may have registered with the network are said to be in idle mode. Mobile devices in idle mode will only send periodic updates indicating that the mobile is still active, thus reducing the signalling load on the network. When a user wishes to make a call, the mobile device will transparently update the network as to its position and move to dedicated mode. In idle mode the location of the mobile device is still known but over a number of cells rather than a single cell. In idle mode the mobile device monitors a certain area spanning a number of cells, known as a Location Area (LA, see below), and sends location update information to the network when:

1. The mobile device physically crosses a boundary between LAs.
2. A certain period of time has elapsed. Even when the mobile device is stationary, after a long period of inactivity it will send an update to allow the network to refresh its stored information regarding the subscriber's location. Devices which do not send this update will be assumed to have left the coverage area and their data may be removed from the network. This interval is network configurable and could be, for example, one hour.

Location area

An LA is a group of neighbouring cells that are controlled by a single MSC. As illustrated in Figure 3.6 an MSC may control a number of LAs.

- The location of a mobile device in dedicated mode (making a call) is known down to the cell level.
- The location of a mobile device in idle mode is only known down to the LA level. In idle mode the mobile device can still listen to cell broadcasts and monitor for paging, so that it can listen for incoming (i.e. mobile terminated) calls.

There is a tradeoff between a small LA and a large grouping of cells. A small LA means that there are many location updates as subscribers move from LA to LA, which may cause signalling congestion. There is also an issue of the power used by the mobile device to transmit these updates, which may eventually cause the battery to go flat. A larger LA, on the other hand, means that when it is required to locate a mobile device (for a mobile terminated call) a page over *all* the cells in the LA is required. This increases the downlink paging signalling, even in those cells within the LA where the mobile device is not located. The size of an LA is very much in the hands of the mobile operator and its network planning process. An LA could be one cell, or it could be all the cells under one MSC. Generally, an urban area will have smaller LAs to reduce the amount of signalling load during subscriber paging. The LA planning is in line with the demographic profile of the country.

One problem that often occurs in practice with LAs is a ping-pong effect, where a subscriber moves a relatively short distance across an LA boundary, and performs an LA update, only to move back again, resulting in another LA update. This problem can be alleviated to some extent by network planning.

3.3 GSM AIR INTERFACE

There is a limited spectrum of frequencies that is both available and suitable for GSM. Cellular operators have to compete for this bandwidth with the likes of the military, broadcast television and broadcast radio. The available electromagnetic spectrum has been split into a number of bands by both national and international regulatory bodies. However, in many cases, the national regulatory bodies of some countries had already allocated the spectrum internally before international standardization had been ratified. The resulting effect is that cellular spectrum bands are not exactly the same worldwide. Fortunately there was much international agreement on the frequencies in the 900 MHz and 1800 MHz bands, which brought in large economies of scale, reducing the price of handsets, and thus enabling GSM to flourish. GSM was originally designed to work in a 900 MHz band but is now used in 1800 MHz, 1900 MHz and a number of others, such as 450 MHz. As shown in Figure 3.7, the 900 MHz range is made up of two separate 25 MHz bands, between 890–915 MHz and 935–960 MHz. The lower 25 MHz is used for the mobile station, or uplink, transmission and the upper 25 MHz of the range is

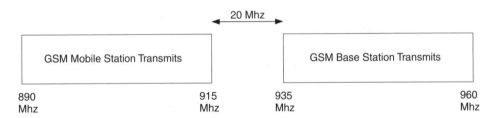

Figure 3.7   Original GSM band

used for base station, or downlink, transmission. There is a gap of 20 MHz between the transmission sub-bands i.e. the GSM base station transmit band starts at $890 + 45$ MHz. The mobile device transmits on the lower frequency since it is a physical property of electromagnetic waves that there will generally be less attenuation on lower frequencies. The base station is not reliant on a small battery and can therefore radiate greater power, thus the greater attenuation in the downlink is not seen as a major problem, allowing the mobile device to avail itself of better transmission characteristics.

As discussed, GSM works on a combination of frequency division multiplexing (FDM), and time division multiplexing (TDM) multiple access schemes. It also uses slotted-Aloha, a contention method which is similar in operation to Ethernet. This contention mechanism is required since it is possible for two mobile subscribers to make a request for resources at exactly the same time. The mobile stations use this contention method to compete with each other to request a traffic channel (TCH), which is required for a call. Like Ethernet, there is a chance that a collision will occur, so mechanisms are implemented to deal with this.

The FDM allocates each GSM channel 200 kHz of bandwidth and therefore there are 25 MHz/200 kHz = 125 channels available in each direction. One of these channels is not used for data transfer but is used as a guard band, leaving 124 channels available for communication. A matching pair of GSM frequency channels, i.e. one uplink and a corresponding downlink, is controlled by a device referred to as a transceiver (TRX). All of the operators in a country using GSM900 have to share these 124 channels and they will be allocated a licence covering a range of them by the national telecommunications regulator. Say there are four mobile operators in a given country. Each of them may be allocated 31 channels (124/4). For example, Operator 1 may be allocated 31 channels starting from 890.0 MHz, 890.2 MHz, 890.4 MHz etc. up to 896.0 MHz in the uplink and 935.0 MHz, 935.2 MHz, 935.4 MHz etc. up to 941.0 MHz in the downlink, as shown in Figure 3.8.

TDM further splits each of these frequency channels into eight separate time slots, each of which may be allocated to a user or used for control purposes. These time slots are individually referred to as slot 0 through to slot 7, and form a *TDM frame*. A single time slot in GSM is also referred to as a *burst*; however, this should not be confused with the term 'error burst'. If a cell is allocated a single frequency (one TRX) then slot 0 on this frequency is reserved as a control channel. If two or more frequencies are employed within the cell then it may require additional control channels to increase the overall efficiency.

The slot 0 control channel always includes the broadcast and control channel (BCCH), which is broadcast from the base station in the downlink to provide information to the mobile devices registered in the cell, such as the cell identifier, network operator etc.,

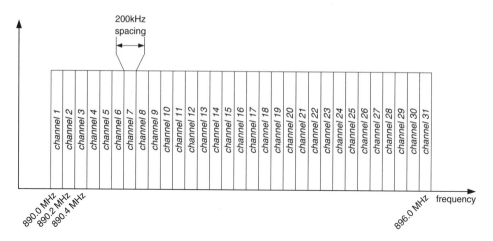

Figure 3.8 200 kHz GSM uplink channels

thus the maximum number of users for the first TRX is actually only seven. If more than
one frequency is used within the cell these additional frequencies can share the control
channel of the first TRX and therefore have eight time slots available for subscriber
traffic channels, allowing slot 0 on these additional frequencies to be used as a normal
traffic channel. A single time slot allocated to control and signalling may not possess a
high enough bit rate, giving rise to congestion on the control channel. This may occur,
for example, in a cell which contains a large number of short message service (SMS)
users, since SMS in GSM is sent over the control channels, or in an airport where many
users may try to access the network at the same time. In this case, the operator may set
aside another time slot to alleviate this congestion. The actual configuration is specific to
the unique requirements of a particular cell. The rapid rise globally of SMS traffic has
presented somewhat of a dimensioning dilemma to operators for this reason.

Figure 3.9 shows an active TRX, where the shaded boxes indicate a particular time
slot currently in use. The top frame indicates the uplink transmission from the mobile
station and the bottom frame indicates the downlink transmission from the base station.
It can also be seen from the diagram that the transmission by the base station is actually
retarded by the duration of three time slots so that the TRXs do not have to listen while
transmitting.[2] Note that the bursts are still paired, e.g. if the mobile device transmits on
time slot 1 then it will receive on time slot 1 because the whole frame has been retarded
by three time slots.

Figure 3.10 shows an example of how the eight time slots can be used. It can be seen
that over time the first time slot (time slot 0) starts off broadcasting system information
across the cell. This same time slot is then used to transfer an SMS message to a single
subscriber and is then used for a location update for a new mobile device, which has
just entered this cell. The other time slots are used by subscribers as and when they wish

---

[2]The mobile station may be transmitting a 0.3 W signal; it is expected to be able to receive a very
weak signal of about 1 millionth of a watt from the base station on the same antenna. Clearly if it
is transmitting and receiving at the same time this is particularly difficult to do.

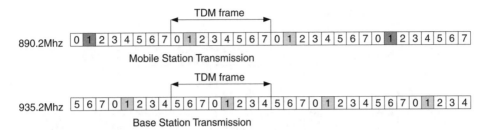

Figure 3.9 TDM channels in GSM

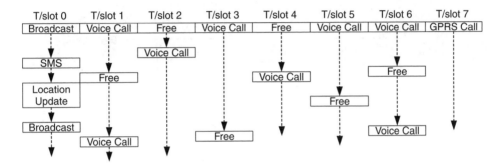

Figure 3.10 Example use of time slots

to make telephone calls (voice or data). The duration of these obviously depends on the call itself.

The network selects which band to use, whether there is an option, for example 900 MHz or 1800 MHz, the frequency, if there is more than one available in this cell, and the time slot the subscriber will use. This selection is completely transparent to the subscriber; in fact, the frequency and time slot will normally change throughout the call, with the user perhaps noticing a small amount of interference or possibly nothing at all. The ability to change the frequency of a subscriber is required when a subscriber roams from one cell into another since adjacent cells cannot use the same frequency.

Within a cell, most GSM systems implement *frequency hopping*, where the BTS and mobile unit transmit consecutive frames on different carrier frequencies across the radio channels. Frequency hopping is a key part of the GSM system and is used to alleviate some of the inherent problems with radio links such as multipath fading by avoiding prolonged use of one frequency. The BCCH channel is not part of the frequency-hopping scheme since it needs to be located by mobile devices wishing to connect to the cell. Figure 3.11 below shows a simplified example of the frequency hopping process. In a practical situation each individual time slot can hop independently to the TDMA frame.

3.3.1 GSM multiframes

The above eight time slot framing structure as shown in Figure 3.9 is part of a much larger multiframe. There are two types of multiframe: the traffic channel and control

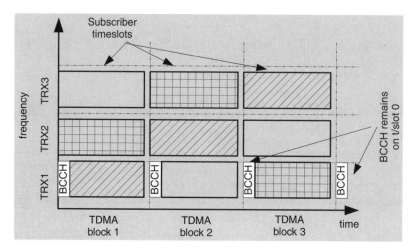

Figure 3.11   Frequency hopping in GSM

channel multiframes. The traffic channel multiframe consists of 26 groups of 8 TDM frames whereas the control multiframe consists of 51 groups of frames.

3.3.2   Traffic channel multiframe

Although this is a designated traffic channel consisting of 26 frames, at most only 24 of these are used for TCH user data such as voice. The frame is also used to carry two logical control channels, the slow associated control channel (SACCH) and the fast associated control channel (FACCH). Time slots 12 and 25 are specifically used for SACCH. In fact, the SACCH may actually alternate between these on different multiframes and so there is only one SACCH channel transmitted per multiframe. During the frame that the mobile device is not transmitting or receiving on its dedicated TCH, it is constantly monitoring the strength of the received signals from the cell it is attached to as well as other cells in the area. The mobile station can actually monitor up to six surrounding cells. The SACCH is then used for sending these measurement results to the network. A SACCH message is 456 bits long (4 bursts × 114 bits per burst); it therefore has a time interval of 480 ms before repeating itself. This information can be used to increase or decrease the transmitted power levels of both the mobile device and the network every 480 ms, or just a little over twice a second. The mobile device uses this information it receives but actually alters its power in steps every 60 ms. Enhanced circuit switched data (ECSD) can also use a fast power (FP) method which is sent within the data stream every 20 ms. Figure 3.12 illustrates the SACCH power control mechanisms: both the mobile device and the BTS send measurement reports to the BSC, which makes the decision to increase or decrease power.

The SACCH may also be used for sending SMS messages to and from the mobile device while a call is in progress. User data frames 0–11 and 13–24 also carry control information in the form of the FACCH channel, as described below. A single user voice

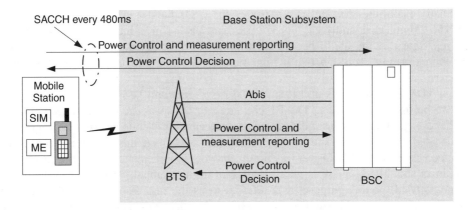

Figure 3.12   SACCH power control

call will be transmitted in frame 0 of the multiframe for a particular amount of time, followed by a particular amount of time in frame 1 etc. Eight such calls can be made on this carrier frequency. A traffic channel is only assigned when the mobile device is in dedicated mode, whereas in idle mode the mobile device does not have a traffic channel assigned to it. Figure 3.13 shows the relationship of the TDM frame within a traffic channel multiframe.

Each of the eight time slots in the TDM frame consists of 148 bits lasting 547 μs and has the structure shown in Figure 3.14.

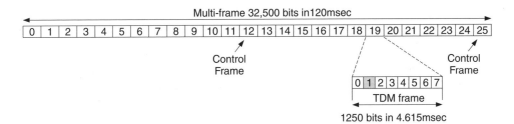

Figure 3.13   GSM multiframe

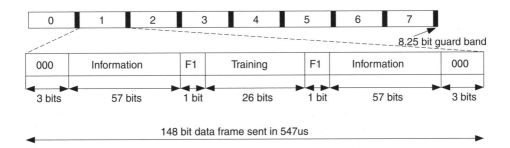

Figure 3.14   GSM data burst

Actually each time slot (burst) consists of the GSM burst and the guard band and is thus 156.25 bits long and takes up 577 μs, with each single bit taking approximately 3.69 μs. Note that the guard band consisting of 8.25 bits is only used as a time interval (approximately 30 μs) between time slots to separate one user transmission from the next.

Each TDM burst consists of 148 bits. The first 3 bits and last 3 bits are always zero and can be used for framing. The F1 bits are referred to as *stealing bits* and indicate whether the burst contains user data or control information. While a call is in progress, if it is decided that a handover to another cell should take place, this handover needs to be executed very quickly. The fastest way for the network to indicate to the mobile device that a handover should take place is via its dedicated traffic channel (TCH). When the traffic channel is used for this purpose, it is referred to as the fast associated control channel (FACCH). The stealing bits indicate to the subscriber that this burst is a control message and not user data. In the uplink, the mobile device will indicate control information to the network in the same manner. Using the traffic channel in such a way exposes the subscriber to a minimal amount of increased interference while the handover is executed. However, in most cases, the subscriber will not notice this extra interference. The training field is used to synchronize the transmitter and receiver. It is also used to 'train' the receiver to the particular characteristics of the channel being used. The receiver knows exactly what the transmitter has sent in this field since this is negotiated beforehand. By using this knowledge it can check for any distortion between the transmitted and received training field signal. This information can then be used to 'equalize' the information fields that are received, reducing the possibility of errors.

3.3.3   Control channel multiframe

In addition to the traffic channel multiframe, there is also a 51-frame multiframe which is employed on the control and signalling channels. This multiframe consists of several logical channels, which incorporate control, timing and signalling. These channels are highlighted below.

Broadcast channel (BCCH)

The BCCH is a continuous stream of data from the base station containing its identity and channel status. All mobile stations can monitor the strength of this signal to ensure that they are still within the cell.

Frequency correction channel (FCCH)

The FCCH is used to ensure that the mobile device adjusts its frequency reference to match that of the base station so that the mobile device does not drift off frequency, reducing the voice quality of the call. The base station simply emits a sine wave on this channel for the duration of a time slot.

Synchronization channel (SCH)

The synchronization channel is used to frame synchronize the mobile device. The base station identity code (BSIC) is also broadcast on the SCH so that the mobile device can tell if the base station it tries to connect to is part of the correct network.

Common control channel (CCCH)

The CCCH is split into three sub-channels; one is used in the uplink and two are used in the downlink:

- *Uplink*: the *random access channel* (RACH), allows a mobile station to request a time slot on the dedicated control channel, which can then be used, for example, to assign a traffic channel for a voice call. The RACH channel utilizes the slotted-Aloha method.
- *Downlink*: the *paging channel* (PCH) is used to alert the mobile station of an incoming call. The base station will announce the assigned slot on the *access grant channel* (AGCH).

Standalone dedicated control channel (SDCCH)

This is used for call setup and location updating of mobiles, as well as for SMS messages to and from mobile devices which are in idle mode. When an initial request for a connection is made by the mobile device on the RACH, the network responds on the AGCH by issuing the mobile device an SDCCH channel rather than directing it immediately to a TCH. By incorporating the SDCCH channel for call setup rather than switching directly to a TCH, the overall efficiency of the GSM system can be increased. This is possible since initial call setup does not require the relatively high bandwidth that a TCH supports. By utilizing the SDCCH, which is a much lower bandwidth channel, the TCH is not tied up and is available for user calls.

Slow associated control channel (SACCH)

As previously mentioned, the SACCH is a bidirectional channel which is used to transfer measurement reports to and from the network. The SACCH may also be used for transfer of SMS data if a traffic channel is allocated to the particular subscriber.

Fast associated control channel (FACCH)

The FACCH is an in-band signalling channel which interrupts user data to transfer system information in both the uplink and downlink. This channel is used when information needs to be transferred quickly, such as when a handover is required.

Figure 3.15 shows the logical channels defined for GSM. The broadcast, common and dedicated control channels together are referred to as the signalling channels.

Figure 3.16 shows a control channel, which consists of the FCCH, SCH, BCCH, CCCH, SDCCH and SACCH channels. This is a single frequency (time slot 0) control channel of 51 frames. However, it actually repeats every 102 frames. Using this method will allow the other seven time slots on this particular frequency to be used by subscribers for calls.

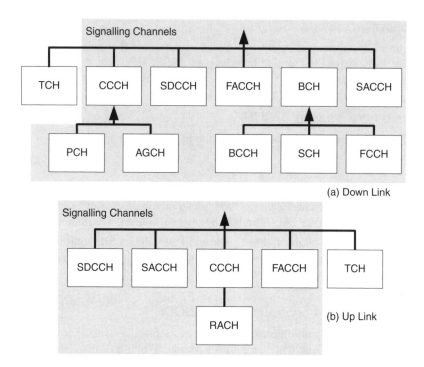

Figure 3.15   GSM logical channels

An alternative method of implementing a control channel is shown in Figure 3.17. Slot 0 comprises FCCH, SCH, BCCH and CCCH. In this example there would be a separate channel (time slot 1) for the SDCCH/SACCH. This method would be used where there is more than one frequency being used in the cell since only six time slots on this particular frequency will be available for subscriber calls.

3.3.4   Frames, multiframes, superframes and hyperframes

Each of the TDMA frames is numbered, and this number is used as an input parameter for the encryption process. To ensure that the number is quite large a hyperframe has been defined. To summarize, a frame consists of 8 TDMA time slots (a total duration of 4.615 ms), a multiframe consists of a block of 26 of these time slots, primarily for data (although some control information is actually transferred) and 51 for control purposes. A superframe consists of 26×51 TDMA frames with a duration of approximately 6.12 s. Since 26 is not a factor of 51, these frames 'slide across' each other so that at the end of the 26×51 period each of the 26 frames has aligned once with every one of the 51 control frames. This sliding process is required to allow the mobile device to monitor the quality of the surrounding cells' BCCH for handover purposes. The hyperframe consists of $2048 \times 26 \times 51$ TDMA frames lasting approximately 3 h 28 min and 54 s.

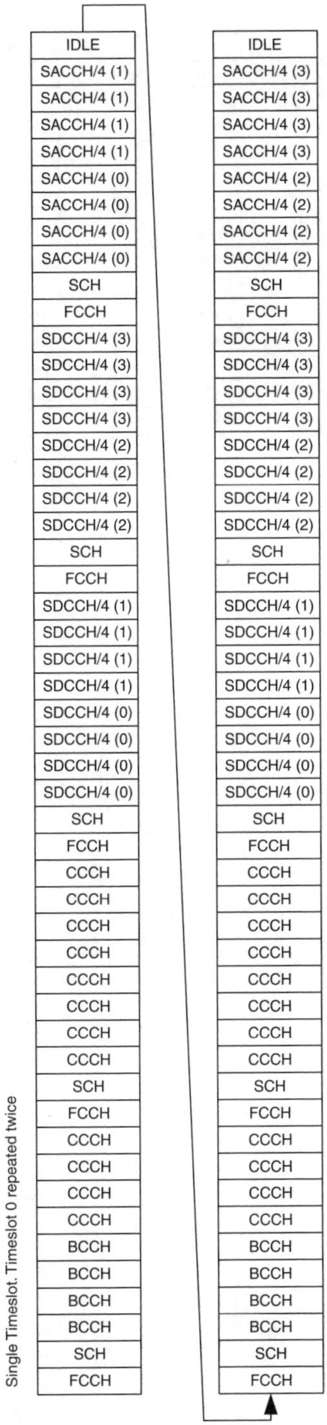

IDLE		IDLE
SACCH/4 (1)		SACCH/4 (3)
SACCH/4 (1)		SACCH/4 (3)
SACCH/4 (1)		SACCH/4 (3)
SACCH/4 (1)		SACCH/4 (3)
SACCH/4 (0)		SACCH/4 (2)
SACCH/4 (0)		SACCH/4 (2)
SACCH/4 (0)		SACCH/4 (2)
SACCH/4 (0)		SACCH/4 (2)
SCH		SCH
FCCH		FCCH
SDCCH/4 (3)		SDCCH/4 (3)
SDCCH/4 (3)		SDCCH/4 (3)
SDCCH/4 (3)		SDCCH/4 (3)
SDCCH/4 (3)		SDCCH/4 (3)
SDCCH/4 (2)		SDCCH/4 (2)
SDCCH/4 (2)		SDCCH/4 (2)
SDCCH/4 (2)		SDCCH/4 (2)
SDCCH/4 (2)		SDCCH/4 (2)
SCH		SCH
FCCH		FCCH
SDCCH/4 (1)		SDCCH/4 (1)
SDCCH/4 (1)		SDCCH/4 (1)
SDCCH/4 (1)		SDCCH/4 (1)
SDCCH/4 (1)		SDCCH/4 (1)
SDCCH/4 (0)		SDCCH/4 (0)
SDCCH/4 (0)		SDCCH/4 (0)
SDCCH/4 (0)		SDCCH/4 (0)
SDCCH/4 (0)		SDCCH/4 (0)
SCH		SCH
FCCH		FCCH
CCCH		CCCH
CCCH		CCCH
CCCH		CCCH
CCCH		CCCH
CCCH		CCCH
CCCH		CCCH
CCCH		CCCH
CCCH		CCCH
SCH		SCH
FCCH		FCCH
CCCH		CCCH
CCCH		CCCH
CCCH		CCCH
CCCH		CCCH
BCCH		BCCH
BCCH		BCCH
BCCH		BCCH
BCCH		BCCH
SCH		SCH
FCCH		FCCH

Single Timeslot. Timeslot 0 repeated twice

Figure 3.16   Single time slot for control

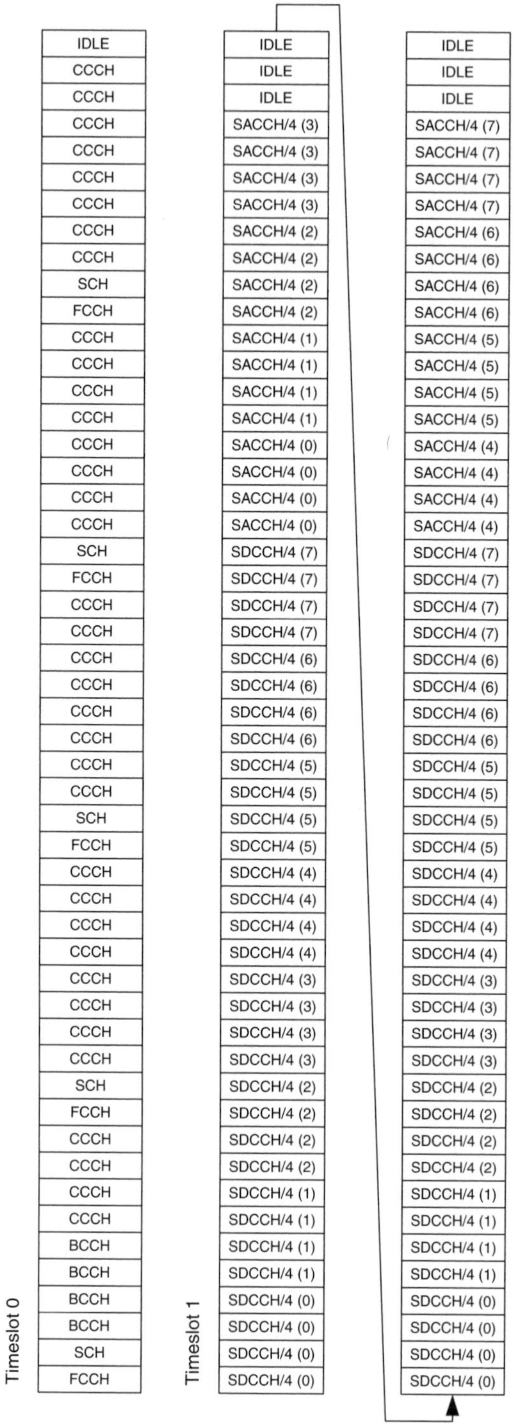

Timeslot 0	Timeslot 1	
IDLE	IDLE	IDLE
CCCH	IDLE	IDLE
CCCH	IDLE	IDLE
CCCH	SACCH/4 (3)	SACCH/4 (7)
CCCH	SACCH/4 (3)	SACCH/4 (7)
CCCH	SACCH/4 (3)	SACCH/4 (7)
CCCH	SACCH/4 (3)	SACCH/4 (7)
CCCH	SACCH/4 (2)	SACCH/4 (6)
CCCH	SACCH/4 (2)	SACCH/4 (6)
SCH	SACCH/4 (2)	SACCH/4 (6)
FCCH	SACCH/4 (2)	SACCH/4 (6)
CCCH	SACCH/4 (1)	SACCH/4 (5)
CCCH	SACCH/4 (1)	SACCH/4 (5)
CCCH	SACCH/4 (1)	SACCH/4 (5)
CCCH	SACCH/4 (1)	SACCH/4 (5)
CCCH	SACCH/4 (0)	SACCH/4 (4)
CCCH	SACCH/4 (0)	SACCH/4 (4)
CCCH	SACCH/4 (0)	SACCH/4 (4)
CCCH	SACCH/4 (0)	SACCH/4 (4)
SCH	SDCCH/4 (7)	SDCCH/4 (7)
FCCH	SDCCH/4 (7)	SDCCH/4 (7)
CCCH	SDCCH/4 (7)	SDCCH/4 (7)
CCCH	SDCCH/4 (7)	SDCCH/4 (7)
CCCH	SDCCH/4 (6)	SDCCH/4 (6)
CCCH	SDCCH/4 (6)	SDCCH/4 (6)
CCCH	SDCCH/4 (6)	SDCCH/4 (6)
CCCH	SDCCH/4 (6)	SDCCH/4 (6)
CCCH	SDCCH/4 (5)	SDCCH/4 (5)
CCCH	SDCCH/4 (5)	SDCCH/4 (5)
SCH	SDCCH/4 (5)	SDCCH/4 (5)
FCCH	SDCCH/4 (5)	SDCCH/4 (5)
CCCH	SDCCH/4 (4)	SDCCH/4 (4)
CCCH	SDCCH/4 (4)	SDCCH/4 (4)
CCCH	SDCCH/4 (4)	SDCCH/4 (4)
CCCH	SDCCH/4 (4)	SDCCH/4 (4)
CCCH	SDCCH/4 (3)	SDCCH/4 (3)
CCCH	SDCCH/4 (3)	SDCCH/4 (3)
CCCH	SDCCH/4 (3)	SDCCH/4 (3)
CCCH	SDCCH/4 (3)	SDCCH/4 (3)
SCH	SDCCH/4 (2)	SDCCH/4 (2)
FCCH	SDCCH/4 (2)	SDCCH/4 (2)
CCCH	SDCCH/4 (2)	SDCCH/4 (2)
CCCH	SDCCH/4 (2)	SDCCH/4 (2)
CCCH	SDCCH/4 (1)	SDCCH/4 (1)
CCCH	SDCCH/4 (1)	SDCCH/4 (1)
BCCH	SDCCH/4 (1)	SDCCH/4 (1)
BCCH	SDCCH/4 (1)	SDCCH/4 (1)
BCCH	SDCCH/4 (0)	SDCCH/4 (0)
BCCH	SDCCH/4 (0)	SDCCH/4 (0)
SCH	SDCCH/4 (0)	SDCCH/4 (0)
FCCH	SDCCH/4 (0)	SDCCH/4 (0)

Figure 3.17 Multiple time slots for control

3.4 TIMING ADVANCE

A mobile device and the BTS have to transmit in specific time slots or bursts. If they transmit too early or too late then they will cause interference to the previous or following call. In an ideal situation all users would be exactly the same distance from the BTS and would transmit at the beginning of the allocated slot. In practice, however, mobile devices are at different distances from the BTS and also each mobile device is free to move closer or further away. Due to the propagation delay over the air interface, mobile devices which are further away actually start to transmit before their allocated time slot. This is known as timing advance and is directly related to the distance from the BTS. The initial timing advance measurement is estimated by monitoring the received signal from a mobile device when it initially sends a burst on the RACH. As the mobile device moves away or towards the BTS, the network informs it of its new timing advance value on the SACCH. Cell sizes in rural areas (also in urban areas if a cell hierarchy is used) can be as large as 35 km in normal circumstances and up to 120 km for GSM400. Note that the timing advance can deal with mobile devices moving at speeds of up to 500 kmph.

3.5 INITIAL CONNECTION PROCEDURE

When the mobile device is switched on, it tries to register with a mobile network. The subscriber's home network will be stored in the SIM module and this will be checked first. If this network is available then the mobile device will request a connection. If the subscriber's home network is not available, the mobile device will try to attach to the last network to which it was connected prior to being switched off. This information is also stored in the SIM module. If neither of the above networks is available, the mobile device begins searching through all of the frequencies in the band to try to find a suitable network. This is the case, for example, when a subscriber arrives at an international airport from a foreign country.

While searching through the various frequencies, the mobile device is looking for a strong BCCH signal. This signal includes a number of channels, including the FCCH and the SCH. The FCCH simply emits a sine wave carrier to enable the mobile device to synchronize its frequency reference with the base station. The SCH contains the base station identity code and a frame number. The BCCH also gives the mobile device information about the network, such as where it is, which LA it falls under and who the operator is. On selecting a strong BCCH, the mobile will try to attach to this network. It does this by sending a request on the RACH channel, to which the base station listens continuously for mobiles wishing to register themselves. The RACH is a shared channel (referred to as a common channel) which works on the slotted-Aloha protocol. If many subscribers try to connect to the network simultaneously their requests will cause interference to each other. This may result in the network receiving the requests in error and discarding them. The mobile devices will wait for a random amount of time before trying to register again. The mobile networks covering an international airport are put under a great deal of strain when a flight arrives as many hundreds of users attempt to make an initial connection.

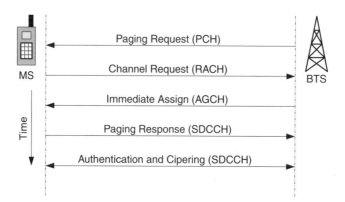

Time

MS

BTS

Paging Request (PCH)

Channel Request (RACH)

Immediate Assign (AGCH)

Paging Response (SDCCH)

Authentication and Cipering (SDCCH)

Figure 3.18 Initial access

Figure 3.18 illustrates how a mobile device attaches to the network. In this example the mobile device has been paged by the network as is the case for a mobile terminated (MT) call. The mobile device continuously monitors the paging channel for such requests and replies on the RACH for a dedicated channel. Once a request is received by the network a response is sent on the access grant channel (AGCH) channel. This response will indicate a dedicated signalling channel which the mobile device should now use to continue its negotiations with the network. A standalone dedicated control channel (SDCCH) is used for this purpose. This channel has a much lower bit rate than a dedicated traffic channel and therefore is more efficient for the small amount of data to be transferred for signalling purposes. The mobile device can now continue with the attach request. It will send its IMSI to the MSC where it will be processed. The MSC will connect to the HLR/AuC of the mobile subscriber's home network to authenticate the SIM module. Authentication triplets will be sent back to the MSC. These include a random number (RAND), a key (Kc) and a result (SRES). The random number is passed to the SIM in the mobile device, which will use its authentication system to also produce a result (SRES′). This result is passed back to the MSC, which will compare it to the result from the AuC. If the results are the same then the SIM is authenticated. The authentication algorithm is rather complex and so an invalid mobile device will reply with a wrong result. The key (Kc) is used to encrypt the data between the mobile station and the BTS. Figure 3.19 illustrates the authentication and encryption procedure.

Once the SIM is authenticated, the MSC may now request the IMEI. Once received from the mobile device this may be checked against the EIR, to see whether or not the mobile device is on a stolen list, not type approved, etc. If it is on such a list then it may not be allowed to register with the network. Once the IMSI and IMEI have been successfully checked, the MSC requests information about the subscriber from the HLR, which will include services available and other details. The MSC will now register the mobile device in the VLR, which will in turn inform the HLR of the current location of this mobile device. The MSC also provides the mobile device with a temporary identifier (TMSI) which is used in any future transactions with the mobile device. Using a temporary identifier, TMSI increases overall security since the IMSI of the mobile device is not sent frequently over the air. The TMSI also consists of a smaller number of digits (4 bytes)

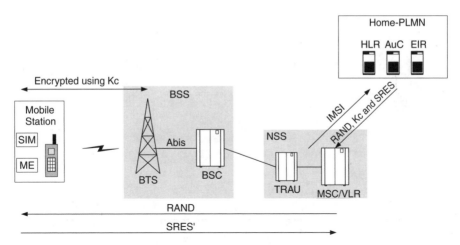

Figure 3.19 GSM authentication

than the IMSI and thus also increases efficiency. The initial signalling procedure is now complete. The mobile device is now assigned an SDCCH or a TCH and its call proceeds.

3.6 PROTOCOLS AND SIGNALLING

GSM has been designed with open interfaces in mind and a simplified diagram of the protocols used over these interfaces is illustrated in Figure 3.20. It can be seen that the air interface consists of the GSM time slots and frequency bands as denoted by TDMA/FDMA. Above this is the point-to-point link access protocol D (LAPD) channel protocol, which is the link layer for traditional ISDN signalling. A modified version of this, LAPDm (TS04.06), is used over the air interface between the BTS TRX and the mobile device. It is modified since the GSM air interface layer 1 already has an FEC mechanism and thus the LAPD CRC error detection at the datalink layer is not required. Also LAPD messages begin and end with an 8-bit synchronization flag, which is not required due to the GSM timing relationship between data and the time slot over the air. The address field includes the 3-bit service access point identifier (SAPI). SAPI 0 is used for call control, mobility management and radio resource signalling. SAPI 3 is used for SMS. All other values are currently reserved for future standardization. This datalink layer also provides the ability to assign three levels of priority, high, normal and low, to messages that are transferred in dedicated mode on SAPI 0. Priority is generally given to radio resource management (RR) messages over both mobility management (MM) and connection management (CM).

The RR layer is used to establish, maintain and release RR connections which allow a point-to-point dialogue between the mobile device and the network. This connection is used for data and user signalling. The procedures include cell selection and reselection as well as the handover procedures and reception of the BCCH and CCCH when no RR connection is established. Figure 3.21 shows a sample of RRM messages.

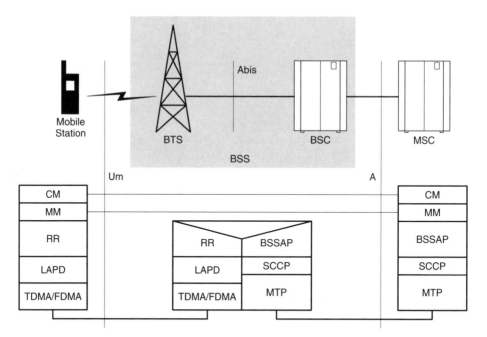

Figure 3.20   GSM protocols

Type	Messages	
Channel establishment	RR INITIALISATION REQUEST	
	IMMEDIATE ASSIGNMENT	
	PACKET ASSIGNMENT	
Ciphering	CIPHERING MODE COMMAND	CIPHERING MODE COMPLETE
Handover	ASSIGNMENT COMMAND	ASSIGNMENT COMPLETE
	HANDOVER COMMAND	HANDOVER COMPLETE
Paging and notification	PAGING REQUEST(1-3)	PAGING RESPONSE
	INTER SYSTEM TO UTRAN HANDOVER	

Figure 3.21   Example RRM messages

The MM layer is required to support the mobility of the mobile device. This includes informing the network of its present location and providing user authentication. It also provides connection management services to the CM layer. Figure 3.22 shows a sample of MM messages.

The CM layer is functionally split into a number of different entities. These include call control (CC), short message service support (SMS), supplementary services support (SS)

Type	Messages	
Registration	IMSI DETACH INDICATION	
	LOCATION UPDATING REQUEST	LOCATION UPDATING ACCEPT
Security	AUTHENTICATION REQUEST	AUTHENTICATION RESPONSE
	IDENTITY REQUEST	IDENTITY RESPONSE
Connection management	CM SERVICE REQUEST	CM SERVICE ACCEPT
	CM RE-ESTABLISHMENT REQUEST	

Figure 3.22 Example MM messages

Type	Messages	
Call establishment	ALERTING	
	SETUP	
	PROGRESS	
	AUTHENTICATION REQUEST	AUTHENTICATION RESPONSE
	CONNECT	CONNECT ACKNOWLEDGE
Call clearing	RELEASE	RELEASE COMPLETE
	DISCONNECT	
Supplementary service control	HOLD	HOLD ACKNOWLEDGE
	RETRIEVE	RETRIEVE ACKNOWLEDGE

Figure 3.23 Example CM messages

and location services support (LCS). The CC procedures have been closely modelled on the ITU-T ISDN Q.931 recommendation. Differences between the two are compared in the 3GPP specification TS24.008 annex E. The elementary procedures for CC are grouped into the following classes:

- call establishment procedures
- call clearing procedures
- call information phase procedures
- miscellaneous procedures.

The CC for GSM differs somewhat from that of Q.931 in one large respect which is that of mobility, i.e. routing to a mobile subscriber. An example of how this routing is performed for a mobile terminated call is highlighted in Section 3.7.4.

The supplementary services include such things as call forwarding, line identification, call waiting, call barring, multiparty calls and closed user groups.

The following protocols are also present in Figure 3.20: BSSAP and SCCP, and are described in Section 3.7.2. Figure 3.23 shows a sample of CM messages.

3.7   GSM AND SIGNALLING SYSTEM 7

SS7 or common channel signalling system 7 (CCS7) is a telecommunications signalling system standardized by the ITU-T under the Q.700 series of recommendations. The USA uses a modified version of SS7, which is standardized by ANSI. The general objective of SS7 was to provide a general purpose signalling system to be used globally. It is optimized to work in digital telecommunication networks operating at 64 kbps but is also suitable for analogue connections and transfer at lower bit rates. SS7 has been designed to provide a reliable system, which transfers information in the correct sequence and without loss or duplication. SS7 is the signalling system on which the vast majority of fixed-line telephone networks are based. It was principally for this reason that it was chosen by the cellular industry for its signalling system, since it would allow simple integration of cellular and fixed-line networks. Many of the existing 2G networks such as GSM use SS7 to connect to a third party through the ISDN or PSTN. Since the core network in 3G R99 has limited changes and reuses much of the GSM core, the 3G network has been designed to integrate with this existing system and a variant of SS7 (3GPP TS24.008) will be used in the UMTS 3G system towards the circuit switched core network (CS-CN). The standard defines the procedures and protocols required for call control such as call establishment, billing, routing, information exchange and modification. Signalling via SS7 is carried in separate channels to the call data, where it uses a *message transfer part* (MTP) to transfer the signalling information around the network. A common channel signalling system such as SS7 enables a number of traffic channels to share a single control channel. The maximum message length for narrowband SS7 signalling messages is 272 bytes (for broadband messages, as used in UMTS, this is extended to 4 kbytes).

Releases 4 and 5 of the 3G specifications introduce new possibilities for signalling within an all-IP network, where the IP transport network is used for signalling message transfer (see Chapter 8). However, some of the specifications are still in draft standard (e.g. M3UA).

To ensure the high reliability required by *carrier class* networks, SS7 has a number of stringent requirements, two of which are highlighted below. These requirements are stipulated under ITU-T Recommendations Q.700 to Q.775.

- The number of messages lost or delivered out of sequence (including duplicate messages) due to transport failure, as well as messages containing an error that is undetected by the transport protocol, should be less than 1 in 10^{10}.

- The complete set of allowed signalling paths from a source to destination, known as the *signalling route set*, should be available for 99.9998% of the time. This equates to a tolerated downtime of around 10 minutes per year.

The following are examples of typical operations handled by SS7:

- *connect +44 0161 980 2323 to +65 789564 using trunk 55*
- *locate 0800 1234 5678* (this is a free-phone number needing to be routed to the correct exchange)
- *the line is busy send back a busy tone to the caller*
- *trunk 88 has failed do not send information on this trunk.*

3.7.1   Signalling points

A typical SS7 network consists of three types of device: service switching points (SSP), signal transfer points (STP) and service control points (SCP). Collectively these devices are referred to as SS7 nodes, or simply signalling points. Although logically they are separate elements, it is common for a single physical device to perform a number of functions. All of the signalling points within an SS7 network are identified by a unique codepoint address.

- The SSP is responsible for originating, terminating and routing the call to the correct destination. When a user wishes to make a call, this request will be passed on to a telephone switch (an SSP). These devices usually have the ability to deal with many calls at the same time and thus have a large switching matrix. Within the GSM network, the MSC is considered to be an SSP.
- The STP acts as a router and is used to provide a path from the source to the destination as well as give alternative paths to a destination, for example, in the case of network failure.
- The SCP is required when there is no actual telephone number for a specific destination but an alias is used instead. This is the case for toll-free numbers where a subscriber may dial, for example, 1-800-FLYDRIVE. In such cases it is necessary to look up the actual telephone number in a database. A similar method may be required when routing a call to a mobile device, since the network needs to query the HLR of this mobile device with the subscriber MSISDN (see later). Within the GSM network the registers HLR, VLR, EIR and AuC are all types of SCP.

It can be seen from Figure 3.24 that to make a signalling connection from user A to user B a variety of network elements are involved. To ensure reliability of the network, it is typical for the SSP to have at least two connections through to the network. These connections are referred to as *access links*, and only signalling which is for the originating point code (OPC) or the destination point code (DPC) traverses these links. STP devices which

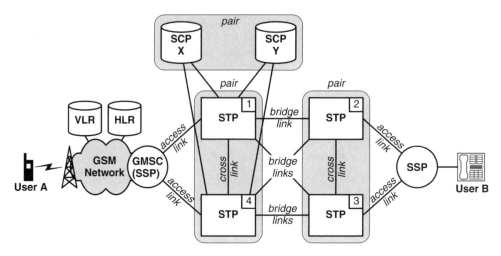

Figure 3.24   SS7 sample logical network

are working in redundant pairs are connected together via a cross-link. In Figure 3.24, STP (1) and STP (4) are paired, as are STP (2) and STP (3), thus improving the reliability of the system. Bridge links are also used to connect one STP to another. In this case the STPs are used as routers to direct the signalling message from the source to destination. The SCPs can also be paired, as is the case for SCP (X) and SCP (Y).

3.7.2   Protocol stack for SS7 signalling over MTP

The standard protocol stack for SS7 is shown in Figure 3.25. The layers are analogous to the layers of the ISO OSI seven-layer model, with the layers in an SS7 stack referred to as parts rather than layers. The lower three layers are collectively referred to as the message transfer part (MTP).

Message transfer part level 1

This is functionally equivalent to the OSI layer 1 and defines the various physical layer interfaces. Messages are usually carried over 56 kbps or 64 kbps links for the narrow-band services. E1 (2048 kbps) consisting of 32×64 kbps channels and DS1 (1544 kbps) consisting of 24×64 kbps channels are also supported. These TDM links ensure that the bits making up the message arrive in the correct order.

Message transfer part level 2

This layer ensures that there is a reliable connection between two datalink network elements. It includes error checking, flow control and sequencing. If an error is detected, MTP2 can ask for a retransmission. Layer 2 is capable of monitoring the state of the link and enables peer devices to communicate link information to one another. This

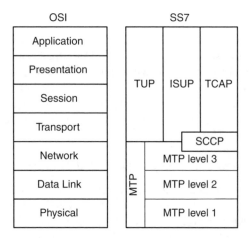

Figure 3.25   SS7 protocol stack

information could be, for example, an indication that the link is congested or has failed. This is functionally equivalent to the OSI layer 2.

Message transfer part level 3

This layer extends the functionality of level 2 so that signalling messages can be transported over a complex network, i.e. there does not need to be any direct connection between signalling points (network elements). This layer deals with routing, re-routing when a link fails and congestion control. It has many similarities with the IP layer on the Internet.

Signalling connection control part (SCCP)

SCCP sits on top of the message transfer part and provides both connectionless and connection-oriented network services. Together with MTP it is referred to as the *network service part* (NSP). Signalling points can have a number of attached applications operating simultaneously; SCCP introduces the subsystem number (SSN) to ensure that the correct application is accessed. This layer can be seen as analogous to TCP in the Internet suite of protocols. SCCP also provides address translation capabilities, known as global title translation (GTT). SCCP is not required for services such as TUP and ISUP and thus in Figure 3.25 it does not extend across the complete stack.

Telephone user part (TUP)

This deals with call setup and release but is designed for the traditional analogue circuits, which are still dominant in some parts of the world. It has been largely superseded by ISUP. TUP was one of the first applications to be defined and as such did not have provision for ISDN services.

ISDN user part (ISUP)

This layer also defines the messages that are to be used for call setup, modification, tear down, etc. It provides the services required by ISDN. ISUP supports basic telephony in a similar manner to TUP; however, it has a greater variety of messages which enable many more services to be offered. Calls that originate and terminate at the same SSP do not require the services of ISUP. Examples of ISUP messages are: Answer, Charge information, Connect, Identification request, Information request and Release.

Transaction capabilities application part (TCAP)

TCAP provides a structured method to request processing at a remote node. It defines the information flow and can report on the result. Queries and responses between SSPs and SCPs are sent via TCAP messages. Typical examples of TCAP services are registration of roaming users in a mobile network, and intelligent network services such as free-phone or calling card. TCAP is employed for the non-circuit-related information exchange, for example, if a subscriber's PIN is required for using a calling card, mobile application part (MAP; see below) messages which are sent between mobile switches and databases to support subscriber authentication, equipment identification and roaming are carried by TCAP. As a mobile subscriber roams into a new MSC area, the VLR requests the subscriber's profile information from the HLR using MAP information carried within TCAP messages. Although SCCP can locate the database, the actual query for data is performed by a TCAP message. TCAP can also be used for transporting billing information.

GSM user part

A cellular network requires additional features to a fixed network; for example, a mobile subscriber needs to be tracked when roaming from location to location. This extra functionality is performed by MAP and BSSAP. Figure 2.26 below shows how these GSM components fit into the SS7 protocol stack.

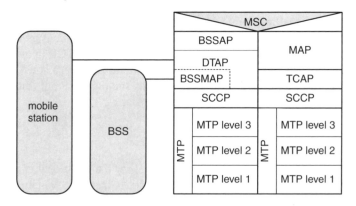

Figure 3.26   GSM positioning in SS7 protocol stack

Message application part (MAP)

MAP is a key protocol in many cellular networks. Both GSM and ANSI-41 networks use MAP for accessing roaming information, paging devices, controlling handover and sending SMS messages. Although GSM and ANSI-41 use MAP, these messages are not directly compatible and interworking functions are required when communication between the two types of MAP is required. MAP messages are carried by TCAP.

Base station system application part (BSSAP)

BSSAP messages are used for signalling both between the mobile device and the MSC and also between the BSC and the MSC. For this reason the BSSAP is divided into two sub-layers:

- The direct transfer application part (DTAP) is used to transfer the messages between the MSC and mobile device. These messages, which include MM and CM, are not interpreted by the BSS and are carried transparently.
- BSS management application part (BSSMAP) messages are transferred between the BSC and the MSC to support procedures such as resource management, handover control and paging of the mobile device.

3.7.3   Address translation

As previously mentioned, an SCP may be required when there is no actual telephone number for a specific destination but an alias is used instead, such as for a toll-free number. Consider the earlier example of a user dialling 1-800-FLYDRIVE. In such cases it is necessary to look up the actual telephone number in a database. A TCAP request is sent to an SCP, which has access to this database. Figure 3.27 illustrates how this translation can be achieved. The SSP will receive the request and look up the destination codepoint (DCP) address of the signal transfer point (STP) that deals with 1-800 requests. The STP will send the request onto a service control point (SCP) with the extra SSN. This process again requires the use of a lookup table, this time located at the STP. The SSN identifier will ensure that the query not only goes to the correct device but, more specifically, to the correct application (a database in this example) within the device. The reply is returned to the SSP, which can then send an initial address message (IAM) to set up the connection to the destination address 12543778659.

3.7.4   Example of routing of a call to a mobile subscriber

When a call is made to a mobile subscriber, the number dialed is the mobile subscriber's MSISDN and not the IMSI. An ISUP IAM, which contains this MSISDN, is routed to the gateway MSC (G-MSC) in the mobile user's home network. The G-MSC will send a MAP request to the HLR containing the MSISDN number, and the HLR will use this to identify the corresponding IMSI for the particular mobile device. In smaller mobile

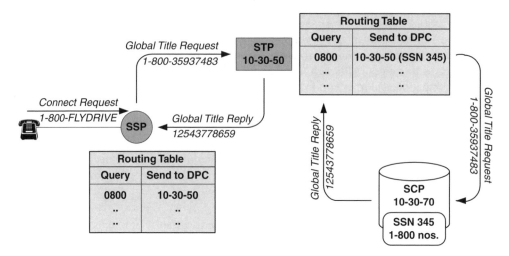

Figure 3.27 Global title translation

networks, there may only be one HLR and a routing information request will be sent to this HLR by the MSC. In larger networks where there are a vast number of subscribers, there may be multiple HLRs. In this case, the MSC needs to determine the correct HLR for this particular subscriber and may utilize global title (GT) translation to assist in this procedure. The HLR record also contains the address of the current VLR serving this subscriber; of course, if the mobile device is switched off or out of coverage there will not be a VLR address entry. The HLR can now contact the VLR, again using the MAP protocol, to request a mobile station roaming number (MSRN). This number, generated by the VLR, is essentially an ISDN number at which the mobile subscriber can be reached. The HLR can now respond to the initial routing information request from the G-MSC. This response will include the MSRN. The G-MSC can now attempt to complete the call to the mobile subscriber using the MSRN, again using an ISUP IAM message. It should be noted that the allocated MSRN is only valid for the duration of the call and the next call to this subscriber may use a completely different MSRN.

The network now has to find which group of cells the mobile is in and send a paging signal asking that particular mobile device to respond. Even though the mobile device is only positioned in a single cell, the page will normally be transmitted across the whole LA, causing extra signalling traffic and noise in the cells that comprise the LA. Once paged, the mobile device will respond on the random access channel (RACH), indicating which specific cell it is in and also requesting resources. The network will respond on the AGCH and offer resources for the mobile device to use. The mechanism used for GPRS is very similar to the above.

Once the call enters the network it is sent over E1 or T1 links using the TDM structure. These links connect the BTSs to the BSC and from the BSC through a transcoder (TRAU) to the MSC. From there the calls are routed via a G-MSC to the PSTN or ISDN. The TDM system is restrictive and can be inefficient since it is circuit switched and the TDM slots are reserved for a user even if no voice/data is being transferred. Figure 3.28 shows an example of the overall procedure.

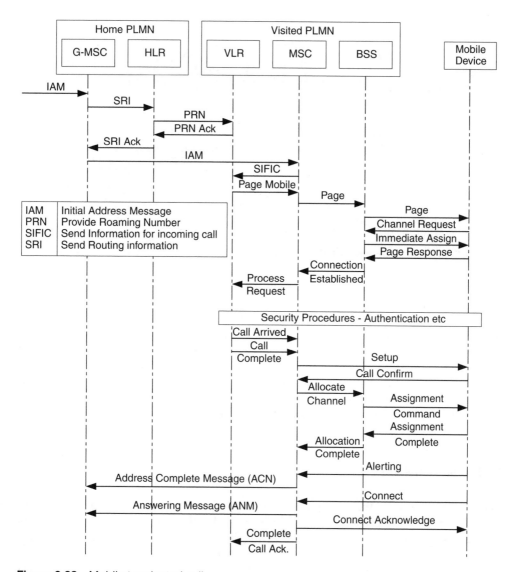

Figure 3.28   Mobile terminated call

3.7.5   Example of routing of an SMS message to a mobile subscriber

When an SMS message is sent to a mobile device it will arrive at a service centre (SC), which is responsible for relaying and store-and-forwarding the SMS to the mobile device. The SC will identify the G-MSC for SMS (SMS-GMSC) for the particular mobile device and will forward the message. The SMS-GMSC will interrogate the HLR using the MSISDN number of the mobile device requesting routing information, and any further

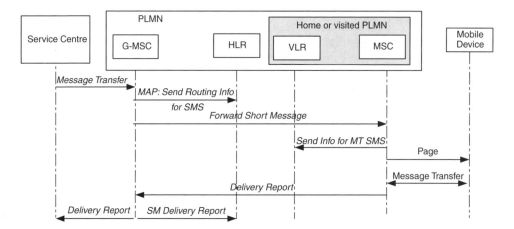

Figure 3.29 Mobile terminated SMS transfer

SMS information such as whether the SMS should be sent via the SGSN or via the MSC. The HLR will return the address of the serving MSC or SGSN and this can then be contacted by the SMS-GMSC. The MSC will query its VLR with the IMSI of the mobile subscriber. The VLR will return the location of the subscriber and the MSC can initiate a page procedure to contact the mobile device. If the mobile device is in idle mode it will perform an IMSI attach and the SMS can be received from the MSC. Once the message has been transferred successfully a delivery report can be sent from the MSC to the SMS-GMSC. The SMS-GMSC can then send an indication to the HLR and also back to the SC. This process is illustrated in Figure 3.29.

3.8 SUMMARY

This chapter describes the principles of operation of the GSM network. For evolution to UMTS, much of the functionality of GSM is retained in the core network, and several key concepts such as the use of the SIM, and security and mobility procedures are also retained. The TDMA/FDMA air interface for GSM is explained in some detail, including the physical, control and transport channels implemented. Each of the key network elements is described, as well as the interfaces between them. The various protocols that comprise GSM are described, including the use and extension of SS7 in the mobile context. Mobility management is explained, along with a description of a mobile terminated call and a short message transfer.

FURTHER READING

D. J. Goodman (1997) *Wireless Personal Communications Systems*. Addison-Wesley, Reading, MA.

3GPP TS03.04: Signalling Requirements Relating to Routing of Calls to Mobile Subscribers.

3GPP TS03.09: Handover Procedures.

3GPP TS04.01: Mobile Station – Base Station System (MS – BSS) Interface General Aspects and Principles.

3GPP TS04.04: Layer 1 – General Requirements.

3GPP TS04.05: Data Link (DL) Layer General Aspects.

3GPP TS08.01: General Aspects on the BSS – MSC Interface.

3GPP TS08.02: Base Station System – Mobile Services Switching Centre (BSS–MSC) Interface – Interface Principles.

3GPP TS08.04: Base Station System – Mobile Services Switching Centre (BSS–MSC) Interface Layer 1 Specification.

3GPP TS23.003: Numbering, Addressing and Identification.

3GPP TS24.008: Mobile radio interface Layer 3 specification; Core network protocols; Stage 3.

A list of the current versions of the specifications can be found at http://www.3gpp.org/specs/web-table_specs-with-titles-and-latest-versions.htm, and the 3GPP ftp site for the individual specification documents is http://www.3gpp.org/ftp/Specs/latest/

4

General Packet Radio Service

4.1 INTRODUCTION TO GPRS

The rapid growth in cellular voice services has led to high user penetration, particularly for the global system for mobile communications (GSM), as has previously been seen. From a user perspective, there is now a growing demand for value-added non-voice services, which the existing GSM infrastructure is not well suited to deliver effectively due to its circuit switched nature. From the cellular operators, perspective, with such high mobile penetration in most developed countries, they are seeing a plateau in revenue streams, and must now turn to non-voice services to create additional revenue.

The general packet radio service (GPRS) is a data service allowing traffic in the form of packets (usually IPv4 or IPv6 packets) to be sent and received across a mobile network. The point-to-point protocol (PPP) has recently been introduced, allowing the transparent transportation of protocols such as AppleTalk and IPX. It is designed to supplement the circuit switched mobile telephone system and the *short message service* (SMS) system, as well as enable new services. In many cases it is seen as an evolutionary step towards 3G, and hence is often referred to as 2.5G. It has been termed *always connected*, since after the initial connection delay subsequent connections are almost instantaneous. This is in contrast to the GSM call or a traditional fixed-line call where every time a new connection is made, there is a considerable delay. Although originally specified by the European Telecommunications Standards Institute (ETSI), in the summer of 2000 the standardization of GPRS was moved to the Third Generation Partnership Project (3GPP).

GPRS works by introducing the services of a packet switched network to the user over the existing GSM network (IS-136 TDMA, as used in North America, also supports the GPRS technology). This enables the user to continue to use the GSM network for voice but if a data transfer is required this can be passed via the GPRS system. In this way the existing base station system (BSS) infrastructure can be reused, as illustrated in Figure 4.1.

Convergence Technologies for 3G Networks: IP, UMTS, EGPRS and ATM   J. Bannister, P. Mather and S. Coope
© 2004 John Wiley & Sons, Ltd   ISBN: 0-470-86091-X

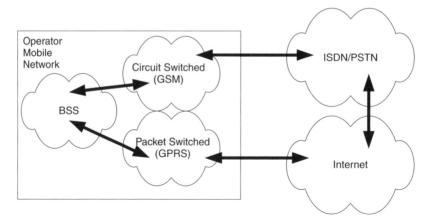

Figure 4.1 Network infrastructure

Voice traffic will continue to be passed between the BSS and the circuit switched GSM core network. This is generally referred to as the CS core network (CS-CN). GPRS traffic will be redirected, usually within the BSC via a new unit called a *packet control unit* (PCU) and be passed on to the packet switched GPRS network. This is generally referred to as the PS core network (PS-CN). It should be noted that the external networks are connected together via gateways. This allows the option of sending Internet protocol (IP) data over the GSM network through the ISDN/PSTN and on to the Internet. This system is in use today and supported by most mobile devices and mobile networks, it is known as circuit switched data (CSD).

GPRS introduces a packet-based core network but still uses much of the GSM functionality, including the home location register (HLR), equipment identity register (EIR) and authentication centre (AuC) (3G systems also use this GSM functionality). What GPRS introduces is the capability to transport different traffic types with more efficiency in network resource usage, and allow the introduction of a wide range of services. However, the general higher-layer functionality does not need to change and can thus be reused. By way of example, a user who is sending an email does not need their location information handled any differently from a user making a phone call. The network is designed to support a number of different quality of service (Qos) classes and these will be gradually implemented throughout the various releases of the standard in order to enable the efficient simultaneous transfer of both real-time and non-real-time traffic. A common GPRS network can be used for both GSM and UMTS; however, some vendors may not support both the GSM Gb interface and the UMTS Iu interface on a single piece of equipment, resulting typically in some hardware changes or additions in migrating to UMTS.

For GPRS/GSM the air interface is allocated in a flexible manner with 1 to 8 of the time division multiplexing (TDM) channels being allocated to GPRS traffic. The active users share the time slots and these are allocated independently in the uplink and downlink. These radio interface resources are shared dynamically between speech and data users, the exact method used being dependent on operator preference and service load of the network. Enhanced GPRS (EGPRS) is an enhancement to the system, which allows higher bit rates through the use of different modulation techniques and coding schemes (see Section 4.11.12).

The concept of a GPRS handover is referred to as a cell reselection procedure and is normally performed by the mobile device. The handover timing for GPRS is not so critical when compared to GSM since the traffic is not real time, and can thus be buffered. In this cell reselection procedure, the mobile device makes measurement reports, as with GSM; however the mobile station (MS) is more involved in the decision process, and can even initiate the procedure for handover. It is, nevertheless, still the responsibility of the network the serving GPRS support node (SGSN) to allow the handover to occur.

Security functions are essentially quite similar to those for GSM services. The SGSN is responsible for authenticating the subscriber as well as encrypting/decrypting of data towards the mobile device (regular GSM encryption is only between the mobile device and the base station). The SGSN and mobile device can also compress data to make more efficient use of the Gb and air interface. A mobile device containing a standard GSM SIM can connect to the GPRS network and use the services, depending on the specifics of the operator network settings.

4.2   GENERAL ARCHITECTURE

Figure 4.2 shows the general architecture of a GPRS network and its interface to other IP-based networks such as the Internet. The GPRS network makes use of much of the existing GSM infrastructure. The HLR, AuC and EIR may require minor modifications to support GPRS, generally in the form of a software upgrade. In the diagram, the different equipment within the GPRS backbone network is connected together using an Ethernet switch. Within the GPRS standard, there is no stipulation as to what Layer 2 technology

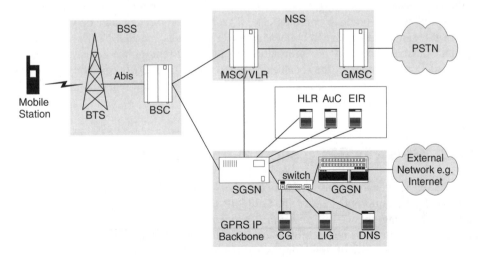

Figure 4.2   GPRS General Architecture

infrastructure should be used to interconnect the IP backbone, since to support roaming, the SGSN and GGSN will be connected through an internetwork. Most current networks are using an Ethernet network to implement this local backbone, as it is a cost-effective architecture choice. This Ethernet switch can be a weak link as it is a single point of failure in the network, so usually it will contain a great deal of redundancy to ensure reliability.

4.3   GPRS NETWORK ELEMENTS

To enable GPRS over an existing second generation network such as GSM, a number of additional network elements are required in the GPRS backbone network. These are described in the following subsections.

4.3.1   Serving GPRS support node (SGSN)

The SGSN serves the mobile devices within its BSS/RAN, and provides authentication, and mobility management, which are derived as much as possible from the GSM mobile application protocol (MAP). It is the connection point between the BSS/RAN and the CN, and at a high level the SGSN provides a similar role for the packet switched network as the MSC/VLR provides to the circuit switched network. When a mobile device is *packet switch attached*, the SGSN is said to provide a *mobility management context* and it then keeps track of the mobile device to a routing area (RA) or specific cell. The SGSN connects to GGSNs and also to other SGSNs via an IP network. When a mobile device is furnished with a *session management context* a connection is established between the SGSN and corresponding GGSN so that the mobile device may transfer data to and from an external network. An SGSN is not restricted to communication with one single GGSN and will in practice communicate with many GGSNs, which may not even be within the same public land mobile network (PLMN) as the SGSN. The SGSN has a dynamic database which stores information about the current mobile devices it is serving. This database will contain the location of the device to an RA or specific cell, security information, such as the ciphering key, charging information, current connections and the QoS being used, etc.

4.3.2   Gateway GPRS support node (GGSN)

The GGSN provides the interface between the mobile and the external packet switched network. Packets are routed across the GPRS IP-based packet network between the SGSN and GGSN using the *GPRS tunnelling protocol* (GTP). Like the SGSN, the GGSN also stores information about mobile devices that have established a session with the SGSN. The database will store the international mobile subscriber identity (IMSI) of the mobile device, QoS negotiated, charging information, as well as the address of the

SGSN serving a particular mobile device. When packets arrive for the mobile device from an external network it is the GGSN which will receive them and route them to the correct SGSN for final delivery to the mobile device. The GGSN does not need to know the location of the mobile device, only the address of the SGSN which is serving the mobile device. The SGSN and GGSN are collectively referred to as GPRS support nodes (GSNs).

4.3.3   Charging gateway (CG)

A charging gateway (CG) is not required in the specifications but is generally implemented since it takes processing load off the SGSN and GGSN. It also introduces a single logical link to the operator's billing system and reduces the number of physical links and connections required to be supported by the billing system, which would otherwise require a separate connection to each of the individual GSNs. It can be used to buffer and consolidate information before passing it on to the billing system.

4.3.4   Lawful interception gateway (LIG)

It is a requirement in many countries for the law enforcement agencies (LEA) to be able to monitor traffic. The LIG is introduced into the network for this purpose. As user traffic traverses the GPRS backbone it is possible to capture their data and forward it to the LEA. However, this interception of user data does normally require a court order.

4.3.5   Domain name system (DNS)

In most cases, when a subscriber wishes to make a connection via GPRS to an external network, they will select an access point name (APN) from a list in the mobile device. A domain name system (DNS) is required so that the SGSN can make a query to resolve the APN to an IP address of the correct GGSN. As with DNS in a standard IP network, this APN is just text, such as 'Internet' or 'home network'. A common scenario would be to define two general access points: *net* and *wap. Net* would indicate a connection directly to the Internet, and *wap* a connection to a wireless access protocol (WAP) gateway. An operator can bill differently based on the access point, therefore many operators have different tariffs for WAP access and Internet access. Further details on the DNS system may be found in chapter 5.

4.3.6   Border gateway (BG)

A border gateway (BG) is used as the gateway to a backbone connecting different network operators together. This backbone is referred to as an inter-PLMN backbone, or global roaming exchange (GRX). The operation and configuration of this connection is according to a roaming agreement between operators. The BG is essentially an IP router and is generally implemented as the same hardware platform as the GGSN.

4.4   NETWORK INTERFACES

GPRS introduces new interface definitions in the network as well as over the air. The interfaces are open standards as described by 3GPP and are shown in Figure 4.3. This enables a multi-vendor network to be constructed with minimum amount of modification. Since GPRS uses much of the GSM network, standardized interfaces are also required between the GSM equipment and the GPRS equipment.

- *Ga*: this is used for transferring the charging records known as *call detail records* (CDRs) from the SGSN and the GGSN to a CG. It uses an enhanced version of GTP known as GTP′.
- *Gb*: the Gb interface resides between the SGSN and the BSS. Its function is to transport both signalling and data traffic. This interface is based on frame relay and is described in more detail in Section 4.7.
- *Gc*: this interface is between the GGSN and the HLR, and provides the GGSN with access to subscriber information. The protocol used here is MAP and the interface is used for signalling purposes only. This interface can be used to activate the mobile device for mobile terminated packet calls. It requires that the mobile device is given a unique IP address, which is often not the case and thus it may not be implemented. The GGSN is essentially an IP device and may not have MAP capabilities, so the specification allows the GGSN to pass requests to the SGSN so that they can be forwarded to the HLR on its behalf. The Gc is an optional interface.
- *Gd*: the Gd interface connects the SGSN to an SMS gateway, thus enabling the SGSN to support SMS services.
- *Gf*: this interface connects the SGSN to the EIR and allows the SGSN to check the status of a particular mobile device, such as whether it has been stolen or is not type approved for connection to the network.

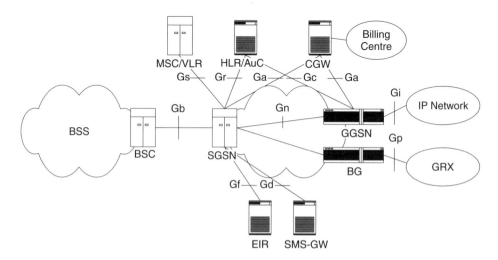

Figure 4.3   GPRS interfaces

- *Gi*: this is a reference point rather than an interface and refers to the connection between the GGSN and some external network. Currently IPv4, IPv6 and PPP[1] are supported by GPRS and the Gi interface simply has to be able to support the required protocol for this particular access point. For example, the access point may be required to transport IPv4 packets; the underlying network is not specified and may be Ethernet, asynchronous transfer node (ATM), frame relay or any other transport protocol.

- *Gn*: the Gn interface resides between the GSNs. It consists of a protocol stack which includes IP and GTP. GTP is explained in detail in Section 4.8. The GTP tunnel is also used between two SGSNs and also between an SGSN via a BG to another operator's GGSN. It is not used between two GGSNs unless they have BG functionality. This tunnel ensures that the operator's IP network is completely separated from the IP used for the mobile device to connect to the external network. The GTP tunnel actually consists of two parts, the GTP-U which is used to carry user data and the GTP-C which is used to carry control data.

- *Gp*: this has similar functionality to the Gn interface and also consists of a GTP protocol. It is required when the SGSN and GGSN are in different PLMNs. It introduces further routing and security functions to the Gn interface. Connection is via BGs and possibly an intermediate inter-PLMN network which may be owned by a third party, hence the increased security functions.

- *Gr*: this interface is between the SGSN and the HLR, providing the SGSN with access to subscriber information. The SGSN and HLR will be in different networks in the case of roaming users. The protocol used here is MAP and the interface is used for signalling purposes only.

- *Gs*: this is another optional interface. It is used for signalling between the SGSN and the visitor location register (VLR), which is usually co-located with the mobile switching centre (MSC) and an SGSN, it uses the BSS application part plus (BSSAP+) protocol. This is a subset of the BSSAP protocol to support signalling between the SGSN and MSC/VLR. To some extent, the SGSN appears to be a BSC when communicating with the MSC/VLR. This interface enables a number of efficiency saving features by coordinating signalling to the mobile device such as combined updates of the location area (LA) and routing area (RA) and IMSI attach/detach which reduces the amount of signalling over the air interface.

In addition to the 'G' interfaces, two relevant interfaces are those across the air, for both GPRS and UMTS:

- *Um*: this is the modified GSM air interface between the mobile device and the fixed network which provides GPRS services.

- *Uu*: this is the UMTS air interface between the mobile device and the fixed network which provides GPRS services.

---

[1]Older networks may support X.25 rather than or in addition to PPP.

4.4.1   Network operation mode

A network can be in three different modes of operation. These modes depend on whether the Gs interface is present and how paging of the mobile device is executed.

- *Network operation mode 1.* A network which has the Gs interface implemented is referred to as being in *network operation mode 1.* CS and PS paging is coordinated in this mode of operation on either the GPRS or the GSM paging channel. If the mobile device has been assigned a data traffic channel then CS paging will take place over this data channel rather than the paging channel (CS or PS).
- *Network operation mode 2.* The Gs interface is not present and there is no GPRS paging channel present. In this case, paging for CS and PS devices will be transferred over the standard GSM common control channel (CCCH) paging channel. Even if the mobile device has been assigned a packet data channel, CS paging will continue to take place over the CCCH paging channel and thus monitoring of this channel is still required.
- *Network operation mode 3.* The Gs interface is not present. CS paging will be transferred over the CCCH paging channel. PS paging will be transferred over the packet CCCH (PCCCH) paging channel, if it exists in the cell. In this case the mobile device needs to monitor both the paging channels.

The *network operation mode* being used is broadcast as *system information* to the mobile devices.

4.5   GPRS AIR INTERFACE

When a mobile network operator introduces GPRS, the service is run in conjunction with GSM. The operator shares the bandwidth allocated to them from the telecommunications regulator between the GSM and GPRS services. GSM and GPRS traffic can also share the same TDM frame, although they cannot share a single burst concurrently. Introducing GPRS may cause higher blocking for GSM calls and the operator may have to redimension the network to counter this problem. GPRS uses the same modulation technique, *Gaussian minimum shift keying* (GMSK), as GSM and since it uses the same time-slots and frame format as GSM there are 114 bits available during a time slot for subscriber data. However, the GSM structure of multiframes consisting of either 26 traffic channel (TCH) frames or 51 control frames has been replaced with a 52-frame format. As illustrated in Figure 4.4, the new multiframe consists of 12 blocks of 4 consecutive frames, which are referred to as radio blocks. Two idle (I) frames and two frames (T) which are used for the packet timing advance control channel (PTCCH) are also components of this multiframe. The time devoted to the idle frames and the PTCCH can be used by the mobile device for signal measurements.

In a GSM system a time slot is dedicated for one user at a time (unless half-rate mode is used). The GPRS system is different in this respect, since each of the radio blocks consisting of 456 bits ($57 \times 2 \times 4$) can actually be used by separate users, where the

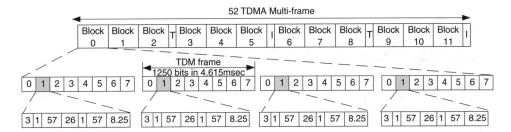

Figure 4.4 52-frame format

users would essentially share the resources of the time slot. This is referred to as a radio block and is a type of TDM within the TDM frame itself. A mobile device is assigned these blocks when it is required to transfer data. This assignment is referred to as a temporary block flow (TBF).

4.5.1 Resource sharing

The air interface is shared between the GSM and GPRS users. It can be considered that GPRS users are utilizing bandwidth that is left over by GSM voice users. Consider a transceiver (TRX) that has five GSM users. There are three time slots remaining that GPRS users can share. When a voice user makes a call, they are dedicated a time slot for the duration of that call. Unless a time slot is exclusively reserved for GPRS users, GSM users generally have priority of resource allocation. However, for GPRS, the remaining time slots should be viewed as a pool of available resources that the data users can share. Therefore the performance experienced by a GPRS user is based on three factors:

1. The number of GSM users in the current cell, which indicates the available time slots for GPRS data traffic.
2. The number of GPRS data users that must share these time slots.
3. The number of time slots that the mobile device can work with.

From the perspective of the mobile operator, it allows them to introduce more services which will, in theory, utilize space that otherwise would be wasted since it is not being used by voice calls. Devoting time slots specifically to GPRS, however, will increase call blocking problems for GSM users since there are only a maximum of eight time slots available per frequency. GPRS users can share a single time slot with each other, each using a single or a number of radio blocks. Since they share the same time slot they also share the time slots bit rate. Currently, in many cases the subscriber demand for data services is rather low and as such dedication of time slots exclusively for GPRS traffic is minimal. However, as the ratio of voice and non-voice traffic becomes more even, operators will re-evaluate this position and redimension resource allocation in the network appropriately. For example, there may be a lot of justification for reserving a

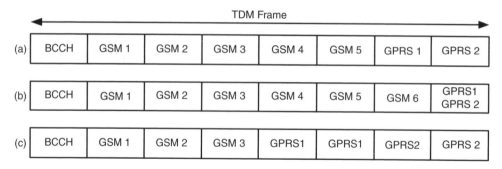

Figure 4.5 TDM frames

larger number of time slots for the exclusive use of GPRS customers in central business districts or city centre areas.

Example

Figure 4.5(a) shows a standard TDM frame with time slot 0 allocated to the broadcast channel (BCCH) and other control channels. The seven time slots remaining are available to GSM and GPRS users. Suppose that there are five GSM users and two GPRS users, which will fill the TDM frame. If an additional GSM user wishes to make a call, the two GPRS users will be moved into a single time slot as in Figure 4.5(b). If three of the GSM users now finish their calls, the two GPRS users will be able to use the vacant time slots, increasing their overall bit rate, as shown in Figure 4.5(c).

The vast majority of packet data transfer of GPRS is expected to be TCP/IP and UDP/IP. TCP/IP is usually used for web browsing, file transfers and email; it is a reliable connection-oriented system that acknowledges all data transferred. When TCP sends a segment of data it starts a timer, known as the *retransmission timer*. If the data arrives and is acknowledged successfully before the timer expires then the next segment is sent. However, if the acknowledgement is not received for any reason the timer may expire. This is referred to as a *timeout*. This timeout is usually dynamic and related to the round trip time. Once a timeout has occurred the connection may close, all data transferred so far may be lost and the transfer will have to start again from the very first byte. By enabling the sharing of time slots, the bit rate may reduce for each GPRS user, and the TCP flow control will slow down the data transfer, but the connection will not timeout.

When a user tries to make a voice call in the busy hour, they may under some circumstances get a *network busy* tone. Usually a customer will immediately try one or two more times to connect. After this, in many cases, the user will leave a long time (2 or 3 minutes) before trying again. When they try this time they may be successful. This situation arises when the number of users simultaneously attempting to make a call in the same cell is greater than the operator has planned for. To a certain extent, network planning is based on educated guesswork, where the planners use tools, tables and past experience to predict the numbers of users in a given geographical area who may wish to access the network simultaneously. However, extenuating circumstances cannot always be planned. Consider, for example, the end of a football match or concert, where a large number of users in the same cell will attempt to make a simultaneous call.

If a connected user had finished their call almost immediately after the first user had given up temporarily, the time slot would have been vacant for 2 or 3 minutes. There is no way that the user trying to connect would know that another user had just finished a call and thus the time slot resources in GSM are not always utilized to maximum efficiency.

To dimension the air interface within GSM, Erlang tables are used. Since GPRS also shares the GSM air interface then this can be taken into account when dimensioning for GPRS. Consider the following example.

Example

If there are 520 calls an hour and the average call duration is 100 seconds how many traffic channels are required?

$$\text{No. of Erlangs} = \text{Calls per hour} \times \text{Average call time}/(60 \text{ min} \times 60 \text{ s})$$

$$= 520 \times 100/3600 \approx 15 \text{ Erlangs}$$

In a GSM cell it is not expected that all calls connect satisfactorily – this is known as grade of service (GoS). Commonly GoS may be 2%, which means that on average two out of every one hundred calls will be blocked and get the network busy tone. By consulting the Erlang tables, it can be seen that to dimension a cell for the above number of GSM calls, 22 TCHs are required (Table 4.1). Adding one TCH for BCCH and one TCH for the stand alone dedicated control channel (SDCCH)/8 means three TRX units (one TRX = eight channels) are required.

To prevent a high level of call blocking at the busy hour we needed 22 traffic channels to supply an average of 15 calls at any one time. Note that at any one time more than 15 channels may be used by GSM subscribers but on average 15 subscribers will connect satisfactorily. Since only an average of 15 channels are used out of 22, this leaves 7 channels vacant. They cannot be used for GSM because of the call blocking. However, they can be used for GPRS since the network can notice that there is a free slot and allocate extra temporary block flows (TBFs) to GPRS subscribers. Using CS-2 (13.4 kbps, see Table 4.2) this provides over 90 kbps of bandwidth without increasing the number of channels.

4.5.2   Air interface coding schemes

Unlike voice, data is very intolerant of errors and generally must arrive error free. This does present some problems, as the air interface introduces a significant number of errors.

Table 4.1   Example of Erlang table

No. of channels	Grade of service		
	2%	3%	5%
20	13.2	14.0	15.2
21	14.0	14.9	16.2
22	14.9	15.8	17.1
23	15.8	16.7	18.1

Table 4.2 GPRS data rates

Coding scheme	Bit rate (kbps)	Raw bit rate (kbps)
CS-1	9.05	8
CS-2	13.4	12
CS-3	15.6	14.4
CS-4	21.4	20

To protect the data, it is necessary to transmit some codes with it which allow either error checking and/or error correction. Four new coding schemes have been specified for the GPRS air interface, known as CS–1-CS–4 and the data rates for these are highlighted in Table 4.2. The column indicating the *bit rate* includes the radio link control/medium access control (RLC/MAC) headers whereas the *raw bit rate* column is for user data.

CS-1 offers the lowest data rate but also offers the most protection on the data with both error detection and error correction. CS-4 offers a much higher data rate but with very little error checking and no error correction. It may initially appear that CS-4 should be used extensively; however, as mentioned, the air interface is notorious at losing data due to interference. If CS-4 is used, there is no error correction and if errored the frame will need to be retransmitted. This introduces transfer delays and also reduces the actual throughput. A transfer using CS-1 will be susceptible to the same interference. However, since it incorporates error correction, frames transmitted between the mobile device and the network can be corrected at the base station without the need for a retransmission. The offerings from CS-2 and CS-3 are somewhere in between. Figure 4.6, illustrates the effectiveness of each of the coding schemes in a noisy environment and in a cell with little interference. C/I is the carrier to interference ratio, which for this purpose can be regarded as similar to a signal-to-noise ratio.

Changes in the coding scheme used may take place during the call. This is done dynamically by the network and is dependent on the current properties of the connection such as number of errors, retransmissions etc., and is transparent to the user. It should be noted that different subscribers in the same cell might be using different coding schemes at the same time.

In practice data rates depend not only on the mobile network but also on the capability of the handset. Currently these are supporting maximum data rates of around 50 kbps.

Higher data rates are possible using advanced modulation techniques, such as eight-phase shift keying (8PSK), which require new TRXs in the base station (BTS) and higher-speed links within the BSS due to the higher data rates possible. This system will be used in enhanced data for global evolution (EDGE) and is known as enhanced GPRS (EGPRS). EDGE also works with high-speed circuit switched data (HSCSD); when used with this system it is known as ECSD.

4.5.3 Classes of devices

GPRS mobile devices can be classified into three general categories, as follows:

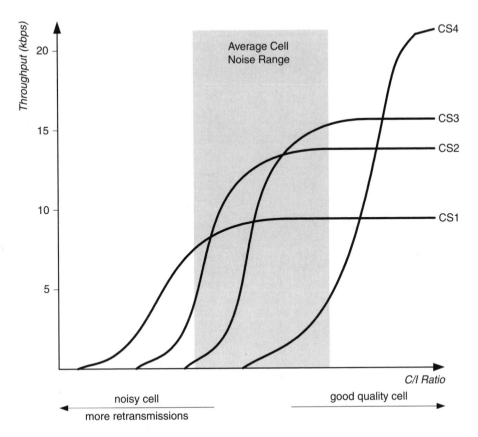

Figure 4.6   GPRS coding scheme throughput

- **Type A** can connect to both the GSM and GPRS cores, i.e. the MSC and SGSN, simultaneously. For example, this would allow a user to talk on the phone while downloading an email.

- **Type B** can also connect to both the GSM and GPRS cores, but the connection can only use one side of the network at any given time. For example, a subscriber who is currently downloading data would be notified of an incoming phone call and must decide whether to accept the call or not. This will put the data transfer on hold.

- **Type C** typically will be a data card for a PC allowing it to send and receive data across the GPRS network. This type of device cannot register with both the GSM and GPRS core networks at the same time.

Due to implementation complexities, currently for 2G systems only types B and C exist and this is likely to be the case for some time.

The device class defines the maximum data rate at which a GPRS device can send or receive. Table 4.3, shows the class number defined for GPRS devices, identifying the time slot combinations allowed. For example, a class 6 device allows the use of three

Table 4.3 GPRS device classes

Class	Downlink	Uplink	Max. slots
1	1	1	2
2	2	1	3
3	2	2	3
4	3	1	4
5	2	2	4
6	3	2	4
7	3	3	5
8	4	1	5
9	3	2	5
10	4	2	5
11	4	3	5
12	4	4	5
13	3	3	Unlimited
14	4	4	Unlimited
15	5	5	Unlimited
16	6	6	Unlimited
17	7	7	Unlimited
18	8	8	Unlimited
19	6	2	Unlimited
20	6	3	Unlimited
21	6	4	Unlimited
22	6	4	Unlimited
23	6	6	Unlimited
24	8	2	Unlimited
25	8	3	Unlimited
26	8	4	Unlimited
27	8	4	Unlimited
28	8	6	Unlimited
29	8	8	Unlimited

time slots in the downlink and two in the uplink. The maximum is four time slots, so a user that takes up three down link time slots can only use one in the uplink.

There are a number of problems with designing devices that can utilize more time slots for data transmission. In GSM, transmission and reception are offset by three time slots to prevent the mobile device from having to transmit and receive at the same time. If GPRS time slots overlap, more complex transceiver circuitry is required. Also, the more time slots used, the more power used since the device is transmitting for longer, and this creates problems for battery life and device cooling. Therefore, GPRS devices are unlikely to support the higher class numbers in the near future. It should be noted that this is the capability of the GPRS terminal. However, the maximum number of time slots available is also highly dependent on the network equipment capability and the operator policy.

4.5.4   Advantages of GPRS over the air

Consider a GSM phone call. The user makes a call setup by dialling the telephone number and is then connected to the call recipient. While the call is connected, the user is charged for that call and the charging is based on the call duration. Many studies have demonstrated that the average user will only talk for approximately 30–40% of the call duration and thus, on average, a telephone call is 60–70% inefficient. In effect, phone subscribers pay for silence.

If data traffic is now examined, for a GSM connection, this problem becomes considerably more pronounced. A typical data transfer is web page access. A user will connect to a website, retrieve its contents and then proceed to read the page. The download may take of the order of 3–4 seconds. However, the user is then idle on the network for a much longer period, while reading. During this long idle period, the user is also being billed, since the charging mechanism is based on duration, and the user is still occupying a fixed resource in the system. Data traffic is generally of this nature, where bursts of data transfer are interspersed among long periods of inactivity.

In this scenario, it makes more sense to bill based on volume rather than duration as is done for voice calls. In this manner, a subscriber will only be billed for actual throughput and not for the periods of network inactivity.

The introduction of the packet switched network approach alleviates this problem by allowing resources to be shared among users, and records kept of how many packets were transferred rather than how long they took.

Another important aspect of GPRS is its *always connected* nature. A mobile device does not take up any of the scarce resources over the air but when a user wishes to transmit data it appears that there is a dedicated channel. In reality the network simply remembers information about the user and thus can give a quick connection. In GSM every time a user wishes to make a call they must dial a number and wait a few seconds for the connection. Unlike GSM, a GPRS user can be 'always connected' to the network, where the user has a logical connection in the form of an IP address allocation (referred to as a packet data protocol, or PDP, context). However, only when the user sends or receives data will they actually use any resources, and consequently be subject to charging. A GPRS user will merely request some data resource, for example, by entering a web location. The network will then service that request in the available time slot space, thus requiring limited setup time. In this way, GPRS separates the connection from the actual resource usage.

4.6   GPRS PROTOCOLS

Similar to any communications protocol, GPRS consists of a layered stack, with different layers providing key functions and communicating with other layers through primitives. This section discusses the GPRS protocol stack and provides an explanation of the key functions of each layer.

Figure 4.7 illustrates the control plane protocol stack for GPRS as it is used within the GSM network across the BSS towards the SGSN. It can be seen that there is little difference between it and the user plane protocol stack, shown in Figure 4.8. The user plane uses an additional layer above the logical link control (LLC) between the mobile device and the SGSN referred to as the subnetwork-dependent convergence protocol (SNDCP). Each of these protocols will be described in detail in the following sections. Unlike Figure 4.7, Figure 4.8 continues the protocol stack across to the GGSN; both the

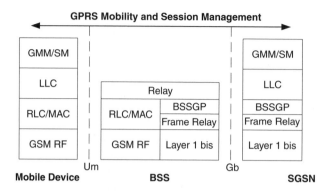

Figure 4.7   Control plane for GPRS/GSM

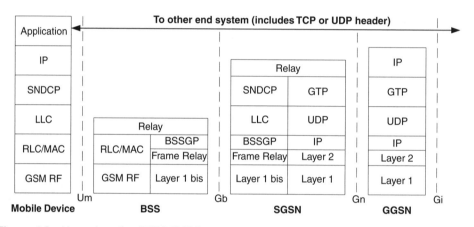

Figure 4.8   User plane for GPRS/GSM

control and user plane protocols use GTP between the SGSN and GGSN. The control plane uses GTP-C and the user plane uses GTP-U.

Figure 4.9 shows a 1500-byte payload passing through the mobile device protocol stack to be transported in the 456 bits available within a radio block over the air interface. This radio block consists of four TDMA frames, each making available 114 bits. The number of actual bits used for user data transfer and for error detection and correction depends on the specific coding scheme being used.

Figure 4.10 shows a generalized diagram of the layered process for GPRS from the perspective of the user equipment (UE). The SGSN stack is similar, but will have a number of differences including the use of BSS GPRS protocol (BSSGP) rather than RLC/MAC to transport the LLC. The left-hand side of this diagram has been presented in Chapter 3. The right-hand side blocks SNDCP, LLC and RLC/MAC are described in the following sections.

A packet received at the UE indicates which upper layer entity the packet is to be routed towards using the 4-bit protocol discriminator field in the Layer 3 header. The protocol discriminator (PD) may for example identify that the received packet mobility management, SMS messages, or a user IP packet.

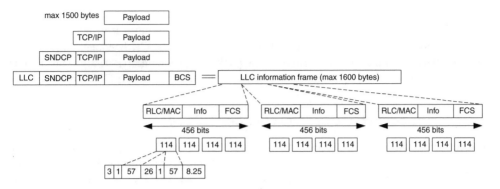

Figure 4.9 Packet fragmentation for transport

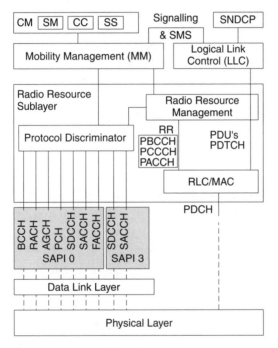

Figure 4.10 Mobile device layered architecture

The RLC/MAC layer supports four radio priority levels as well as an additional level for signalling messages. This information is used by the BSS to determine the access priority and the transfer priority under heavy load.

4.6.1 Physical and logical channels

Information about a cell's ability to deal with GPRS subscribers is broadcast on the GSM broadcast channel (BCCH). GPRS introduces a number of additional control channels to the air interface; some of these are mandatory and some are optional. When the new channels were introduced, the naming scheme was to simply put a P in front of the old channel name, if such a channel existed; for example, the BCCH becomes the PBCCH. Each of these channels is transferred via a packet data channel (PDCH). This PDCH equates to a physical channel taken from the total pool of GSM and GPRS resources.

4.6.1.1 Broadcast and control channel (BCCH)

The broadcast and control channel (BCCH) transmits general information from the base station to all mobile devices in the cell. One small part of this information is to indicate whether or not GPRS is supported in this particular cell. If GPRS is supported and the optional packet broadcast and control channel (PBCCH) is configured, the position of this channel is also indicated on the BCCH. The PBCCH is then used to broadcast information to mobile devices which is required for packet transmission operations. The information transmitted on the BCCH is also reproduced so that mobile devices connected in packet switched mode can listen to a single channel (PBCCH) for all general cell information. As mentioned, the PBCCH is optional; if GPRS is supported but the PBCCH is not, then information for GPRS devices is broadcast on the BCCH.

4.6.1.2 Common control channel (CCCH)

This is a GSM channel, but can be used for GPRS if PCCCH does not exist.
    The CCCH is actual constructed from a number of channels, including the following:

- *paging channel (PCH)*: a downlink channel used to page mobile devices
- *random access channel (RACH)*: an uplink channel used to request a SDCCH channel
- *access grant channel (AGCH)*: a downlink channel used to allocate the requested SDCCH. It can also allocate a traffic channel (TCH) directly
- *notification channel (NCH)*: this is used to notify mobile devices of voice group or voice broadcast calls.

4.6.1.3 Packet common control channel (PCCCH)

The PCCCH is an optional channel that is transported on a PDCH; if it is not allocated then information required for packet switch operation is transmitted on the CCCH. The PCCCH may be implemented if the demand for packet data transfer warrants this or if there is enough spare capacity within the cell since this will increase the QoS for packet data access. It consists of the following:

- *packet paging channel (PPCH)*: a downlink channel used to page mobile devices prior to packet transfer. This paging can be used for both circuit switched and packet switched paging
- *packet random access channel (PRACH)*: an uplink channel used to request one or more packet data traffic channels (PDTCH).

- *packet access grant channel (PAGCH)*: a downlink channel used to assign the requested PDTCH channels

- *packet notification channel (PNCH)*: this downlink channel is used to notify a group of mobile devices of a point-to-multipoint packet transfer.

The actual GPRS traffic is transferred over the packet data traffic channel (PDTCH). This corresponds to the actual resources that have been made available for this transfer. It may be a single time slot, part of a time slot or a number of time slots up to the maximum of eight, all of which must be on a single frequency.

The packet associated control channel (PACCH) is a signalling channel which is dedicated to a particular mobile device. It is required in both the uplink and downlink. The information on this channel may consist of resource assignment information, power control information or acknowledgements.

Figure 4.11 illustrates the downlink (a) and uplink (b) channels.

As discussed, it is not necessary for a cell with GPRS to implement the PCCCH channels since the mobile device will be able to use the existing GSM control channels. However, if there are any PDCHs that contain PCCCH the network will broadcast this information on the BCCH. On the downlink the first radio block (B0) will be used as a PBCCH, if a PBCCH exists. If required, up to three more blocks on the same PDCH can be used for additional PBCCH information.

Since GSM and GPRS may dynamically share the same radio resources in the cell, it is important that GPRS releases resources as soon as possible without introducing too much signalling. The following situations for resource release are possible:

- wait for the assignment on the PDCH to terminate

- notify all users that have an assignment on the PDCH individually

- broadcast a message to deallocate the PDCH.

In practice a combination of the above may be used.

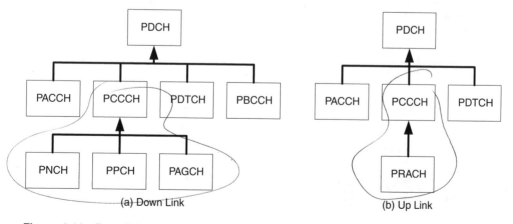

Figure 4.11   Downlink and uplink channels

4.6.2 Subnetwork-dependent convergence protocol (SNDCP)

The SNDCP is only used in the user plane to indicate a specific PDP context and not used by the mobility and session management. A subscriber may have a number of PDP contexts open and each one of these is associated at this layer to a network services access point identifier (NSAPI). The main functions of the SNDCP layer are to provide:

- multiplexing of PDPs;
- compression of user data (including IP header compression);
- segmentation of data packets to be passed to the LLC layer.

The LLC stipulates the maximum size of protocol data unit (PDU) it can carry within a single segment. If the Layer 3 packet (e.g. IP packet), referred to as a network layer PDU (N-PDU), cannot fit into this size then it is a requirement of the SNDCP layer to break the Layer 3 packet into smaller segments (SN-PDU) that can be carried within one LLC frame and then at the receiving end reassemble the N-PDU packet. The SNDCP layer compression capability for the IP header conforms to the IETF header compression RFCs: RFC 1144 and RFC 2507. It is also possible to compress the data in compliance with ITU-T V.42bis. Although compressing the user data reduces the number of data bytes transmitted over the air, it does have the negative effect of increasing the processing power required by the MS and the SGSN.

Figure 4.12 illustrates how two different data connections can be active on a single mobile device, both of these active connections sharing the same logical link between the mobile device and the SGSN. The PDP connection types do not have to be the same; for example, one session may be an IPv4 connection to the Internet while the other may be a PPP or IPv6 connection to the home network. They are identified by their different NSAPI identifiers. If a number of PDP contexts require the same QoS, they will have

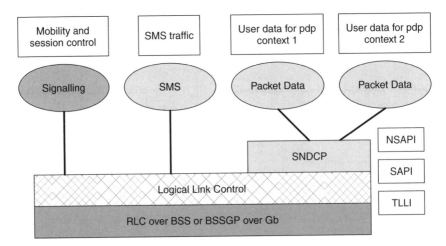

Figure 4.12 Subnetwork-dependent convergence protocol

different NSAPIs; however, it is possible for them to be associated with a single SAPI at the LLC layer.

As can be seen from Figure 4.13, there are two separate frame formats, Figure 4.13(a) shows the frame for the acknowledged mode transfer while Figure 4.13(b) shows the frame for the unacknowledged mode. When using acknowledged mode the SNDCP layer will buffer network layer N-PDU packets and keep them until all segments (SN-PDU) of the N-PDU packet have been acknowledged by the receiving entity. In unacknowledged mode, as soon as the N-PDU packets have been passed to the LLC layer for transmission the packet is immediately discarded. The SNDCP will request the lower LLC layer to transfer acknowledged data using the acknowledged frame format; it is the responsibility of the LLC to ensure that the data arrives in the correct order. During this request the SNDCP can request QoS requirements such as precedence class, peak throughput and the delay class in the SGSN as well as indicate the peak throughput of the mobile device. A radio priority class can also be established for the mobile device, which will be used by the RLC/MAC layer for data transfer. When unacknowledged data transfer is requested, the QoS parameters at both the SGSN and mobile device will also include a reliability classification which indicates whether the LLC layer should use protected or unprotected mode[2] and whether the RLC/MAC layer should use acknowledged or unacknowledged mode.

The fields of the frames shown in Figure 4.13 are defined as follows:

- *Spare bit (X)*: this is set to 0 by the transmitting SNDCP entity and ignored at the receiver.
- *First segment indicator (F)*: this bit is 1 if this is the first segment part of the N-PDU. It indicates that the DCOMP, PCOMP and N-PDU number fields are included in the packet. It also indicates that there are more segments to follow. If this bit is 0 it indicates that this is *not* the first segment part of the N-PDU and that data is to follow directly, i.e. there are no DCOMP, PCOMP and N-PDU number fields.

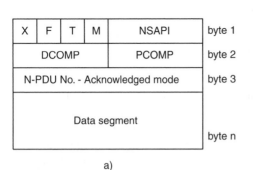

a)   b)

Figure 4.13   SNDCP frame formats

---

[2]The LLC layer can detect errors in frames and depending on whether protected or unprotected mode is being used will either discard or deliver the erroneous frame.

- *SN-PDU type (T)*: a 0 indicates that this is an SN-DATA (acknowledged) frame while a 1 indicates that this is an SN-UNITDATA (unacknowledged) frame.

- *More bit (M)*: since an N-PDU may consist of a number of SN-PDUs (segments), the last one of these segments has to be indicated so that the N-PDU can be reassembled. A 0 indicates that this segment is the last one of the N-PDU, while a 1 indicates that there are more segments to follow.

- *NSAPI*: indicates which PDP context the SN-PDU is associated with since a number of PDP contexts may share the same logical link connection.
  0 This is an escape mechanism for future extensions
  1 Point-to-multipoint multicast (PTM-M) information
  2–4 Reserved for future use
  5–15 Dynamically allocated NSAPI values, allowing for a maximum of 11 contexts.

- *Data compression coding (DCOMP)*: this is only present in the first segment.
  0 No compression
  1–14 This points to the data compression identifier that has been dynamically negotiated
  15 Reserved for future extensions.

- *Protocol control information compression coding (PCOMP)*: this is only present in the first segment.
  1 No compression
  1–14 This points to the protocol control information compression identifier that has been dynamically negotiated
  15 Reserved for future extensions.

- *Segment number*: only required in unacknowledged mode. Since the LLC mechanism will ensure delivery order in ack mode
  0–15 Sequence number for segments carrying an N-PDU.

- *N-PDU number – acknowledged mode*
  0–255 number of the N-PDU.

- *N-PDU number – unacknowledged mode*
  0–255 number of the N-PDU.

4.6.3   Logical link control (LLC)

The LLC provides a reliable link between the mobile device and the SGSN for both control and user data. It supports variable length information fields from 140 bytes up to a maximum of 1520 bytes for the payload and can transfer data and control messages, which may or may not be encrypted. The LLC protocol supports both acknowledged and unacknowledged modes of operation with the ability to also reorder frames that have been received out of sequence. This could occur, for example, in the case of retransmission of frames with errors. It is designed to be independent of the underlying radio protocols to enable different radio solutions to be adopted. In the control plane LLC carries GPRS mobility management (GMM) messages, such as attach, authentication, and session management (SM) information, such as PDP context activation, as well as also transporting SMS messages to the higher layers. In the user plane, LLC frames carry the

SNDCP packets which will contain the user data such as IP packets. Figure 4.9 shows how a single LLC is used for transporting signalling, SMS and different data connections concurrently. An LLC connection is identified by a datalink connection identifier (DLCI). This consists of the service access point identifier (SAPI) and the temporary logical link identifier (TLLI) of the mobile device.

As was seen previously, in GSM for security reasons, the transfer of the user IMSI is minimized, and thus a TMSI, assigned by the VLR, is used instead. In GPRS, the SGSN will assign a user the equivalent for the packet network, known as a P-TMSI. Like a TMSI, this is a 4-byte number, with local significance only. The mobile device will build a TLLI from the P-TMSI and use this TLLI to refer to itself in subsequent transactions with the network. This TLLI is a 32-bit number and uniquely defines the logical link between the mobile device and the SGSN. The mobile can build three different types of TLLI: local, foreign and random. A local TLLI is derived from the P-TMSI allocated by the SGSN in the current routing area (see later) and is therefore valid only within that area. A foreign TLLI is derived from a P-TMSI from a different routing area. A random TLLI is built if there is no P-TMSI available. Figure 4.14, shows how each is built.

The SAPI is carried within the LLC frame header and defines the particular SAP within the mobile device and SGSN with which it is associated. Since the TLLI is used to uniquely identify the mobile device, it is therefore required to be present. However, it can be seen from Figure 4.15 that there is no TLLI field within the LLC header. The underlying BSSGP (between the BSC and SGSN) and the RLC/MAC (between mobile device and BSC) during contention periods are required to transport the TLLI information that uniquely identifies the mobile device.

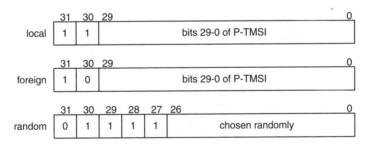

Figure 4.14 Format of the TLLI

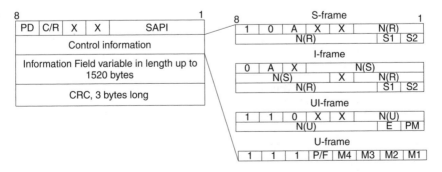

Figure 4.15 LLC frame format

4.6.3.1 Unacknowledged mode

For this mode of operation an entity may initiate transmissions to a peer entity without having established a logical connection. The LLC does not guarantee in-order delivery of the frames and no error recovery procedures are defined. The LLC can however detect errors in a received frame, and depending on whether protected or unprotected mode has been used for the transmission will either discard or deliver the frame with errors. Recall that data has no tolerance of errors so an indication of errors being present is generally what is required, rather than a measure of how many errors are present. In protected mode, a frame check sequence (FCS) in the form of a CRC covers both the frame header and information field. In unprotected mode, on the other hand, the CRC check covers the header and only the first 4 bytes of the information field, corresponding to the maximum length of a SNDCP segment PDU header. The rest of the information is unprotected and permits applications which can tolerate errors to receive frames that may have errors. There is also no flow control for this transfer option. The frame format used is the numbered unconfirmed information (UI) frame. This mode of operation is referred to as asynchronous disconnected mode (ADM). Unacknowledged mode of operation is specified for all SAPIs that are not reserved, as indicated in Table 4.4. Both GPRS mobility management and SMS messages use this mode for transfer.

4.6.3.2 Acknowledged mode

In the acknowledged mode, each sending entity is responsible for the organization of its data flow and for error recovery procedures. To enable this, the link needs to be first established using the set asynchronous balanced mode (SABM) command. The frame format used is the numbered information (I) frame and the frames are acknowledged at the LLC layer. Procedures are specified for both retransmission of any frames that are

Table 4.4 SAPI identifiers

Value	SAPI	Service description	SAP name	Mode
0	0000	Reserved		
1	0001	GPRS mobility management	LLGMM	Unack
2	0010	Tunnelling of messages 2	TOM2	Unack
3	0011	User data 3	LL3	Ack/Unack
4	0100	Reserved		
5	0101	User data 5	LL5	Ack/Unack
6	0110	Reserved		
7	0111	SMS	LLSMS	Unack
8	1000	Tunnelling of messages 8	TOM8	Unack
9	1001	User data 9	LL9	Ack/Unack
10	1010	Reserved		
11	1011	User data 11	LL11	Ack/Unack
12	1100	Reserved		
13	1101	Reserved		
14	1110	Reserved		
15	1111	Reserved		

unacknowledged as well as for flow control. This mode of operation is referred to as asynchronous balanced mode (ABM) and provides a reliable in-order delivery service. This mode of operation is allowed for all SAPIs that are not reserved except for SAPIs 1, 2, 7 and 8 (see Table 4.4).

4.6.3.3   LLC frame formats and procedures

The LLC frames that are received will have different SAPIs to indicate what is being transported. For example, SAPI 1 is reserved for mobility management as shown in Table 4.4.

Figure 4.15 shows the format of the LLC header, and its fields are now described.

The **protocol discriminator (PD)** bit in the frame address header should be set to a logic 0 to indicate that this is an LLC frame; if it is set to 1 the frame will be treated as an invalid frame.

The **command/response (C/R)** bit indicates whether this is a command or a response to a command. The options are highlighted in Table 4.5.

The SAPI identifies the address of the higher layer services, to which this frame should be sent. There are 4 bits set aside for SAPI addresses and thus there can be a maximum of 16 SAPIs. A number of these are reserved and Table 4.4, correlates the addresses to the specific SAPI entities. The service access point (SAP) name identifies the actual service that is associated with the particular logical link frame.

It can be seen that there are four separate SAPs for user data. Each one of these can be assigned a different QoS. There are also two SAPs for tunnelling of messages. These give high and low priority to the messages being transferred. The tunnelling of messages is an optional procedure which uses the LLC unacknowledged mode of operation to tunnel non-GSM messages such as EIA/TIA-136 messages.

There are four types of control field:

- *Supervisory (S frame)* frames are used to perform LLC functions. They can acknowledge I frames using the sequence number of the received frame, N(R), and also temporarily suspend I frame transmission. The acknowledge request bit (A) is set to logic 1 by the sender if an acknowledgement is required, and to 0 if no acknowledgement is required. The S frame is sent if there is no information field data that needs to be transferred.

- *Confirmed information transfer (I frame)* where there is a sequence number for both the sent frames, N(S), and the received frames, N(R). Each I frame also contains a supervisory (S frame) and is sometimes referred to as an I + S frame.

Table 4.5   C/R values

Type	Direction	C/R
Command	SGSN to UE	1
Command	UE to SGSN	0
Response	SGSN to UE	0
Response	UE to SGSN	1

- *Unconfirmed information transfer (UI frame)* is used to transfer information to higher layers that does not need acknowledgements and in this respect no verification of the sequence numbers, N(U), is performed. This means that a frame could be lost with no notification given to the higher layer. The information may or may not be encrypted. This is indicated by the E bit. If it is set to logic 1 then the information is encrypted. The protected mode (PM) bit indicates whether the CRC checks across the header and payload or just the header. If it is set to logic 1 it indicates that the payload is also protected.

- *Unnumbered (U frame)* provides additional LLC functions. It does not contain any sequence numbers. The poll/final (P/F) bit is referred to as the poll bit if this is a command frame and the final bit if this is a response frame. The P bit is set to logic 1 to request a response frame from the receiver. The F bit is set to a logic 1 to indicate that this is a reply to a poll request command.

It can be seen from Figure 4.15 that the I and S frames have two further bits, S1 and S2, and also that the U frame has four M bits. These bits are used for control and response functions and Figure 4.16 highlights the values these fields can have. If a field has a value which is not defined, then the frame is rejected.

The function of the I frame is to transfer sequentially numbered frames containing information for higher layers across the logical link between the SGSN and the mobile station. Referring to Figure 4.16(a), it can be seen that there are four command/responses and these are briefly described below:

- The *receive ready (RR)* command/response is used by a receiver to indicate that it is ready to receive an I frame. It is also used to acknowledge received I frames.

- The *acknowledgement (ACK)* command/response is used to acknowledge a single or multiple I frames up to and including the last frame received, N(R) −1.

- The *selective acknowledgement (SACK)* command/response supervisory frame is also used to acknowledge I frames. In this case the information field contains a list of the frames that have been received successfully so that the sender needs to retransmit only the missing frames.

- The *receive not ready (RNR)* command/response is used to indicate that the entity is unable to accept incoming I frames at this time.

Command	Response	S1	S2
RR	RR	0	0
ACK	ACK	0	1
RNR	RNR	1	0
SACK	SACK	1	1

a) I and S Frames

Command	Response	M4	M3	M2	M1
	DM	0	0	0	1
DISC		0	1	0	0
	UA	0	1	1	0
SABM		0	1	1	1
	FRMR	1	0	0	0
XID	XID	1	0	1	1
NULL		0	0	0	0

b) U Frames

Figure 4.16   I frame command/response

Referring to Figure 4.16(b), for the unnumbered frames it can be seen that there are four M bits and seven combinations. Unlike the I and S frames above, these are not identical for command and response. The combinations are briefly described below:

- *Set asynchronous balanced mode (SABM)* is used to place the addressed mobile device or SGSN into ABM (i.e. acknowledged) mode of operation. ABM mode of operation ensures that each entity assumes responsibility for its own data flow and error recovery. Both of the entities act as data source and data sink, allowing information to flow in two directions.

- The *disconnect (DISC)* command is used to terminate an ABM connection.

- The *unnumbered acknowledgement (UA)* response is used to acknowledge receipt of the SABM or DISC commands.

- The *disconnect mode (DM)* response informs its peer that it is unable to perform ABM mode of operation at this time.

- The *frame reject (FRMR)* response is used to indicate to the sender that a frame has been rejected and simply resending the particular frame will not suffice. This may occur if a control field is undefined or not implemented.

- The *exchange identification (XID)* command/response is used to negotiate and renegotiate LLC layer parameters. An entity will send an XID command with a set of parameters for a particular application; the receiver can accept the values or offer different values in the XID response. The parameters may include such things as maximum information size for U and UI frames, timer timeouts, maximum number of retransmissions and window size.

- The *NULL* command is used to indicate a cell update.

Figure 4.17, illustrates how data is transferred in both unacknowledged and acknowledged modes. In unacknowledged mode the information is simply transferred in a UI frame. When information is transferred in the acknowledged mode, the I+S frame is used and this will be acknowledged by either an S frame if information is not being sent in the reverse direction, or an I+S frame, which is piggybacking an acknowledgement for data received with the data being sent. For acknowledged mode the data transfer is preceded by the SABM command which is required to actually establish the acknowledged mode of operation. Figure 4.18 shows examples of the signalling required for establishing, releasing and renegotiating LLC procedures. Standard primitives are used between the Layer 3 and the logical control layer. These primitives are request (REQ), indication (IND), response (RES) and confirm (CNF).

Figure 4.19 shows both an uplink and downlink LLC trace across the Abis, which is carrying an IP packet. The 03 C2 0D is the LLC header and indicates that this particular frame is using SAPI 3 and the UI header. The 65 00 06 83 indicates that this is an unacknowledged single segment SN-PDU which is destined for NSAPI 5. Also highlighted in the figure are the source and destination IP addresses (0A 01 02 7B C0 A8 01 02) and the TCP port numbers (04 0F 00 14), indicating that this is an FTP session.

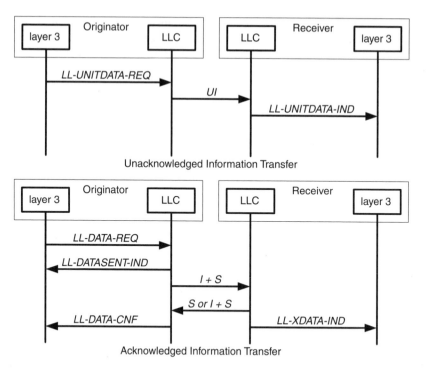

Figure 4.17   LLC data transfer

4.6.3.4   Ciphering algorithm

The ciphering algorithm shown in Figure 4.20 is used to provide both integrity and confidentiality for user data between the mobile device and the SGSN. It can be used in both the downlink and uplink as well as in the downlink for point-to-multipoint group transmissions. The algorithm provides security over the information and the CRC but does not cover the actual header. Both I and UI frames can be encrypted. Encryption of the I frame is based on whether ciphering has been assigned to the TLLI whereas encryption of the UI frame is based on this assignment as well as the setting of the E bit in the UI header by upper layer protocols. The *key* (Kc) is 64 bits in length and is generated in the GPRS authentication procedure. The *direction* is 1 bit and indicates whether this is an uplink or downlink packet. The *input* is 32 bits in length and derived from the LLC frame number, which is N(S) for I frames and N(U) for UI frames. This protects against replay of previously sent frames.

4.6.4   Radio Link Control/Media Access Control (RLC/MAC)

The main role of the RLC/MAC layer is to segment uplink LLC packets for transfer over the radio link from the mobile device through to the BSC. At the BSC these LLC packets are then reassembled and relayed through to the BSSGP for transfer to the SGSN. In

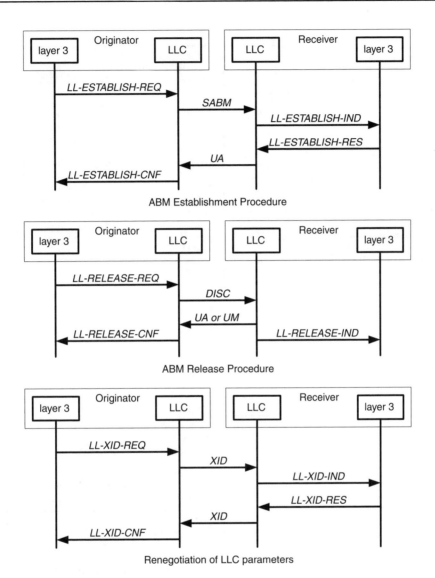

Figure 4.18   Example LLC control procedures

the downlink a similar mechanism occurs, with the BSC receiving the LLC packets via the BSSGP and segmenting them into the RLC/MAC blocks for transfer to the mobile device. The RLC/MAC layer can transfer the RLC data blocks in both acknowledged and unacknowledged mode. The packet control unit (PCU), which is usually located within the BSC, is responsible for the RLC/MAC tasks. These include the segmentation and reassembly of the LLC frames.

- *Acknowledged mode*: when acknowledged mode is used a selective ARQ mechanism coupled with block sequence numbering (BSN) is used to ensure correct transfer of

Uplink

```
Conn:4 Line:1 TS:2 Subch:3
LLC UI format (58)
03 C2 0D 65 00 06 83 45 00 00 30 1F 07 40
00 80 06 FB 23 0A 01 02 7B C0 A8 01 02 04
0F 00 14 00 17 A5 1D 48 48 C9 CE 70 12 21
80 C9 1F 00 00 02 04 02 18 01 01 04 02 5F
13 9F
SN-UNITDATA PDU
Spare (X)
- ok
First Segment Ind. (F)
- First segment of N-PDU
SN-PDU Type (T)
- SN-UNITDATA PDU
More (M)
- Last segment of N-PDU
NSAPI
- Dynamically allocated NSAPI: 5 (5h)
DCOMP
- No data compression
PCOMP
- No protocol control info compression
Segment Number
- Current segment: 0 (0h)
N-PDU Number
- UNACK mode: 1667 (683h)
Data Segment
IP DATAGRAM
Source address : 10.1.2.123
Destination address : 192.168.1.2
```

Downlink

```
Conn:4 Line:1 TS:2 Subch:3
LLC UI format (586)
43 C6 45 65 00 09 91 45 00 02 40 C0 C4 40
00 7E 06 59 56 C0 A8 01 02 0A 01 02 7B 00
14 04 0F 48 48 C9 CE 00 17 A5 1E 50 10 40
E8 CC 88 00 00 D0 CF 11 E0 A1 B1 1A E1 00
00 00 00 00 00 00 00 00 00 00 00 00 00 00
00 3E 00 03
                    [abridged]
00 FE FF 09 00 06 00 00 00 00 00 00 00 00
00 00 00 09 00 00 00 7B 02 00 00 00 00 00
00 00 10 00 00 89 02 00 00 01 00 00 00 FE
FF FF FF 00 00 00 00 76 02 EC A5 C1 00 71
00 09 04 00 00 08 12 BF 00 00 00 00 00 00
10 00 00 00 00 51 42 3B
SN-UNITDATA PDU
Spare (X)
- ok
First segment Ind. (F)
- First segment of N-PDU
SN-PDU Type (T)
- SN-UNITDATA PDU
More (M)
- Last segment of N-PDU
NSAPI
- Dynamically allocated NSAPI: 5 (5h)
DCOMP
- No data compression
PCOMP
- No protocol control info compression
Segment Number
- Current segment: 0 (0h)
N-PDU Number
- UNACK mode: 2449 (991h)
Data Segment
IP DATAGRAM
Source address : 192.168.1.2
Destination address : 10.1.2.123
```

Figure 4.19   LLC trace. Reproduced by permission of NetHawk Oyj

the blocks. The sending side will transmit the blocks within a preset window size and will receive ACK/NACK messages when required. The ACK/NACK will acknowledge all blocks up to an indicated BSN. If there are any blocks missing these can be indicated within a bitmap field and selective retransmission of these blocks can take place. The acknowledgement procedures for the LLC (mobile device to SGSN) and RLC/MAC (mobile device to BSC) layers should be seen as independent of each other.

- *Unacknowledged mode*: the transfer of blocks in the unacknowledged mode is controlled by block sequence numbering, and ACK/NACK messages are sent by the receiver. However, this mode of operation does not include retransmissions and thus the ACK/NACK information can only be used for assessing the quality of the link. The length of the user field is preserved by inserting *dummy* information.

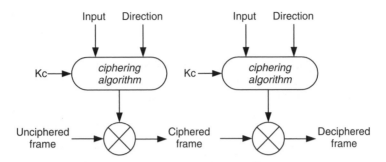

Figure 4.20 Ciphering algorithm

RLC/MAC *data blocks* are used to transport LLC PDUs containing both user data and upper layer signalling. RLC/MAC *control blocks* are used to transport control messages related to TBF management, paging and system information, etc. These RLC/MAC control blocks have a higher priority than the data blocks.

4.6.4.1 Temporary block flow (TBF)

The temporary block flow mechanism is a unidirectional physical connection to support the transfer of LLC frames. It enables a number of mobile devices to share either a single time slot or occupy a number of time slots in both the uplink and downlink directions. The uplink and downlink are independently assigned, giving the opportunity for asymmetric transfer. This is more suitable for data traffic, where in general there is a greater volume of traffic in the downlink. Each of the mobile devices is assigned a single or multiple radio blocks at a time, therefore there could be a number of mobile devices sharing the same 9.05 kbps channel. This would give a very slow data rate but would, as previously discussed, be sufficient to keep a TCP/IP connection from timing out. A mobile device is given an assignment consisting of the channel, time slot and radio block on which they can transmit. This is required as a means of contention control since there may be a number of mobile devices wishing to transmit on the uplink at the same time. This situation does not occur in the downlink since there is only one base station. There are three techniques of working this system: the dynamic, extended dynamic and fixed allocation methods. Both the dynamic and fixed allocation methods are required to be supported by all GPRS networks whereas the extended dynamic allocation is an optional feature.

- *Dynamic allocation*: an uplink state flag (USF) is transmitted on the downlink data channel to allow the multiplexing of a number of mobile devices to send data on the uplink in the correct radio blocks. The USF comprises three bits. Each mobile device is allocated an individual number and thus a maximum of 8 mobile devices can be multiplexed at any one time even though there are 12 radio blocks.

- *Extended dynamic allocation*: this method is similar to the above but eliminates the need to receive a USF on each of the time slots. In this case when a mobile device receives a USF on a PDCH downlink channel time slot it will assume that it can send data not only on this time slot but on all the others that have been allocated to it. For example, if a mobile device has been allocated blocks on time slots 0, 2 and 3 and it receives a USF on time slot 0, it will transmit on time slots 0, 2 and 3, not just on time slot 0.

- *Fixed allocation*: a packet uplink assignment message is used to communicate to the mobile device which resources (radio blocks) it has been allocated. The fixed allocation method does not include the USF and the mobile device is free to transmit on the uplink without monitoring the downlink for the USF.

Each TBF is allocated a temporary flow indicator (TFI) which is assigned by the network. The TFI is a 5-bit field contained within the RLC header and is unique within the cell. It is used in both uplink and downlink to ensure that the received radio blocks are associated with the correct LLC frame and SGSN/mobile device. Figure 4.21, illustrates how four subscribers can share a single time slot using the dynamic allocation method. The 52-frame multiframe is split into 12 blocks each consisting of four GSM bursts. Mobile device 1 has been assigned USF = 1. It monitors the base station downlink signal and notices that blocks 2 and 3 have USF = 1 in their header information, it knows therefore that it has been allocated these blocks to send information. In the uplink to the base station. Notice that the mobile device which has been assigned USF = 2 has been assigned a larger number of blocks. This may be because the subscriber has paid a premium for higher QoS.

4.6.4.2 Establishment of TBF

The mobile station can establish a TBF for the transfer of LLC packets in the uplink (both signalling[3] and data packets) by sending the *packet channel request* message on the

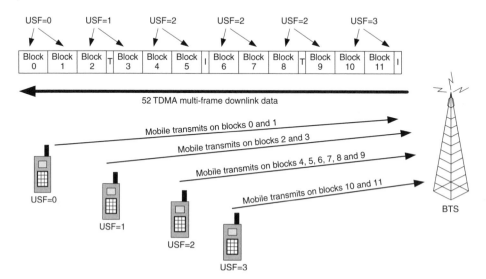

Figure 4.21 GPRS users sharing time slots

---

[3]When signalling data is sent this is transferred using the RLC acknowledged mode of operation.

(P)RACH channel to the network. The mobile device may use a *one-* or *two-phase access method*. It is possible under certain circumstances, highlighted below, that the mobile device will use channels other than the PRACH.

- If a TBF has already been established in the downlink direction from the network to the mobile device then the PACCH can be used for the initial access.
- If the mobile device is operating in dual-transfer mode[4] (DTM) and already has a dedicated mode connection established then the DCCH should be used for the TBF initiation.

In the one- and two-phase access methods the network will respond with a *packet uplink assignment* message on the (P)AGCH. If the mobile station does not receive a *packet queuing notification* or a *packet uplink assignment* message from the network it will wait for a certain time period before resending the message again. The *packet queuing notification* message may be sent if there are more packet channel requests than can be handled. This message indicates to the mobile device that its packet channel request message has been received correctly and a *packet uplink assignment* message will be forthcoming. For the single-phase access, reservation of resources is accomplished with use of the information within the *packet channel request* message. On the RACH there are only two *cause* values available for GPRS which can be used, to request limited resources or two-phase access. The PRACH channel (if available), which is designed for GPRS, can contain more adequate information about the requested resources than the RACH.

In the two-phase access method the *packet uplink assignment* message sent on the (P)AGCH will reserve the limited resources required for the *packet resource request* message (not for data transfer) which is sent on the PACCH. This *packet resource request* message carries a complete description of the requested resources for the data transfer. The network will respond on the PACCH with a *packet uplink assignment* message and will reserve the resources required and define the actual parameters for the data transfer. These parameters may include the power control information, number of blocks allocated, TFI to be used, USF issued and channel coding type (e.g. CS-2). This procedure is highlighted in Figure 4.22 and data transfer would follow this procedure. The resources are usually released by the mobile device as it counts down (see *countdown value* in uplink frame format; Section 4.6.4.3) the last few blocks it wishes to send. The network is then free to reallocate the USF to some other user. To initiate downlink data transfer to a mobile device in standby state the network would page the device as illustrated in Figure 4.23. As can be seen, the mobile device sends a *packet channel request* message in the same way as before.

This procedure enables the mobile device to send a packet paging response to the network. Once this is received, the mobile device will be in ready mode, where it can send and receive data (see later). The network can now initiate a *packet downlink assignment* message. If there is already an uplink packet transfer in progress this message will be sent on the PACCH; otherwise it will be sent on the PCCCH. If there is no PCCCH allocated

---

[4]The mobile device can be simultaneously connected to the network in dedicated mode and packet transfer mode. DTM is a subset of the class A mode of operation.

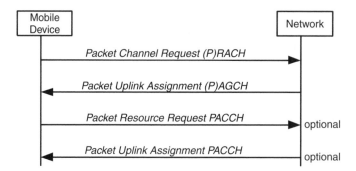

Figure 4.22 Packet uplink assignment

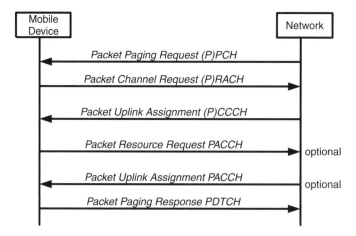

Figure 4.23 Paging for downlink packet transfer

in the cell then an *immediate assign* message will be sent on the CCCH. This message will include the PDCHs that will be used for the data transfer.

There are two message formats that can be used for the *packet channel request* message, containing 8 bits or 11 bits of information. The correct one to use for the cell for a particular purpose is broadcast on the (P)BCCH. The 11-bit format includes two bits which define four levels of priority. Both formats have the same access types specified and these are shown in Table 4.6. The one-phase access request has five bits set aside to indicate the multislot capability of the mobile device.

- If the mobile device wishes to establish the TBF for user data using the RLC unacknowledged mode the two-phase access method will be used.
- If acknowledged mode and the amount of data consists of eight or fewer RLC/MAC blocks then the short access method will be used.
- If acknowledged mode (>eight blocks) is to be transferred then either the single or two-stage access can be used.

Table 4.6 Packet channel request access types

Access type
One-phase access request
Short access request
Two-phase access request
Paging response
Cell update
Mobility management procedure
Single block without TBF establishment

- If the purpose of the packet access procedure is to respond to a page then page response will be used.
- If the purpose of the packet access procedure is to send a cell update then the cell update will be used.
- If the purpose of the packet access procedure is for any mobility management procedure then the mobility management procedure will be used.
- If the purpose of the packet access procedure is to send a measurement report or a packet pause message then the single block without TBF establishment will be used.

4.6.4.3 Contention resolution

Contention resolution is an important part of the RLC/MAC operation since a single channel can be used for the transfer of a number of LLC frames, and thus two mobile devices may perceive that a channel is allocated to both of them. It applies to both dynamic as well as fixed allocation methods of operation. Since there are two basic access possibilities, one-phase and two-phase, they are dealt with separately. The two-phase access is inherently immune from contention since the second access phase, packet resource request, addresses the mobile device by its TLLI. Since the same TLLI is included in the downlink *packet uplink/downlink assignment* message, no mistake can be made. The one-phase access method, however, is more insecure in its action and an effective contention resolution mechanism needs to be introduced. The first part of the solution requires identification of the mobile device. This is already necessary to establish a TBF on the network side and also to take into consideration the multislot capability of the device. By including the TLLI in the uplink ACK/NACK messages, the network is notified of to whom the allocation belongs. This message needs to be sent early (even before the receive window for RLC/MAC is full) as contention is resolved after the first occurrence of this ACK/NACK message.

While contention resolution is ongoing, the mobile device will use its TLLI to uniquely identify itself within every RLC data block that is sent within the TBF. The TLLI is 32 bits long and as such introduces significant overhead. However, to alleviate this, once the contention resolution has completed, the TLLI need not be transmitted within the RLC data block. The network will reply with a *packet uplink ACK/NACK* message, which will include the same TLLI value that the mobile device has used.

For one-phase access, contention resolution on the network side is assumed to be successfully completed once the network receives an RLC data block which contains both the TLLI identifying the mobile device and the TFI value that has been associated with the TBF. The mobile device assumes contention resolution is successfully completed when it receives a *packet uplink ACK/NACK* message which includes the same TLLI value that the mobile device has sent in the initial RLC block and the TFI associated with the uplink TBF.

For two-phase access the contention resolution is assumed to be complete when the network receives a TLLI value identifying the mobile device as part of the contention resolution procedure. The mobile device assumes contention resolution is successfully completed when it receives a *packet uplink assignment* message which includes the same TLLI value that the mobile device has sent included in the *packet resource request* and its *additional MS capabilities* messages.

4.6.4.4 RLC/MAC frame format

Figures 4.24 and Figure 4.25, show the frame format for RLC/MAC data and control frames. The fields of each are explained below.

Data fields

- *Payload type (PT)*: this 2-bit field indicates whether this is a data (00) or a control (01) block. A value of 10 in the downlink indicates a control block with the inclusion of the optional header. All other values are reserved.

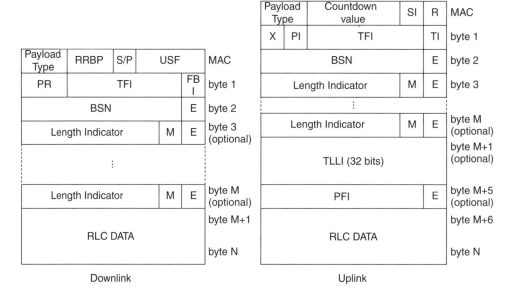

Downlink   Uplink

Figure 4.24 RLC/MAC data block

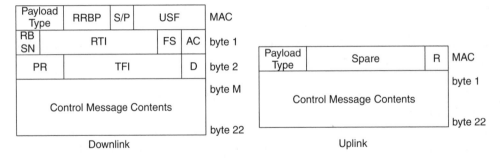

Figure 4.25 RLC/MAC control block

- *Relative reserved block period (RRBP)*: this 2-bit field specifies a reserved block that the mobile device may use for a *packet control acknowledgement* message or a PACCH block.

- *Supplementary/polling (S/P)*: a 0 in this single-bit field indicates that the RRBP field is not valid and a 1 indicates that it is.

- *Uplink status flag (USF)*: the USF consists of 3 bits and is sent in all downlink RLC blocks to indicate the owner of the next uplink radio block on the same time slot.

- *Power reduction (PR)*: this 2-bit field indicates the power level reduction of the current RLC block as compared to the power of the BCCH. A value of 0 indicates 0–2 dB, 1 indicates 4–6 dB and 2 indicates 8–10 dB less than the BCCH. A value of 3 is unused.

- *Temporary flow identity (TFI)*: this 5-bit field identifies the TBF within which this block belongs.

- *Final block indicator (FBI)*: this single-bit field indicates the last downlink RLC data block. A 0 indicates that there are more blocks to come whereas a 1 indicates that this is the last block in this TBF.

- *Block sequence number (BSN)*: this 7-bit field carries the sequence number of the RLC block (mod 128) within the TBF.

- *Extension (E)*: this single-bit field indicates the presence of the optional byte in the data block header. A 0 indicates that an extension byte follows.

- *Length indicator (LI)*: this 6-bit field is used to delimit LLC PDUs within a single RLC data block by identifying the last byte of the LLC PDU.

- *More (M)*: this single bit, along with the E bit and the LI, is used to delimit LLC PDUs within a TBF. It identifies whether or not another LLC PDU follows the current one within a single RLC data block.

- *Countdown value (CV)*: this 4-bit field is sent by the mobile device to allow the network to calculate the number of RLC blocks remaining in the current uplink TBF.

- *Stall indicator (SI)*: this single-bit field indicates whether the mobile station's RLC transmit window can advance or not. A 0 indicates that the window is not stalled.

- *PFI indicator (PI)*: this single-bit field indicates the presence of the optional PFI field. A 0 indicates that it is not present.

- *TLLI indicator (TI)*: this bit indicates whether the TLLI field is present or not. A 0 indicates that the TLLI is not present.

- *Temporary logical link identifier (TLLI)*: this 32-bit field is optional and is used while contention resolution is required.

- *Packet flow identifier (PFI)*: this 7-bit field is assigned by the SGSN and used to identify a particular flow and QoS value. Legitimate identifiers can be: best effort, signalling, SMS or dynamically assigned.

- *RLC data*: this field contains the LLC PDU or part of it if it has been segmented. The amount of data transferred depends on whether there are any optional RLC headers in place and also on the coding scheme used. Table 4.7 indicates the number of bytes each coding scheme can carry.

Control fields

- *Reduced block sequence number (RBSN)*: used to indicate the sequence number of the downlink RLC control blocks.

- *Radio transaction identifier (RTI)*: this 5-bit field is used to group downlink control blocks which make up a single message.

- *Final segment (FS)*: this single-bit field is used in downlink control blocks to indicate the final segment of a control message. A 0 indicates that this is not the final segment.

- *Address control (AC)*: this single bit indicates the presence of the optional PR/TFI/D byte in the downlink control block. A 0 indicates that this byte is not present.

- *Direction (D)*: this single-bit field indicates whether the TBF identified in the downlink control block TFI field is for an uplink (0) or downlink (1) TBF.

- *Retry (R)*: this single-bit field indicates whether the mobile station has sent the *channel packet request* message more than once. A 0 indicates that it was sent once and a 1 indicates that it was sent multiple times.

4.6.4.5 *Summary of roles of SNDCP, LLC and RLC/MAC*

The SNDCP is used to transfer data packets between the mobile device and the SGSN. It is capable of multiplexing a number of channels onto the underlying LLC. It also provides for compression of both the header and data.

Table 4.7 Number of bytes each coding scheme can carry

Coding scheme	Bytes of LLC data
CS-1	20
CS-2	30
CS-3	36
CS-4	50

The datalink between the mobile device and the network is split into two separate sublayers, the LLC between the mobile device and the SGSN and the RLC/MAC between the mobile device and the BSS. The LLC provides both acknowledged and unacknowledged modes of operation between the mobile device and the SGSN. It includes flow control, error detection/correction through ARQ retransmissions and encryption of the data.

The RLC/MAC layer uses the services of the physical layer for information transfer. It handles the segmentation and reassembly of LLC PDUs between the mobile device and the BSC. Its functions include backward error correction through the use of selective retransmissions and arbitration of resources between multiple mobile devices across the shared medium. The MAC enables a single mobile device to use several physical channels in parallel. The RLC/MAC provides two methods of operation, acknowledged and unacknowledged modes.

4.6.5 GPRS radio protocol

Since the air interface is particularly unreliable, extra error checking and correction bits are added to the data. However, since this plus the data needs to fit into a transfer size of 456 bits, the amount of coding reduces the space remaining for data, and hence the throughput. In Table 4.8, the radio block (DATA) indicates the amount of subscriber data transferred. As can be seen from the table, when a subscriber is using CS-1, there are 181 bits of *user data* being transferred in the 456-bit block. This *user data* is data at the RLC/MAC layer and therefore includes their associated headers. The BCS bits are a block check sequence which checks for errors, and the size of this field depends on the coding scheme used.

CS-1, CS-2, and CS-3 all use 1/2 rate convolution coding to increase the reliability of the data transfer and are therefore multiplied by 2; CS-4 does not use convolution coding, but merely error checking using a CRC field. The error checking and correction for CS-2 and CS-3 introduce too many bits for the RLC frame, therefore puncturing is required. Puncturing simply reduces the number of bits to 456 so that the frame will fit into the four bursts. It does this by removal of non-critical error correction bits. The code rate gives an indication of the amount of subscriber data within the frame; for example, for CS-1, two bits of information are required at this layer to transfer one bit of subscriber data. It is therefore 50% efficient. For CS-4 we can see that the efficiency has risen to 100% at this particular layer. However, there is a lack of error checking/correction and in cells with high interference, using CS-4 will most likely reduce overall throughput.

Table 4.8 Coding scheme bit details

Coding scheme	Code rate	USF	Pre-code USF	Radio block (DATA)	BCS	Tail	Coded bits	Punctured bits	Data rate (kbps)
CS-1	$\frac{1}{2}$	3	3	181	40	4	456	0	9.05
CS-2	$\approx\frac{2}{3}$	3	6	268	16	4	588	132	13.4
CS-3	$\approx\frac{3}{4}$	3	6	312	16	4	676	220	15.6
CS-4	1	3	12	428	16	–	456	0	21.4

Table 4.9 Physical layer interaction with coding schemes

Coding scheme	Layer		
	Information	Block coding	Convolution code
CS-1	184	$40 + 0 + 4 = 228$	456
CS-2	271	$16 + 3 + 4 = 294$	588
CS-3	315	$16 + 3 + 4 = 338$	676
CS-4	431	$16 + 9 + 0 = 456$	456

Table 4.9 shows the information bits (including the RLC/MAC and USF) and how extra bits are added at the various entities within Layer 1. As can be seen, the block entity adds on a CRC which is 40 bits for CS-1 and 16 bits for CS-2, CS-3, and CS-4. For CS-2, CS-3 and CS-4 the USF (3 bits) is modified to give better protection. This is not required with CS-1 due to its lower data rate. CS-1, CS-2 and CS-3 also have four tail bits added. The resulting bits are passed to the convolution entity, which performs 1/2 rate convolution coding on CS-1, CS-2 and CS-3. The output of the convolution coder is required to be 456 bits, therefore puncturing as described above may be necessary.

4.6.6 Layer 1

Layer 1 is divided into two separate distinct sub-layers, the physical radio frequency (RF) layer and the physical link layer. The physical RF layer performs the modulation based on the data it receives from the physical link layer and at the receiver demodulates the signal. The physical link layer provides framing, data coding and the detection and

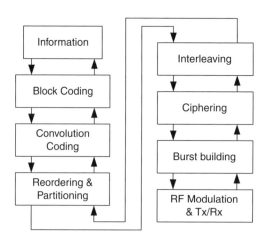

Figure 4.26 Physical layer procedures

possible correction of physical medium transmission errors. Figure 4.26 illustrates the procedures that are executed at the physical layer. These functions include forward error correction through the use of convolution coding and interleaving of RLC/MAC radio blocks over four consecutive bursts.

4.7   Gb INTERFACE PROTOCOLS

The Gb interface connects the BSS to the SGSN and is used for both signalling and data. It is designed to allow many users to be multiplexed over the same physical resources and allows different data rates to be available depending on the user requirements. Resources are allocated to a user when data is either transmitted or received. This is in contrast to the A interface used for circuit switched connections where a user has the sole use of a dedicated resource throughout the call, irrespective of the amount of activity.

4.7.1   Layer 1 bis

There are a number of physical layer configurations possible, and the actual physical connection of this interface is subject to negotiation between the equipment vendor and operators. The specification allows for point-to-point connections between the SGSN and the BSS, but also allows for an intermediate frame relay network to be used. In situations where an intermediate network is used, a number of physical layer technologies may be used over the different links between switches, and between switches and the SGSN and the BSS. Since the standards allow for the intermediate network, this does increase the complexity of this interface and unfortunately introduces a large number of identifiers. In situations where the MSC and SGSN are co-located, it may be advantageous to multiplex channels within the same E1 (2.048 Mbps) or T1 (1.544 Mbps) link for both circuit switched (A interface) and packet switched (Gb interface) connections. When multiple 64 kbps channels are used for this interface it is recommended that they are aggregated into a single $n \times 64$ kbps channel since this will take advantage of the statistical multiplexing at the upper layer.

4.7.2   Frame relay

The network link layer is based on frame relay as defined in GSM08.16. Virtual circuits (VC) are established between the BSS and SGSN and the transmissions from a number of users can be multiplexed over these VCs. In many cases it is expected that there will be a direct link between the BSS and the SGSN; however, frame relay will permit an intermediate network in between the SGSN and BSS.

The frame relay connection will allow different frame sizes with a maximum of 1600 bytes. The frame relay header is 2 bytes long. A number of permanent virtual circuits (PVC) are expected to be used between the SGSN and the various BSS to transport the BSSGP data packets. These links are to be set up using administrative procedures. The actual PVC between the SGSN and the BSS is referred to as the network services virtual connection (NS-VC). Thus the NS-VC identifier (NS-VCI) has Gb end-to-end significance, and uniquely identifies this connection between an SGSN and a particular BSS. If an intermediate network is used, each of the frame relay links is referred to as a network services virtual link (NS-VL). It is often the case that there are a number of paths between the SGSN and a particular BSS. This can be useful for redundancy and load sharing. It is therefore necessary to combine all of the NS-VCs between an SGSN and a specific BSS into a group, which is referred to as the network services virtual connection group (NS-VCG). This group of connections is identified by a network services entity identifier (NSEI), which identifies the actual BSS to the SGSN for routing purposes. It is important to note that frame relay guarantees in-order delivery of frames. By introducing different paths between the SGSN and the particular BSS, there needs to be a mechanism introduced to maintain this ordering of frames.

Frame relay is a packet switched technology that was developed to provide high-speed connectivity. It takes advantage of the low error rates on modern networks by leaving retransmission to the end stations. By stopping all point-to-point error correction and flow control within the network itself, the nodes do not have to wait for acknowledgements or negative acknowledgement. This can increase the throughput tremendously since with acknowledgements, end-to-end delays are amplified. Point-to-point error checking is still performed; however, if a frame is found to contain errors it is simply discarded. The end nodes will typically use higher-layer protocols, such as TCP, to perform their own error control mechanism. Frame relay is generally regarded as a successor to X.25 and a precursor to ATM.

4.7.3 Base station system GPRS protocol (BSSGP)

The base station system GPRS protocol (BSSGP) resides above the frame relay network and is used to transport both control and user data over the Gb interface. The primary function of this layer is to introduce and provide the required QoS for the user as well as routing information between the BSS and the SGSN. On the uplink, the BSC will take RLC/MAC frames from the mobile device and reassemble a complete LLC packet from these to be passed within a single BSSGP packet to the SGSN. On the downlink the BSC will extract the LLC frame from the BSSGP packet and segment it into the required number of RLC/MAC frames to be transported to the mobile device. To complete this task, the BSC makes use of the TLLI which is provided by the SGSN and is carried within the BSSGP packet header. It uses this to identify the correct resources provided to the RLC/MAC that this particular packet corresponds to. Each RLC/MAC–BSSGP association is linked via the TLLI. As well as the TLLI, the SGSN provides further information to the BSSGP protocol for the specific mobile device. This information includes:

- The radio capability of the mobile device indicating the simultaneous number of time slots the device is capable of handling.

- The QoS profile which defines the peak bit rate, whether the BSSGP packet is Layer 3 signalling or data (signalling may be transferred with higher protection), whether the LLC frame being carried is ACK/SACK or not (it may be transferred with higher priority if it is ACK/SACK), the precedence class as well as the transmission mode (RLC/MAC acknowledged mode using ARQ or unacknowledged transfer) to be used when transmitting the LLC frame between the BSC and the mobile device.

- A time period for which the packet is valid within the BSS. Any packets held up for longer than this period are to be discarded locally.

The precedence class, lifetime and peak bit rate may be incorporated into the BSC radio resource scheduling algorithm for efficient transfer of LLC frames. In periods of congestion the BSS may initiate a *network controlled cell reselection* for a particular mobile device to ensure efficiency and maintain the maximum number of service requests. If such an event occurs, the BSS will update any internal references to the location of the mobile device and inform the SGSN. It is, however, the responsibility of the SGSN to cope with any LLC packets that have been discarded.

Figure 4.27, shows the format of the BSSGP frames; the various fields are explained below, and follow the format shown in the bottom of the figure.

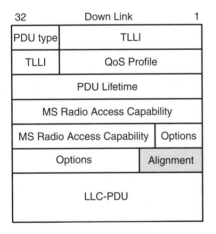

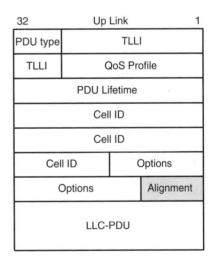

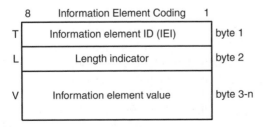

Figure 4.27 BSSGP data frames

Table 4.10   Examples of BSSGP
PDU types

Value (hex)	PDU type
x00	DL unit data
x01	UL unit data
x02	RA capability
x03	DPTM unit data
x05	Paging packet switched
x06	Paging circuit switched
x0b	Suspend
x0c	Suspend ack
x28	Flow control MS
x29	Flow control MS ack

- *PDU type*: identifies the type of PDU and thus the frame format to follow. Table 4.10 is a sample list of assigned PDU types.

- *QoS profile*: defines the peak bit rate, whether the SDU is signalling or data, the type of LLC frame (ACK/SACK or not), the precedence class and the transmission mode to use over the air.

- *MS radio access capability*: defines the radio capability of the mobile device. This field is optional and only present if the SGSN is aware of the mobile device capability.

- *PDU lifetime*: defines the time period that the PDU is considered valid within the BSS. This period is set by upper layers in the SGSN.

- *Cell identifier*: to support location-based services, the uplink PDU includes the cell identity where the LLC was received.

- *Localized service area (LSA)*: this is an optional field as it is an operator-defined group of cells for which specific access conditions apply. This may, for example, be used for negotiating cell reselection.

Unlike the RLC/MAC between the mobile station and the BSC, the BSSGP does not provide error correction, and if a retransmission is required this is performed between the mobile station and the SGSN at the LLC layer. This is because, like frame relay, it assumes that the link is reliable.

There is a one-to-one mapping of the BSSGP protocol between an SGSN and a particular BSS. If the SGSN controls more than one BSS, then there will be additional separate mappings to each of these BSSs from the SGSN. The BSSGP virtual connection (BVC) is carried over a single NS-VC group, which is a collection of frame relay links between an SGSN and a particular BSS; the NS-VC group can carry more than one BVC. Each cell supporting GPRS is allocated and identified by a BVCI. The BVCI and the NSEI are used within the SGSN to identify the cell where a mobile device in *ready mode* resides. The SGSN does not need to know the BVCI of a mobile device in *standby mode*, it simply needs to know which NSEIs relate to the routing area that the mobile device is in for paging purposes. If a mobile device requires to send data or is paged for downlink data then the SGSN will be informed of the cell (BVCI) where the mobile device is located.

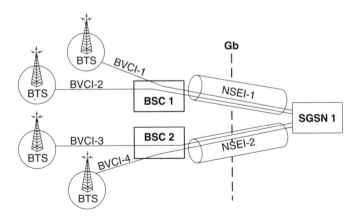

Figure 4.28   Example use of BVCI and NSEI identifiers

Figure 4.28 shows how the BVCI and NSEI can be used to successfully transport data to and from a cell where a mobile device resides. As discussed in Section 4.7.2, the NSEI identifies a group of frame relay permanent virtual circuits. The BVCI is used to identify the particular cell. It should be noted that the BVCI has relevance only over the Gb interface and that it will be mapped to the correct RLC/MAC identifier across the Abis interface. As well as being used to identify a cell, at least one BVC is also required for signalling purposes such as paging.

The BSSGP protocol actually provides a connectionless link between the SGSN and BSS. The flow control mechanism based on the 'leaky bucket' introduces QoS profiles, and a cell (BVC) may have a queue and context in the SGSN and BSS. This queue may be further subdivided for particular subscribers (identified by the TLLI). There are three types of context predefined: best effort, SMS and signalling, where data can be transported over either the best effort or the SMS context. If it is transported over the SMS context, then it will be transported with the QoS profile which has been established for SMS. The flow control mechanism is only required in the downlink direction, since buffers and link capacity in the uplink should be over-dimensioned to avoid loss of data in this direction.

Figure 4.29 shows an example trace from a GPRS network. In this particular trace a mobile device is requesting attachment to the network. Both the uplink *attach request* and the downlink *attach accept* are shown. It can be seen that the mobile device provides a great deal of information, including the QoS required and the encryption algorithms that can be supported. It can also be seen that the *attach request* message sent with RLC/MAC ARQ functionality (acknowledged mode) whereas the downlink *attach accept* message is using the unacknowledged UNITDATA transfer over the RLC/MAC.

The trace in Figure 4.30 is again taken across the Gb interface and is a session management *activate PDP context request*. Again, it can be seen that in the uplink over the RLC/MAC ARQ functionality is used whereas the downlink *pdp context accept* message is using the UNITDATA transfer. Also transferred are the NSAPI and SAPI identifiers as well as the requested QoS and the access point to which this request is for.

UPLINK

DOWNLINK

```
Conn:1 Line:1 TS:1 Hyperch:1984 kb/s
113 13:42:59.502
UL-UNITDATA
TLLI
- Random TLLI
- Value: 52 67 34 DE
QoS Profile
- bit rate : best effort
-  precedence: Reserved/Unknown
- Radio Interface uses RLC/MAC ARQ
functionality
- The SDU contains signalling
- LLC ACK/SAC present
Cell Identifier
- MCC : 211
- MNC : 77
- LAC : 29 (1Dh)
- RAC : 161 (A1h)
- cell id value: 2 (2h)
Alignment Octets      :  FF FF FF
LLC-PDU
ATTACH REQUEST  (GPRS MM)
MS Network Capability
- Encryption algorithm GEA/1 available
- SM capabilities via dedicated channels
supported
- SM capabilities via GPRS channels
supported
- Use of default alphabet over UCS2
prefered
- Default value of phase 1
- The ME does not support SoLSA
- Mobile station supporting earlier
versions of the protocol
Attach Type
- GPRS attach
- No follow-on request pending
GPRS Ciphering Key Seq. Nr
- value: 7 (07h)
DRX Parameter
- no DRX used by the MS
- DRX cycle length coefficient not
specified by the MS
- Split pg cycle on CCCH not supported
- no non-DRX mode after transfer state
Mobile identity
- length:  8 (08h)
- IMSI: 211772000012345
Routing Area Id.
- Mobile Country Code: 000
- Mobile Network Code: 000
- Location Area Code: 0 (0h)
- Routing Area Code: 0 (00h)
MS Radio Access Capability
GSM 900-P Access Technology Type
- Length: 33 bits
- RF Power class: 1
- Encrypt. algorithm A5/1 available
- Encrypt. algorithm A5/2 available
- Encrypt. algorithm A5/3 not available
- Encrypt. algorithm A5/4 not available
- Encrypt. algorithm A5/5 not available
- Encrypt. algorithm A5/6 not available
- Encrypt. algorithm A5/7 not available
- Controlled early Classmark Sending not
implemented
- PS capability not present
- VGCS capability and notifications not
wanted
- VBS capability and notifications not
wanted
- GPRS Multislot Class: 6
- GPRS Ext. Dynamic Alloc. Capability not
implemented
- SMS Value: 5/4 timeslot
- SM Value: 2/4 timeslot
- Spare ok
```

```
Conn:1 Line:1 TS:1 Hyperch:1984 kb/s
114 13:42:59.679
DL-UNITDATA
TLLI
- Random TLLI
- Value: 52 67 34 DE
QoS Profile
- bit rate : best effort
- precedence: Reserved(Low/Priority 3)
- Radio Interface uses RLC/MAC-UNITDATA
functionality
- The SDU contains data
- LLC ACK/SACK not present
PDU Lifetime
- 8.00 sec
MS Radio Access Capab.
GSM 900-P Access Technology Type
- Length: 33 bits
- RF Power class: 1
- Encrypt. algorithm A5/1 available
- Encrypt. algorithm A5/2 available
- Encrypt. algorithm A5/3 not available
- Encrypt. algorithm A5/4 not available
- Encrypt. algorithm A5/5 not available
- Encrypt. algorithm A5/6 not available
- Encrypt. algorithm A5/7 not available
- Controlled early Classmark Sending not
implemented
- PS capability not present
- VGCS capability and notifications not
wanted
- VBS capability and notifications not
wanted
- GPRS Multislot Class: 6
- GPRS Ext. Dynamic Alloc. Capability not
implemented
- SMS Value: 5/4 timeslot
- SM Value: 2/4 timeslot
- Spare ok
DRX Parameters
- no DRX used by the MS
- DRX cycle length coefficient not
specified by the MS
- Split pg cycle on CCCH not supported
- no non-DRX mode after transfer state
TLLI
- Random TLLI
- Value: 52 67 34 DE
LLC-PDU
ATTACH ACCEPT  (GPRS MM)
Attach Result
- GPRS only attached
Force to Standby
- not indicated
GPRS Timer
- Value: 54 min
Radio Priority
- Priority level 3
Routing Area Id.
- Mobile Country Code: 211
- Mobile Network Code: 77
- Location Area Code: 29 (1Dh)
- Routing Area Code: 161 (A1h)
GPRS Timer
- Value: 44 sec
P-TMSI
- length:  5 (05h)
- TMSI/P-TMSI: c00003b6
```

Figure 4.29   Attach trace across Gb. Reproduced by permission of NetHawk Oyj

UPLINK	DOWNLINK
Conn:1 Line:1 TS:1 Hyperch:1984 kb/s 120 13:43:10.092 UL-UNITDATA TLLI - Local TLLI - Value: 52 67 34 DE QoS Profile - bit rate : best effort - precedence: Reserved(Low/Priority 3) - Radio Interface uses RLC/MAC ARQ functionality - The SDU contains signalling - LLC ACK/SAC present Cell Identifier - MCC : 211 - MNC : 77 - LAC : 29 (1Dh) - RAC : 161 (A1h) - cell id value: 2 (2h) Alignment Octets : FF FF FF LLC-PDU ACT. PDP CONTEXT REQUEST (GPRS SM) Network Serv. Access Point Id - NSAPI 5 LLC Serv. Access point Id - SAPI 3 Quality of Service - length: 3 (03h) - Reliab. class: Unack. GTP and LLC Ack. RLC Protected data - Delay class: Delay class 1 - Precedence class: High priority - Peak throughput: Up to 2 000 octet/s - Mean throughput: 200 octet/h Packet Data Protocol Address - Length: 6 (06h) - PDP type organisation: IETF allocated address - PDP type number: IPv4 - Address: 10.1.21.2 Access Point Name - length : 06 (06h) - value (hex) : 05 47 47 53 4E 31	Conn:1 Line:1 TS:1 Hyperch:1984 kb/s 121 13:43:10.119 DL-UNITDATA TLLI - Local TLLI - Value: 52 67 34 DE QoS Profile - bit rate : best effort - precedence: Normal/Priority 2 - Radio Interface uses RLC/MAC-UNITDATA functionality - The SDU contains data - LLC ACK/SACK not present PDU Lifetime - 8.00 sec MS Radio Access Capab. GSM 900-P Access Technology Type - Length: 33 bits - RF Power class: 1 - Encrypt. algorithm A5/1 available - Encrypt. algorithm A5/2 available - Encrypt. algorithm A5/3 not available - Encrypt. algorithm A5/4 not available - Encrypt. algorithm A5/5 not available - Encrypt. algorithm A5/6 not available - Encrypt. algorithm A5/7 not available - Controlled early Classmark Sending not implemented - PS capability not present - VGCS capability and notifications not wanted - VBS capability and notifications not wanted - GPRS Multislot Class: 6 - GPRS Ext. Dynamic Alloc. Capability not implemented - SMS Value: 5/4 timeslot - SM Value: 2/4 timeslot - Spare ok DRX Parameters - no DRX used by the MS - DRX cycle length coefficient not specified by the MS - Split pg cycle on CCCH not supported - no non-DRX mode after transfer state TLLI - Random TLLI - Value: 52 67 34 DE LLC-PDU: ACT. PDP CONTEXT ACCEPT (GPRS SM) LLC Serv. Access point Id - SAPI 3 Quality of Service - length: 3 (03h) - Reliab. class: Unack. GTP and LLC Ack. RLC Protected data - Delay class: Delay class 4 (best effort) - Precedence class: High priority - Peak throughput: Up to 2 000 octet/s - Mean throughput: 200 octet/h Radio Priority - Priority level 3

Figure 4.30   PDP context request trace across Gb. Reproduced by permission of NetHawk
Oyj

4.8   GPRS TUNNELLING PROTOCOL (GTP)

Figure 4.31 shows a layered model of the Gn interface. Layers 1 and 2 are the physical and datalink protocols. This is not specified and can be an Ethernet 100baseTX connection, an ATM connection, frame relay or other transport mechanism. In many cases the SGSN and the GGSNs will be in the same room or building, thus Ethernet, with its limited distance but high speed and simple network configuration, is commonly used. In other cases the GGSNs may be remote, and there may be a leased line or ATM network connecting the GGSN to the SGSN over a vast distance. Above this layer is the network layer which runs the IP. Above this is the connectionless UDP protocol, which is the transport mechanism for the GPRS tunnelling protocol (currently GTP version 1, also referred to as Release 99). Since UDP is connectionless it does not require acknowledgements and can therefore provide higher throughput than TCP; it is assumed that the core network is over-dimensioned and very reliable. Support has recently been introduced (in Release 99) for a PPP PDP context which allows transfer of packets other than IP, such as AppleTalk or IPX, over the cellular network (the SDU size for these is 1502 bytes; whereas the SDU size for IP is 1500 bytes). In the original specification (also referred to as Release 97) there was support for X.25 protocol PDP contexts. In this case, it was required that the GTP protocol be carried over TCP as opposed to UDP as packet order consistency is a requirement for this type of traffic. The support for X.25 has been removed from GTP version 1.0; this has resulted in GTP being carried only over UDP. The single port number 3386/udp was originally used for both control and user data in GTP version 0 (Release 97). This is as originally used for GPRS/GSM. However, for GTP 1, as used in GPRS/UMTS, the destination device will be designated port number 2152/udp for GTP-U and 2123/udp for GTP-C. The source port number can be dynamically assigned. Clearly, a GGSN will need to have a daemon listening to both ports to make it compatible with both 2G and 3G GPRS networks.

Mobile devices such as those used in a GSM network are uniquely identified worldwide by their International Mobile Equipment Identity (IMEI). The user of such a device is also uniquely identified by their IMSI, which is stored on the SIM card. If the user chooses to use another mobile device, they can simply place their SIM card in the new mobile device, thus transferring the IMSI. The user can be contacted at the same telephone number and the charges will go on the standard bill. When the mobile device requests a connection to the network, its IMSI is transferred across the air interface and onwards to the core

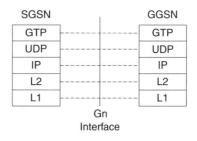

Figure 4.31   SGSN to GGSN interface

network (CN) to authenticate the user. Once a connection is established, a t[
number, the TMSI, is now used for security purposes rather than transmitting the IMSI
continuously over the air interface where it can easily be captured. In direct parallel with
the TMSI used on the GSM core network, another temporary number, the packet-TMSI
(P-TMSI) is used on the GPRS CN. Note that usually GSM (and GPRS) transmissions
are encrypted and thus merely obtaining the IMSI will not be sufficient to allow a hacker
to intercept a call in progress over the air interface.

Figure 4.32 indicates the points of encryption on both the GSM and GPRS networks
and also the identifiers (P-TMSI, TLLI and tunnel end point identifier, or TEID) used to
transport the user IP packets at the lower layers.

In the case of a GPRS connection, where the mobile device will connect to an IP
network, it will need to have an IP address with which it can also be identified. This
address is used at Layer 3 of the OSI and is not currently used as an end system identifier
(ESI) for the mobile device. On a fixed network, the ESI corresponds to the MAC address,
which is globally unique and generally burned into the hardware of a network interface.
On the GPRS network, this ESI corresponds to the IMSI. On a fixed IP network, a link is
required between a device's IP and MAC addresses. This binding is generally achieved via
a dynamic mapping protocol known as the address resolution protocol (ARP). Similarly,
on a GPRS network, a link between the IP address and the TEID (or TMSI or TLLI)
is also required. The GPRS equivalent of ARP is usually achieved by mapping the IP
address to the TEID in the GGSN. Therefore, when data is sent to the mobile device, it
will be routed over the various intermediate networks, for example the Internet, based on
the IP address. Once it reaches the mobile network GGSN, a lookup of the associated
TEID is required, and this can be seen as analogous to an ARP entry lookup. The GGSN
can now route the packet to the correct SGSN based on the SGSN's IP address. When

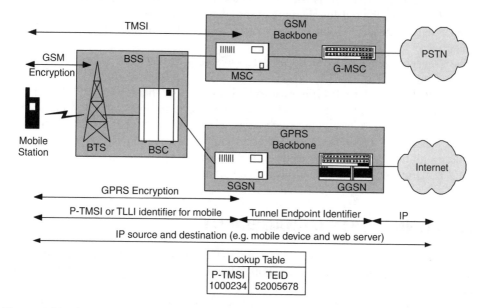

Figure 4.32   GPRS network identifiers

8					1
Version	PT	0	E	S	PN
Message Type					
Length 1st octet					
Length 2nd octet					
Tunnel Endpoint Identifier 1st octet					
Tunnel Endpoint Identifier 2nd octet					
Tunnel Endpoint Identifier 3rd octet					
Tunnel Endpoint Identifier 4th octet					
Sequence Number 1st octet					
Sequence Number 2nd octet					
N-PDU Number					
Next Extension Header Type					

Figure 4.33   GTP header

the packet arrives at the SGSN, the TEID is associated with a P-TMSI[5] or TLLI and this is used for the onward journey across the BSS.

The GTP protocol actually consists of two parts, the GTP-C and the GTP-U. The GTP-C carries control data for creating, modifying and deleting GTP tunnels, while the GTP-U transports user data and some control information. The header for both of these is of variable length, the minimum length being 8 bytes. Three separate bits (E, S and PN) indicate the presence of additional fields. The format of the header is shown in Figure 4.33.

The version number is used to determine the version of this header; the GTP protocol is backward compatible. The current version is version 1 (Release 99) but there is a version 0 (Release 97).

The protocol type (PT) bit is used to indicate whether this is standard GTP or GTP'. The GTP' protocol is used for charging purposes in the Ga interface.

The extension header (E) bit is used to indicate whether there is an extension header present.

The sequence number (S) bit is used to indicate whether there is a GTP sequence number field.

The N-PDU number (PN) bit is used to indicate whether there is an N-PDU number.

The message type field indicates what type of message is being carried in this packet. Examples of message types include echo request, node alive request, create PDP context request, delete PDP context request, sending routing information etc.

The length field indicates the length of the payload in bytes.

The TEID unambiguously identifies the end points of the tunnel.

4.9   CONNECTION MANAGEMENT

The initial connection for GPRS is essentially the same as for GSM. The main difference between the two is that for GPRS registration the mobile device will be dealt with by the

---

[5]Further information is also required for the journey across the BSS which will identify the actual process in the mobile device. This will include the SAPI and NSAPI identifiers.

SGSN rather than the MSC which is used for GSM. A mobile device that has both GSM and GPRS capabilities will generally make a connection to a GSM network via the MSC followed by a connection to the GPRS network via the SGSN. This requires that the mobile device be authenticated twice and that requests back to the home network's HLR are also made twice. As well as this, requests by the mobile device have to be made twice over the air interface. Since the air interface is a scarce resource and a major bottleneck with regard to bandwidth, this can present a problem. The GPRS standards allow an optional interface between the MSC and SGSN known as the Gs interface. If a network supports this option then a single attach is required. Details of the Gs interface are presented in Section 4.4. In this case, the mobile device will register with the SGSN, which will authenticate the SIM and check the status of the mobile device. The HLR is updated with the details of the SGSN that is serving the mobile device for GPRS connections. Using the Gs interface, it can now update the MSC/VLR of the mobile device's location and explain that authentication has already been done. The VLR can update the HLR that is serving the mobile device for GSM calls. This method reduces the amount of signalling over the air interface. A mobile device will know that a cell has GPRS capability and that it should contact the SGSN rather than the MSC, since a cell broadcasts its GPRS capability on the (P)BCCH.

4.9.1 Mobility management

In 2G systems, mobility management has been performed by the core network. The location of the mobile device is known within the MSC/VLR for those devices which are circuit-switched-connected, and in the SGSN for those that are packet-switched-connected. For the packet switched network, GPRS introduces a new location entity, known as a *routing area* (RA). Any LA (circuit switched) or RA (for packet switched) updates from the mobile device are passed transparently over the BSS to the core network to be stored in the correct device (MSC or SGSN). In this way, paging for a mobile device is achieved by first finding the correct MSC/VLR or the correct SGSN. Either of these devices will know the location of a mobile device to a precise cell, or to a number of cells (the LA or RA), depending on the state the mobile device is in. This can introduce a lot of signalling traffic.

As shown in Figure 4.34, an RA is a subset of an LA, and is defined within the SGSN and not the MSC. An RA can be the same size as an LA but it cannot be larger, i.e. an RA cannot overlap separate LAs. There is no direct correlation between MSCs and SGSNs and the RA is identified by the following formula:

Routing area identifier (RAI) = Mobile country code (MCC)

+ Mobile network code (MNC)

+ Location area code (LAC) + Routing area code (RAC)

There are three basic mobility states that a GPRS device can be in: idle, standby and ready state, as shown in Figure 4.35.

Idle mode

In the idle state the mobile device is not attached to the network and therefore the network holds no valid location or routing information for the device. Since the network does not

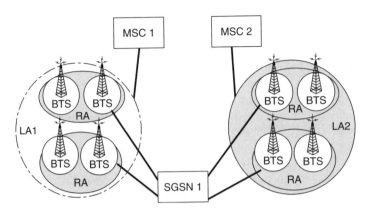

Figure 4.34 GPRS routing area

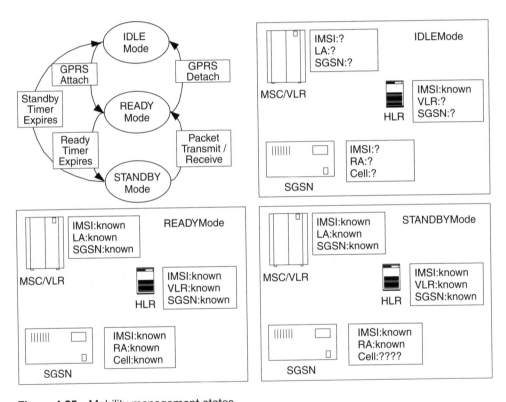

Figure 4.35 Mobility management states

know where the mobile device is, it cannot be reached by paging. The mobile device must perform an attach procedure to establish a mobility management context so that it is registered within the network. This procedure will move the mobile device from the idle state to the ready state. It is usually done automatically when the device is switched on. Idle state for GPRS is not the same as idle state for GSM.

Standby state

In the standby state the subscriber is attached to the mobility management of the network and information is stored about the location of the mobile device. The location of the mobile device is known to the RA level, which usually includes a number of cells, i.e. the actual cell is not identified. By only updating the network when it moves from one RA to another rather than one cell to another, the mobile device sends fewer signalling messages and also conserves its battery power. In this state the mobile device can activate or deactivate the session management PDP context; in practice it automatically moves to the ready state to do this. A PDP context is required to transfer data over the GPRS network; however data transmission and reception are not possible until the mobile device moves to the ready state. The mobile device can be paged in the standby state and asked to move to the ready state. This state is very similar to the GSM idle state.

Ready state

In the ready state the location of the mobile device is known to the exact cell. Every time the mobile device changes cell it will update the network with its new location. In this state of operation the mobile device is capable of transferring data across the GPRS network if there is a valid PDP context. The mobile device will remain in this state while data transfer takes place. If no transfer takes place for a particular length of time, the mobile device may move to the standby state when the *ready timer* expires. The timer value is set by the operator, with the specifications having a default setting of 44 seconds after this time when no cell updates are performed. When the mobile device first switches on it will normally move from idle to ready mode. If the subscriber does not initiate a PDP context within the time allowed then the mobile device will move to the standby mode. A mobile device will move from standby to idle at the expiration of another timer; again, this is operator dependent but the specifications set a default value of 54 minutes. After this time the mobile device will perform a location update (even if it has not moved to another cell) or it will be moved to idle mode.

4.9.1.1   GPRS attach procedure

Figure 4.36, shows the signalling flow for a combined GPRS/IMSI attach procedure. Each step in the procedure is explained below.

1. The mobile device initiates the attach procedure by sending the *attach request* message to the HLR. This message will include either the IMSI or the P-TMSI and the RAI that it was valid in (the IMSI will be sent if the mobile device does not already have a valid P-TMSI), the attach type and the classmark of the mobile device indicating its multislot capabilities. The attach type will indicate whether this is simply a GPRS attach or a combined GPRS/IMSI attach.

2. The security functions can now be executed as described in Section 3.5. The SGSN will query the HLR with the IMSI and be given the triplets back which include the ciphering key (Kc), random number (RAND) and signed response (SRES). The RAND will be passed to the mobile device and it will reply with an SRES'. The

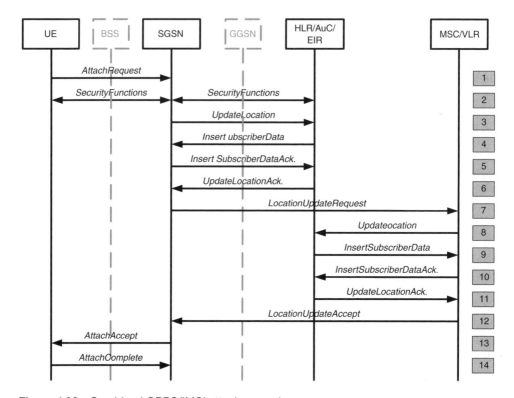

Figure 4.36 Combined GPRS/IMSI attach procedure

SGSN will compare the SRES from the AuC and the mobile device and if they are the same then the IMSI is authenticated. The mobile device (IMEI) can now be checked against the EIR; however, this is not compulsory.

3. The SGSN sends an *update location* message to the HLR which will include the SGSN identifier and the IMSI of the mobile device.

4. The HLR will send an *update subscriber data* message back to the SGSN which will contain the subscription data associated with the IMSI.

5. The SGSN validates the mobile device's presence in the new RA and acknowledges receipt of the *update subscriber data* message.

6. The HLR acknowledges the *update location* message from the SGSN in step 5.

7. If the attach type in the initial attach request indicated *combined GPRS IMSI attach* then the VLR will be updated across the Gs interface. The specific VLR can be ascertained from the RAI.

8. The VLR send an *update location* message to the HLR which will include the VLR identifier and the IMSI of the mobile device.

9. The HLR will send an *update subscriber data* message back to the VLR which will contain the subscription data associated with the IMSI.

10. The VLR acknowledges receipt of the *update subscriber data* message.

11. The HLR acknowledges the *update location* message from the VLR in step 10.

12. The VLR will now reply to the SGSN with the *location update accept* message which includes the VLR identification and the TMSI of the mobile device.

13. The SGSN sends an *attach accept* message back to the mobile device which will include the P-TMSI and the TMSI.

14. The mobile device acknowledges receipt of the P-TMSI and TMSI.

4.9.1.2 Location updating

Figure 4.37 illustrates how mobility management signalling is routed to the core network. In GSM dedicated (connected) mode and GPRS ready state, cell updates are performed; in GSM idle or GPRS standby states LA and RA updates are performed.

An RA update can take place because a subscriber moves into a new RA and the mobile device detects the new RA identifier, or the RA timer has expired. This latter case is referred to as a *periodic routing area update*. The RA update can take two forms:

- **Intra SGSN RA update** where the new RA is controlled by the same SGSN as the current RA. The SGSN will have all information pertaining to the mobile device and it does not need to inform the HLR or the GGSN. The mobile device will probably be given a new P-TMSI identifier. All periodic RA updates are of this type.

- **Inter SGSN RA update**. In this case the mobile device has noticed a change from one RA to another; however, these RAs are administered by different SGSNs. This procedure is illustrated in Figure 4.38 and described below.

1. The mobile device sends a *routing area update request* which includes the old RAI, the update type and the capability of the mobile device to the new SGSN.

2. The new SGSN sends the *SGSN context request* message to the old SGSN which includes the TLLI, old RAI and new SGSN address. The old SGSN now knows where to send packets that arrive from a GGSN over the Gn interface.

3. The old SGSN replies with the mobility context of the mobile device and any PDP context information it has. It stops sending packets directly to the mobile device and starts a timer. If the timer expires before the RA procedure is completed successfully then the old SGSN will remain in control of the mobile device. This

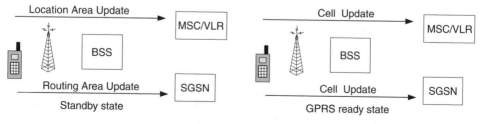

Figure 4.37 GPRS mobility management

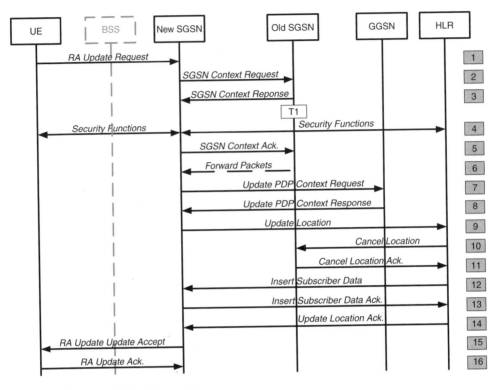

Figure 4.38   Inter SGSN RA update

may be the case when a mobile subscriber changes geographical direction and moves back into the old RA.

4. Standard security functions can now be performed.

5. The new SGSN sends an acknowledgement to the old SGSN. Its purpose is to inform the old SGSN that it is now ready to receive packets destined for the mobile device from any active PDP contexts.

6. The old SGSN starts to transfer packets via the new SGSN to the mobile device. Note that the packets to and from a GGSN are tunnelled to the old SGSN and then tunnelled from the old SGSN to the new SGSN.

7. The new SGSN informs each of the GGSNs that it is now handling the mobile device and requests an update PDP context. This information includes the new SGSN address, TEID and QoS that has been negotiated.

8. The GGSNs will update their databases and reply with an update PDP context response.

9. The new SGSN now sends a *location update* message to inform the HLR that it is now controlling the mobile device. This message will include the SGSN address and the IMSI of the mobile device.

10. The HLR will send a *cancel location* to the old SGSN. The mobile device will be identified by its IMSI and all mobility management and PDP context information in the old SGSN will be removed after the timer expires.

11. The old SGSN acknowledges the *cancel location* message.

12. The HLR sends the *insert subscriber data* message to the new SGSN. This includes the IMSI and GPRS subscription data.

13. The new SGSN acknowledges the *insert subscriber data* message.

14. The HLR now acknowledges the *update location* message sent by the new SGSN in step 9.

15. The new SGSN now establishes a logical link connection to the mobile device and sends the *routing area update accept* message.

16. The mobile device acknowledges the message by sending the *routing area update compete* message back to the new SGSN.

Figure 4.39 is an actual trace which shows the SGSN context request, SGSN context response and SGSN context acknowledge messages that are identified in Figure 4.38.

4.9.2 Session management

Once the mobile device has registered with the mobile network, it cannot start to send and receive data over the packet core until it has established a session, which is known as having a packet data protocol context (PDP context). A mobile device has to be in the ready mode to activate a PDP context; however, the PDP context will continue to be present if the mobile device moves to the standby state. This is to allow a subscriber to retain the same IP address, even if they have not generated or received any traffic for some time.

This process of PDP context activation results in the mobile device obtaining an IP address. This IP address can be given either by the operator's network or by an external network. To obtain an IP address the user may select a particular service on the mobile device by scrolling through a menu. For example, the menu may have 'Internet' or 'home Intranet' as an item. This request for a connection will be passed through the network to the SGSN. The SGSN has to locate the correct GGSN, which will be used to route the data to the correct external network. Note that there may be more than one GGSN connected to the external network for load sharing and redundancy. To find the GGSN, the SGSN will contact a DNS server within the operator's private network; the DNS server will return the IP address of the GGSN where the particular connection (known as an *access point*) is located. A GGSN can have a number of access points, i.e. *connections*, to different external networks. The SGSN can now contact the GGSN and ask for a PDP context to be granted for this user. Figure 4.40, illustrates this process.

The PDP context ensures that a GPRS Tunnel is set up between the SGSN and GGSN for the sole use of this user. An IP address for the mobile device will also be given to the mobile station at this time. It should be noted that the GGSN may not be in the same network as the SGSN. This is the case when the mobile is in a visited or V-PLMN and may wish to connect to an access point in its H-PLMN. Figure 4.41 illustrates the end points

NEW SGSN

```
Source address : 10.1.2.123
Dest address : 10.1.2.231
SGSN CONTEXT REQUEST
- Version number: 1 (1h)
- Protocol type: 1 (1h)
- Extension header flag: 0
(0h)
- Sequence number flag: 1 (1h)
- N-PDU number flag: 0 (0h)
- Message Type: 50 (SGSN
CONTEXT REQUEST)
- Header Length: 30 (1eh)
- Tunnel endpoint id: 0
(00000000h)
- Sequence number: 13 (dh)
- No extension header
Routeing Area Identity
- MCC : 511
- MNC : 044
- LAC : 38 (0026h)
- RAC : 1 (01h)
TLLI
- TLLI : foreign TLLI
2325850897(8AA1AB11h)
MS Validated  : no
Tunnel Endpoint Iden CP
- id: 0 (00000000h)
SGSN Address for Control Plane
- length: 4 (4h)
- Ipv4 Address: 10.1.2.123
```

NEW SGSN

```
Source address : 10.1.2.123
Dest address : 10.1.2.231
SGSN CONTEXT ACKNOWLEDGE
- Version number: 1 (1h)
- Protocol type: 1 (1h)
- Extension header flag: 0
(0h)
- Sequence number flag: 1 (1h)
- N-PDU number flag: 0 (0h)
- Message Type: 52 (SGSN
CONTEXT ACKNOWLEDGE)
- Header Length: 19 (13h)
- Tunnel endpoint id: 13
(0000000Dh)
- Sequence number: 13 (dh)
- No extension header
Cause
- acceptance: request accepted
Tunnel Endpoint Iden Data II
- NSAPI: 5(5h)
- id: 11 (0000000Bh)
SGSN Address for user traffic
- length: 4 (4h)
- Ipv4 Address: 10.1.2.123
```

OLD SGSN

```
Source address : 10.1.2.231
Dest address : 10.1.2.123
SGSN CONTEXT RESPONSE
- Version number: 1 (1h)
- Protocol type: 1 (1h)
- Extension header flag: 0 (0h)
- Sequence number flag: 1 (1h)
- N-PDU number flag: 0 (0h)
- Message Type: 51 (SGSN CONTEXT
RESPONSE)
- Header Length: 117 (75h)
- Tunnel endpoint id: 0
(00000000h)
- Sequence number: 13 (dh)
- No extension header
Cause
- acceptance: request accepted
IMSI
- MCC : 511
- NMSI : 232134628909
Tunnel Endpoint Iden CP
- id: 13 (0000000Dh)
MM Context
- length : 17 (0011h)
- security mode : GSM key and
triplets
- CKSN : 1 (01h)
- Used Cipher : GEA/2
- KC :  FF 03 91 AC 02 8F E5 54
- split PG Cycle value: 7
- DRX cycle length coefficient not
specified
- split pg cycle on CCCH not
supported
- max. 2 sec non-DRX mode after
transfer state
- MS Network capability:
- Encryption algorithm GEA/1
available
- SM capabilities via dedicated
channels supported
- SM capabilities via GPRS
channels supported
- Use of default alphabet over
UCS2 prefered
- Capab. of ellipsis notation and
phase 2 error handling
- The ME does not support SoLSA
- Mobile station supporting
earlier versions of the protocol
- Mobile station does not support
BSS packet flow procedures
- encryption algorithm GEA/2
available
- encryption algorithm GEA/3 not
available
- encryption algorithm GEA/4 not
available
- encryption algorithm GEA/5 not
available
- encryption algorithm GEA/6 not
available
- encryption algorithm GEA/7 not
available
- container length : 0 (00h)
PDP Context
- length: 67 (0043h)
- NSAPI: 5 (05h)
- order : No
- VAA : No
- SAPI: 5 (05h)
- QoS Sub
- length 12 (0Ch)
- allocation/retention priority 2
(02h)
- unacknowledged GTP; acknowledged
```

```
LLC and RLC, protected data
- delay class 4 (best effort)
- precedence class: normal
priority
- peak throughput : up to 256
000 octet/s
- mean throughput : best effort
- traffic class : subscribed
- delivery order: with delivery
order ('yes')
- delivery of erroneous SDUs:
not delivered
- maximum SDU size: 1500
- maximum bit rate for uplink:
2048 kbps
- maximum bit rate for downlink:
2048 kbps
- residual BER: 1*10-5
- SDU error ratio: 1*10-6
- transfer delay: subscribed (MS
only)
- traffic handling priority:
subscribed
- guaranteed bit rate for
uplink: subscribed (MS only)
- guaranteed bit rate for
downlink: subscribed (MS only)
- QoS REQ
- length 4 (04h)
- allocation/retention priority
0 (00h)
- subscribed reliability class
- subscribed delay class
- precedence class: subscribed
- peak throughput : subscribed
peak throughput
- mean throughput : Subscribed
- QoS NEG
- length 4 (04h)
- allocation/retention priority
0 (00h)
- unacknowledged GTP;
acknowledged LLC and RLC,
protected data
- delay class 4 (best effort)
- precedence class: normal
priority
- peak throughput : up to 256
000 octet/s
- mean throughput : best effort
- SND : 0 (0000h)
- SNU : 4 (0004h)
- send N-PDU : 0 (00h)
- receive N-PDU : 4 (04h)
- uplink TEI Control Plane:
16777296 (01000050h)
- uplink TEI Data I: 15802628
(00F12104h)
- PDP Context Identifier: 10
(0Ah)
- PDP type organisation :
reserved value 10 (0Ah)
- PDP type number : 3 (03h)
- PDP address length: 2 (02h)
- PDP address :  040A
- GGSN address for control
plane:
- length 10 (0Ah)
- address: :  [Abridged]
- GGSN address for User Traffic:
- length 108 (6Ch)
- address: :  [Abridged]
- transaction identifier 0
(000h)
Unknown IE: 0 (0h)
```

Figure 4.39   Inter SGSN trace. Reproduced by permission of NetHawk Oyj

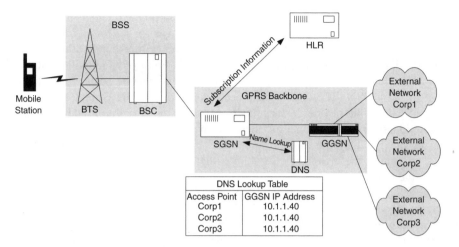

Figure 4.40 PDP context activation

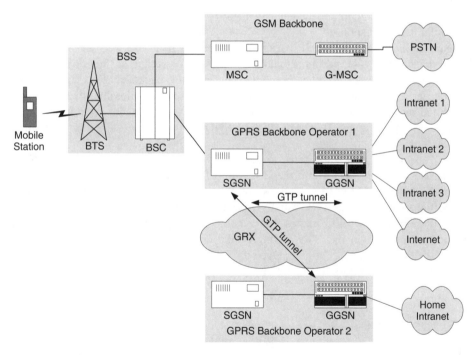

Figure 4.41 GTP tunnel

of the tunnel between SGSN and GGSN in the local network, and also a tunnel through the global roaming exchange (GRX) to another PLMN network in a different country. Note also that the GGSNs can have a number of access points to external networks.

Once a PDP context has been activated, the user can then use the services provided by the particular access point. For example, if the user with a device possessing the appropriate application software connected to the access point 'Internet' then they would be able to 'surf the web'. Suppose the user requests a web page from a server, 'get page request www.orbitage.com/default.htm'. The web server will respond with the web page, and the destination IP address from the web server's point of view will be the mobile device which made the initial request. This IP address will be routed back across the Internet to the GGSN. The GGSN now needs to send it to the correct SGSN for final delivery to the mobile device. It does this by cross-referencing the IP address with all of the tunnel identifiers it has; hopefully there will be a match. The tunnel identifier actually comprises the IMSI of the mobile device and is therefore unique. As long as the IP address arriving at the GGSN is also unique, there should be no problem with mapping one onto the other.

The GPRS tunnel now transfers the IP packet to the SGSN. Once the data arrives at the SGSN, the IP packet is removed from the GPRS tunnel. The SGSN now needs to resolve the IP address into the P-TMSI address of the mobile device. The unique connection from the SGSN to the mobile device is known as the *temporary logical link identifier* (TLLI) and comprises the P-TMSI of the mobile device. The TLLI uniquely identifies the mobile device within an RA. It is sent in all packet transfers. The SGSN has a database of P-TMSI to IMSI mapping which will identify the correct mobile device to which it should deliver the response.

Figure 4.42 shows how the IP packet is transported from the web server back to the mobile device. The GTP header (containing the IMSI) is routed to the correct SGSN via the underlying protocol (another IP network). Once the packet reaches the SGSN, the GTP header is discarded and a new TLL header is added. It is possible that between the SGSN and mobile device, the IP packet may be broken into smaller sections and carried by separate LLC frames. It is actually the SNDCP that deals with this segmentation and reassembly.

The IP packet is now transported to the mobile station using a number of LLC layer packets. Between the SGSN and the BSC, the LLC is itself carried over a frame relay network. At the BSC the LLC packets are extracted from the frame relay packets and transported to the correct BTS, usually over the TDM slots of an E1 or T1 line. The BSC will segment the LLC frames into a number of smaller RLC/MAC frames. These will be transported via the BTS to the mobile device, where they will be grouped to reconstruct the original LLC frame. Once the BTS has received the RLC/MAC frames, it then transmits them across the air interface to the mobile device. When the LLC packets

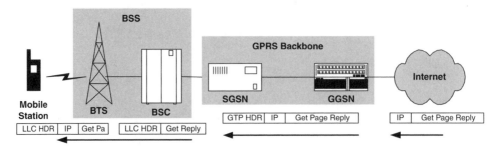

Figure 4.42   IP transport across GPRS infrastructure

arrive at the mobile device, the IP packet is reassembled, and processed by the application on the device.

A mobile device may have a number of PDP contexts active at any one time. These may or may not be to the same GGSN, but are given separate tunnel identities, and the mobile device will have a number of addresses associated with it, for example IP addresses. It will have one address for each primary PDP context activated. When a PDP context exists and another context is defined which uses the same address, and possibly some of the same context information, this is referred to as a secondary context. The secondary PDP context procedure can be used to make an alternative connection to an already active context but with different QoS. The PDP address and other PDP context information already established are reused. Since an access point must have been defined for the primary context, procedures for access point name (APN) selection and PDP address negotiation are not executed. The secondary PDP context will use the same IP address as the primary context; it will be identified through a unique transaction identifier and its own NSAPI. As an example, consider a videoconferencing application. This can consist of a video stream and a whiteboarding application. Here, the video traffic presents quite stringent QoS requirements, whereas the whiteboarding is non-real-time and thus has looser QoS demands, since it can tolerate delays.

Figure 4.43 illustrates the procedures for a mobile-originated PDP context activation.

1. The mobile device sends the PDP request, which contains the NSAPI, PDP type, PDP address, APN, QoS requirements and any PDP configuration options. The NSAPI identifies the service access point in the mobile device, which will be dealing with this specific PDP context. The PDP type indicates if this is an X.25 or IP connection[6] and the PDP address indicates the actual address. In the case of IP this will be an IP address for the mobile device. If the mobile device does not have a static IP address

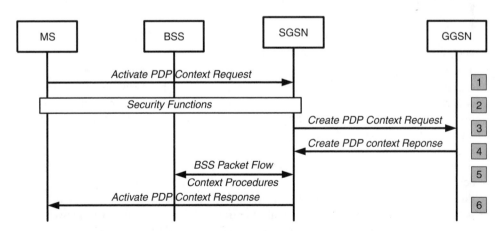

Figure 4.43 UE-originated PDP context activation

---

[6]X.25 was originally supported in GTP version 0 but has been removed from GTP version 1. It has been replaced with PPP, which can carry many types of protocol transparently.

then this field will consist of 0.0.0.0 and the GGSN will allocate the IP address. It practical situations it is more common that dynamic addresses are allocated since this provides greater flexibility. The APN, described in Section 4.9.4, is the mobile operator's external connection point. This is a text name and has local significance only. Examples of this could be Internet, Corporation-1 etc. Each PDP context may have a different QoS requirement; perhaps higher QoS may mean that the subscriber has to pay a premium price. The options field is transparently transported through the SGSN to the GGSN where optional parameters may be implemented.

2. Once the SGSN has received the request it will most likely invoke the security procedures to authenticate the mobile subscriber and device. These are very similar to the GSM security procedures. The SGSN will validate the subscriber's request with those permitted through the subscription records in the HLR. For example, if the subscriber has asked for a higher level of QoS than they have subscribed for in their service level agreement, then this QoS value may be reduced. Also checked is the use of static IP addresses, as well as whether the requested APN is defined in the HLR for that subscriber.

3. The SGSN generates a TEID consisting of the IMSI and NSAPI or a link to it, and checks whether it has the resources available to allow the requested QoS. If not, it may restrict the requested QoS. The SGSN locates a GGSN which has access to the access point and sends a *create PDP context request*. This message consists of the TEID, MSISDN, selection mode (as described in Section 4.9.4 for the APN), PDP type, PDP address, APN, negotiated QoS, charging characteristics and any PDP configuration options. The charging characteristics are checked with the HLR to ascertain how the subscriber will be billed for this connection. The GGSN will create a new entry in its PDP context table and generate a charging ID. The GGSN may also restrict the QoS for this connection if it does not have the required resources available. If the mobile device has been allocated an IP address then this may be configured by the GGSN (see Section 4.9.3).

4. The GGSN will send a *create PDP context response* to the SGSN including the charging ID and any modifications to the QoS.

5. The BSS packet flow procedures may be executed to ensure that the QoS requirements are met. All the packet flows related to a single mobile device are stored in a specific BSS context. Individual contexts are identified by a unique packet flow identifier which is allocated by the SGSN. As was discussed, there are three packet flows that are predefined: *SMS, signalling* and *best effort*. In these cases, no negotiation of BSS packet flow contexts is required. Both the best effort and SMS packet flows can be assigned to a PDP context and PDP data packets will be carried across the BSS with the same QoS as the predefined flows. To ensure QoS is guaranteed, the SGSN divides the transfer delay between the BSS and core network part. It can do this since it can estimate the transfer delay over the core network to the GGSN. The SGSN can then convert the maximum SDU size, SDU error ratio, residual bit error ratio, maximum bit rate, guaranteed bit rate and thus the transfer delay to that applicable for this transfer. The SGSN inserts the NSAPI and the GGSN address into its PDP context and selects the radio priority and packet flow ID based on the QoS negotiated.

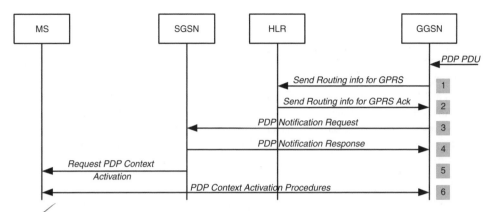

Figure 4.44   GGSN-originated PDP context activation

6. Finally, the *activate PDP context accept* message is sent to the mobile device from
   the SGSN, indicating that the request has been accepted.

A network-requested PDP context activation procedure is illustrated in Figure 4.44. Here
a PDU arrives at the GGSN from an external source for a mobile device. This system
only works when the GGSN knows the IP address of the mobile device, i.e. the mobile
device has a static IP address.

1. If there is no PDP context activated then the GGSN may query the HLR for
   connection information. To limit the number of queries to the HLR, the *mobile
   station not reachable for GPRS flag (MNRG)* message may be set in the GGSN. In
   this case the GGSN will not query the HLR. The *send routing information for GPRS*
   message will request the IMSI of the intended recipient.

2. The HLR will respond with the IMSI and the SGSN address of the mobile device in
   the *send routing information for GPRS acknowledge* message.

3. Once the GGSN has this information, it can send a *PDU notification request*
   message to the SGSN which contains the IMSI number of the mobile device, the
   APN, PDP type and PDP address of the mobile device.

4. The *PDU notification response* message is returned to the GGSN to indicate that it
   will try to reach the mobile device. If the mobile device is not known then an
   unsuccessful response such as *IMSI not known* may be returned to the GGSN.

5. The SGSN can send a *request PDP context activation* message to the mobile device
   to request the device to activate the indicated PDP context.

6. This is followed by the regular PDP context activation procedures as highlighted in
   Figure 4.43.

The issues with regard to this type of mobile-terminated data access stretch beyond
the technical capabilities of the network. Allowing external sources to activate a PDP
context presents a number of consequences, some of which may be undesirable for both

subscriber and network operator. In comparison, the Internet is essentially a *pull* medium, meaning that information is accessed by the user, and can not generally be network originated. One exception to that is unsolicited email, or spam, which is already presenting problems in both fixed-line and mobile networks. If information *push* is permitted, it is difficult to identify which traffic is wanted and which is not. The central issue is with regard to payment. A subscriber would understandably be unhappy if not only had they received unsolicited data, perhaps advertising, but also that they had been billed for this receipt. This issue is non-trivial if one considers that blocking of particular IP addresses or domain names in the context of the Internet is a laborious task with arguable effectiveness.

Figure 4.45 shows the full trace of the PDP context request from the SGSN and the response from the GGSN.

Figure 4.46 illustrates how the PDP context can be deactivated. The diagram shows how this is achieved from the mobile device; however, the SGSN and GGSN can also deactivate the PDP context.

1. The mobile device sends a *deactivate PDP context request* to the SGSN.
2. The SGSN may request for security functions to be executed.
3. The SGSN sends a *delete PDP context request* to the GGSN indicating the TEID. The GGSN deletes the context at its end and returns any dynamic IP addresses to the pool so that they can be allocated to other subscribers.
4. The GGSN then responds with the *delete PDP context response* indicating the TEID that has been released.
5. The SGSN can now send a reply to the mobile device stating that the deactivation has been complete.

Figure 4.47 shows a trace of the PDP context deletion. This procedure can be executed by either the SGSN or the GGSN, hence the identifier GSN above the traces.

4.9.3   Transparent and non-transparent mode

Figure 4.48, shows how a mobile device may obtain an IP address directly from the GGSN. This is referred to as transparent mode. Alternatively, the GGSN may ask an external dynamic host configuration protocol (DHCP) or remote authentication dial-in user service (RADIUS) server from the subscriber's home intranet for an IP address for the mobile device. This is referred to as non-transparent mode.

4.9.4   Access point name (APN)

The access point is the external connection point for the mobile operator's network. A subscriber will indicate the access point to which they would like to connect. This is typically performed either through selecting an item on a list presented on the mobile

SGSN

```
IP1                      1 15:58:14.241
IP DATAGRAM
Source address : 172.26.143.1
Destination address : 172.26.131.118
UDP DATAGRAM
Source port: 2123
Destination port: 2123
CREATE PDP CONTEXT REQUEST
- Version number: 1 (1h)
- Protocol type: 1 (1h)
- Extension header flag: 0 (0h)
- Sequence number flag: 1 (1h)
- N-PDU number flag: 0 (0h)
- Message Type: 16 (CREATE PDP CONTEXT REQUEST)
- Header Length: 106 (6ah)
- Tunnel endpoint id: 0 (00000000h)
- Sequence number: 49 (31h)
- No extension header
IMSI
- MCC : 310
- NMSI : 010000000030
Recovery
- restart counter: 73 (49h)
Selection mode
- MS or network provided APN, subscribed
Tunnel Endpoint Iden Data I
- Tunnel endpoint id: 84 (00000054h)
Tunnel Endpoint Iden CP
- id: 85 (00000055h)
NSAPI
- NSAPI: 5(5h)
End User Address
- length : 6 (0006h)
- PDP type organisation : IETF
- PDP type number : 33 (21h)
- IPv4 address : 135.2.70.123
Access Point Name
- length : 24 (0018h)
- APN :  .apn1.mnc001.mcc310.gprs
SGSN Address for signalling
- length: 4 (4h)
- Ipv4 Address: 172.26.143.1
SGSN Address for user traffic
- length: 4 (4h)
- Ipv4 Address: 172.26.143.1
MSISDN
- length : 9 (0009h)
- MSISDN : A1 13 00 01 00 00 00 30 F0
Quality of Service Profile
- length 12 (000Ch)
- allocation/retention priority 3 (03h)
- unacknowledged GTP, LLC, and RLC, unprotected
data
- delay class 4 (best effort)
- precedence class: low priority
- peak throughput : up to 8 000 octet/s
- mean throughput : 100 octet/h
- traffic class : background class
- delivery order: without delivery order
- delivery of erroneous SDUs: no detect
- maximum SDU size: 1500
- maximum bit rate for uplink: 64 kbps
- maximum bit rate for downlink: 64 kbps
- residual BER: 1*10-5
- SDU error ratio: 1*10-4
- transfer delay: 10 ms
- traffic handling priority: level 1
- guaranteed bit rate for uplink: 64 kbps
- guaranteed bit rate for downlink:64 kbps
```

GGSN

```
IP1                      2 15:58:14.251
IP DATAGRAM
Source address : 172.26.131.118
Destination address : 172.26.143.1
UDP DATAGRAM
Source port: 2123
Destination port: 2123
CREATE PDP CONTEXT RESPONSE
- Version number: 1 (1h)
- Protocol type: 1 (1h)
- Extension header flag: 0 (0h)
- Sequence number flag: 1 (1h)
- N-PDU number flag: 0 (0h)
- Message Type: 17 (CREATE PDP CONTEXT
RESPONSE)
- Header Length: 61 (3dh)
- Tunnel endpoint id: 85 (00000055h)
- Sequence number: 49 (31h)
- No extension header
Cause
- acceptance: request accepted
Reordering Required  :no
Recovery
- restart counter: 27 (1bh)
Tunnel Endpoint Iden Data I
- Tunnel endpoint id: 180 (000000B4h)
Tunnel Endpoint Iden CP
- id: 179 (000000B3h)
Charging ID
- value : 11829536 (00B48120h)
GGSN Address for Control Plane
- length: 4 (4h)
- Ipv4 Address: 172.26.131.118
GGSN Address for user traffic
- length: 4 (4h)
- Ipv4 Address: 172.26.131.118
Quality of Service Profile
- length 12 (000Ch)
- allocation/retention priority 3 (03h)
- unacknowledged GTP, LLC, and RLC,
unprotected data
- delay class 4 (best effort)
- precedence class: low priority
- peak throughput : up to 8 000 octet/s
- mean throughput : best effort
- traffic class : background class
- delivery order: without delivery order
- delivery of erroneous SDUs: no detect
- maximum SDU size: 1500
- maximum bit rate for uplink: 64 kbps
- maximum bit rate for downlink: 64 kbps
- residual BER: 1*10-5
- SDU error ratio: 1*10-4
- transfer delay: 10 ms
- traffic handling priority: level 1
- guaranteed bit rate for uplink: 64 kbps
- guaranteed bit rate for downlink: 64 kbps
Charging Gateway Address
- length: 4 (04h)
- Ipv4 Address: 172.26.146.1
```

Figure 4.45 Trace PDP context creation. Reproduced by permission of NetHawk Oyj

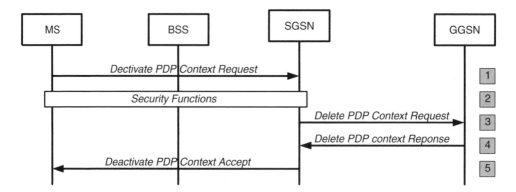

Figure 4.46   PDP context deactivation

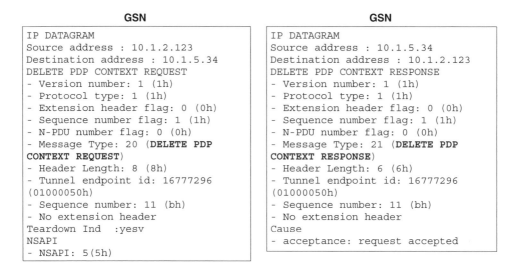

Figure 4.47   Trace PDP context deletion. Reproduced by permission Oyj

device, or by the subscriber typing the name in manually. At the time of writing, GPRS configuration on a mobile device is still often manual and rather tedious; however, gradually the process is becoming more automated, with a script file containing the necessary settings being loaded to the mobile device from the operator's network.

When the PDP context is requested, the SGSN will check with the HLR whether or not the APN is defined there. This information is passed to the GGSN (through the selection mode field), which will make a decision about whether or not to allow the request. The GGSN may require that the HLR validate that the subscriber is allowed to try to gain access to an external network via the access point. This adds further protection, so that if the subscriber has simply typed in the APN then this request for a PDP context activation may be rejected.

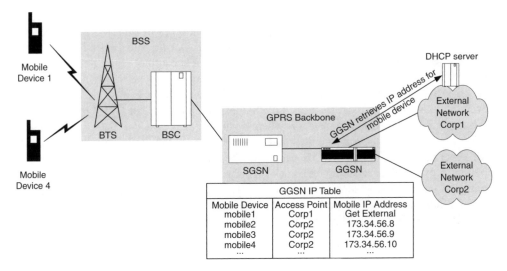

Figure 4.48 Transparent and non-transparent modes

An APN is composed of two parts:

- *APN network identity.* This is mandatory and is a label for the AP. It has to conform to DNS naming. Examples are Internet, Corp1.com, HRServer.Org5.com. The AP network identity does not end in **.gprs**.
- *APN operator identity.* This is optional and has also to comply with the DNS fully qualified domain name convention. It consists of three labels and does end in **.gprs**. The first two labels are the mobile network code followed by the mobile country code. An example is MCCMalaysia.MNCOperator.gprs. The whole AP name may then be Corp1.com. MCCMalaysia.MNCOperator.gprs. Note: wildcards (*) may also be accepted as APNs but may present a security risk.

4.9.5 Charging and billing

Charging information for each of the mobile devices is collected by the SGSN and the GGSN. The SGSN can collect charging information for the mobile device for the radio network usage and SMS transfers via the SGSN. Both the SGSN and GGSN collect charging information on the actual data transfer, including whether it is uplink or downlink, QoS and access point. The SGSN and GGSN generate charging data records (CDRs) and pass them to the charging gateway, which will then consolidate these records and pass them to the billing system. It is then the responsibility of the billing system to process the CDRs and apply the charging policy to produce a correct charge for each transaction. There are a huge number of different ways in which an operator can structure its charging model. One of the challenges is then processing and presenting this information quickly and efficiently. Unlike a phone call, one data call may generate several CDRs. The complexity is further increased due to the variety of different types of transaction that may be occurring at any given time. The operator must ensure that the subscriber is billed only for the information they have solicited. For example, a subscriber should clearly not be charged for receipt of an advertisement. Since the operator may be involving third-party service providers, records of these transactions also need to be kept. This may involve several parties, and also more than one GPRS operator, in the case of GPRS roaming. This situation is highlighted in Figure 4.49, where a subscriber is in a V-PLMN and wishes to connect to the home network via their H-PLMN. Here both the SGSN and remote GGSN will generate CDRs, which will be passed to the charging gateway and on towards the operator's own billing system. At some stage the operators need to compare the billing records to work out who owes what to whom. This may be performed for example at the end of a monthly period. It can be seen that the subscriber's traffic also traverses a GRX. The GRX is probably a third party which will also require revenue for use of the network.

4.9.6 QoS over the GPRS network

To guarantee provision of QoS it may initially appear that acknowledgements between the SGSN and GGSN should be necessary, requiring the use of TCP instead of UDP. However, this may have an adverse effect on the network by actually slowing down the response time if every packet is checked on this point-to-point link (SGSN to corresponding GGSN), and then checked again on the end-to-end link (e.g. mobile device to website). However, leaving the error checking and correction to the end-to-end stations means that the packets may travel across the Internet to the other side of the world before errors are noticed. Any request for retransmission will slow down the data transfer and not be at all useful for many real-time applications. It is often the case that the core network is built to be robust and is also over-dimensioned. In many cases, the equipment will be located in the same area, perhaps even the same room, and thus it is inexpensive to do this when compared to the cost of the BSS/RAN network. Building a robust infrastructure and over-dimensioning the core network is expected to introduce sufficient reliability.

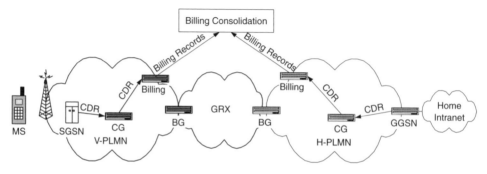

Figure 4.49 Charging for roaming subscribers

4.9.6.1 QoS mechanisms

Each flow of data (PDP context) has a QoS associated with it, which defines the following classes:

- precedence
- delay
- reliability
- peak throughput
- mean throughput.

The QoS is requested during the activation but this may be modified by various elements within the network in accordance with the resources available. The RLC/MAC layer supports four radio priority levels for signalling messages. The mobile device can indicate one of these levels and whether the access request is for data or signalling. This information is used by the BSS to determine the priority (precedence) of the information. The radio priority level for user data is determined by the SGSN during the PDP context activation/modification procedures. The precedence class prioritizes signalling and data flows. Under normal operating conditions the network will attempt to meet the requirements of all service commitments. However, under abnormal conditions it may be necessary to discard packets etc. The precedence class priority level determines which flows will be maintained and which flows will be affected first. There are three precedence classes, with class 1 offering the highest priority and class 3 the lowest.

Delay

There are four delay classes defined and these are outlined in Table 4.11. A delay class of best-effort makes no guarantees about the delay experienced.

Reliability

This is defined in terms of residual error rates, which include probability of loss of data, probability of data delivery out of sequence, probability of duplicate data and probability

Table 4.11 GPRS delay classes

Delay class	Delay maximum value (s)			
	SDU size 128 bytes		SDU size 1024 bytes	
	Average transfer	95th percentile	Average transfer	95th percentile
1	<0.5	<1.5	<2	<7
2	<5	<25	<15	<75
3	<50	<250	<75	<375
4	Best effort (unspecified)			

Table 4.12   GPRS reliability classes

Reliability class	Modes of operation				Traffic type
	GTP	LLC frame	LLC protection	RLC block	
1	ACK	ACK	Protected	ACK	Non-real-time. Application cannot deal with errors
2	UNACK	ACK	Protected	ACK	Non-real-time. Application can deal with minimum errors
3	UNACK	UNACK	Protected	ACK	Non-real-time. Application can deal with errors. GMM/SM and SMS
4	UNACK	UNACK	Protected	UNACK	Real-time. Application can deal with minimum errors
5	UNACK	UNACK	Unprotected	UNACK	Non-real-time. Application can deal with errors

of corrupted data. Table 4.12, indicates the different classes and the appropriate modes of operation to transport different types of traffic through the GPRS network.

Each different flow of data will require specific reliability classification according to its particular requirements. The reliability classes state whether data is transferred in acknowledged or unacknowledged mode across various lower-level protocols throughout the network.

Throughput

Throughput of used data is characterized in terms of bandwidth, split into a peak and mean throughput class. As the name suggests, the peak throughput class defines the maximum rate at which the data is expected to be transferred. The class provides no guarantee that the defined rate will be reached, as this depends on both the resources available in the network, and the capabilities of the mobile device. Table 4.13 shows the peak throughput classes and the related data rate for each.

The mean throughput class defines the average data rate in bits per second. The actual definition is in terms of bytes per hour, as the bursty nature of most data applications

Table 4.13   GPRS peak throughput classes

Class	Peak throughput (kbps)
1	8
2	16
3	32
4	64
5	128
6	256
7	514
8	1024
9	2048

Table 4.14 GPRS mean throughput classes

Class	Mean throughput (bps)	Class	Mean throughput (bps)
1	Best effort	11	220
2	0.22	12	440
3	0.44	13	1110
4	1.11	14	2200
5	2.2	15	4400
6	4.4	16	11100
7	11.1	17	22000
8	22	18	44000
9	44	19	11100
10	111		

means that although there may be high instantaneous demand for resources, on average the throughput can be quite low. Table 4.14, shows the defined classes. As with delay, a class of best effort makes no guarantees.

4.9.6.2 Traffic flow templates

It is possible within GPRS to have a number of PDP contexts on a single device and these can be connected via different GGSNs to completely independent networks. One of these could be the Internet and another may be the subscriber's home network. Thus the mobile device will have two IP addresses, one for each context, and these are both known as primary PDP contexts. Within R99 it is also possible to have two PDP contexts that connect to the same network and share a single IP address. As an example, consider that a subscriber may be connected to the Internet and simultaneously browsing the web and downloading a large file. When used in this method by sharing an IP address, the second PDP context established is referred to as a *secondary PDP context*.

The traffic flow template (TFT) is used to distinguish and separate packets arriving at the GGSN from the same external network under more than one PDP context, but sharing an IP address. The TFT will ensure that these are routed to the mobile device over the correct tunnel for the PDP context. This enables the network to allocate different profiles for each tunnel, to provide, for example, different levels of QoS or security.

Relating back to the example, by using separate contexts, the packets of data that are associated with the interactive browsing session can be allocated a higher priority when compared to the download session. Since the packets for both the browsing session and the download session share the same IP address, the TFT is used to differentiate them at the GGSN. The TFT is defined using between one and eight packet filters, which are comprised of different sets of fields within the IP header. The header attributes that may be used are as follows:

- source address and subnet mask (since this is incoming from the external network, the source is not the mobile device IP address, but rather the source address of the incoming external IP packet)
- protocol number for IPv4 or next header for IPv6

- destination port range
- source port range
- IPSec security parameter index (SPI)
- type of service for IPv4 and traffic class and mask for IPv6
- flow label for IPv6.

The above attributes are described in the IP description in Chapter 5. An example filter could be:

Packet Filter Identifier = 1

IPv4 Source Address = 168.223.10.9/24

Destination Port = 4000

Packets arriving at the GGSN from the external network with the above values can be filtered by the GGSN and passed to the correct GTP tunnel associated with the specific PDP context

4.10   CONNECTION SCENARIOS

There are many factors governing how the core network is actually implemented. Each has an impact on the QoS that the subscriber receives. The simplest scenario, illustrated in Figure 4.50, shows a user wishing to connect to a home network which is in close proximity to the mobile operator's network. In this situation, the SGSN and GGSN may be located in the same physical site, probably the same building. An Ethernet network may be employed between the two units. However, if there is a large amount of data or if a high level of QoS is required, then this could be some other high-speed technology such as ATM or multi-protocol label switching (MPLS).

If there is a great deal of traffic between a mobile user and the home network, then a leased line may be used as a connection between the GGSN and home router. This will give maximum reliability and higher data rates but the cost will also be high. Another option may be the use of a microwave link between the two sites. The initial expense

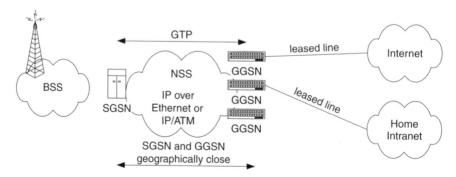

Figure 4.50   Simple connection scenario

may be high but the running costs of such a link can be very low, especially if it is using the free-band on a point-to-point link and the power is kept below 1 W. This type of link can give very high data rates at a fraction of the cost of a wired leased line. In many cases, however, a great deal of error checking and error correction is required since these radio links may not be so reliable.

If the user is in a different region to the home network, the connection implementation has even more factors in the equation. It is also possible for lower traffic rates to use a public network such as the Internet, or use a frame relay or ATM network. To introduce security over the public Internet, it is likely that a virtual private network (VPN) will be constructed between the mobile operator and the home Intranet (Figure 4.51). This will probably be the most cost-effective solution and may be suitable for low volumes of data transfers, which are not real time and can absorb delays.

An alternative would be to let the mobile network operator deal with the problem of the subscriber being geographically remote. In this case, the SGSN (where the subscriber is located) and the GGSN (where the leased line to the home network is located) may be many hundreds of kilometres apart (Figure 4.52). In this situation, the mobile operator has an estimate of how many roaming subscribers there are, and what sort of data rates they expect. Since the SGSN and GGSN are geographically remote, Ethernet cannot be used for this connection. The operator may use a public network such as the Internet to

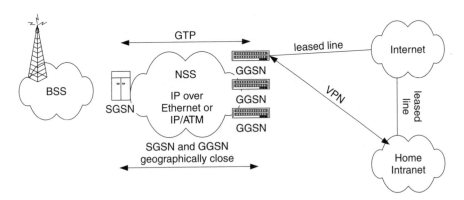

Figure 4.51 Connection using a leased line or VPN

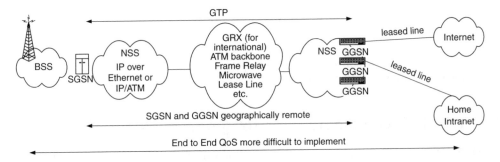

Figure 4.52 GPRS roaming configuration

connect the SGSN to GGSN but the transfer delays may not be constant and the data rates attained will generally be low. It is possible for the mobile operator to come to an arrangement with the ISP whereby the operator's traffic has priority over standard Internet user traffic. This in effect gives the appearance of a leased line but should be much cheaper. An ATM, frame relay or X.25 network may also be implemented, or a number of radio towers (which the operator already has) may be used as point-to-point microwave links.

It can be seen from the above that the core network can get more and more complex, depending on the actual implementation. The scenario of over-dimensioning the core network is only cost effective when the SGSN and GGSN are physically located in close proximity. To over-dimension a network covering many kilometres may not be economically viable, but microwave links using existing infrastructure may be used. Since the GTP packets are carried between the SGSN and GGSN via UDP and not TCP, reliability is a concern. Any error checking and retransmission will only be end-to-end (with possibly very high delays) unless a method for doing this is implemented transparently under the GTP protocol.

For international roaming it is likely that the operator will enter into a roaming agreement, as is currently done with GSM, allowing a user to roam transparently. A GPRS roaming exchange (GRX) network is used for this inter-PLMN connection.

4.11   OTHER CELLULAR HIGH-SPEED DATA TECHNOLOGIES

4.11.1   High-speed circuit-switched data (HSCSD)

HSCSD is based around the existing GSM interface, allows a user to occupy multiple time slots for data transfer and guarantees a certain number of bits per second across the air interface and through the operator's mobile network. Currently HSCSD mobile devices support two or three GSM time slots; each of these time slots will offer 14.4 kbps, giving a total throughput of around 40 kbps. It is possible to provide up to eight time slots to a single HSCSD user; however, this will have an adverse effect on other GSM users, introducing increased call blocking etc. The user is usually billed on time in a similar method to a normal dial-up Internet connection. If a user downloads a web page in 10 seconds and then spends 5 minutes reading that page while remaining connected, they will be charged for 5 minutes 10 seconds of connected time. The resources reserved for the user can be asymmetric, allowing faster transfer in the downlink than in the uplink. This is because it is anticipated that users will be downloading more information than they upload. Since it is still based on a circuit switched system the HSCSD system does not utilize the resources of the operator efficiently for bursty packet data.

4.11.2   Enhanced data rates for global evolution (EDGE)

EDGE was originally seen as an evolutionary step for GSM and the acronym stood for enhanced data for GSM evolution. However, the TDMA community has since embraced

EDGE as a 3G solution under the UWC-136 proposal and it now stands for enhanced data for global evolution. It is seen as both a migratory step towards 3G standards such as UMTS and also as a complementary technology which will support such networks in the future. Like GPRS, EDGE uses much of the underlying GSM system, including the existing frequency band that an operator has been allocated. UMTS, on the other hand, does require new frequencies, which in many cases have been very expensive for the operator to purchase. This section describes the new features and attributes that are gained from EDGE. Figure 4.53 illustrates how EDGE, with its higher transfer rates, can be used to efficiently free up time slots, which can then be reallocated to other GSM users.

EDGE can be used in conjunction with both HSCSD and GPRS to provide higher data rates for the subscriber. EDGE introduces nine new modulation and coding schemes over the air interface, as shown in Table 4.15.

Four of these coding schemes use the standard GSM GMSK modulation technique and the other five use 8 PSK. The introduction of the new modulation technique enables data rates up to three times that possible with standard GMSK. It can theoretically support 384 kbps using all eight time slots. However, to take advantage of this system the BSS has to be upgraded: generally this requires new transcoders in the BTS and software upgrades to both the BTS and the BSC. The mobile devices also have to be capable of operating with the EDGE air interface and signalling. When used with GPRS, EDGE is known as E-GPRS and when used with HSCSD it is referred to as ECSD. The Abis interface for

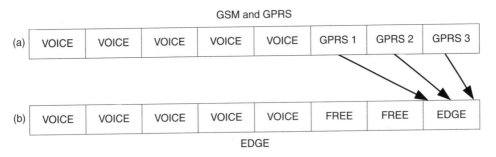

Figure 4.53   Improved data capacity using EDGE

Table 4.15   EDGE modulation schemes

Coding scheme	Modulation type	Data rate (kbps)
MCS-1	GMSK	8.8
MCS-2	GMSK	11.2
MCS-3	GMSK	14.8
MCS-4	GMSK	17.6
MCS-5	8PSK	22.4
MCS-6	8PSK	29.6
MCS-7	8PSK	44.8
MCS-8	8PSK	54.4
MCS-9	8PSK	59.2

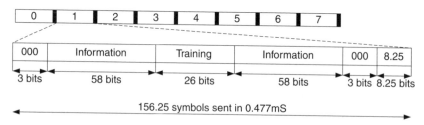

Figure 4.54 EDGE data burst

GSM consists of $n \times 64$ kbps links that are broken up into 16 kbps time slots, so there is a one-to-one correspondence with a time slot on the air interface. For an EDGE solution there may be more data in one air time slot than will fit into one 16 kbps slot, therefore the Abis now needs to allow dynamic allocation of bandwidth to support the air upgrades.

An EDGE burst structure is the same as GSM/GPRS consisting of 200 kHz spacing and 156.25 bits per burst and taking 0.577 ms. However, one modulated bit (referred to as a symbol) can actually represent three bits due to the introduction of an enhanced modulation technique. The data burst is shown in Figure 4.54. A simple GSM burst consists of 114 bits of information. This can be compared to EDGE, which with the same burst can transport up to 348 bits due to the 8PSK modulation technique. When compared to the GSM burst, it can be seen that there are no stealing bits, which are required in GSM/GPRS to indicate that the burst contains either control or data. This is because the stealing bits are encoded as part of the 58 symbol information parts. It can therefore be seen that whereas a GPRS radio block can transfer 456 bits of data, an EDGE radio block using GMSK (MCS-1, 2, 3 and 4) can support up to 464 bits and EDGE using 8PSK (MCS-5, 6, 7, 8 and 9) can support up to 1392 bits. This data consists of higher-layer data after convolution coding, CRC addition etc., and is not pure user data.

4.11.3 Modification to RLC/MAC

When compared to the GPRS RLC/MAC, it can be seen that there are slight modifications to the RLC/MAC header. In fact, there are actually three variations to the header format in both the uplink and downlink for transferring RLC data blocks. However the RLC/MAC header for control messages is the same as that for GPRS. The three header formats are referred to as types 1, 2 and 3 and are used for the different coding schemes as listed below:

- type 1 is used for MCS-7, 8 and 9
- type 2 is used for MCS-5 and 6
- type 3 is used for MCS-1, 2, 3 and 4.

There are a number of reasons for this, one being that with EDGE it is possible to transmit two RLC blocks within a single radio block using MCS-7, 8 and 9 and the location of these separate RLC blocks needs to be identified.

The RLC/MAC window size has also been modified with EDGE. In a GPRS transfer the maximum window size is 64 blocks. This means that if a block has been sent in error then only 63 more blocks can be outstanding before a successful retransmission of the errored block. This maximum can be easily reached using a mobile with multislot capability and when the data transfer is stalled. Taking the higher data rates available with EDGE into consideration this problem would become more acute. The window size has therefore been increased to a maximum of 1024 blocks depending on the multislot allocation being used.

By comparing the EGPRS RLC/MAC header formats for data transfer (Figures 4.55 and 4.56) with those of GPRS (Figure 4.24), it can be seen that there are a number of modifications. The header fields are described below:

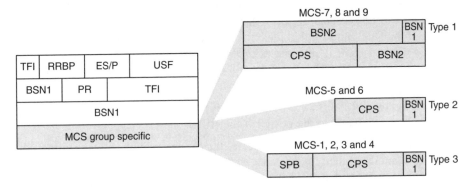

Figure 4.55   EDGE downlink RLC/MAC headers

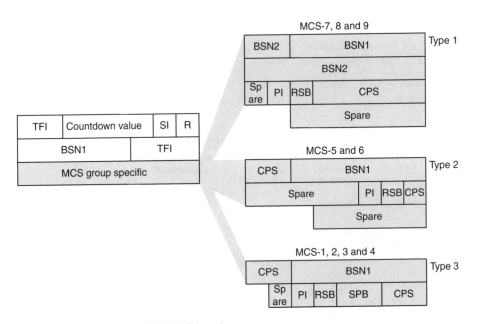

Figure 4.56   EDGE uplink RLC/MAC headers

- *Temporary flow identity (TFI)*: this 5-bit field identifies the TBF within which this block belongs.

- *Relative reserved block period (RRBP)*: this 2-bit field specifies a reserved block that the mobile device may use for a *packet control acknowledgement* message or a PACCH block.

- *EGPRS supplementary/polling (ES/P)*: this 2-bit field indicates whether the RRBP field is valid or not. If this is 00 then the RRBP is not valid.

- *Uplink status flag (USF)*: the USF consists of 3 bits and is sent in all downlink RLC blocks to indicate the owner of the next uplink radio block on the same time slot.

- *Block sequence numbers 1 and 2 (BSN1 and BSN2)*: this works in a similar fashion to GPRS. However, it is extended from 7 bits to be an 11-bit field in EGPRS, allowing more blocks to be outstanding, i.e. a larger window size. It consists of the sequence number of the RLC block within the TBF to identify missing blocks. As discussed and shown in Table 4.16, it is possible using MCS-7, 8 and 9 for the EGPRS RLC/MAC frame to carry two RLC blocks. Each of these will have its own individual block sequence number. This is why the *type 1* header format has BSN1 and BSN2.

- *Power reduction (PR)*: this 2-bit field indicates the power level reduction of the current RLC block as compared to the power of the BCCH.

- *Coding and puncturing (CPS)*: this indicates the channel coding and puncturing scheme used. For example, if the type 1 header was employed and the CPS field had the value 8, this would indicate that MCS-9 was used for both blocks 1 and 2 and that block 1 would use puncturing scheme 3 whereas block 2 would use puncturing scheme 1.

- *Split block indicator (SPB)*: this is only required in header type 3 and is used for identifying retransmissions.

- *Countdown value (CV)*: this 4-bit field is sent by the mobile device to allow the network to calculate the number of RLC blocks remaining in the current uplink TBF.

- *Stall indicator (SI)*: this single-bit field indicates whether the mobile station's RLC transmit window can advance or not. A 0 indicates that the window is not stalled.

Table 4.16 Coding scheme

Family	Coding scheme	Payload size (bits)	Payload units	Data bits in each RLC block
A	MCS-3	296	1	296
	MCS-6	296	2	592
	MCS-8	272	4	2×544
	MCS-9	296	4	2×592
B	MCS-2	224	1	224
	MCS-5	224	2	448
	MCS-7	224	4	2×448
C	MCS-1	176	1	176
	MCS-4	176	2	352

- *Retry (R)*: this single-bit field indicates whether the mobile station had sent the *channel packet request* message more than once. A 0 indicates that it was sent once and a 1 indicates that it was sent multiple times.

- *PFI indicator (PI)*: this single-bit field indicates the presence of the optional PFI field. A 0 indicates that it is not present.

- *Resent block bit (RSB)*: this field indicates whether there are any RLC data blocks within the radio block that have already been transmitted before.

The EGPRS RLC/MAC data block is illustrated in Figure 4.57. This data block fits inside the RLC/MAC block and the size of the actual data block for each of the coding schemes is highlighted in Table 4.16.

The following identify the fields within the RLC data block illustrated in Figure 4.57:

- *Final block indicator (FBI)*: this single-bit field indicates the last downlink RLC data block. A 0 indicates that there are more blocks to come whereas a 1 indicates that this is the last block in this TBF.

- *Extension (E)*: this single-bit field indicates the presence of the optional byte in the data block header. A 0 indicates that an extension byte follows.

- *Length indicator (LI)*: this optional 7-bit field is used to delimit LLC PDUs within a single RLC data block by identifying the last byte of the LLC PDU. If there are a number of LLC PDUs within an RLC data block the first LI will indicate the length of the first LLC PDU, the second LI will indicate the length of the second LLC PDU etc. Only the last segment of a LLC PDU has an associated LI.

- *TLLI Indicator (TI)*: this bit, which is only present in the uplink, indicates whether the TLLI field is present or not. A 1 indicates that the TLLI is present.

- *Temporary logical link identifier (TLLI)*: this 32-bit field is optional and is used while contention resolution is required.

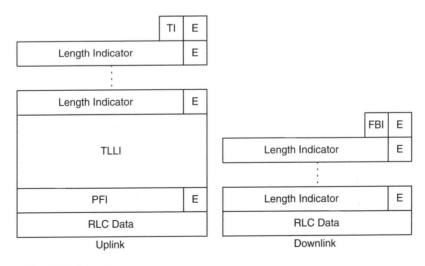

Figure 4.57   RLC data block

- *Packet flow identifier (PFI)*: this 7-bit field is assigned by the SGSN and used to identify a particular flow context and QoS value. Legitimate identifiers can be best effort, signalling and SMS, which are predefined, or it can be dynamically assigned and offer a particular QoS for this specific context.
- *RLC data*: this field contains the LLC PDU or part of it if it has been segmented. The amount of data transferred depends on whether there are any optional RLC headers in place and also on the coding scheme used. Figure 4.57 indicates the number of bits each coding scheme can carry.

4.11.4   Channel coding for PDTCH

The nine different modulation and coding schemes (MCS) shown in Table 4.15 are divided into different families as indicated below:

- family A consists of MCS-3, MCS-6, MCS-8 and MCS-9
- family B consists of MCS-2, MCS-5 and MCS-7
- family C consists of MCS-1 and MCS-4.

Each of these families has a fixed payload size: family A is 296 (and 272[7]) bits, B is 224 bits and C is 176 bits. This allows different code rates within a family to be achieved through transmitting a number of blocks (known as payload units) within a radio block. MCS-7, 8 and 9 each consist of four payload units, MCS-5 and 6 each consist of two payload units and MCS-1, 2 and 3 each consist of one payload unit. This is highlighted in Table 4.16. It can also be seen from this table that MCS-7, 8 and 9 actually transfer two RLC blocks within a single radio block.

Figure 4.58, is a simplified view of how data is transferred in the downlink using MCS-9. The stealing bit (SB) field is used to indicate the header format used. The numbers in the diagram indicate the number of bits associated with each field. It can be seen that there are actually two RLC blocks that are carried. Since MCS-9 uses 8PSK the four radio bursts at the physical layer across the air transfer 1392 bits of data. MCS-6, which is in the same family, will transfer one RLC block (612 bits) with more protection and since it also uses 8PSK will transfer 1392 bits at the physical layer. MCS-3, which is also in the same family, actually carries 316 bits of data within a block. It does not use 8PSK, therefore at the physical layer 464 bits are transferred. The HCS is a header checksum and the BCS is a checksum over the data.

Figure 4.59 shows another example; this time MCS-1 is being used. In both this and the preceding diagram each of the payload data blocks (612 bits in Figure 4.58 and 372 bits in Figure 4.59) is punctured with an individual puncturing scheme. There are up to three puncturing schemes available and in the case where two RLC blocks are transmitted within a single radio block (MCS-7, 8 and 9) they may be subjected to

---

[7]MCS-8 is only 272 bits, therefore when switching between MCS-3 or MCS-6 padding is required to ensure that the length of 296 bits is consistent throughout the family.

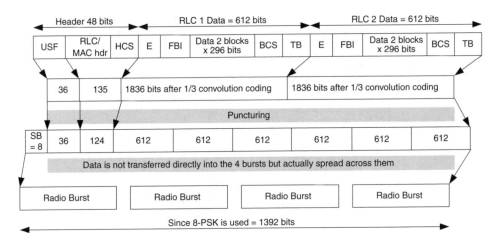

Figure 4.58   Coding and puncturing for MCS-9

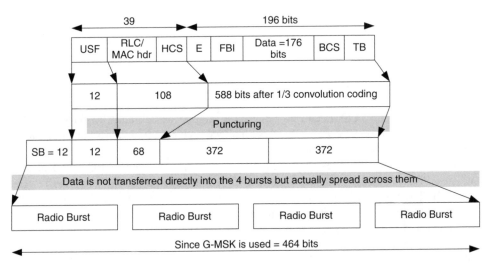

Figure 4.59   Coding and puncturing for MCS-1

different puncturing schemes. The coding and puncturing scheme (CPS) indicator field in the RLC/MAC header indicates to the receiver which scheme has been implemented.

4.11.5   Link adaptation and incremental redundancy

As discussed, EDGE introduces nine modulation and coding schemes, each of which is designed to deliver the optimal throughput under different radio conditions. As the radio environment changes during a connection, then the coding scheme may change, enabling more efficient use of the air resources and improving performance and robustness. The selection of this MCS is determined by the network and takes into consideration the link

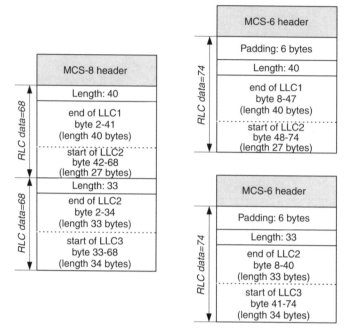

Figure 4.60 Example of IR retransmission

quality measurements (LQM) carried out by the mobile device and the BTS. This link quality may be based on a number of parameters such as block error rate (BLER) or channel to interference ratio (CIR).

When acknowledged mode is selected, EDGE may use the same ARQ mechanism as used in GPRS but it can also use incremental redundancy (IR), which is also referred to as hybrid ARQ type 2. In the traditional ARQ mechanism the data is segmented at the transmitter, convolution encoded and a checksum added. At the receiver the block is decoded and the CRC checked. If the computed checksum does not equal the CRC transmitted then the data is discarded and a retransmission is requested.

When IR is selected, the block of data received in error is not discarded but retained. There are a number of alternative puncturing schemes and each block of data is transmitted with redundant information using a different puncturing scheme. In the case of an erroneous block being received, additional coded bits are retransmitted using an alternative puncturing scheme and together with the previously errored block are used to reconstruct the data. If the block is received in error then it can be resent using the same MCS or an alternative MCS within the same family. For example, if MCS-6 had been initially selected and the block was received in error, then any MCS in family A including MCS-6 can be used for the retransmission.

Figure 4.60 shows an example of IR retransmission, the original transmission using MCS-8 and the retransmission using MCS-6. In this example there are three LLC partial or full frames to be transferred; LLC data 1 consists of the last 40 bytes of the LLC, LLC data 2 consists of a complete LLC of 60 bytes and LLC data 3 consists of the first 34 bytes of the LLC.

The MCS-8 block can carry two RLC data blocks each consisting of 68 bytes. This value includes the length indicator and not just the actual data. The first length indicator field with the value 40 indicates where LLC1 data ends. The LLC2 data follows this directly within the same RLC, i.e. the RLC block is not specifically for one single LLC frame and can span multiple frames. The LLC2 data does not fit completely into the first RLC block and thus the second length indicator indicates the last byte of this LLC frame. Again, LLC3 data follows on directly.

In the case of the retransmission, it can be seen that MCS-6 does not have such a large payload capacity and as such two RLC/MAC blocks have to be sent. If MCS-3 had been chosen for the retransmission then four RLC/MAC blocks would have been required, introducing a large transfer overhead due to the four headers required. It can be seen from Figure 4.60 that padding has been introduced for alignment purposes. This padding takes up the initial part of the RLC data and consists of 6 bytes. Since the RLC data unit can be 74 bytes this can now transport the first RLC data unit of the MCS-8 RLC/MAC block. The second MCS-6 RLC/MAC block can also transport the second RLC data unit of the MCS-8 RLC/MAC block completely. At the receiver, both the MCS-8 block and the MCS-6 blocks can be used to reassemble the LLC data frames.

4.11.6   Compact EDGE

UWC-136 EDGE classic was specified by ETSI and uses the standard GSM carriers and control channels. A BCCH is located in each cell; it transmits continuously and unlike data channels does not hop between frequencies. This system typically requires 2.4 MHz of spectrum in both directions. Currently it is typical for a GSM system to adopt a frequency reuse pattern of 3/9 (three BTS and three sectors each) or 4/12 (four BTS and three sectors each). It is expected that EDGE systems will be co-located and thus also use the same reuse pattern.

In the USA, where there is limited spectrum available, a deployment that uses 1 MHz in both directions is being considered which puts considerable constraints on the system implementation. This will require a frequency reuse pattern that is optimal and spacing of carriers will be much closer than with the classic model. In fact, it requires a frequency reuse of 1/3 whereby a BTS that has three cells using F1, F2 and F3 will be adjacent to other BTSs using the same three frequencies. This reuse pattern is seen as possible with the dedicated channels which can withstand lower signal to noise ratios. However, to ensure that the system works effectively, control signalling such as system broadcast, paging and packet access will require to be extremely robust. The robustness cannot come from frequency separation due to the limited spectrum but rather this is achieved through the time domain. This system is known as compact EDGE. To achieve the reliability and robustness of the common signalling channels through the time domain requires that all the BTSs are frame synchronized. This is not a requirement for classic EDGE. This can be achieved through the use of GPS.

In compact EDGE, for control signalling purposes, each cell is not only identified by its frequency but also by the time group to which it belongs and there are four time groups specified (although not all of these need to be implemented). Essentially this means that although cells using the exact same frequency will be in close proximity geographically and will cause a certain amount of interference, it is a design consideration to ensure that they belong to different time groups. When it is one time group's turn to use the common channel signalling channel other cells that are *not* in the same group but are at the same frequency will not transmit at all, thus reducing the interference. This mechanism is illustrated in Figure 4.61. It can be seen that there are three frequencies and that there are four time groups. Also highlighted is a group of twelve cells which includes only one cell using frequency 1 and timing group 4.

The actual time groups are identified by the time slot number in the GSM/EDGE frame. There are eight time slots in a GSM frame and time slots 1, 3, 5 and 7 are used for these four time groups. The time group is not fixed to a particular time slot and is rotated over the time slots, so at any one time, for example, time group 2 may use slot 1, 3, 5 or 7. The time groups are synchronized so that the other time groups will use the other slots that are available. The time groups are also only transmitted at certain times within the hyperframe (52-frame); these are frames 0–3, 21–24, 34–37 and 47–50. When they are not being used for common channel signalling these slots (1, 3, 5 and 7) and the other even time slots can be used for data transfer (DTCH) or dedicated signalling (PACCH). The only exception as mentioned is when another time group is using the common control channel and to reduce interference all other slots at that time will be idle and will not transmit. This is highlighted in Figure 4.62, where the synchronized downlink time slots of four time groups all using the same frequency are illustrated.

4.11.7 GSM/EDGE radio access network (GERAN)

The next phase of EDGE for R4/5 is the GSM EDGE radio access network (GERAN), which is closely aligned with UTRAN. It is designed to give better QoS than the existing

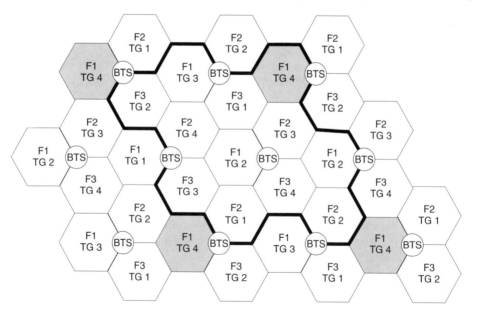

Figure 4.61 Compact EDGE frequency and time reuse

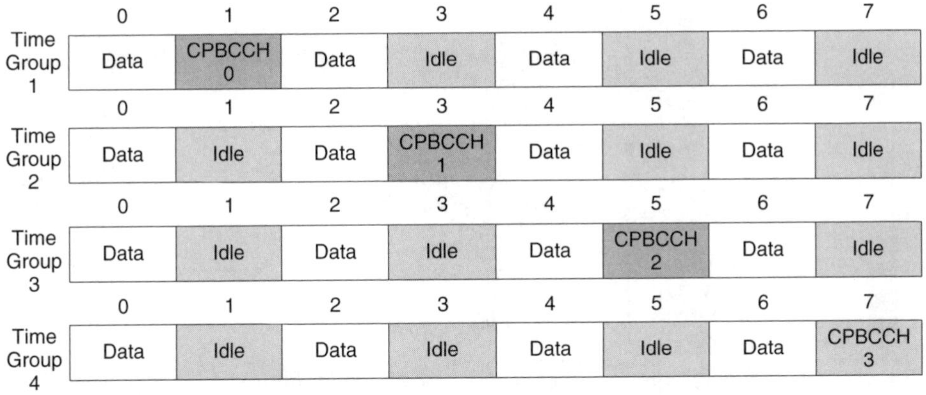

	0	1	2	3	4	5	6	7
Time Group 1	Data	CPBCCH 0	Data	Idle	Data	Idle	Data	Idle

	0	1	2	3	4	5	6	7
Time Group 2	Data	Idle	Data	CPBCCH 1	Data	Idle	Data	Idle

	0	1	2	3	4	5	6	7
Time Group 3	Data	Idle	Data	Idle	Data	CPBCCH 2	Data	Idle

	0	1	2	3	4	5	6	7
Time Group 4	Data	Idle	Data	Idle	Data	Idle	Data	CPBCCH 3

Figure 4.62 Compact EDGE downlink time slots

EDGE systems and will enable such services as VoIP. Since in this case the voice data will be transported over the Iu-PS, it will be possible to multiplex other data services within the same time slot as long as they do not interfere with the real-time QoS requirements of the voice. Figure 4.63 illustrates a generalized GERAN architecture. It can be seen that there are options for using both the A/Gb interface and/or the Iu interface for circuit- and packet switched connections to the core network. Which of these interfaces is supported is indicated in the broadcast system information messages. The actual user and control plane protocol stacks depend on which interfaces (A/Gb or Iu) are implemented within the GERAN. For example, on the user plane for the packet core, if the Gb interface is implemented then SNDCP/LLC will be used and if Iu-PS is implemented then PDCP will be used. The protocol stacks for the control plane and user plane for the Gb mode of operation are illustrated in Figures 4.7 and 4.8, respectively. For the Iu-PS mode of operation please refer to Chapter 6.

It is anticipated that additional transport options rather than just ATM will be utilized, such as IP.

4.11.7.1   *Iur-g interface*[8]

The Iur-g interface is an open standard allowing for multi-vendor interoperability. Implementing this optional Iur-g interface to connect different RANs together, or even to other BSSs, enables the GERAN routing area (GRA) to be introduced. This has a similar function to the URA within UTRAN and enables the RAN to assist the core network in the mobility management related to each of the mobile devices. It is used if the Iu mode is implemented but is not used for the A/Gb mode of operation. The GRA consists of one or more cells which may overlap BSS/RNS or LA/RAs. To enable this overlapping feature, a single cell may broadcast a number of GRA identifiers.

The RNSAP protocol will be used across the interface but will support only a subset of those procedures that are incorporated in the Iur in UTRAN. For example, there is no soft

---

[8]The Iur-g is similar to the Iur interface, which is described in detail in Section 6.19.3.

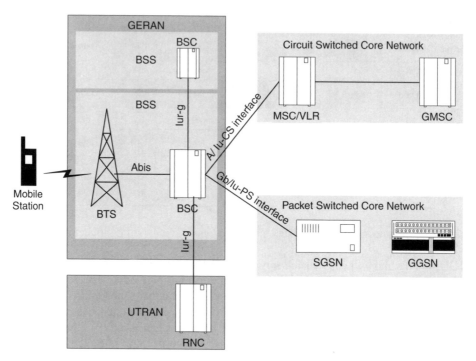

Figure 4.63   Generalized GERAN architecture

handover in GERAN, therefore this does not need to be supported. The interface allows for paging and SRNS relocation, as well as uplink and downlink signalling transfers. The protocol stack for RNSAP is the same as used in UMTS later releases. It has the option of being carried over MTP3-B or M3UA/SCTP.

A mobile device which has an RRC connection will be issued a G-RNTI, which consists of the serving BSC identifier (SBSC-id) as well as a unique identifier, within the BSC serving area, for the mobile device (S-RNTI).

4.11.7.2   Radio resource connection modes

This is when the Iu interface is being used and the mobile device can be in either of two modes: idle or connected.

In idle mode the CN knows the location of the mobile device to the current LA or RA. There is no knowledge of the mobile device within the BSS, and for downlink transfers the mobile device needs to be paged. The mobile device will return to this state from connected mode when the RRC connection is released.

The RRC connected mode is entered when the mobile device is assigned a G-RNTI. There are three states defined for this mode:

• *RRC-cell shared*: no dedicated basic physical subchannel is allocated and the mobile device is known to the cell by GERAN.

- *RRC-cell dedicated*: one or more dedicated basic physical subchannels has been allocated in the uplink and downlink, which it can use at any time. The mobile device may also be assigned resources on one or more shared channels. Again, the mobile device is known to the cell by GERAN.

- *RRC-GRA_PCH*: no channels are allocated and no uplink activity is possible. The location of the mobile device is known to the GRA level by GERAN.

In both *RRC-cell shared* and *RRC-GRA_PCH* mode, the core network will know the location of the mobile to the serving BSC. GERAN will know the location to the cell and GRA, respectively.

4.12 SUMMARY

This chapter explains the evolution of the basic GSM network to effectively handle non-real-time data traffic with the introduction of the GPRS infrastructure. The required changes and additions throughout the network are described, as well as the new protocols introduced. The dynamic allocation of resources on the air interface and the new coding schemes introduced are also explained. Like GSM, the GPRS system forms a central part of the UMTS core network. The key procedures are explained with use of examples of network traces to illustrate their operation in a real environment. The evolution of GPRS may be to an EDGE network, so, in this context, the operation of EDGE is described in some detail, along with its use in the evolved GSM network, GERAN, which will eventually be integrated into a ubiquitous 3G cellular environment.

FURTHER READING

T. Halonen, J. Romero, J. Melero (2002) *GSM, GPRS and EDGE Performance*. John Wiley & Sons, Chichester.

3GPP TS03.64: General Packet Radio Service (GPRS); Overall description of the GPRS radio interface; Stage 2.

3GPP TS04.60: General Packet Radio Service (GPRS); Mobile Station (MS)–Base Station System (BSS) interface; Radio Link Control/Medium Access Control (RLC/MAC) protocol.

3GPP TS04.64: General Packet Radio Service (GPRS); Mobile Station–Serving GPRS Support Node (MS–SGSN) Logical Link Control (LLC) layer specification.

3GPP TS04.65: General Packet Radio Service (GPRS); Mobile Station (MS)–Serving GPRS Support Node (SGSN); Subnetwork Dependent Convergence Protocol (SNDCP).

3GPP TS05.01: Physical Layer on the Radio Path (General Description).

3GPP TS05.02: Multiplexing and Multiple Access on the Radio Path.

3GPP TS05.03: Channel coding.

3GPP TS08.14: General Packet Radio Service (GPRS); Base Station System (BSS)–
Serving GPRS Support Node (SGSN) interface; Gb Interface Layer 1.

3GPP TS08.16: General Packet Radio Service (GPRS); Base Station System (BSS)–
Serving GPRS Support Node (SGSN) Interface; Network Service.

3GPP TS08.18: General Packet Radio Service (GPRS); Base Station System (BSS)–
Serving GPRS Support Node (SGSN); BSS GPRS Protocol.

3GPP TS22.60U: Mobile multimedia services including mobile Intranet and Internet services.

3GPP TS23.060: General Packet Radio Service (GPRS) Service description; Stage 2.

3GPP TS23.107: Quality of Service (QoS) concept and architecture.

3GPP TS24.008: Mobile radio interface Layer 3 specification; Core network protocols;
Stage 3.

3GPP TS29.060: General Packet Radio Service (GPRS); GPRS Tunnelling Protocol (GTP)
across the Gn and Gp interface.

3GPP TS43.051: GSM/EDGE Radio Access Network (GERAN) overall description;
Stage 2.

A list of the current versions of the specifications can be found at http://www.3gpp.org/
specs/web-table_specs-with-titles-and-latest-versions.htm, and the 3GPP ftp site for the
individual specification documents is http://www.3gpp.org/ftp/Specs/latest/

5

IP Applications for GPRS/UMTS

5.1   INTRODUCTION

From its earliest inception, the Universal Mobile Telecommunications System (UMTS) (Release 99) makes use of the Internet Protocol (IP) for data transport within the core network. In this first release of UMTS, IP is used to route general packet radio service (GPRS) traffic between the user equipment (UE) and external IP networks (for example, the Internet) and across the UMTS core network using the GPRS tunnelling protocol (GTP). This role of IP expands through release 4, culminating in release 5, where IP may be used for the entirety of the transport (traffic and control) across the UMTS terrestrial radio access network (UTRAN) through the core network and beyond. The use of IP provides a number of benefits. The IP protocol suite is a mature, well-tested technology that has proven to be a highly robust and scalable architecture supporting many millions of nodes on the Internet. IP has a vast range of applications (e.g. WWW and email) and services (e.g. e-commerce and security) already developed for it. With the use of IP as a transport mechanism these applications and services can be ported directly to the UMTS environment. Finally, IP has the capability to carry mixed voice and data traffic efficiently through the use of a range of IP quality of service (QoS) protocols. To facilitate a smooth progression to IP version 6 (IPv6) the UMTS specifications state that both IP version 4 (IPv4) and IPv6 must be supported within each and every network element.

This chapter and Chapters 8 and 9 examine how IP is applied in the various releases of UMTS. This chapter focuses on the IP protocol suite and describes how IP is used within the UMTS R99 network for GPRS service. In particular, the provisioning of QoS and security for GPRS networks is discussed in some detail. Later chapters show how the role of IP is expanded within the network for releases R4, R5 and R6. With R4 all of the UMTS core network will use packet switching but is still split into separate domains for

Convergence Technologies for 3G Networks: IP, UMTS, EGPRS and ATM   J. Bannister, P. Mather and S. Coope
© 2004 John Wiley & Sons, Ltd   ISBN: 0-470-86091-X

circuit- and packet-type service. The voice traffic within the core for R4 will be moved in packets using technologies such as voice over IP (VoIP) or voice over ATM. In R5 and R6 the UMTS network evolves to an all-IP architecture. In this, both the UTRAN and the core network can use IP transport. The core network at this stage will consist of only one packet switch domain and all services will be converged over a single network transport.

5.2   IP PROTOCOL SUITE OVERVIEW

The history of IP dates back to the 1970s, when its development was part of the US Advanced Research Projects Agency (ARPA). At that time the primary objective was the development of a network which was robust against a single point of failure. The original network interconnected the US military with several research centres throughout the US, and was known as ARPANET. In 1983, as the network development was coming more from the research community, ARPANET split into MILNET for the military, with the academic community retaining the rest of ARPANET. As the scope of ARPANET increased, it eventually became known as the Internet.

This led to the development of a technology called packet switching, in which data is broken up into variable size chunks called packets, marked with a header containing a source and destination address, then sent into the network for delivery (Figure 5.1). As the packet is sent between the internal network switches (called routers), each router makes a decision on where to send the packet next to get it to its final destination based on the destination address in the packet, the beauty of the scheme being that if a given router in the network fails, packets can be re-routed via another path, thus keeping the

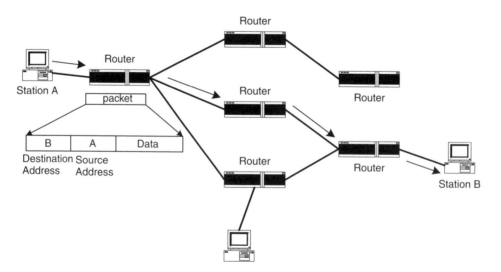

Figure 5.1   IP network operation

network in operation. Note in the context of IP, the term *gateway* is commonly used, and is synonymous with the term *router*.

IP is an open standard, and its standardization process is under the coordination of the Internet Engineering Task Force (IETF, www.ietf.org). The standards are published through a process of Internet drafts, Request for Comment (RFC) documents and standards (STD) documents.

5.2.1 IP protocol

The IP protocol suite was originally developed to support transport of data across the Internet. Each packet in an IP network is referred to as a datagram, with the primary purpose of the IP protocol being to route these datagrams across the network. Other functions provided include such services as fragmentation and reassembly of packets that are too large to be transported over a given network transport technology and the appropriate handling of traffic with different priorities within the network. Figure 5.2 shows a diagram of an IP header, which is added to each packet before entering the network.

The header starts with the version number of the protocol. Currently version 4 is used but the newer version, v6, is already being implemented by many manufacturers. The following description actually applies to IPv4 (IPv6 is covered later in the chapter). Following the version number is the Internet header length (IHL) field giving the length of the header in 32-bit words. The usual value for this field is 5, resulting in a minimum header length of 20 octets. The type of service field specifies reliability, precedence, delay and throughput parameters. Historically, many routers have ignored this field; however, it has been redefined as the differentiated services (DiffServ) codepoint and can be used to help provide differentiated QoS. The importance of this field will be seen when QoS within the IP network is covered later in the chapter.

The total length field indicates the total length of the datagram before fragmentation. The next line of the header consists of fields which handle fragmentation within the network. When fragmenting a packet, IP sets the identification field to the same value

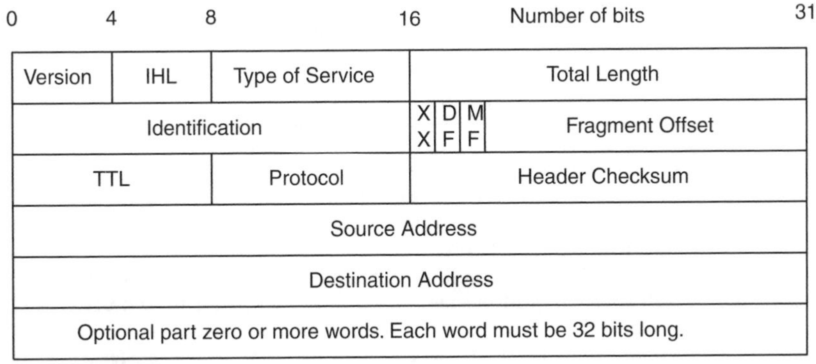

0	4	8		16	Number of bits	31
Version	IHL	Type of Service			Total Length	
Identification			X X	D F	M F	Fragment Offset
TTL		Protocol			Header Checksum	
Source Address						
Destination Address						
Optional part zero or more words. Each word must be 32 bits long.						

Figure 5.2 IP header

for each fragment of the original datagram. This allows the receiver to work out which fragment belongs to which datagram when reassembling.

The fragment offset, on the other hand, is used to indicate where an individual fragment belongs in the whole datagram. This field is used to help put the fragments back together in the correct order when they reach their final destination. As well as these two fields there are a couple of fragmentation flags. DF stands for 'don't fragment' and can be set by a host that wishes to send data without fragmentation. The MF (more fragment) flag is set to 1 if there are more fragments to come; a value of 0 indicates the end of a list of fragments.

If the router itself has to perform fragmentation this will introduce an extra overhead in terms of processing when forwarding the packet across the network. The use of fragmentation headers also leads to an increased number of headers, decreasing the bandwidth available for the payload. For these reasons fragmentation is only offered as an optional service for IPv6, with the fragmentation information being relegated to the extension headers. In this case each host is expected to send only datagrams no longer than the maximum transfer unit (MTU) of the path across the network from source to destination.

The time to live (TTL) field is used to prevent misrouted packets circulating forever on the network. Before forwarding a packet the router decrements this field by one; if the value in the TTL field is zero then the packet will be discarded. The protocol field indicates the higher layer protocol that is using IP to move its packets; in most cases this will be TCP or UDP (see Sections 5.2.4 and 5.2.5). The header checksum protects the header against corruption, and packets with an incorrect header checksum will be discarded.

The base header shown in Figure 5.2 can be followed by a number of extra optional headers. The presence of these extra headers is indicated by an IHL greater than 5. Table 5.1 describes some of the IP options available.

5.2.2   IP addressing and routing

IP addressing is based on the concept of hosts and networks. A host is anything on a network that is capable of transmitting and receiving IP packets, such as a workstation or a router. Hosts are connected together through one or more networks, therefore to send a datagram to a host both its network and host address must be known. IP addressing

Table 5.1   IP options

Name	Description
Record route	Each IP router that forwards a packet adds its address to packets marked with the record route option. This allows a packet's route to be traced across the IP network
MTU probe/reply	Used to discovers the MTU for a given path across the network
Internet timestamp	Allows router to timestamp packets as their route is recorded across the network
Source routing	Allows the source host to constrain the route of the packet across the network

combines the network and the host into one single data structure called an IP address. A mechanism called subnet masking is used to determine which part is the network address and which part is the host address. This subnet mask can vary from situation to situation.

An IP address consists of 32 bits. By convention, it is expressed as four decimal numbers separated by full stops, for example:

IP address 152.226.86.23

Subnet mask 255.255.240.0

Since there are four bytes, addresses range from 0.0.0.0 to 255.255.255.255, which gives a total of over 4 billion unique addresses. In fact, because of the way IP addresses are allocated in practice, fewer addresses than this are actually available.

There are various classes of address, as shown in Table 5.2. Note for each class there is a limit on the total number of network and host addresses available.

When trying to identify an address as class A, B, C, D or E, the simplest way is to look at the address ranges given in Table 5.2. Class A addresses start with 1–126, class B addresses start with 128–191, class C addresses with 192–224 and multicast addresses with 224–239.

For example, 152.226.0.0 is easily identified as a class B address and therefore supports addresses in the range 152.226.0.0 to 152.226.255.255.

Subnetting is the division of the single network with many hosts into smaller subnetworks with fewer hosts. We do this by using part of the 32-bit IP address to indicate the subnetwork and another part for the actual host address. For example, with class C we could have 6 subnets each with 30 hosts. Figure 5.3 shows how we can increase the number of subnetworks at the expense of the number of hosts on those subnetworks.

It is possible to increase the number of 1s to generate a subnetwork, as shown in Table 5.3 (a class B network is assumed in this example). This mechanism of using

Table 5.2 IP address classification

Class	Network address field size	Host address field size	Total network addresses	Total host addresses	Address range, dot decimal form
A	7	24	126	16 777 214	1.0.0.0–126.0.0.0
B	14	16	16 383	65 534	128.0.0.1–191.255.0.0
C	21	8	2 097 151	254	192.0.1.0–223.255.255.0
D (multicast)		28			224.0.0.0–239.255.255.255
E (reserved)					240.0.0.0–255.255.255.255

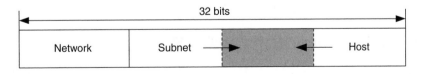

Figure 5.3 Subnet and host tradeoff

Table 5.3 Subnet mask tradeoff

Subnet mask	Bit pattern	Subnets	Hosts
255.255.0.0	11111111.11111111.00000000.00000000	0	65 534
255.255.192.0	11111111.11111111.11000000.00000000	2	16 382
255.255.224.0	11111111.11111111.11100000.00000000	6	8190
255.255.240.0	11111111.11111111.11110000.00000000	14	4094
255.255.248.0	11111111.11111111.11111000.00000000	30	2046
255.255.252.0	11111111.11111111.11111100.00000000	62	1022
255.255.254.0	11111111.11111111.11111110.00000000	126	510
255.255.192.0	11111111.11111111.11111111.00000000	254	254

different subnet masks to identify different subnetworks is referred to as variable-length subnet masking (VLSNM).

Since there are only 126 class A network addresses these were originally allocated on the Internet to only big organizations with very large internal networks. Class B addressing allocates 14 bits to the network portion and therefore provides more networks at a cost to host space (only 65 534 hosts per network). Class C networks have the smallest allocation of hosts per network at only 254 but have the largest number available. Due to many organizations having more than 254 hosts, demand for class B addresses has traditionally been high. This has led to a depletion of the class B network address space and problems finding suitable network address allocations for medium size organizations (more than 254 hosts). This problem and others will be finally solved by the introduction of IPv6, which uses 128 bits for each IP address and therefore provides an almost inexhaustible supply.

Class D addresses are for the support of multicasting. Class E addresses are reserved, and thus far their use has not been standardized.

There are some special addresses that cannot be used. To take these exceptions into account, the general formulas to compute the number of possible subnets and hosts using a given subnet mask are:

$$\text{Possible subnets} = 2^{(\text{number of masked bits})} - 2$$

$$\text{Possible hosts} = 2^{(\text{number of unmasked bits})} - 2$$

Consider a class B network using the subnet mask of 255.255.240.0. This has 4 additional masked bits representing the value 240 and therefore the number of subnets = $2^4 - 2 = 14$ and the number of hosts is = $2^{12} - 2 = 4094$.

Table 5.4 lists the addresses that have special significance within the IP protocol and should not be used when assigning addresses.

In addition, the IETF has defined the addresses shown in Table 5.5 for use internally in private networks, and these are invalid on the Internet. An application of these addresses is described in Section 5.6.6.

5.2.3 Address depletion and CIDR

Since IPv4 addresses are 32 bits long, this allows a theoretical maximum of 4 294 967 296 hosts to be addressed, as has been noted. The class structure introduced in the previous

Table 5.4   IP addresses with special significance

Address	Function
0.0.0.0	Refers to the default network
127.0.0.0	Reserved for loopback. Usually 127.0.0.1 is used to refer to the local host. This address is used for testing purposes
An address with all network bits = 0	Refers to 'this' network
An address with all host bits = 0	Refers to the network itself, e.g. the address 135.34.0.0 refers to network 135.34. This notation is used in routing tables
A network or host address of all 1s	Refers to 'all' hosts
255.255.255.255 (all 1s)	Broadcast address

Table 5.5   Reusable private addresses

Class A	10.0.0.0
Class B	172.16.0.0–172.31.0.0
Class C	192.168.0.0–192.168.255.0

section reduces this host count to actually about 3.7 billion. In fact the address space is depleted still further by the wasting of unused address space within the class B space.

For most organizations, a class A network with 16 million addresses is too big but conversely a class C network with 254 hosts is too small.[1] Therefore most organizations would like a class B network address assigned. The implication of this is that an organization with 2000 hosts will be assigned a class B address with its 65 534 addresses, even though this is far more than it requires. It will in fact have wasted 97% of its allocated address space. The other problem is that there are only 16 383 class B network address available. If every operator that asks for one is assigned one they will soon run out. Another possibility is to assign eight class C addresses to the operator. This helps in the address allocation problem but causes a problem on the Internet in terms of an explosion in the size of routing tables. Now this organization will require eight entries in each routing table instead of the one required for the class B allocation.

To allow multiple IP addresses (called address blocks) to be assigned without causing the increase in routing table sizes, a new scheme called classless inter-domain routing (CIDR) was devised (RFC 1519). With CIDR the network address is expressed as follows:

IP address/mask size

The IP address is expressed in the conventional dot decimal notation, for example 175.6.0.0. The mask size that follows represents the number of bits in the network portion of the address, i.e. the network part of the address that is to be used for routing. This allows numbers of class C addresses to be blocked together for routing purposes. For example, if an organization requires 1000 addresses, then it will be allocated four blocks of class C (4×254) = 1016 addresses. This is known as *supernetting*. For CIDR

---

[1]Known as the 'Goldilocks' problem.

to work, these blocks must be contiguous. The operator can then be assigned the following four class C addresses:

196.220.0.0, 196.220.1.0, 196.220.2.0 and 196.220.3.0

These can be written as 196.220.0.0/22, which will allow any packet on the Internet with the same first 22 bits to be routed to this network. This will provide the operator with 1016 host addresses but only add one entry to the routing tables within the Internet (instead of four).

As a final note CIDR is merely intended as a stopgap until the widespread introduction of IPv6 since the address size will be expanded to 128 bits, providing a vast address space.

5.2.4 Transmission control protocol (TCP)

The transmission control protocol (TCP) is designed to provide a reliable connection between communicating hosts. This is a connection-oriented service but the underlying internetworking protocol is the unreliable IP datagram service. To the hosts it appears that they have a dedicated circuit whereas in reality it is based on a datagram service. TCP not only guarantees the delivery of packets across the network, it also puts the stream of packets back in the correct order at the receiving host. Each packet is checked for errors by the application of a checksum that covers both the TCP header (Figure 5.4) and the data payload. To guarantee the delivery of data TCP uses a technique called automatic repeat request (ARQ). When packets are received at the destination an acknowledgement is sent back to the source. If the acknowledgement is not received within a certain time (retransmit timeout) the packet is sent again. To make sure that repeated packets are not mistaken for new transmissions each new packet (not repeated) is marked with a unique

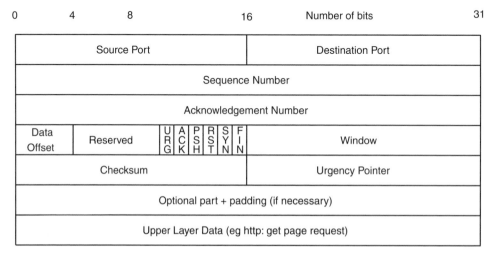

Figure 5.4 TCP header

32-bit number called its sequence number. This number is also used to make sure that data is received in the correct order. Also note that in the TCP header (see Figure 5.4) there is a 32-bit acknowledgement number. This is used to match up each acknowledgement with a block of transmitted data marked with a particular sequence number and length.

Apart from guaranteeing the sequenced flow of data from sender to receiver, TCP also handles issues such as flow and congestion control. Flow control is used to allow a receiver that is becoming overloaded with data to slow down a fast sender. The receiver sets a window size that determines the maximum amount of data the sender can send. By setting a window size of zero the receiver can stop the flow of data entirely. Congestion control protects the network itself from getting overloaded. With TCP it is assumed that packet loss is due to congestion on the network. This is because a highly congested router will lose packets due to buffer overflow. Every time TCP has to retransmit a packet it slows down its transmission, assuming that the lost packet was due to congestion on the network. Note that this technique does not make much sense for TCP running over a wireless link since packet loss is likely due to interference on the link rather than congestion.

TCP is a connection-oriented protocol. Before sending data a connection must be made with the remote host. The connection set up consists of three messages and is therefore called a three-way handshake. This process is shown in Figure 5.5. The handshake serves two basic purposes: first, ensuring that each end is ready to receive data and, second, allowing each party to agree on the *initial sequence numbers* each end will expect to receive.

Once the connection has been made data can be transferred in both directions between the two hosts. When the hosts have finished data transfer they terminate the connection. In general terms clients initiate TCP connections and servers wait (or listen) for incoming connections. This relationship is of particular significance when considering the use of port numbers.

The TCP header starts with two 16-bit fields called port numbers, which are used to route data to a particular application or service (Figure 5.6). This technique is called *port multiplexing*. The port numbers are needed because there can be many different software applications using the IP connection to the network running on a particular machine (for example email, web browsing, file transfer). To handle these different streams of data each application is allocated a particular port number that is used to route the packets within the client or server device. Port numbers are allocated differently at the client and server end of the connection. At the server end each service is defined by a different well-known

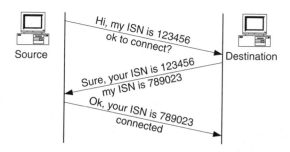

Figure 5.5 TCP three-way handshake

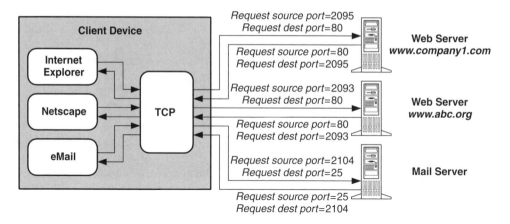

Figure 5.6 TCP port multiplexing

port number (for example decimal 80 for HTTP web access). When the client opens a connection with the server the TCP software in the client will allocate a local port number from its free list. It then sends a TCP connection request with a source port set to the locally allocated port and the destination port set to the well-known port number. When the server receives the connection request it sends back a connection confirm with the source and destination ports reversed. This routes the confirm back to the originating application. When the confirmation is received at the client it is acknowledged, completing the three-way handshake. Figure 5.6 shows a client running three applications, two browsers (Netscape and Internet Explorer) connected to different websites and an email client.

In summary, TCP provides a reliable connection stream between two end points on the network. It controls the flow of data to avoid receiver overrun (flow control) and also protect against overloading the network (congestion control). TCP together with IP provides a basic end-to-end reliable data transfer and for this reason the IP network service is commonly referred to by the conjugation of these protocol acronyms, i.e. TCP/IP.

5.2.5 User datagram protocol (UDP)

TCP provides a very good service where reliability is paramount but is not generally suitable when delay is critical. The problem occurs if a packet is lost and TCP retransmits data: the retransmission takes time and this introduces delay into the overall transmission. There are applications such as voice and video transfer where the occasional loss of data is much less important than its timely delivery. For example, with voice if a single voice sample is lost this may not be perceivable to the listener (or at worst be heard as a slight click) whereas increased delay can result in the voice call being unusable.

For these reasons another protocol called UDP was developed (see Figure 5.7), which provides the port numbering facility and error checking facilities of TCP with none of the other services. UDP is often the preferred choice for delay-sensitive traffic since packets that are lost are not retransmitted.

0   4   8   16   Number of bits   31

Source Port	Destination Port
Length	Checksum
Upper Layer Data	

Figure 5.7   UDP header

Another difference between TCP and UDP is the possibility of using multicast addressing. UDP supports multicast; TCP does not, since with TCP the transmitter cannot be sure of who is receiving packets within the multicast group, and therefore cannot run ARQ not knowing if and when to retransmit.

5.2.6   Domain name service (DNS)

This protocol provides a mapping service for the IP protocol stack. DNS maps text names to IP addresses and vice versa. It removes the requirement for a user to remember an IP address to access a particular service by providing a lookup between a text name and the IP address. This echoes the situation in a cellular network, where users can now address people by their name, and the phone book stored in the device provides the mapping to mobile phone number. For example, with DNS, a company's website can normally be found by placing 'www' and 'com' around the company name. A user merely types www.companyname.com into the browser. This address is then passed to a DNS server, which in turn is responsible for translating it to an IP address.

Domain names themselves are arranged and managed as a hierarchical tree (Figure 5.8 shows a small subset of the DNS naming space). Moving from the right-hand side of the domain name to the left moves one further down the tree. The namespace for DNS is partitioned and administered according to this tree. Looking at Figure 5.8, one can see the domain cam.ac.uk. This is the domain for Cambridge University and is therefore administered by Cambridge University's network administrator. This domain is contained within ac.uk, the domain allocated for all academic institutions in the UK. The domain ac.uk in turn is contained with the UK domain, which is the domain allocated for all DNS names within the UK. This partitioning of the address space makes it possible for any given operator to allocate new DNS names without worrying about name collisions (e.g. two organizations having the same name for their web server) since their namespaces are separated by the postfix part of their DNS addresses. For example, Cambridge University can be assured their address will always be distinct from Oxford's because their domain names will be postfixed cam.ac.uk and ox.ac.uk, respectively.

When resolving DNS names into IP address the client machine must refer to a DNS server. Most clients have the address of the default DNS server allocated on configuration. If the local DNS server does not have the mapping for the DNS address then it will use the DNS name to work out where to look for the mapping. For example, when resolving the DNS address www.3com.com, a local DNS server will first interrogate the Internet

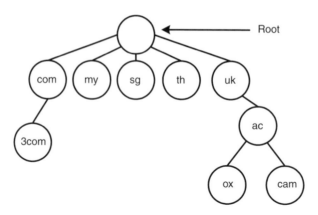

Figure 5.8   DNS namespace hierarchy

DNS root servers for the address of the name server which handles the **.com** domain. When this address has been returned then the **.com** name server is asked for the address of the 3com name server. Once the 3com name server is found then it can be asked for the address of www.3com.com. This process of working down the tree from the root through the whole of the DNS name can either be done by the client itself, interrogating each name server in turn, or done on its behalf by another DNS server. When an external name server does all the work for the client as a proxy this is called a *recursive DNS lookup*. This is advantageous for the client since it does not have to do so much work and reduces DNS traffic on the local loop. This is particularly important for clients connected over a wireless link. Note that each name server knows the address of its children but not necessarily its children's children. Also each name server must know the address of its parent so that it can forward unknown requests up the DNS hierarchy.

As was discussed in Chapter 4, DNS is used in GPRS for resolution of access point names (APN), which define the interface on the GGSN where the user will connect to an external network. The packet data protocol (PDP) context activation will inform the network of the required APN, and this will be resolved to the IP address of the correct GGSN.

5.2.7   Address resolution protocol (ARP)

This protocol is responsible for mapping from IP addresses to media access control (MAC) addresses. It is important in IP since frames are actually routed to destinations on a local area network (LAN) using MAC and not IP addresses. A MAC address is a unique 48-bit (6-byte) address, contained in the hardware of a network interface, so no two interface cards can have the same MAC address. The first three bytes of the address indicate the manufacturer and the remaining three bytes are allocated by that manufacturer as each new card is produced. For example, when sending a packet on an Ethernet network both the Ethernet MAC address and the IP address of the destination must be known beforehand. The operation of the protocol is as follows. An ARP request packet is sent out on the LAN with the destination MAC address set to *broadcast*. It also contains the IP address of the required mapping. The MAC broadcast address consists of the destination address being set to all 1s, i.e. FFFFFFFFFFFF. All stations on the local network will receive and analyse the frame but only one will recognize its own IP address in the request and reply to the sender. In the reply, the recipient host inserts its own MAC address. Now that the sender has a copy of the hardware address (MAC address) it can send packets to that IP address directly. ARP actually works in conjunction with a local cache which stores copies of the recent ARP mappings. This cache reduces network traffic and increases performance but is cleared out on a regular basis to allow for the fact that the LAN hardware may have changed (or even the IP address might have been reallocated).

Figure 5.9 shows an Ethernet network connected to the Internet. If a station with IP address 192.10.1.100 needs to send a packet to 192.10.1.101, the packets will be sent as shown in Table 5.6.

If station 192.10.1.100 is required to send a packet to address 165.10.10.1, this address is not on the same subnet. The packet will have to be forwarded via the router at 192.10.1.254, therefore the ARP request is for the Ethernet address of the router. The messages will be as shown in Table 5.7.

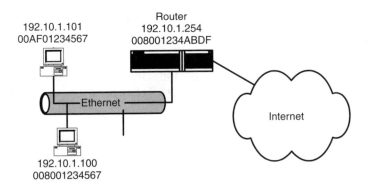

Figure 5.9   ARP example

Table 5.6   ARP example (1)

Packet type	Destination IP address	Destination MAC address	Source IP address	Source MAC address
ARP request	192.10.1.101	FFFFFFFFFFFF	192.10.1.100	008001234567
ARP response	192.10.1.100	008001234567	192.10.1.101	00AF01234567

Table 5.7   ARP example (2)

Packet type	Destination IP address	Destination MAC address	Source IP address	Source MAC address
ARP request	192.10.1.254	FFFFFFFFFFFF	192.10.1.100	008001234567
ARP response	192.10.1.100	008001234567	192.10.1.254	008001234ABD

The station will now forward the packet to Ethernet address 008001234ABD. Note that when 192.10.1.100 forwards the packet to the destination, although it forwards to the Ethernet address of the router, the IP address will be set to 165.10.10.1 (the final destination) and not the IP address of the router.

5.2.8 IP summary

The IP protocol suite provides a range of data transport services (reliable or unreliable expedited), it is robust against a single point of failure, independent of network hardware and capable of some degree of automated network management. It was designed originally with robustness in mind and not quality of service (which is needed for voice and video). However, IP QoS protocols have since been developed and a number of these will be discussed in some depth later in this chapter.

5.3 IP ROUTING

When forwarding packets the routers first look at the network address of the destination address. If the network address of the destination is directly attached to the router then the packet is forwarded straight to the host via the attached network.

If the network address refers to a network not directly attached then the packet is forwarded to another router which is closer to the destination network.

This process is illustrated in Figure 5.10. A packet to be sent from host 4.0.0.195 (i.e. network ID = 4, host ID = 0.0.195) to host 7.0.0.234, is sent first over network 4 to router A. Since router A is not directly attached to the destination network (network ID 7) it forwards the packet to router B. Router B in turn forwards the packet to router C, which delivers it to the destination. Each router has a table (the routing table) telling it the next hop where to forward a packet to deliver it to a given destination. If the

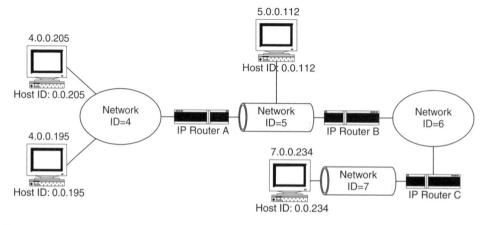

Figure 5.10 IP routing

network address of the packet is not present in the routing table then most routers will have a routing entry called the *default* which is used as a forward for the packet in this circumstance. For instance, router A in Figure 5.10 could have a default entry of router B. Many Intranet[2] routers will have a default entry pointing to their external router (connected to the Internet), assuming that traffic that cannot be routed internally must be bound for the Internet.

Routing tables can be updated using two basic techniques, static and dynamic routing. With static routing the entries are configured manually into the router. This is only suitable for simple network configurations where the routing tables are small and easy to configure. With dynamic routing the routers will 'talk' to each other, exchanging information about the topology of the network. Using this topology and a technique called a *routing algorithm* the routers calculate the optimum path for packets to flow to their destinations. The combination of a routing algorithm plus the protocol to exchange routing information is called a *routing protocol*. Routing protocols are classified into two groups: internal and external. Internal routing protocols such as routing information protocol (RIP) and open shortest path first (OSPF) carry out routing within a network managed by a single operator. In IP terminology, this is referred to as an *autonomous system* (AS). External routing protocols such as border gateway protocol (BGP) handle routing between autonomous systems and are used, for example, to connect the gateways belonging to different Internet service providers (ISPs) on the Internet. This is illustrated in Figure 5.11.

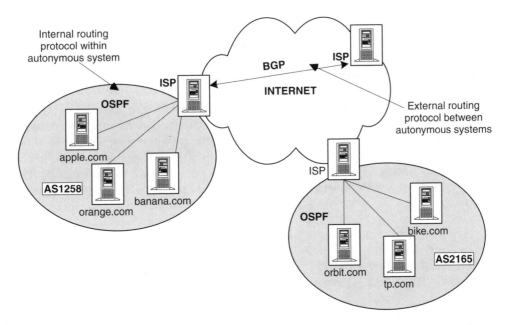

Figure 5.11   Internal and external routing protocols

---

[2]The term 'Intranet' is used to denote a private network which is based on the IP protocol. Intranets are connected to the Internet via an external router. Normally, the private network will use some security mechanisms to protect the internal network, such as the implementation of a firewall.

Routers and hosts forwarding packets using the IP protocol do not make any guarantees about the delivery of datagrams. If the underlying physical transmission channel does not deliver a packet then IP will not compensate for this. Also IP does not guarantee the order in which packets will be delivered: since datagrams between two hosts can take different paths through the network, they can be delivered in variable order dependent on the time of transmission through each path. These services, if required, are provided by layers above the IP protocol.

5.3.1   Dynamic routing algorithms

Dynamic routers maintain and update the routing table dynamically. To keep an up-to-date picture of the network, the routers periodically exchange information with each other. Each of the routes will have *cost*[3] *information* associated with it. This is the *metric* by which one route will be chosen over another. A metric is a standard measurement, such as path length, to determine the optimal path to a destination. This metric can be very simple such as 'the best route is the one with the least number of routers to cross'. It could, however, be more complex, taking into consideration a number of factors such as bandwidth, latency and reliability. The two most common types of protocol using routing algorithms to do this are:

- distance vector routing protocol
- link state routing protocol.

5.3.2   Distance vector routing protocol

Distance vector routers compile and send network route tables to other routers which are attached to the same segment. Each router builds up its own table by broadcasting its entire table and receiving broadcast tables from other neighbouring routers, to allow it to maintain its own table. Broadcasts can occur as often as every 30 seconds, which can cause congestion on the network. Cost information is relative to other routers. For example, if router B tells router A that it can reach network 22 in 3 hops, then router A will assume that it (i.e. router A) will be able to reach network 22 in 4 hops. An example showing how the hop count is incremented is shown in Figure 5.12. The information gathered by a router will be broadcast to neighbouring routers.

The process of updating all of the routing tables so that they all have the same information is called *convergence*. The most common implementation of the distance vector algorithm is RIP, which is described below.

5.3.2.1   *Routing information protocol (RIP)*

RIP has been a popular routing protocol for a number of years and it is used in IP networks and also by Novell LANs. However, note that the implementations for the two

---

[3]This is not a monetary cost, but an indication of the speed, bandwidth or latency, etc. of the link.

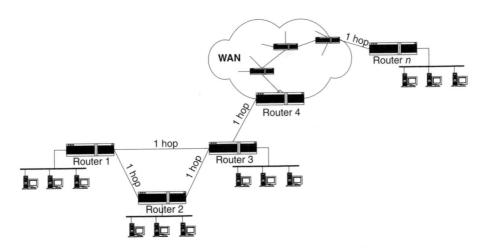

Figure 5.12 Example of distance vector routing

are different (Novell's implementation also takes into consideration the time of a route in 'ticks'). The RIP protocol has a hop field limit of 15 hops, which means that there can be a maximum of 15 hops (i.e. 15 routers) from source to destination. If a computer cannot be reached within this limit then a message is usually sent back to the source, saying 'network unreachable'. RIP is defined under the IETF's RFCs 1508 and 1388. It was introduced in 1988 and was the original routing protocol of the Internet.

The following factors are taken into account by a router when it is selecting the optimum path for a packet through a network:

- The number of hops. A *hop* is a jump across a router, a *metric* count defines the number of hops for a given route.
- For Novell, the time in 'ticks' (approximately 1/18 second) each route takes.
- The cost of the path.

To understand how RIP works, refer to Figure 5.13 as an example. Suppose that routers A, C and D have been exchanging routing information for some time. Router A's routing tables will appear as illustrated in Figure 5.14. The 'next hop' value of 0.0.0.0 indicates the default network.

Router B is now switched on and broadcasts its availability. Router A will update its routing table to include B as one hop away. Router A will then broadcast to routers C and D that it can reach router B with one hop. Both routers C and D will update their routing tables to include the new information that they can reach router B in two hops (via router A). Eventually the routing tables will appear as in Figure 5.15.

Now consider that router C becomes faulty. Both routers A and D will notice that router C has not broadcast its routing table. All routers should broadcast their routing tables to their neighbours every 30 seconds. After waiting for a period of 6 times the broadcast interval (i.e. $6 \times 30 = 180$ seconds) the routers will delete router C's information from their routing tables. Router C and its associated network will now be regarded as unreachable via routers A and D.

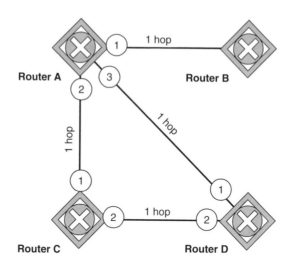

Figure 5.13   Routing information protocol

Destination	Next Hop	Metric	Interface
Router C	0.0.0.0	1	2
Router D	0.0.0.0	1	3

Figure 5.14   RIP router A routing table

Router A

Destination	Next Hop	Metric	Interface
Router C	0.0.0.0	1	2
Router D	0.0.0.0	1	3
Router B	0.0.0.0	1	1

Router B

Destination	Next Hop	Metric	Interface
Router A	0.0.0.0	1	1
Router D	Router A	2	1
Router C	Router A	2	1

Router C

Destination	Next Hop	Metric	Interface
Router A	0.0.0.0	1	1
Router D	0.0.0.0	1	2
Router B	Router A	2	1

Router D

Destination	Next Hop	Metric	Interface
Router A	0.0.0.0	1	1
Router C	0.0.0.0	1	2
Router B	Router A	2	1

Figure 5.15   RIP routing tables

In a second example, suppose that the link between C and D is faulty. Both C and D will advertise the link loss and each router will update its tables with the latest information. In this case C would advertise that it can reach D via A in two hops and vice versa.

Count to infinity

There is a problem with RIP known as *count to infinity*. This can be explained through the first example above, where router C is faulty. Routers A and D have updated their routing tables; router A now tells router B that router C is unreachable. Router B will look at its own routing table and see that it can reach router C in two hops. Router B broadcasts this information to router A. Router A now believes that there is an alternative path to router

C via router B. Router A broadcasts this information to both routers B and D. They both update their routing tables, adding 1 to the hop count for router C, believing that they can reach router C through A. This continues with each router believing that it can reach router C via another router. Eventually the maximum hop count of 15 is reached, where router C is considered unreachable. However, this is only after a considerable period of time.

To overcome this problem of slow convergence, a solution called *split horizon* was proposed. Split horizon prevents routers from sending routing information back along paths where they first received that information. For example, in Figure 5.13, router B learns of router C's existence via router A. Using the split horizon technique, router B will not be allowed to broadcast information on its path connected to router A about router C. An example of this is illustrated in Figure 5.16. Router 2 has informed router 3 that it can reach router 1 in one hop. Router 3 informs the other routers that it can reach router 1 in two hops but does not inform router 2 of this.

Another method to overcome the slow convergence problem is called *poison reverse*. In this method routers send routing information back along paths where they first received that information, but will always send a metric count (number of hops) of 16, which means that the destination is unreachable. This essentially performs the same function, but is simpler to implement.

An example of this is shown in Figure 5.17. Router 3 learns of router 1's existence via router 2. Using the poison reverse technique, router 3 will broadcast on its path to router 2 that it will take 16 hops to reach router 1, i.e. router 1 is unreachable via router 3.

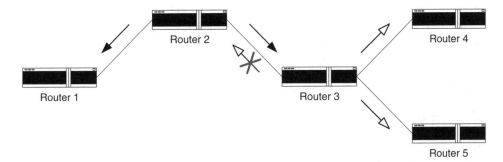

Figure 5.16 Example of split horizon

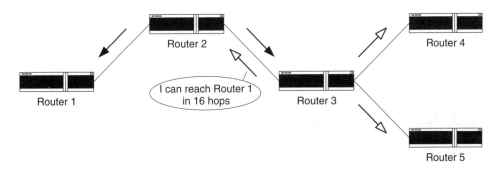

Figure 5.17 Example of poison reverse

There are several problems associated with RIP. One is that by allowing the routers to talk to each other very frequently (i.e. to update tables) a lot of traffic is generated on the network and this overhead can reduce performance. Another problem is *slow convergence*, which is caused by routers not being kept up to date by distant routers because of delays within these distant routers while they recalculate their own routing tables. For example, using RIP in an IP environment where the update interval is 30 seconds, it will take around 7 minutes to configure a large network (30 seconds × 15 hops). In a Novell environment where the default update is 60 seconds it will take even longer. The 30-second interval, or 60 in the case of Novell, can be reduced (or increased) by the network administrator but more RIP traffic will be sent, causing more congestion on the network. Link state protocols such as OSPF and NetWare Link State Protocol (NLSP) can overcome the limitations of distance vector routing and are becoming more widely implemented.

5.3.2.2 Triggered updates

An extension to RIP is the triggered update. The triggered update only sends information about the routing of a network when something substantial happens rather than broadcasting every 30 seconds or so. This can be used to reduce convergence time on a LAN. However, it is particularly suitable for public data networks, where the corporation may have to pay for packets sent over the wide area network (WAN). Note also that as WAN links are usually much slower than the LANs they are connecting, sending periodic RIP updates can decrease the bandwidth available for data traffic. To ensure the link is still up, RIP packets across the WAN are repeated until acknowledged, whereas on the LAN interfaces, RIP is not acknowledged.

5.3.2.3 Routing information protocol version 2 (RIP2)

RIP version 2 (RIP2) is defined in RFC 2453. It has been introduced to address some of the shortcomings of RIP. RIP version 1 does not consider interior gateway protocol/exterior gateway protocol (IGP/EGP) interaction and variable length subnetting. This is because RIP predates the implementation of these protocols. This means that RIP cannot process addresses other than based on the classful address, i.e. class A, B or C. This reduces its effectiveness on modern networks containing subnets. RIP2 understands variable length subnet masks and will work with CIDR networks. RIP is also limited to 15 hops and does not provide any authentication.

Figure 5.18 shows an RIP message advertising the three links available on a router, with the metric for each.

5.3.3 Link state protocols

Link state protocols are more complicated than distance vector algorithms but have many significant advantages. Many organizations on the Internet are now using an implementation of a link state protocol known as OSPF, in preference to RIP.

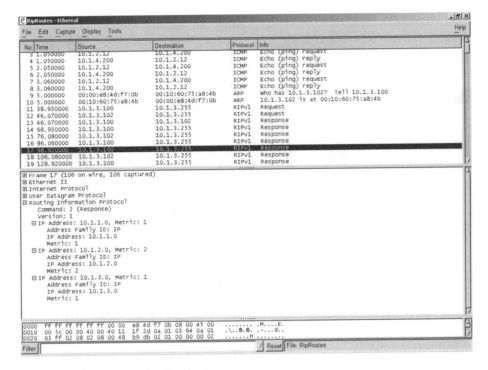

Figure 5.18 RIP router advertisement

Assume a new router is attached to the network. It will send out a broadcast *hello* packet to the other routers. An existing router will discover its new neighbour and will learn its network address. This router will send an *echo* packet back to the new router and the new router will reply immediately to this packet. It is now possible to calculate the delay in reaching the new router. Knowing the delay is important because on a complicated internetwork there may be different paths through numerous routers to a destination, and the quickest route is usually sought (there are other factors which may result in a preferred route such as bandwidth and cost).

Once information about this new router has been established a link state packet is constructed and flooded to all other routers on the network (not just immediate neighbours as in RIP). This is shown in Figure 5.19. All routers on the network will now have a full picture of the internetwork in their link state protocol table rather than just a table of directly connected routers as is the case with RIP.

The above can be broken down into five simple steps:

1. Discover neighbours and learn their network addresses.

2. Measure the delay or cost to reach each neighbour.

3. Build a packet with the information it has received.

4. Send this packet to all routers (not just neighbours as in RIP).

5. Compute the shortest path to every other router.

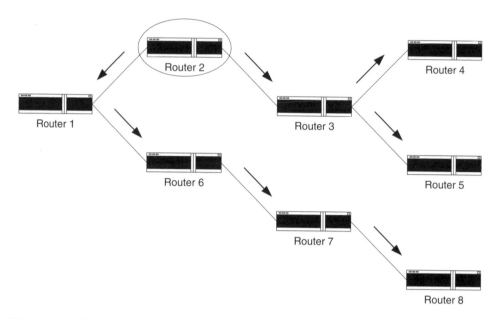

Figure 5.19   Example of link state protocol flooding

5.3.3.1   *Open shortest path first (OSPF)*

OSPF is a commonly used link state routing protocol on larger IP networks. It is a complicated routing system and only a simplified view is presented here. For further information on OSPF the reader is referred to RFCs 1245, 1246 and 1247. The OSPF *hello* packet is responsible for establishing and maintaining a relationship between neighbours. Figure 5.20 shows how different paths through an internetwork have different costs associated with them. Unlike the RIP routing protocol, where each path has a metric of 1,

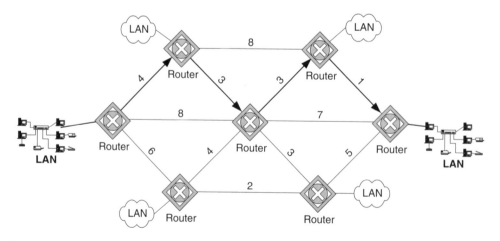

Figure 5.20   OSPF routing metrics

OSPF has many factors which can influence the cost of a route, e.g. bandwidth, reliability and latency. As such, the 'cost' for each route is worked out individually by combining the above factors.

The name, OSPF, comes from the fact that it uses an implementation of Djikstra's 'shortest path' algorithm for quickly calculating the best route based on cost factors. This calculation is not done each time a packet is routed, but rather a table of best routes is generated as routes are added or removed. One problem that has been experienced with OSPF is that it will always choose the best route, so 'popular' routes become heavily loaded, with no option to choose non-ideal routes for load balancing. One could consider this analogous to a highway to the seaside becoming congested on a public holiday, whereas a back road, although longer and normally slower, would work out quicker in these circumstances.

The *hello* packet:

- announces the router's availability, including its address and subnet mask, to the other routers on the network;
- is used to determine the router's neighbours;
- establishes the interval at which each router will send *hello* packets. This is used to determine when a neighbouring router is down;
- identifies the designated router (DR, see below) and backup DR (BDR).

Each router will periodically *multicast hello* packets to its neighbouring routers to inform them that it is still functioning. These are small packets which do not cause too much congestion on the network. Figure 5.21 shows the structure of the OSPF header. To indicate a *hello* packet, the packet type field will contain 1.

Unlike RIP, where there is one type of packet, OSPF has five different types (Table 5.8). The other fields in the OSPF header are listed in Table 5.9

Each router constructs its own database representing a map of the whole internetwork from information it receives from other routers via these packets. When a router detects that one of its interfaces has changed, it will propagate this update information to all other

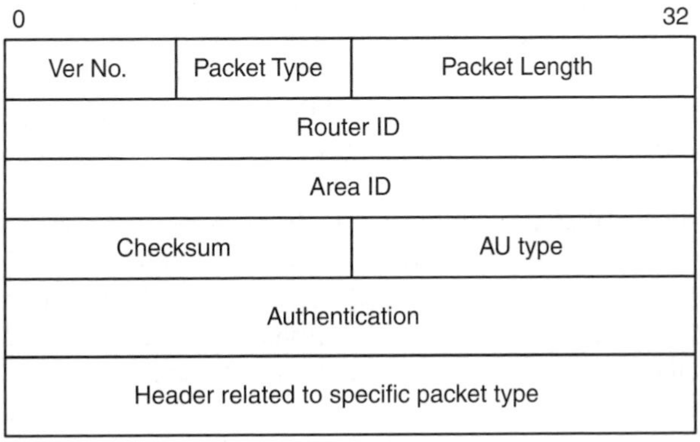

Figure 5.21 OSPF header structure

Table 5.8   OSPF packet types

Packet type	Description
1 Hello	See text
2 Database description	Sends sequence number of data, to show if it is up to date
3 Link state request	Requests information from another router
4 Link state update	Sends a router's costs to its neighbour
5 Link state acknowledge	Acknowledges link state update

Table 5.9   OSPF header fields

Field	Description
Version number	The version number of the OSPF protocol used by this router, currently version 1
Packet length	The length of the protocol packet in bytes, including the header
Router ID	The address of the source router
Area ID	All OSPF packets are associated with an area that the packet belongs to
Checksum	The standard IP checksum of the entire contents of the packet including the header but excluding the 64-bit authentication field
AU type	Identifies the authentication scheme used in this packet
Authentication	A 64-bit field to ensure that the packet is genuine

routers through a process known as *flooding*. Flooding is achieved through a *designated router* (DR) and allows other routers to update their databases.

Designated router (DR)

It is inefficient to have every router on the network send large link state packets to every other router on the network as this causes congestion. To avoid this situation, one router is elected as the *designated router*. One of the tasks of the *hello* packets is to inform routers which one is the DR. The DR is the router that has the highest priority number in its priority field in the *hello* header. This is a parameter set by the network administrator to elect the DR. In addition to the DR, a backup DR (BDR) is also elected in case of failure of the DR, which is the router with the next highest priority. Both of these routers (DR and BDR) are said to be *adjacent* to the other routers, and exchange link state information with them.

Neighbouring routers that are not adjacent only exchange the small *hello* packets with each other, so that they know the router is not faulty. When a route changes, the router affected will inform the DR and the DR will broadcast this information to the other routers. This is illustrated in Figure 5.22, where R22 sends the initial packet (1) to the DR. The DR checks the authentication, checksum and sequence number before multicasting it (2) to the other routers in the area.

During normal operation, each router periodically floods *link state update* packets to its adjacent routers. The default period for this is 30 minutes or when a route comes up

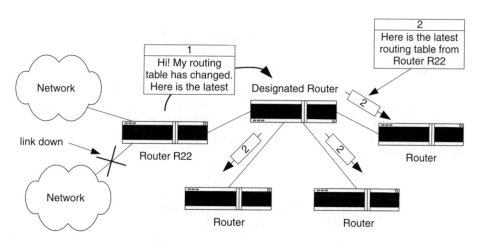

Figure 5.22 DR updating routers

or goes down, or its cost changes. These packets contain its state and the costs used in its database. The flooded messages are acknowledged, to make them reliable (*link state acknowledge packet*). Each packet has a sequence number, so that a router can check whether an incoming *link state update* is older or newer than the one that it currently has.

Database description packets provide the sequence numbers of all the link state entries currently held by the sending router. By comparing its own values with those of the sender, a receiving router can determine who has the most recent values. As well as a sequence number, entries also have a time associated with them. The database has an ageing timer, which keeps track of all entries. When the ageing timer reaches a value four times the *hello* interval (usually 4×10 seconds), entries are discarded. This is known as the *router dead interval*.

Routers can request link state information from one another using *link state request* packets. Using this method, *adjacent* routers can check to see who has the most recent data. The latest information can be spread throughout the network in this way.

By using the information received through flooding, each router can construct a graph for its network and compute the shortest paths. In this way they can maintain synchronization with the DR.

OSPF terminology

The following are some of the common terms relating to OSPF:

- autonomous system
- autonomous system border routers
- areas
- backbone.

Autonomous system

An autonomous system (AS) is a group of routers that exchange routing information within a single administrative unit. On the Internet, they link to other ASs through *autonomous system border routers* (ASBRs) to allow data to be transferred from one network to another. (ASs are described in greater detail in Section 5.3.5.1.)

Autonomous system border routers

An autonomous system border router (ASBR) is a router that exchanges routing information with routers from other autonomous systems. ASBRs communicate routing information with each other through an EGP. ASBRs must be able to translate between the IGP, e.g. OSPF, and the EGP, e.g. border gateway protocol 4 (BGP4). Figure 5.23 shows how an ASBR is used to connect to other autonomous systems.

Area

For small or medium-sized networks, distributing data through the internetwork and maintaining topological databases at each router is not a problem. However, in larger networks that include hundreds of routers, maintaining the topological database can require several megabytes of RAM for routing information and heavily utilize the CPU.

For this reason, large networks are often logically divided into smaller networks called *areas*. An area usually corresponds to an administrative domain such as a department, a building or a geographic site. In this way much of the routing information can remain hidden, thus reducing the burden on the routers. This is shown in Figure 5.24.

Backbone

A backbone is a logical area to which all other areas of the network are connected. This special area must be *directly* connected to all other areas of the network (either physically or virtually).

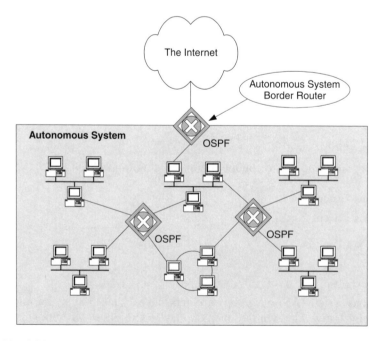

Figure 5.23 OSPF border router

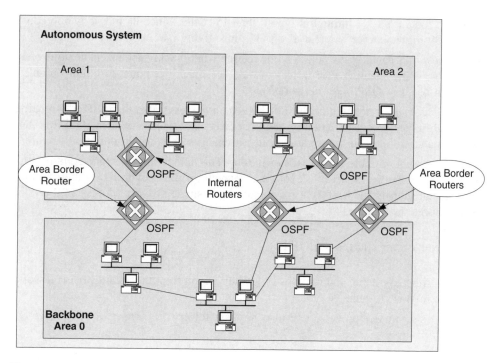

Figure 5.24   OSPF autonomous system connectivity

Routers that attach an area to the backbone are called *area border routers* (ABRs). Routers within a particular area receive routing information about the rest of the network from the ABRs. In this way the internal routers can have their workload reduced. The ABRs are also responsible for exchanging information about the area with the backbone.

Non-backbone areas can be classified as either of the following:

- *Transit area*: an area containing more than one ABR (e.g. area 2 in Figure 5.24).
- *Stub area*: an area where there is only one ABR (e.g. area 1 in Figure 7.27): all routes to destinations outside the area must pass through this router.

The backbone routers accept information from the area border routers to compute the best route from each backbone router to every other router. This information is transmitted back to the area border routers, which advertise it within their areas. Using this information, a router can select the best route to the backbone for an inter-area packet. Figure 5.24 shows how the different areas connect together.

5.3.3.2   A comparison of OSPF and RIP

Although it is much more complex, OSPF is considered superior to the RIP routing protocol for larger networks for the following reasons:

- RIP can route a packet through no more than 15 routes, since its metric is hop count. An OSPF metric can be as great as 65 535, thus giving more diversity for routing.

- OSPF networks can detect changes in the internetwork quickly and calculate new routes faster than RIP, resulting in a faster convergence time. The count-to-infinity problem does not occur in OSPF internetworks.

- OSPF reduces the amount of congestion by generating less traffic than RIP. RIP requires that each router broadcast its entire database every 30 seconds whereas OSPF routers only broadcast link state information when it changes or every 30 minutes.

- OSPF supports variable-length subnet masks. This allows the network administrator to assign a different subnet mask for each segment of the network. This increases the numbers of subnets and hosts that are possible for a single network address. RIP version 2 introduces this feature.

RIP does have advantages over OSPF:

- RIP is a simple protocol and requires little intervention from the administrator; it works well in small environments.

- Since it is a simpler process, RIP also requires fewer CPU cycles and less memory than OSPF.

5.3.4   Other routing protocols

There are a number of other routing protocols which are widely used. These are described briefly here.

5.3.4.1   Interior gateway routing protocol (IGRP)

IGRP is a distance vector protocol developed by Cisco in the early 1980s. It was devised to solve the problems associated with using RIP to route datagrams between interior routers. As such, IGRP converges faster than RIP and does not share RIP's hop count limitation. However, like all distance vector protocols, IGRP routers broadcast their complete routing table periodically, regardless of whether the routing table has changed. IGRP routers do this by default every 90 seconds. If a router has not received an update for a given path for 180 seconds, it removes the route from its table. This periodic sending of routing table information wastes bandwidth on the network. IGRP determines the best path through an internetwork by examining the bandwidth (deduced from the interface type), delay, reliability and load of the networks between routers.

A more advanced version of IGRP has been developed by Cisco and is known as enhanced IGRP (EIGRP). It is still a distance vector protocol and uses the same metrics as IGRP to work out the best route. However, it combines some of the properties of link state protocols to address the limitations of conventional distance vector routing protocols, i.e. slow convergence and high bandwidth consumption in a steady-state network. Also, whereas IGRP does not support variable-length subnet masking, EIGRP does support this function.

Instead of using hop count as a metric, it used various link measurements such as:

- bandwidth on the link;
- delay on the link (not measured but defined as a constant dependent on the technology);
- current load on the link (measured at regular time intervals);
- reliability of the link.

By default generally a Cisco router only uses the first two of these parameters.

The advantage of using this more complex metric over hop count is that IGRP will work well if some routes are slower than others, forwarding traffic down the faster link. RIP, for example, is only really well suited to more homogeneous networks where alternate links have the same bandwidth. IGRP can be used to load balance traffic between two alternative routes to the same destination.

IGRP uses the following techniques to help with convergence:

- split horizon
- poison reverse updates
- triggered updates
- holddowns.

A holddown is where a route is not allowed to be reinstated for a given time after it has just been removed. This stops the route being reinstated incorrectly by an erroneous routing table update.

5.3.5   Exterior routing protocols

These have historically been called exterior gateway protocols (EGPs), since gateway is just another term for router on IP networks such as the Internet. They are used to connect discrete ASs together. There are two main exterior routing protocols, exterior gateway protocol (EGP) and border gateway protocol (BGP). EGP has been largely superseded by BGP, a newer and more versatile protocol. It is rather unfortunate that EGP is the collective term for these protocols as well as being the term for the specific exterior gateway protocol. Unlike IGPs, EGPs do not have a single administrative unit and as such routing decisions are more complicated. Routes may pass through different countries, each with its own laws on importing and exporting data.

5.3.5.1   Border gateway protocol (BGP)

Routing through an internetwork is relatively straightforward. However, choosing the best path in a certain set of circumstances can be a difficult problem. EGP was the original routing protocol for the Internet and worked satisfactorily in the early days. However, recently its lack of policy-based route selection has become a big issue. This is because of the increasingly complicated nature of the Internet, technically, socially, and politically. BGP was designed to overcome EGP's problems. Like EGP, BGP is an inter-AS routing protocol created for use in the Internet's core routers. As Figure 5.25 shows, BGP enables routing between different ASs which have internally different IGPs, such as RIP, OSPF or intermediate system–intermediate system (IS–IS). Only the ASBR needs to understand BGP.

The current EGP on the Internet is BGP4, defined in 1994 and documented in RFC 1771. BGP requires reliable transportation of updates and as such uses TCP as its transport protocol, on port 179. When a connection is established, BGP peers exchange complete copies of their routing tables, which can be quite large. After synchronization only changes are then exchanged, which makes long-running BGP sessions more efficient than shorter ones.

The primary function of a BGP system is to exchange network reachability information with other BGP systems (known as BGP speakers). This is achieved through small 'keep alive' packets. If a router does not receive a 'keep alive' message from its neighbour within a certain 'hold time' (RFC 1774 suggests these should be sent every 30 seconds) then it will update its routing table to reflect the loss of the route. BGP is a *path vector protocol*, which has many characteristics that are similar to a distance vector routing protocol. However, whereas a distance vector routing protocol simply uses the number of hops to determine the best path, the path vector protocol uses a more sophisticated policy-based route selection.

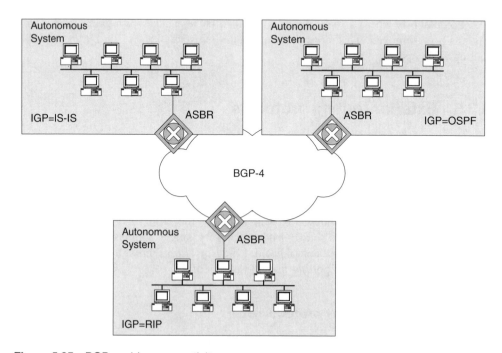

Figure 5.25 BGP enables connectivity

5.4   TCP AND CONGESTION CONTROL

Congestion is caused by network overload. In extreme cases congestion can cause packet loss due to the overflow of buffers within the network. This fact is exploited by the TCP protocol to ensure that congestion is limited within the Internet. Four different algorithms have been specified to control congestion for a TCP connection (RFC 2581).

5.4.1   Slow start/congestion avoidance

Each sender uses a value called the congestion window to determine how many outstanding bytes it can send that are unacknowledged. This value provides a limit on the total number of bytes the sender can have in transit at any one time. This is because as soon as data is acknowledged it must have been received at the far end, and therefore finished its journey across the network. The initial value of the congestion window is set to a value less than twice the maximum sender segment size. This algorithm is split into two phases, slow start and congestion avoidance.

During the slow-start phase, for each segment successfully acknowledged the congestion window is increased by the maximum sender segment size. Consider that a communication starts with a congestion window (CW) of 1500 bytes and a maximum segment size (MSS) of 1500 bytes.

After 1500 bytes are sent and received, the congestion window becomes 3000. This means the sender can now send two 1500-byte segments and the window will increase to 6000 bytes (1500 for each segment acknowledged). The sender can now send four 1500-byte segments and the window will increase to 12 000 bytes.

This process continues until the congestion window reaches a value called the slow-start threshold (SST). This second phase of the algorithm is called congestion avoidance. Now the congestion window is increased by one maximum sender segment size for each round trip time (RTT). For many implementations the following formula is used to update the window for each non-duplicate acknowledgement received:

$$CW = CW + MSS \times MSS/CW$$

This gives an acceptable approximation to the MSS per RTT. This is because the difference between the delivery times for each whole congestion window's worth of data will approximate to the RTT time, since the next block cannot be sent until this block is acknowledged.

At any time within the slow-start or congestion avoidance phases, the network may become congested and a packet will get lost in transit (due to buffer overflow). The sender will not receive an acknowledgement, timeout and proceed to retransmit. When this happens the congestion control algorithm adjusts the congestion window and slow-start threshold using the following formula:

$$CW = MSS$$

$$SST = MAX \text{ (flight size, MSS/2)}$$

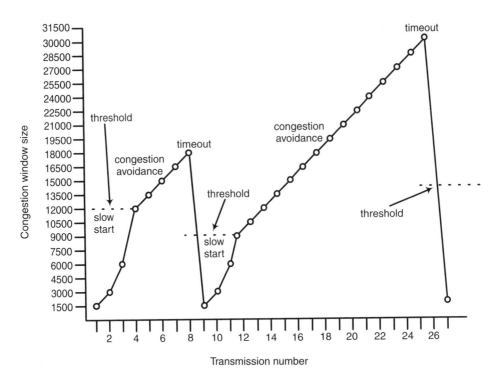

Figure 5.26 Slow-start/congestion avoidance algorithm

where flight size is the current amount of data outstanding in the network (sent but not received). At this point the slow-start algorithm begins again. Note that the threshold can be adjusted upwards or downwards depending on the value of flight size. Figure 5.26 shows an example of the slow-start/congestion avoidance algorithm in operation.

At the start, the threshold is set to 12 000 and the congestion window is at 1500. In the first slow-start phase, for each transmission burst the congestion window doubles in size until it reaches the threshold of 12 000. At this point the window grows linearly until transmission number 8, when a timeout occurs due to a packet loss. The threshold is now set to half the data in flight (from 18 000 to 9000), the window is set back to 1500 and the slow start begins again. This time the window grows until transmission number 26, where the timeout occurs again and the new threshold is set to 15 000.

One can see that the value of the slow-start threshold is the sender's approximation of the maximum value that the congestion window can be set to without causing congestion. All the time this value is adjusted to reflect prevailing network conditions.

5.4.2 Fast retransmit/fast recovery (RENO TCP)

If a TCP segment is lost in transmission the receiver will lose part of the stream of data. All of the following TCP segments cannot be delivered to the high layers until the missing segment is retransmitted. After a while the sender will timeout and then retransmit the missing segment. The problem with this procedure is that it introduces delay as the receiver waits for the missing segment.

When a TCP receiver receives a segment out of sequence, it should send an acknowledgement number back to the sender indicating which sequence number is expected next. This means if a segment is lost, the sender will receive duplicate ACKs for all the segments sent after the lost one due to the *go back N* ARQ mechanism regularly employed by TCP/IP. It is also possible for duplicate ACKs to be sent if segments are reordered within the network. A sender receiving three duplicate ACKs assumes this is caused by a lost segment and retransmits the segment indicated by the ACK number. This is called fast retransmit and serves to repair the loss.

When the missing segment has been acknowledged, instead of using slow start, the congestion window and threshold are set as follows.

$$SST = MAX \text{ (flight size, MSS/2)}$$

$$CW = SST$$

This procedure is called fast recovery. This is used because the sender receiving duplicate ACKs is an indication that TCP segments are being delivered to the receiver and therefore congestion has only been indicated for the missing segment.

5.4.3   Drop tail buffer management

In this case the router only drops packets when its buffers are full, where it drops the incoming packets that cannot be stored in the buffer. This scheme is called drop tail and has the advantage of being extremely simple as no complexity is needed at the router. It does, however, have a number of drawbacks. It is possible that a number of transmitters are increasing their speeds together until a time when they observe that the network becomes heavily congested. At this point of congestion all of these flows could lose packets and the senders slow down together. This phenomenon, called global synchronization, has the undesirable effect of reducing overall throughput, since the network load oscillates from being highly congested to highly under-utilized. Drop tail is also not particularly fair since flows are curtailed regardless of their data rate.

5.4.4   Random early detection (RED)

With RED, the router tries to avoid the congestion before it starts. With this scheme buffer lengths are monitored on a continuous basis. As the buffer starts to grow and get nearer its limit, a packet is selected out of the buffer on a random basis for dropping. The probability of picking a packet for dropping increases as the buffer size increases. Since the packets are dropped at random from the queue, the chance of a stream having a packet dropped is in exact proportion to the number of packets it has in the queue. Faster streams will have more packets in the queue and be dropped first. RED provides a more gentle form of flow control and avoids problems such as global synchronization. It is also somewhat fairer than drop tail since the flows that are more responsible for the congestion are likely to be curtailed first.

5.5 TCP OPTIMIZATION FOR THE AIR

TCP as a protocol assumes that packet loss is due to congestion within the network. This is reasonable in a wired network since the error rates of copper and fibre optic cables are very low. When a packet gets lost, the TCP stack slows down the flow of data so that the network can recover from the congestion, which naturally is good if the loss was as the result of congestion. However, it is undesirable if the loss was due to genuine packet loss since the slow down will just reduce the utilization of the link. As such, with a wireless link it makes better sense to send the packet again as soon as possible. To allow the TCP protocol to handle the wireless link correctly, a number of different options have been proposed. Two of them, wireless profiled TCP (proposed in the WAP 2.0 specification) and the Berkeley snoop module, are discussed here.

Figure 5.27 shows the configuration with the wireless profiled TCP. The wireless profiled TCP stack is a modified TCP stack optimized for the wireless link. An example of this modification is the way retransmissions are handled. With standard TCP if a transmitted packet is lost and ten more packets are transmitted, these will arrive out of order and all eleven packets must be transmitted again.

This standard mechanism is a version of ARQ, referred to as *go back N*, and prevents the receiver having to buffer out-of-order packet transmissions. This process is adequate for a high-bandwidth wired link which loses very few packets, but presents a major problem for a link with low initial bandwidth that has a tendency to lose packets. With wireless profiled TCP, the stack is able to send only the packet that was lost (this is called *selective repeat ARQ*), saving on link bandwidth and improving the efficiency of the transmission. Note that with this configuration, the link is broken into two pieces, a TCP link between the WAP device and the WAP proxy, which implements the wireless profiled TCP, and another standard TCP link between the WAP proxy and the server on the wired network. This means that wireless profiled TCP is only required beyond the proxy, with no changes required to the servers being accessed.

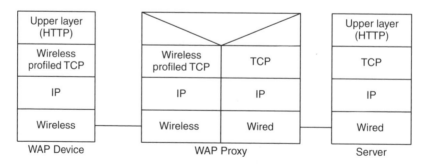

Figure 5.27 Wireless profiled TCP

The Berkeley snoop module implements a cache in the IP software at the base station. As packets come from the wired host they are stored in the cache and then forwarded to the wireless host. If the base station at some time in the future receives a duplicate acknowledgement from the wireless host this means that a packet has been lost in transmission. If the packet is in the cache it is forwarded to the wireless host and the acknowledgement is discarded. If the packet is not in the cache then the acknowledgement is forwarded to the wired host for standard TCP handling. The snoop module is implemented transparently, such that no modification is required for the two end points. Also, the optimization is only provided for downlink traffic to the wireless station from the wired host.

5.6   IP FOR GPRS AND UMTS R99

IP is used for GPRS/UMTS in two ways. First, it is used as a transport mechanism by mobile users to carry their data between the UE and hosts on the Internet or other network. Second, it is used between network devices to support the GPRS tunnelling protocol (GTP). This is shown in Figure 5.28; the advantage of this configuration is that it isolates the operator's network, keeping it secure from hackers operating from both the mobile and the fixed network end. This diagram is modified on the UTRAN side from that used in GSM/GPRS as described in Chapter 4.

In R99 of UMTS the role of IP is limited to the packet switched domain. The basic network configuration is shown in Figure 5.29. The UTRAN network transport is based on asynchronous transfer mode (ATM) and uses ATM Adaptation layer 2 (AAL2) for user plane traffic with signalling carried over AAL5. Within the circuit switched core, the network is based on a GSM core with the transport being circuit switched (i.e. ISDN, SDH etc.). The IP protocol is used as transport in the packet switched domain between the serving GPRS support node (SGSN) and the gateway GPRS support node (GGSN) and over the Iu interface between the radio network controller (RNC) and the SGSN (i.e. the Iu-PS).

Figure 5.30(a) shows how IP is used as a transport mechanism in the user plane. The GTP protocol is used to carry user GPRS traffic between the RNC and the external data network. Notice that for R99 all user plane traffic between the circuit core network and

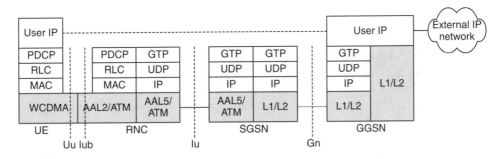

Figure 5.28   GTP tunnelling for user IP (release 99)

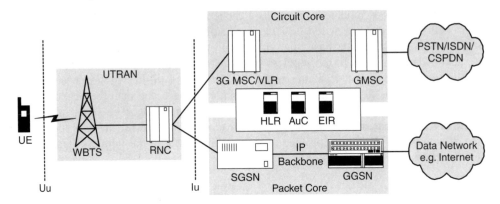

Figure 5.29   UMTS release 99

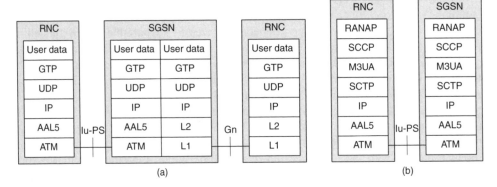

Figure 5.30   IP transport user and control plane (R99)

the UTRAN is expected to use ATM at layers 1 and 2. In later releases this constraint is removed as the network moves to an all-IP architecture independent of L2/L1 transport.

Figure 5.30(b) shows how IP may be used as transport for radio access network application part (RANAP) signalling over the Iu packet switched interface, although ATM is currently the most popular choice for this release. Since the signalling is a signalling system no. 7 (SS7) SS7 application protocol it is usually transported using the message transfer part (MTP) protocols (i.e. MTP3, MTP2 and MTP1). To carry SS7 over IP (SS7/IP) two protocols were developed: SCTP and M3UA. The addressing schemes of IP and SS7 are different (SS7 uses a three-level address called a point code instead of IP addresses). The MTP3 user adaptation layer (M3UA) provides routing for SS7 messages across an IP network. It provides translation between the SS7 codepoint addressing and the IP address and allows the IP end point to be seen effectively as part of the SS7 network. Using this mechanism the SS7 user part protocols are not aware that they are not running over MTP3 as is usual.

The streaming control transmission protocol (SCTP) is a replacement for TCP and provides a service more suitable for SS7 messaging. SCTP provides extra reliability by allowing multiple IP end points to be used for each end of the connection (fail safe redundancy). It also provides protection against some denial of service attacks that TCP can be vulnerable to. SCTP in terms of data service allows a number of streams to be delivered independently across one SCTP association. It also allows individual packets to be delivered out of order without having to wait for a single packet that was lost in transit. This is useful for protocols such as SS7, which need the data to be delivered expeditiously but also require a guaranteed service. The architecture for transporting SS7 over IP, commonly referred to as sigtran (signalling transport), is covered in more detail in Chapter 8.

5.6.1 Reliability and virtual router redundancy protocol (VRRP)

Even though IP network components have gained a reputation for reliability and robustness they are not yet comparable to traditional circuit switch telephony based equipment (e.g. PSTN exchange equipment) in terms of fail safe operation. To overcome these problems a number of solutions have been used over the years, most of them based on system redundancy. With this type of scheme parts of the hardware (and software) are duplicated and if one part fails the others can take over. Examples of these are redundant power supplies, disk RAID systems and backup links.

When operating IP in a carrier network environment it is essential to provide backup for the routing components. If an access router on an internal network fails this can result in the network being denied connectivity. Looking at Figure 5.31, it can be seen that the failure of the GGSN will deny GRPS access to the Internet.

Traditionally, with IP, routing failure has been handled via the use of dynamic routing protocols. These will re-route the packets if a router crashes or is taken out of service. This works well with routers within the core of the network but it does pose a number of difficulties with edge routers, for the following reasons:

- only one connection may be available to the outside network;
- commonly hosts are configured to use a default gateway to the outside world; if that gateway fails they must be reconfigured to use a different gateway;
- it will take some time for the new route to be established (convergence time), depending on the routing protocol;

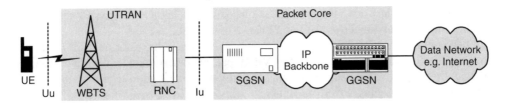

Figure 5.31   GGSN single point of failure

- many networks rely on the use of static routes when dealing with routes to other domains and in this case dynamic re-routing is not available.

With SS7 networks, signalling transfer points (STPs) are arranged in what is called a mated pair configuration in which each is connected to the same set of incoming and outgoing components.

With IP networks a redundancy protocol has been proposed called virtual router redundancy protocol (VRRP; RFC 2338). Using this mechanism a router can be backed up with one or more alternative routers which will take over in the case of failure. When a router takes over the functions of its companion it not only uses the same IP address but also the same MAC address. This means that other components on the IP network can continue working without knowing about the configuration change; even the ARP cache within the other routers will still contain valid information.

VRRP is designed to provide backup extremely quickly so that network service is disrupted only for a very short period of time.

The protocol operates as follows. Each physical router running VRRP can be configured to provide services for a number of virtual routers, which act as the default routers for sets of hosts on the network. The virtual routers have a virtual router identifier (VRID) and one or more IP addresses (and corresponding MAC addresses) to which they are expected to respond. Within the physical router each virtual router is allocated a priority from 0 to 255. A priority of 255 means that the physical router is expected in normal operation to perform that virtual router's routing function. A priority of less than 255 means that router will act as backup. By taking over ownership of a virtual router's set of IP and MAC addresses a router can act as a backup to a companion that has failed. To keep each other up to date, each router advertises the current virtual routers it is master of by sending VRRP packets to its neighbours. If a router fails to receive updates for a given length of time from the master of a virtual router that it is backing up then it takes over the role of master and starts advertising itself as the new master. An example of this is illustrated in Figure 5.32.

Router A is configured as the master of the virtual router with ID 1 and IP address 128.7.0.1 and backup for virtual router ID 2 with IP address 128.7.0.2 and a priority of 200. Router B's configuration is the reverse (master of 2 and backup for 1), making them a mated pair configuration. In normal operation, shown in Figure 5.32(a), each router sends VRRP packets advertising the virtual routers it supports to multicast address 224.0.0.18 (VRRP allocated multicast). If router A crashes or is taken out of service router B will detect it has a backup ID from which it has not seen VRRP packets for some time. After a given timeout it will take over virtual router with ID 1 and start to advertise itself as the master for 1. If some time later router A comes back online it can take over as master of virtual ID 1 by just advertising itself as owner with a higher priority. Router B will see that router A has higher priority and release ownership.

When configuring VRRP it is important to understand how the priorities will affect the timeouts. Each router is expected to send VRRP packets determined by the advertisement interval, the default value being 1 second. The time for a backup to take over the role of a master when not seeing advertisements is:

$$((256 - backup\_priority)/256) + 3 \times advertisement\ interval$$

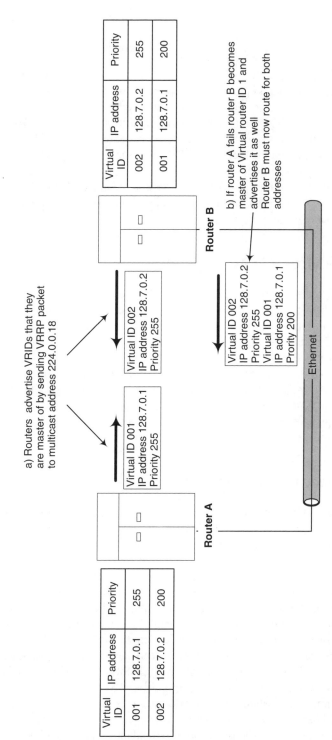

Figure 5.32   VRRP operation

So for the previous example with an advertising interval of 1 second this time will be:

$$(256 - 200) + 3 = 59 \text{ seconds}$$

The shortest timeout available will be $(255 - 254) + 3 = 4$ advertisement intervals.

This means to achieve a backup time of less than 0.1 seconds the advertisement interval would have to be 0.02 seconds or 50 times a second. There is evidently a tradeoff between advertising frequently and generation of a considerable volume of network traffic for fast response times against less network traffic with poorer response. It should also be noted that the destination address of the VRRP packet is multicast and therefore will be forwarded by bridges and switches across the LAN.

While not directly part of the specification, VRRP is recommended by the Third Generation Partnership Project (3GPP) for the backup of GGSN services within a GPRS/UMTS network environment.

5.6.2 VRRP virtual MAC addresses

Since VRRP allows the router's MAC address to be modified, there is an issue in terms of what MAC address to use. Routers could in theory use the original MAC address of their neighbour's interface card. This, however, makes configuration complex since all the original MAC addresses would have to be discovered and distributed between the VRRP routers. In practice virtual MAC addresses are used instead, each with a 5-byte prefix of 00-00-5E-00-01. The 00-00-5E is set aside for virtual addresses by the IEEE and the 00–01 is especially for VRRP. The last byte of the MAC address is the virtual router ID. This means that up to 255 VRRP routers can be supported on a single LAN segment. For the example given previously, the MAC addresses would be 00-00-5E-00-01-01 and 00-00-5E-00-01-02.

5.6.3 IP header compression

When sending data over the air, optimal use of the available bandwidth is essential since the obtainable data rate is somewhat limited. Also of note is the fact that when sending packets of data containing interactive voice, since transmission delay increases with packet size, shorter packets are preferred. The problem with short packets is that they tend to be less efficient because the overhead suffered due to fixed header components will increase as the packet size shrinks. If D is the data length and H the header length, the overhead percentage is $100 \times H/(D + H)$. Hence, as D decreases the percentage overhead tends towards 100%. For example, consider voice samples carried using the real-time transport protocol (RTP) over UDP/IP. The total header length will be RTP $(12) + \text{UDP}(8) + \text{IP}(20) = 40$ bytes. Since many voice schemes send data in 20-byte samples, then this will result in an overhead of 40/60, or 66%, with only 33% of the bandwidth being available for voice data.

To help overcome this problem, a number of schemes have been devised to compress the IP and UDP header parts of the packets (and therefore reduce the size of H). All compression schemes work on the principle of redundancy. The redundancy of a message is that which can be removed and the message still reconstructed at the far end. For example, if shown the sentence 'The cat s t on the m' an average reader would have little problem filling in the blanks even though the message has had some data removed. Examining a stream of packets moving from one host to another and looking at the contents of the headers, a great deal of similarity is seen from packet to packet. For example, the IP source and destination address may not change often over a given length of time, and fields such as the initial header length (IHL) and version number will probably never change for the whole of the transmission time.

Most header compression schemes work roughly as follows:

- never send the contents of fields that are not changing;
- for those fields that are changing, try to send only the *change* in the field and not the whole field again.

At the beginning of transmission, the whole of the header is sent uncompressed over the link and then subsequently, only the changes are sent. This is the basis for a popular TCP/IP header compression scheme developed by Van Jacobson and described in RFC 1144. Van Jacobson noticed that for any given TCP connection, about half of the 40 bytes of header information (TCP 20 + IP 20) does not change for the lifetime of the connection. So instead of sending the data repeatedly, the uncompressed header is sent once and then for subsequent packets, an 8-bit connection identifier is transmitted with the changing fields. For the fields that might change, a bit mask is sent indicating which fields have changed since the previous header was sent, and instead of sending the whole fields again, only incremental value are sent. This method is particularly relevant for sequence and acknowledgement numbers, since these will only be incrementing.

In the case of a packet being lost on the link due to errors, then the TCP header of new packets will be reconstructed incorrectly due to the loss of the compressed header. This corruption of the packet header will cause the sequence number to be reconstructed incorrectly and the receiver will discard the packet. When the sender times out the packet will be sent again, the compressor will detect the duplicate sequence number and will send an uncompressed header. At this point the sender and receiver will be back in sync and the compression can proceed as normal.

IP header compression is used between the UE and the RNC, and the compression is performed by the packet data convergence protocol (PDCP).

5.6.3.1   PDCP IP header compression

For PDCP R99 the compression scheme used is specified in RFC 2507. This provides support and defines compressed header formats for all of the following:

- IPv4 and IPv6 (including IPv6 extension headers)
- TCP and UDP transport options
- IP Security (IPSec) header, i.e. encapsulating security payload (ESP) and authentication header (AH).

The compression uses a scheme loosely based around the Van Jacobson technique but modified to allow it to cope with non-reliable protocols such as UDP. Each compressed packet is sent prefixed with a context identifier (CID), which indicates the compression context to be used at the decompressor. The context contains a full copy of the header of the last packet received by the decompressor (or at the send side, the last packet header compressed by the compressor). The compression is achieved by sending only differences between this context and the next header to send. Packets are grouped into streams, where a packet for a given stream will be compressed using a particular context. The grouping of packets into streams is dependent on the headers present. For IPv4/UDP the packets are grouped using source and destination IP address, source and destination UDP port number, IP version, IHL and type of service.

Five types of packet are supported:

- Full header: packet with uncompressed header
- Compressed non-TCP: non-TCP packet with compressed header
- Compressed TCP: TCP packet with compressed header
- Compressed TCP non-delta: TCP packet but not using differential coding
- Context state: lists of CIDs for contexts needing repair.

The full header is sent when a new context is created or when the receiver's context is in need of repair (for example when a packet is lost). These headers have the same format as a standard IP header except that they also contain the CID and generation field (for non-TCP headers). To make sure that the length of the packet is not increased, the CID and generation field are carried in the length field. The length field is then reconstructed using the Layer 2 frame length information at the receiver.

The compressed non-TCP packet is made up of a CID, a generation ID and the header fields that change from packet to packet. The format of the compressed non-TCP packet is fixed for a given set of headers. Figure 5.33 shows the format for a UDP/IPv4 compressed header. The figure shows an 8-bit CID. It is possible to use 16-bit CIDs by setting the extension bit, X, to 1. In this case the generation field will be followed by the least significant byte of the CID. The D bit allows for the addition of an optional data field. Its use must be negotiated before the compression session begins. One possible application is as a sequence number when compressing streams such as RTP.

The generation value is used to indicate the version of the context to use to decompress the packet. It is incremented each time a full packet header is sent. If a packet is received for a given context with an incorrect generation value, it cannot be compressed and must be discarded, or saved until a full header establishes the correct context.

CID	X	D	Generation	IP fields	UDP fields
8 bits	0	0	6 bits	Fragment identification	UDP Checksum

Figure 5.33   IPv4/UDP compressed header

The only header fields that have to be sent are the IP fragment identification field and the UDP checksum. All the other fields are expected to be read from the context or calculated when the packet is received (e.g. datagram length or IP header checksum).

The compressed TCP packet is similar to the RFC 1144 format. The fields are only sent if they have changed from the previous packet. Flags are used to indicate which fields have changed since the last packet sent. Some of the fields are sent as they are, whereas others are sent as delta changes (i.e. the difference between this value and the previous one). If a compressed TCP packet is lost, then all following compressed TCP packets cannot be decompressed because the context will be invalid.

The compressed TCP non-delta packet contains a compressed TCP/IP header but all the fields that are usually sent as delta values are sent as is. The TCP header is sent in this form when a repair of the context is required, for example when a compressed TCP packet is lost.

Finally, the context state is sent to the sender to tell it which CIDs require refreshing. The sender is expected to send full headers for all the packets. This allows for optimization of the repair mechanism since the receiver does not have to wait for the TCP retransmit timer to timeout before receiving the context update.

5.6.3.2 Robust header compression

Because of the slow turnaround time and high error rates of wireless links, many packets can be lost in the face of loss of context. To help overcome these problems, a complex scheme called robust header compression (ROHC; RFC 3095) has been developed. This scheme is offered as an enhancement over RFC 2507 for UMTS R4 and beyond.

This technique is similar to RFC 2507 but implements a more complex algorithm to try to reconstruct a valid header in the face of errors. A cyclic redundancy check (CRC) is also sent with each compressed header, which is valid for the uncompressed header. If the header CRC fails when uncompressing the header, then the decompressor tries to interpolate from previous headers the missing data and checks the result again using the CRC (Figure 5.34). This process of trying to repair the decompressor's context will be

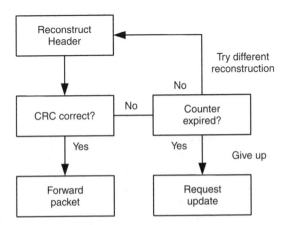

Figure 5.34 ROHC decompressor

tried a number of times and if this process fails the decompressor will ask for a context update from the compressor, which will then send it enough information to fix the context.

ROHC uses a number of different message types between compressor and decompressor. A static update message is sent at the beginning of a transmission and contains fields that are not expected to change throughout the lifetime of the packet stream (for example, IP source and destination addresses). To identify each packet stream, an 8-bit CID is used. Dynamic updates are used to update field addresses that are not covered in the static update, and need to be sent in an uncompressed form. In general, dynamic updates will be sent once at the start of the connection and then only if the fields cannot be sent in a compressed form. The final type is the compressed packet, which transfers the minimal data required to update the decompressor context so that the header can be reconstructed. Robust header compression is of such importance to wireless transmission schemes that it has been assigned its own IETF working group. For more information please refer to www.ietf.org/html.charters/rohc-charter.html.

5.6.4   IP address depletion and GPRS

With the advent of GPRS and the fact that there are over 800 million mobile subscribers globally there will be renewed pressure on the IPv4 limited address space. Currently a large number of IP network addresses have already been allocated to the major ISPs, leaving little space for future expansion. Eventually with the move to IPv6 this will cease to be a problem but until then other solutions must be sought. Two solutions currently already used by organizations providing connections for their intranet users to the Internet are dynamic host configuration protocol (DHCP) and network address translation (NAT).

5.6.5   Dynamic host configuration protocol (DHCP)

Saving address space by using dynamic host configuration relies on the fact that not every user will be using IP services at any given time. For example on a mobile network the phone may be switched off or even if on may not be making a data call. The ISP or operator starts with a pool of addresses. When a user requires to connect to the IP service he or she is given an IP address temporarily from the pool. This is called a *lease* of the address. This system works for as long as the number of users online at any one time does not exceed the size of the pool. This is very much how a PABX supports many phone users with only limited outside lines. The protocol that supports dynamic host configuration is called DHCP and is used globally by ISPs to assign IP addresses to their user base. DHCP also assigns other important addresses to the host such as the default DNS server and default gateway. In fact, the great benefit for the ISP is that it relieves them and their customers from the onerous task of having to manage the address space and configure each host manually. DHCP is the successor to another dynamic configuration protocol, called bootstrap protocol (BOOTP). BOOTP also allocates an IP and gateway address to a host but does not use the concept of a lease (the IP address is allocated for an indefinite period). For this reason it is not suitable if IP address reuse is needed. DHCP is used within GPRS to dynamically allocate an IP address to the mobile device once it has requested a PDP context activation. It is used for any mobile devices that do not have a static mapping defined in the home location register (HLR) for a particular access point. If a user is connecting to an external network such as the Internet, then the DHCP allocation is performed within the GPRS IP backbone. The DHCP service normally resides within the GGSN. This allocation method is referred to as transparent mode. If, however, the user is connecting to their corporate intranet, the IP address may be allocated from within that network. This is known as non-transparent mode.

5.6.6   Network address translation (NAT)

DHCP works well at assigning a limited pool of addresses to a number of users but does nothing to expand the total address space available. A network operator working with a class C network address would only be able to support a maximum of 254 IP addresses. Some system clearly needs to be introduced to allow a larger number of hosts to connect to the Internet, without the requirement for large numbers of public IP addresses. This is particularly the case with GPRS/UMTS, where the network is often expected to support many millions of users simultaneously. As was seen earlier, there are a range of addresses set aside for internal use, as shown in Table 5.10.

The addresses in Table 5.10 can be used internally within the operator's private network as long as they are not seen by any external hosts on the Internet. Any user that needs to connect *directly* to the Internet must have a public address so that IP packets can be correctly routed back to the user. However, consider that the operator has a class C address available. This clearly will not support the number of subscribers that could potentially be connecting to the Internet simultaneously.

To solve this dilemma, the operator will allocate *private* internal IP addresses to users when they establish their PDP context, usually using DHCP. When a user connects to the Internet, network address translation (NAT) performs the necessary translation to an external IP address.

In the simplest form of NAT, referred to as *static translation*, there is a fixed mapping between an internal IP address and the public IP address. This is typically used for devices such as a publicly accessed server (e.g. a WAP gateway), which needs to be able to accept connections from an external source.

Static translation is well suited for use with servers that are permanently situated and accepting connections. However, it does not provide flexibility for users who are being dynamically allocated an IP address on demand when they connect. For this general situation, *dynamic translation* is where many internal users who have their own private

Table 5.10   Reusable private addresses

Class	Addresses
A	10.0.0.0
B	172.16.0.0–172.31.0.0
C	192.168.0.0–192.168.255.0

IP addresses share the external address. In this case, to ensure that the correct translation takes place, the different internal IP addresses are mapped to different port numbers in the router. This NAT translation often resides in the GGSN, but may be situated in a separate unit, possibly incorporated into the firewall. Dynamic translation is sometimes referred to as NAPT (network address and port translation) but more commonly as simply NAT. Recall that a TCP or UDP packet has a 16-bit number for the port value, which allows $2^{16} = 65\,536$ unique ports per IP address. Those below 1024 are referred to as well-known ports and are used for specific services. However, many of those ports above this range are available for NAT. It is therefore possible to have a large number of TCP/IP connections and thus many internal devices sharing a single external IP address.

Consider the example configuration shown in Figure 5.35. Here, the operator is using the class A private address 10.0.0.0 internally and has a public class C address 212.56.65.0 and the UE is allocated the IP address 10.1.1.102 by DHCP. The user connects to the external web server 135.237.78.6:80, where port 80 identifies the HTTP service. The request from the UE uses its allocated IP address 10.1.1.102 and port number 1345. At the NAT service, the internal request is translated to an external class C address 212.56.65.10 with port number 16456. It is to this address and port number that the web server replies, and again the NAT service translates the destination IP address back to the internal address 10.1.1.102 and port number 1345.

The port mapping entry can be made in the lookup table once the TCP connection is established, and then deleted at the end of the TCP transaction. There is another form of NAT, referred to as *load balancing translation*, where a router can spread the connections of a very busy server to a number of identical servers, each of which will have its own

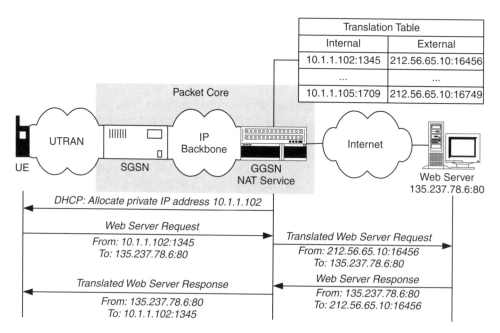

Figure 5.35 NAT address translation

unique IP address. This scenario is often the case for access to a busy web server on the Internet. The router will check to see which of the servers is the least busy and map IP translations to that machine. It should be noted that load balancing is commonly achieved through proprietary methods and the servers need to communicate their availability to the router in a format which it can understand.

As a solution to IP addressing constraints and the problems of host configuration NAT is very effective. An operator working with one class C network address using dynamic NAT translation can easily support over 16 million IP hosts. One critical limiting factor will be the processing power of the NAT service doing the address translation, particularly if the device where the service is located has a number of other tasks to perform, for example the GGSN or firewall. Another problem is the support for UDP within NAT. Since UDP is connectionless the NAT service does not have any session information to tell it when to delete a translation entry. In the case of UDP some sort of timeout mechanism is generally used to determine when translations are idle and can be removed.

5.7   IP-BASED QoS FOR UMTS NETWORKS

To provide a UMTS packet user with QoS, this must be an end-to-end consideration and must therefore cover the Uu air interface, the Iub/Iu-PS connection through the UTRAN, and the packet core IP backbone Gn interface. For each interface QoS must be provided and possibly negotiated. Essentially QoS must be provided in three domains: air, radio access and core network. For the air and radio access domains the RNC is responsible for managing the QoS. For the PS core network the QoS is negotiated between the SGSN and GGSN and is provided over the IP backbone. For the CS core network from R4, QoS must be provided on the new packet switched core between the incoming and outgoing media gateways (see Chapter 8). The core network and the UTRAN network for R5 and beyond are both IP based and will use technologies such as DiffServ, integrated services (IntServ) and resource reservation protocol (RSVP) to deliver QoS. For the UTRAN prior to R5, the transport is based on AAL2, and ATM protocols will be used to negotiate QoS for the radio access bearers.

5.7.1   QoS negotiation in UMTS

The basic operation of the QoS negotiation process for a GPRS service within UMTS is illustrated in Figure 5.36. Initially the UE makes a request to the SGSN for GPRS service (PDP context request) with a given QoS. In the example the request is for a streaming service class with a data rate of 192 kbps. This request is passed to SGSN transparently using radio resource control (RRC) and RANAP direct transfer messages. Next the SGSN determines that it can provide the requested QoS and then sends a *create PDP context request* to the GGSN. The GGSN at this point downgrades the QoS request from 192 kbps to 128 kbps due to lack of bandwidth available on its connected network and replies to the SGSN with a *create PDP context response* containing the adjusted QoS.

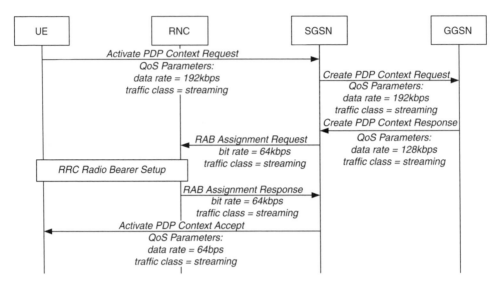

Figure 5.36 QoS negotiation in GPRS (R99)

At this point QoS has been negotiated for the backbone and the GGSN connection to the final destination. However, it is the SGSN that is responsible for the packet switched core network QoS, and, based on its available resources, as shown in Figure 5.36, it makes a decision to further downgrade the bandwidth allocation to 64 kbps.

To transfer data between the SGSN and the UE a radio access bearer (RAB) must be provided. The SGSN sends a *RAB assignment request* to the RNC. The RNC will determine if it has enough resources available over the Uu interface and within the UTRAN (i.e. over the Iub and Iu interfaces) to fulfil the request. It does this through its radio resource management procedures of load control and packet scheduling. If so, it instructs the base transceiver station (BTS) using the node B application part (NBAP) to reconfigure the radio link to support this new bearer and then sends a *radio bearer setup* message to the UE via its existing signalling connection, as discussed in Chapter 6. The UE confirms to the RNC that it has received allocation for the new bearer (*radio bearer setup complete*), and the RNC then confirms the RAB assignment to the SGSN (*RAB assignment response*). Now the SGSN replies back to the UE that the PDP context has been activated with the new data rate of 64 kbps.

All the way along in this process each element responsible for controlling a different part of network resources can renegotiate the QoS parameters. If the RNC for example does not have enough radio frequency (RF) resources within the target cell to provide the requested data rate then it is free to renegotiate the QoS downwards.

5.7.2   GPRS QoS parameters

Table 5.11 shows the QoS attributes that can be negotiated for a GPRS link over UMTS (see 3GPP TS24.008 and TS23.107). In practice, many of the QoS parameters are not

Table 5.11   UMTS QoS parameters

Parameter	Description
Traffic class	Options: conversational, streaming, interactive, background
Delivery order	Options: ensure order, don't ensure order
Delivery of erroneous service data unit (SDU)	Options: do not detect, deliver them, don't deliver them
Maximum SDU size	Range from 10 to 1520 bytes
Maximum bit rate for uplink	Range 0–8640 kbps
Maximum bit rate for downlink	Range 0–8640 kbps
Residual bit error rate (BER)	Bit error ratio range $5 \times 10^{-2} - 6 \times 10^{-8}$
SDU error ratio	SDU error ratio range $1 \times 10^{-2} - 1 \times 10^{-1}$
Transfer delay	Delay in range 10–4000 ms only for streaming and conversational class
Traffic handling priority	1–3 (only relevant for interactive class)
Guaranteed bit rate uplink	Range 0–8640 kbps (streaming and conversational)
Guaranteed bit rate downlink	Range 0–8640 kbps (streaming and conversational)

negotiated, but are taken directly from those assigned in the user's profile, which is part of their service agreement. For example, a user may have opted for a budget package, where a connection to the packet core is only permitted to be a maximum of 64 kbps. In this case, the PDP context activation request will show 'subscribed' for these parameters.

For GPRS/GSM (R97/98) networks a range of QoS attributes, including classes for delay, reliability and precedence as well as maximum and average throughput, were defined. The QoS specifications available for UMTS provide a higher degree of control. TS23.107 has recommendations on how to map the R97/98 attributes to the R99 attributes in case of inter-system handover.

5.8   QOS FOR THE GPRS CORE NETWORK

Networks are less capable of providing QoS as they become more highly loaded but the provision of a lightly loaded network will not always guarantee good QoS to the user. If the traffic is assumed to follow the Poisson distribution then one can apply queuing theory. For a normalized traffic rate of A Erlangs the chance of the router having N frames waiting to be transmitted is:

$$P(N) = (1 - A) \times A^N$$

An Erlang is a normalized expression of traffic and is calculated as follows for a packet switch domain:

$$\text{Traffic in Erlangs} = \frac{\text{Arrival rate in packets per second}}{\text{Service time for packet}}$$

It is also directly convertible from the network percentage load:

$$\text{Erlang rate} = \frac{\text{Load } (\%)}{100}$$

For a traffic load of only 10% or 0.1 Erlangs the probability of five frames being queued at the router at any given time would be 0.000009. This percentage seems small but if the router is receiving and transmitting frames each of 1500 bytes in length down a gigabit link then it will receive 83 333 frames $(1 \times 10^9/(1500 \times 8))$ a second. The average number of these frames that will find the routing queue contains five packets will be 0.75 $(83\,333 \times 0.000009)$, therefore the condition will occur 2700 times in a one-hour period.

Since the traffic is serviced at the router quickly this may not be a problem since the delay involved in forwarding five 1500-byte packets will be only:

$$5 \times 1500 \times 8/1\,000\,000\,000 = 0.00006 \text{ s or } 60 \text{ } \mu s$$

If the incoming buffer to the router does not store more than five frames then the next incoming frame could be lost and may require retransmission. In this case the increase in delay can be considerable since the packet will have to be transferred over the slower UTRAN network.

Notice that the formula for the probability is unbounded by the queue length, i.e. there is a chance (albeit a small one) that there will be any number of packets queued at the router at any given time:

$$(1 - A) \times A^N \text{ is non-zero for any value of } N \text{ as long as } 0 < A < 1$$

There is another important formula that can be used to help predict the behaviour of network switching components, which gives the average length of the queue at a given router:

$$L = A/(1 - A)$$

As the value of A approaches 1 the network becomes saturated and the average queue length (L) moves to infinity. It is prudent to check the average queue length against the buffering capability of the router; if the network is at a high load and the buffer overflows then packets will be lost and the network is deemed to be congested.

Another issue with QoS is the effect that low priority data traffic (e.g. email) can have on the traffic which has tighter QoS requirements (e.g. packetized voice). This is best illustrated by an example. Consider a router has started to forward a 1500-byte email packet over a 64 kbps link. A newly arrived voice packet would be delayed for:

$$1500 \times 8/64\,000 = 0.1875 \text{ or } 187.5 \text{ ms}$$

This is regardless of the allocation of bandwidth for the email. The solution in this case is to fragment the data packets before they are forwarded on the link.

It can be seen that controlling average traffic load is a necessary but not always sufficient condition for providing QoS. Apart from total load control, QoS techniques may include the following:

- individual flow control (limiting flow data rate);
- shaping the traffic (e.g. limit on maximum burst size over a given time);

- reserving resources such as bandwidth or buffer space at routers or hosts for a given flow;
- handling different traffic flows with different priorities;
- breaking long packets up to avoid head-of-line blocking.

The next section examines two types of service that can be used to provide QoS in the GPRS core network, DiffServ and IntServ. In fact with R4 and R5 these techniques can be extended into the UTRAN and circuit switched domain as the whole of the UMTS transport evolves to an all-IP architecture.

5.8.1 Differentiated services (DiffServ)

DiffServ (RFC 2474) works by separating traffic into classes and provides an appropriate treatment for each class. For example all voice traffic might be run in one class and be given a higher priority than another class of packets, such as those containing email data. DiffServ makes use of the 8-bit type of service (TOS) field in the IP header (Figure 5.37), which is referred to as the *differentiated services* (DS) byte. Use of the TOS field will not in general affect how the field has historically been used. In particular, the three precedence bits (used for priority) which are used by IP will still maintain their function. However, no attempt is made to ensure backwards compatibility with the traffic control bits (DTR, i.e. delay (D), throughput (T) and reliability (R)) as defined in RFCs 791 and 1349. Figure 5.37 illustrates how the original ToS field in the IP header is modified by DiffServ.

DiffServ relies on boundary devices at the edge of the network (Figure 5.38) to provide an indication of the packet's service requirements. These boundary devices need to examine the packet and set the DS byte codepoint value. The DS byte codepoint value specifies the *per-hop forwarding behaviour* (PHB). Within a GPRS core network the boundary function could be hosted within the SGSN and GGSN.

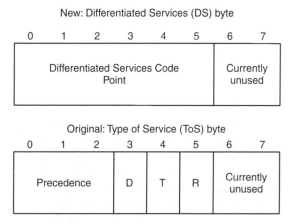

Figure 5.37   ToS to DS mapping

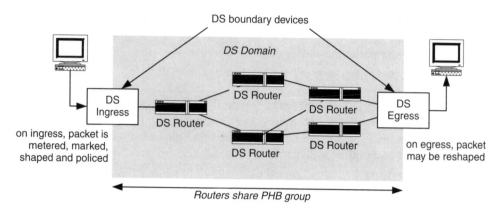

Figure 5.38 DS domain

The packet is then forwarded across the DS domain with a profile based on this value. A simple example of a PHB is where the PHB guarantees a minimal bandwidth allocation of X% of a link to a collection of packets with the same DS codepoint. A more complex PHB would guarantee a minimal bandwidth allocation of X% of a link, with proportional fair sharing of any excess link capacity. The ATM Forum is studying a proposal to map DiffServ PHBs to the ATM available bit rate (ABR) service to provide end-to-end QoS for IP traffic. One of the codepoints is reserved to provide a PHB of best try; this will be for non-real-time traffic with no QoS requirements.

It is important with DiffServ that traffic entering the network is managed correctly. If all hosts are allowed to send traffic into the network with a codepoint providing a high priority then the network will become congested and the QoS provision has failed. The boundary device must control which traffic coming into the network is marked with which codepoint and possibly shape the traffic coming in to ensure the network will not suffer from congestion.

5.8.2 Expedited forwarding

Expedited forwarding (EF) is a PHB behaviour defined in RFC 2598 where packets are expected to be treated with highest priority. Each router must allocate a fixed minimum bandwidth on each outgoing interface for the expedited traffic. Packets marked with the EF codepoint must be given preferential treatment over any other traffic and suffer minimal delay and jitter. In this case, packets are expected to be buffered for a minimum amount of time. For EF behaviour to operate successfully the allocation of available flows into the network must be managed so as not to overload the total expedited bandwidth available. This is because EF traffic will typically be time-sensitive interactive media traffic such as voice or video carried over UDP/IP. For this type of traffic it is inappropriate to throttle back on the sender since this will result in unacceptable delay. A router suffering congestion within an EF flow will have no alternative than to drop packets hence the need to shape expedited traffic as it enters the DS domain. The DS codepoint for EF is defined as 101110.

5.8.2.1 Assured forwarding

The assured forwarding (AF) PHB is defined within RFC 2597. In this case four different classes of PHB are specified each of which corresponds to a different level of service in terms of minimum bandwidth available. Even though bandwidth is allocated for each class this allocation is not guaranteed in the face of congestion. If buffers allocated for a given class fill up, packets will be discarded and congestion control will come into operation.

Within each class packets can be classified with low, medium or high drop probability. In the face of congestion the router will choose packets from the buffer for dropping based on this drop precedence, packets with a higher drop precedence having a higher probability of being dropped. One possible use of the drop probability is with non-conforming flows.

Senders that transmit above their allocated data rate (non-conforming flows) could have their packets marked with high drop precedence so that they will be more likely to be discarded in times of congestion. Table 5.12 shows the DS codepoints for the AF PHB groups.

5.8.2.2 Default PHB

The default PHB maps to a best effort service. In this case the packet will be delivered as quickly as possible but with no guarantees. The recommended codepoint for default PHB is 000000 but any packet not covered by a standardized or locally defined PHB will also be mapped to default PHB.

5.8.2.3 DiffServ for GPRS

DiffServ is ideally suited to provide QoS within the GPRS core network. This is because:

- the core network is a closed network;
- the core network is usually short span and therefore can be over-scaled using gigabit technologies;
- ingress and egress flows to the network are controlled by the SGSN and GGSN.

The 3GPP does not stipulate how UMTS traffic classes should be mapped to differentiated service codepoints. TS 23.107 (QoS Concept and Architecture) states: 'The mapping

Table 5.12 AF PHB DS codepoints

Drop precedence	Class 1	Class 2	Class 3	Class 4
Low	001010	010010	011010	100010
Medium	001100	010100	011100	100100
High	001110	010110	011110	100110

Table 5.13 Mapping UTMS class to DS codepoint

UMTS QoS class	DS codepoint
Conversational	Expedited Forwarding EF 101110
Streaming	Assured forwarding class 1 AF1 (001xxx)
Interactive	Assured forwarding class 2 AF2 (010xxx)
Background	000000 (default behaviour)

from UMTS QoS classes to DiffServ codepoints will be controlled by the operator'. Table 5.13 shows a possible mapping using both the EF and AF PHB classifications.

Conversational class using EF has the highest precedence. This type of traffic will typically carry VoIP telephone conversations. The bandwidth requirements for each traffic flow may be relatively modest (typically < 20 kbps) but the delay and jitter constraints will be very tight. The next class, streaming, will be traffic such as streamed video (one-way). In this case the bandwidth requirements may be high but the delay and jitter requirements are not so crucial. This is because the packets can be buffered at the receiver. The interactive traffic class representing traffic flows such as web access needs some reasonable service in terms of delay but with more modest bandwidth requirements than streaming. Finally, all other traffic will be handled as background. It is important to note that each router must have bandwidth allocated for every PHB group including the default, otherwise services such as FTP may fail in the face of network congestion.

5.8.3 QoS and the integrated services (IntServ)

The IntServ model for IP networks defines a single network service (based on IP) capable of carrying audio, video, and real-time and non-real-time data traffic. This is very much akin to the service definition capabilities of the Integrated services digital network (ISDN) but for a packet switching environment. The IETF IntServ working group has been responsible for defining and describing the service capability and interfaces to the underlying network but does not work on the details of how this might be achieved. For example, it is not part of their remit to look at the modification of routing or transport protocols. However, certain recommendations on the use of protocols developed elsewhere to provide an integrated service have been produced. IntServ defines two basic types of service: controlled load and guaranteed service. Both types of services can be requested by a host from the network for a given traffic specification (TSpec). The TSpec defines parameters such as average required data rate and maximum burst size. The network will be expected to exercise some type of admission control to determine if it can provide the service.

5.8.3.1 Controlled load

Controlled load service is defined as being the same QoS that would be achieved when sending data on an unloaded network element with other traffic within the network not significantly impinging on the data flow. The network must be working well within its limits and each packet must suffer little or no average queuing delay or packet loss due to congestion at each router. These requirements are expected to be met within timescales larger than the burst time, therefore short-term delays (less than the burst size) will be considered to be statistically anomalous. In terms of end-to-end service the following must be met:

- A very high percentage of packets are expected to be delivered successfully.
- The transit delay suffered by a very high percentage of packets will not be significantly more than the minimum transit delay, this minimum delay being the total processing time within each of the routers.

Notice how the controlled load service is somewhat vague in its definitions. Its purpose is to define a network in which service levels will not be degraded significantly as network load is increased. Applications can be tested at low loads and if working successfully can be expected to keep working as the network traffic increases since new load into the network will be policed effectively.

5.8.3.2   Guaranteed service

Guaranteed service carries packets across the network within a given maximum end-to-end queuing delay. To request the guaranteed service the applications must provide both a traffic specification and a requested maximum delay.

5.8.3.3   RSVP and IntServ

Both the controlled load and guaranteed services for IntServ imply some degree of admission control and resource reservation to make sure that sufficient bandwidth and buffer space are available within the network. The exact mechanism to carry out this control is beyond the scope of the IntServ working group but they have produced a recommendation on how to use RSVP to provide integrated services (RFC 2210).

5.8.4   Resource reservation protocol (RSVP)

RSVP (RFC 2205) was defined to address the issues of sending time-sensitive traffic over IP networks. For example, when a host has real-time video to send, it can use RSVP to request an appropriate level of service from the network to support this request. If a part of the network does not support RSVP, it cannot reserve the level of service. In this case RSVP can tunnel the packets through this section of the network until an RSVP router is again reached, which will continue the level of service. This enables RSVP to be implemented progressively as is required on the Internet. RSVP requests network resources in a single direction, therefore for a full duplex path two separate reservations need to be established. RSVP can support QoS for both unicast and multicast traffic.

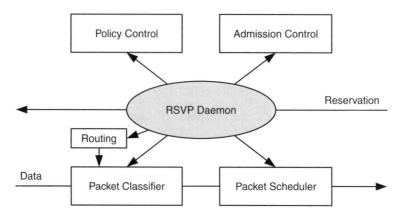

Figure 5.39   Resource reservation

An overview of RSPV operation within a network node is shown in Figure 5.39. Each node is capable of resource reservation and has a number of procedures to support it. *Policy control* ensures that a user has the required authority to make the reservation. *Admission control* keeps track of the resources to ensure that the resources are available to provide the requested QoS. If either of these controls fail, an error is returned to the originator of the RSVP request. The *packet classifier* determines the QoS for each packet by checking to see if it belongs to a reserved flow and marking it appropriately. The packets once classified can be fed into the *packet scheduler*, which orders the transmission to achieve the required QoS for each data flow.

Reservations are implemented using two message types: *path message* and *resv message*. A path message comes from the sender and contains flow information, describing the data that it wishes to send, such as data format, source address and source port, as well as the traffic characteristics. This *path message* is used to set up a set of virtual paths between the source and destination stations. When an RSVP-aware router or host receives the path message it stores the immediate upstream address of the node it received the path message from, as well as information about the traffic flow that is to be sent from the source. This information, called the path state, is used when reservation requests are sent from the destination hosts. When a receiving host wishes to make a reservation to receive data for the particular flow it sends a *resv message* to its immediate upstream RSVP neighbour. The upstream neighbour will check to see if the request is allowed. If the request is permitted then it will forward the request to its upstream neighbour (as identified in the path state) until the request reaches the source or is rejected due to admission failure.

A good example of RSVP operation is when providing QoS for a multicast session, as illustrated in Figure 5.40. Computer A sends a path message to all members of the multicast group, in this case computers B and C. The path message is received by the router, which stores the information in a table. The router now knows that when B or C replies to the message, the next hop back to the source is router to A. Note that the path message does not set up the reservation requirements. Now suppose that B wishes to receive information from the multicast group. It will now send a resv message back to the

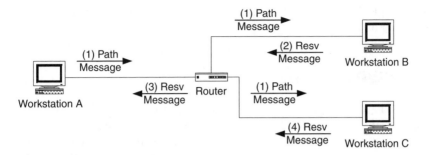

Figure 5.40   An example of RSVP

router asking for the required QoS. When the router gets the resv message it will reserve that requested amount of resources for B. The router also forwards the resv message to A (using the path state information) to ensure that resources are also reserved on that link. When C wishes to join the group, it will also send a resv message to the router. The router will reserve the required resources for the link router to C. However, in this case the router does not need to forward the request to A since a link has already been established (through B's request). However, if C's resv request is larger than B's then the router will have to forward the request to A, so that A can increase the reservation of the link A to router. As mentioned before, there are two main points to note:

1. The required resources must physically be available or the link will not be established.
2. The stations must have the required authority to request the desired resources.

RSVP supports three traffic types:

1. Best effort: this is traditional IP traffic. Applications include file transfer (such as email) and disk mounting.
2. Rate sensitive: this type of traffic will give up timeliness for throughput. For example, an application requests 100 kbps of bandwidth and then wishes to send 200 kbps for an extended time. In this implementation, the routers can delay the traffic. H.323 videoconferencing is such an application since it encodes at almost a constant rate.
3. Delay sensitive: this type of application requires data to arrive quickly and will allow the data rate to change accordingly. MPEG-2 video requires this system since the amount of data transfer depends on the amount of change in the picture from one frame to the next.

   In practice RSVP has not been used widely to support QoS over the Internet and large-scale IP networks but has been used to some limited extent within intranets to support VoIP and video streaming. The problem with RSVP is that it does not scale well for large networks for routers. To keep an update path and reservation state information for thousands of connections takes up a lot of processing power and storage space. Another

problem with RSVP is that is generates a lot of extra messages, which in turn may lead
to congestion on the network.

5.8.5 RSVP for GPRS

Provision of QoS within the GRPS core network will be largely based on DiffServ (and
controlled by PDP), the reason for this being scalability. When handling many thousands
or tens of thousands of sessions, RSVP is not well suited for the reasons stated above.
The place where RSVP can play a role, however, is between the UE, GGSN and the
external network. To provide QoS end-to-end, QoS must be supported in every domain
between sender and receiver. Since the PDP context creation goes only as far as the
GGSN, separate end-to-end QoS mechanisms are required for the domain external to
UMTS. This is illustrated in Figure 5.41.

Figure 5.42 shows the signalling flow for a UE originated call (3GPP TS 23.207
V5.7.0). The UE first establishes a PDP context across the UMTS network by sending a *create PDP context request* to the SGSN. The SGSN will then establish the PDP
context with the GGSN and request the creation of radio access bearers between it and
the UE. Assuming the request is satisfied, the SGSN sends a reply to the UE accepting
the request. The UE will then start RSVP reservation procedures end-to-end. For data in
the uplink direction, first a path message is sent from the UE, then the receiver makes a
reservation request, and finally the UE acknowledges the reservation. In Figure 5.42 the
GGSN is RSVP-aware and uses the request to reserve bandwidth on its external network
interface. In the second request, the reservation of bandwidth for the downlink direction
is shown. The path message from the external host with traffic to send is received by
the UE. This contains the traffic specification for the sender. Assuming the UE wants to
establish end-to-end QoS, it must first request the appropriate QoS characteristics from
its UMTS bearer. It does this by sending a modify request PDP context to the SGSN.
The SGSN will then attempt to satisfy the request and signal back to the UE. Now that
the UE has a bearer which can handle the traffic, a reservation request is sent back to the
sender. Assuming the reservation succeeds, a reservation confirm message is sent back to
the UE.

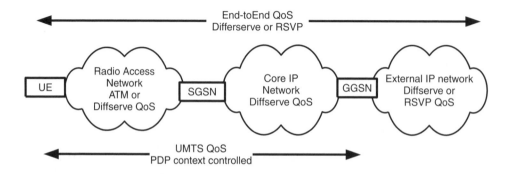

Figure 5.41 End-to-end GPRS QoS

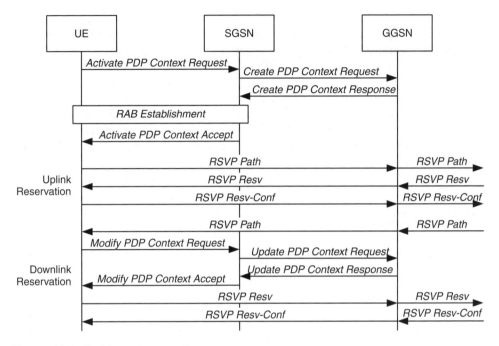

Figure 5.42   End-to-end reservation

All the RSVP messages in this case are tunnelled across the UMTS network, which will be unaware of their contents. The only UMTS components using RSVP in this case are the UE and the GGSN. It is also possible to have a GGSN which is not RSVP-aware. In this case the GGSN will forward the RSVP message but not enforce or perform resource or policy control for the QoS requests.

5.8.6   IntServ versus DiffServ

The type of services provided by IntServ and DiffServ are quite different. Whereas IntServ looks to provide separate QoS service for each flow, DiffServ only distinguishes traffic into flow classes. IntServ provides a greater degree of control to the application in terms of requesting QoS but at a price in terms of complexity of processing within the router and extra protocol information packets to be sent. DiffServ relies on the network being very carefully scaled to provide enough bandwidth for a particular class, and practical applications would likely require the network to be somewhat over-dimensioned to take into account the variances in the load it may suffer. With DiffServ the traffic can be policed at the network boundary device, which is considerably more scalable a solution than provision of a controlling mechanism within each of the routers.

Given all these factors and the fact that bandwidth on networks is becoming more readily available at lower costs, DiffServ will be the preferred choice over RSVP and IntServ for voice and data integration for networks with a small geographical span. For example, it is not uncommon to have the GGSN and SGSN in the same building or even room. In this case all components can be networked using gigabit or even 10-gigabit, both available for Ethernet, and use DiffServ and 802.1p[4] to differentiate between real-time and non-real time traffic.

Of course the two types of service are not mutually exclusive and it would be possible to have a differentiated service to run voice and data over a network and use RSVP to allocate bandwidth when streaming video from a server. Another application of RSVP is the reservation of bandwidth for the Internet connection leaving the operator's premises. As this is a likely bottleneck in communications, careful control of access to this resource is desirable.

5.9   IP SECURITY

Many of the applications that are run over networks require security. Internet banking, e-commerce or email access all may require the message transfer to be protected against snooping or forgery. Since data services for UMTS will be provided over IP networks there are a number of tried and tested security protocols available to choose from. The two most popular security protocols are transport layer security (TLS) and IP security (IPSec). The former provides privacy and authentication for World Wide Web users and through its use security is provided for 99% of all e-commerce applications. IPSec is the preferred choice for virtual private networks, a system that allows a corporation to use the Internet (or any other public IP network) as if it is private by encrypting the messages sent.

5.9.1   Transport layer security (TLS) and WAP security (WTLS)

TLS and WTLS were designed to provide secure access to Web and WAP services, respectively. Using WTLS provides weaker security than TLS and was only introduced as a stop gap to provide encryption for WAP devices that had limited processing power. TLS will eventually be able to replace WTLS as the preferred choice for wireless security with the advent of WAP 2.0 and GPRS.

5.9.1.1   *Transport layer security (TLS)*

TLS is an open protocol, largely based on the secure sockets layer (SSL) version 3.0 protocol, which was developed by Netscape. TLS has now been standardized by the IETF. TLS supersedes and incorporates all the functionality of SSL; however, the two are not interoperable. The rationale for TLS/SSL, and its widest application, is for the provision of a secure connection between a web browser and server for the transfer of

---

[4]802.1p is a datalink layer protocol that allows for prioritization of traffic at this layer.

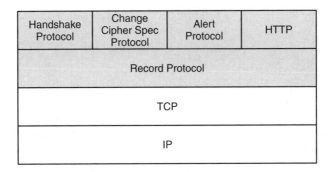

Figure 5.43   TLS protocol stack

sensitive data, such as credit card numbers, through the HTTP protocol. A session that is protected by TLS/SSL uses port 443, instead of the usual HTTP port of 80, and is identified by using https instead of http in the URL, for example:

https://www.orbitage.com/

Figure 5.43, shows that the TLS/SSL protocol stack consists of a number of protocols.

It utilizes TCP to provide reliable and secure end-to-end connections and defines four different protocols to support this service, namely the record, change cipher spec, alert and handshake protocols. The record protocol provides basic security services to the protocols above, significantly HTTP. The other three protocols are for management of the secure session.

The record protocol provides confidentiality and integrity through encryption and authentication. The keys for both are defined by the handshake protocol when the session is started.

Application data to be transferred securely is first split into blocks of a maximum size of 16 kB, which may then be compressed. Next, a *message authentication code* (MAC) is calculated for the data, based on the shared secret authentication key. The resulting message including the MAC is encrypted using a symmetric encryption algorithm, again based on the shared encryption key. There are several encryption algorithms permitted, including the data encryption standard (DES) and triple DES (TDES).

Finally, a record header is added to the result to provide some information about the compressed data.

Change cipher spec protocol

This is a 1-byte message. Its only function is to update the cipher suite being used on the connection.

Alert protocol

This passes alert messages, which are errors with regard to the current connection, for example, if an incorrect MAC was received, or an inappropriate message was received. Alert messages are classified as either level 1, which is a warning, or level 2, which is fatal and causes the connection to be terminated. The alert messages are also compressed and encrypted.

Handshake protocol

The functions of the handshake protocol are:

- to provide two-way authentication between the client and server, which is done through digital certificates;
- to negotiate the encryption and authentication algorithms to be used;
- to negotiate the encryption and authentication keys to be used.

The process of establishing a secure session involves four phases. In phase one, the client and server establish each other's security capabilities. The client first sends a *client_hello*, message which contains:

- the TLS/SSL version it is using;
- an identifier for the session;
- a list of the encryption, authentication and compression algorithms supported.

The server then replies with a *server_hello* containing the same information, but with the server having picked an encryption, authentication and compression algorithm from the list supplied by the client. This also contains the chosen method of key exchange.

The second phase involves server authentication and key exchange. First, the server sends its *certificate* message, such as issued by Verisign or Digicert. If required, a *server_key_exchange* message is now sent, depending on the key exchange mechanism. The server may also optionally request a certificate from the client, using a *certificate_request* message. The server closes this phase with a *server_hello_done* message. Note that both Internet Explorer and Netscape Navigator, as well as WAP browsers from vendors such as Openwave, have the public keys for Verisign and others hardwired in. Therefore, servers furnishing certificates from Verisign can be immediately authenticated by a client.

In phase three, once the client is satisfied that the certificate from the server is valid, the client sends its certificate in a *certificate* message. Next, the client sends a *client_key_exchange* message, which is dependent on the type of key exchange being used. For example, with the RSA scheme, the client generates a 48-byte key, encrypts it using the public key enclosed in the server certificate, and sends it to the server. This key is then used to generate a master key. Finally, the client may send a *certificate_verify* message to provide explicit verification for the client certificate.

The closing phase, phase four, is the final part of the establishment of the secure connection. The *change_cipher_spec* message, as previously described, and the *finished* message are exchanged by the client and server. Once this is completed, the client and server may now exchange encrypted data.

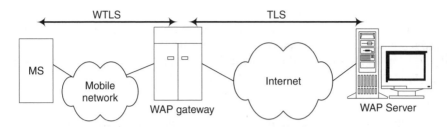

Figure 5.44   WTLS security model

5.9.1.2 *Wireless transport layer security (WTLS)*

The protocol was developed to allow WAP devices to gain secure access to a WAP gateway. The security model for WTLS is shown in Figure 5.44.

The data transfer between the mobile station and the WAP gateway is secured using WTLS. From the WAP gateway to the WAP server the connection is secured using TLS. It is the WAP gateway's responsibility to authenticate the WAP server and make sure it has a correct digital certificate that generates the appropriate signatures. WTLS operates in a similar manner to TLS in terms of messages sent and services provided but is not as secure for the following reasons. WTLS allows support for weak encryption algorithms and even allows the user to switch off encryption. This is due to the limited processing capability of some WAP handsets. Also, the configuration for WTLS is vulnerable to being attacked at the WAP gateway, since in the change from one encryption mechanism to another, the information is momentarily in plain text. With the introduction of WAP 2.0 a new model for security has been implemented. This is illustrated in Figure 5.45.

In this case the WAP transport protocols – wireless session protocol (WSP), wireless transaction protocol (WTP) and wireless datagram protocol (WDD) – have been replaced by standard IP protocols (HTTP, TCP and IP). This is possible because of the introduction of GPRS. The original WAP protocols were designed to be replacements for IP protocols and optimized for a wireless link.

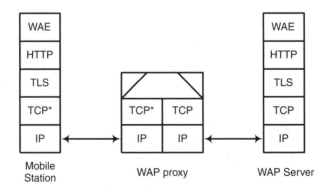

Figure 5.45   WAP 2.0 security

Since GPRS is designed to manage data transfer over the wireless link, many of the original WAP transport functions are somewhat redundant. Notice on the connection between the mobile station and the WAP proxy the TCP protocol has been replaced with TCP*. This is the version of the standard TCP protocol that has been optimized for wireless transmission, as was explained in Section 5.5.

5.9.2   Virtual private networks and IP security (IPSec)

Looking at Figure 5.46, one can see that the two networks are connected together via the Internet. The data leaving network 1 is encrypted by the firewall, carried securely across the Internet and then decrypted at firewall 2. This type of configuration is called a virtual private network (VPN) since even though the data is being transferred over a public network, its privacy is maintained through the use of encryption. It is also possible to connect a mobile user onto the VPN. In the diagram the mobile terminal is connected via the VPN to networks 1 and 2. The applications running on the mobile terminal and within workstations A and B see no difference in terms of their operation: it is as if they are connected to the network directly. This is not possible with SSL because protocols such as UDP are not supported. This would, for example, preclude the use of protocols such as DNS being carried over the VPN. The whole idea of VPN access is that it is transparent and the two networks can be connected without modifying the software in the client or server machines.

VPN solutions are also used for operators to provide secure links between the GGSN and a corporate client's intranet. This allows the operator to offer its GPRS network as a mobile extension to the existing fixed-line services of the corporate customer, while providing secure access to the corporate network. In addition to the VPN, only permitted users would be allowed to connect to the defined access point for the corporate by configuration in both the HLR and GGSN. Initially this GPRS corporate connection will usually

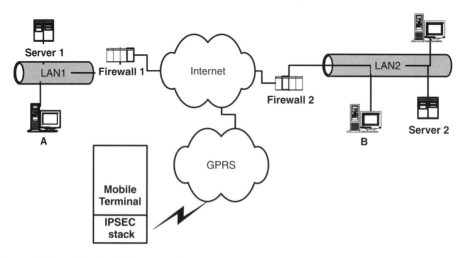

Figure 5.46   Virtual private networks

terminate at a laptop, using the GPRS device essentially as a data 'modem'. However, this situation will gradually change as GPRS devices become more powerful, supporting a wider range of standard IP applications, and also as GPRS access becomes more routinely incorporated into PDAs. With programmable and Java-enabled GPRS devices emerging, this expands the scope of corporate mobile services to include client–server and other enterprise applications.

5.9.2.1   IP security (IPSec)

To allow for encryption at the IP layer a protocol called IPSec was developed. IPSec is not simply a VPN solution since it also allows for encrypted and authenticated traffic to be sent end-to-end as well as within VPN tunnels. Its main application, however, is VPNs and it is supported by a number of vendors, including Checkpoint in its VPN-1/Firewall-1 product. IPSec has been designed to provide security and negotiated compression for IPv4 and IPv6. In common with most security solutions, the following services are defined:

- authentication
- integrity
- privacy
- protection against replays
- compression.

IPSec provides security services within IP packets and therefore can provide security for all IP application protocols. There are two protocols defined within IPSec: *authentication header* (AH) and the *encapsulating security payload* (ESP). The protocols are designed to be algorithm-independent and this modularity permits selection of different sets of algorithms without affecting the other parts of the implementation. However, a standard set of default algorithms has been specified to enable interoperability on the Internet.

5.9.2.2   Security associations

IPSec services are based upon a mechanism called security associations. A security association (SA) is a simplex 'connection' that provides security services to the traffic carried by it. SAs are fundamental to IPsec, and both AH and ESP make use of them. Since an SA is simplex, to secure bidirectional communication between two hosts, or between two security gateways, two SAs (one in each direction) are required.

An SA is uniquely identified by the following:

- security parameter index (SPI). This in turn can be associated with key data, encryption and authentication algorithms;
- IP destination address;
- a security protocol (AH or ESP) identifier.

An SA provides security services to a traffic stream through the use of AH or ESP but not both together. If both AH and ESP protection is applied to a traffic stream, then two SAs are created to provide protection to the traffic stream. Although the destination address may be a unicast address, an IP broadcast address or a multicast group address, currently IPSec SA management mechanisms are only defined for unicast SAs.

5.9.2.3 Authentication header

The IP authentication header (AH) is defined in RFC 2402 and provides data integrity, source address authentication and an anti-replay service. AH provides authentication for the upper-layer protocols and for some of the IP header. It is not possible to provide authentication for all of the IP header fields since some of these will change en route and the sender cannot always predict what these values will be. AH can be applied on its own, in conjunction with ESP, or in a nested fashion through the use of tunnel mode.

5.9.2.4 Encapsulating security payload protocol

The ESP protocol is defined in RFC 2406 and provides confidentiality through encryption. Like AH, ESP can also provide authentication; however, ESP does not cover the IP header fields unless those fields are first encapsulated using ESP (tunnel mode). Both confidentiality and authentication are optional, but at least one of them must be selected. The ESP header is inserted after the IP header and before the upper-layer protocol header (transport mode) or before an encapsulated IP header (tunnel mode). Both of these modes are described below.

5.9.2.5 AH and ESP modes of operation

AH and ESP can be applied individually or in combination with each other to provide a set of security services in IPv4 and IPv6. Each protocol can also support two modes of operation:

- transport mode
- tunnel mode.

In transport mode the protocols provide protection primarily for upper-layer protocols; in tunnel mode, the protocols are applied to tunnelled IP packets.

Transport mode

A transport mode security association is usually between two host machines as shown in Figure 5.47.

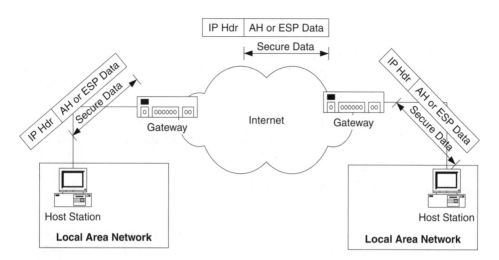

Figure 5.47   IPSec transport mode security association

- *Scenario 1 – using AH*: the host on a secure LAN digitally signs the packet to authenticate it. The receiving host will check the signature and either accept or reject the packet. If the packet has been altered during its journey, then the digital signature will not concur with the packet contents. The contents are not encrypted as the packet travels across the Internet.

- *Scenario 2 – using ESP*: the host on a secure LAN may encrypt the packet for its journey across the Internet. Anybody that happens to listening to packets using a network analyser will be able to receive the packet but will not be able to decipher its contents. The receiver, however, will have the correct key to unlock the encrypted data.

In IPv4, the transport mode security protocol header appears immediately after the IP header and any options, but before any higher-layer protocols (e.g. TCP or UDP), as shown in Figure 5.48.

In IPv6, the security protocol header appears after the standard IP header and extensions, with the exception of the destination options extension, which may appear before or after

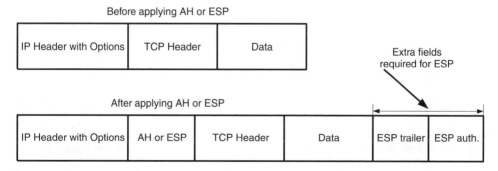

Figure 5.48   Transport mode IPSec protocol header (IPv4)

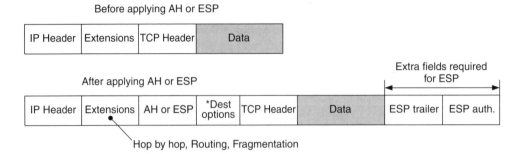

Figure 5.49   Inclusion of IPSec header (IPv6)

the security protocol header (Figure 5.49). In the case of AH, the protection includes higher-layer protocols and selected portions of the IP header, selected portions of extension headers and selected options (contained in the IPv4 header, IPv6 hop-by-hop extension header, or IPv6 destination extension headers). In the case of ESP, a transport mode SA provides security services only for higher-layer protocols, not for the IP header or any extension headers preceding the ESP header.

Tunnel mode (VPN mode)

A tunnel mode security association can be between two host machines, two gateways or a gateway and a host, as shown in Figure 5.50. Here the *secure gateway* receives a packet from the secure LAN. This packet is then encapsulated within another IP packet to be sent to the destination. The destination in this example happens to be another *secure gateway* (it could have been the receiving host). This device will decapsulate the packet for its

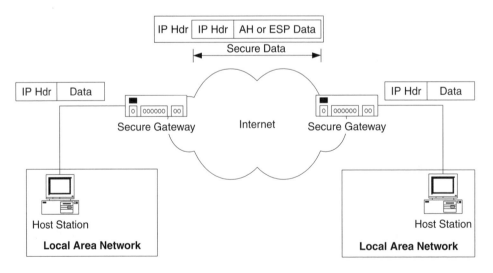

Figure 5.50   Tunnel mode

onward journey. During its time on the insecure Internet all details about the original packet will remain secure. If AH is used this means that the receiver will be able to check for any inconsistencies, and if ESP is used then this could mean (if encryption was selected) that it was encrypted throughout this leg of the journey. A tunnel provides a VPN service between two networks (if secure gateways encrypt the data). If IPSec encryption is executed at the host this can be used to provide secure remote access via the VPN to the network for a remote client.

Whenever either end of a security association is a security gateway, the SA must be in tunnel mode and not transport mode. An SA between two security gateways is therefore always a tunnel mode, as is an SA between a host and a security gateway.

For a tunnel-mode SA, there are two IP headers:

- an *outer* IP header that specifies the IPSec processing destination;
- an *inner* IP header that specifies the ultimate destination for the packet.

The security protocol header is placed after the *outer* IP header and before the *inner* IP header. If AH is employed in the tunnel mode, portions of the outer IP header are provided protection, in addition to the entire tunnelled IP packet (i.e. all of the inner IP header is protected, as are the higher-layer protocols), as shown in Figure 5.51.

If ESP is employed, protection is only supplied to the tunnelled packet, not to the outer header, as shown in Figure 5.52.

5.9.2.6 Establishing security associations

The setting up of a security association between two entities on the network is not defined within IPSec and can be performed in a number of different ways. Essentially there are two types of establishment:

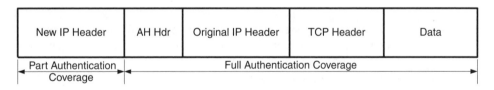

Figure 5.51 AH tunnelled IP packet

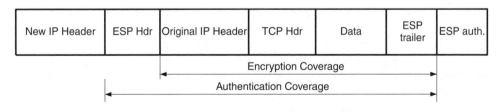

Figure 5.52 ESP tunnelled IP packet

- *Manual*: where both machines are configured 'by hand', including the key distribution, encryption and authentication algorithms.
- *Dynamic*: the key exchange and algorithm negotiation is carried out using a key exchange protocol.

The first option is realistic in cases where there are not too many different secure connections that are required to be made, the set-up is relatively static and travel between sites feasible. For all other cases some form of key exchange protocol is required.

5.9.3 Internet key exchange (IKE)

This mechanism specifies a particular public-key-based approach called the *Internet key exchange* (IKE), which is defined in RFC 2409, for automatic key management. However, other automated key distribution techniques such as Kerberos and SKIP may be used. IKE is the amalgamation of the IP security association key management protocol (ISAKMP), defined under RFC 2408, and Oakley (RFC 2412), a key determination protocol. IKE manages the key exchanges between the sender and recipient through the Diffie–Hellman key exchange protocol.

Authentication can be done through pre-shared secrets or digital certificates issued through a certificate authority (CA). Using this system it is possible to control the level of trust since there are a number of algorithms that can be used within IPSec. Thus for example if a high level of security is required then Triple DES could be used with a new key being negotiated every 5 minutes.

5.9.4 Security and GPRS

GPRS provides IP transport as an end-to-end solution, that is, all the way from a mobile device to another IP host on the Internet or private IP network. This means that the IP security services such as authentication and privacy can be provided end-to-end using mechanisms such as encryption and digital signatures. This is in fact the preferred model for securing transmissions, encrypting them as they leave the original source host and only decrypting them on arrival at their final destination. GPRS does in fact provide encryption for transmissions as they are carried from the mobile device to the SGSN via the LLC protocol. However, this cannot be consider secure as the data is transported in plain text across the GPRS core network and on to the Internet. Figure 5.53 shows how IPSec could be used in tunnel mode to provide VPN services to a mobile user. The user data in this case is encrypted all the way from the mobile device as far as the enterprise firewall. This configuration will also allow external users to be authenticated, blocking out unauthorized access to the enterprise network. Note that the operator's network is not involved in providing the security but merely a service providing data transport.

For users requiring secure access to a server then it is likely that TLS end-to-end will be the preferred solution. In this case the server usually has to provide authentication but

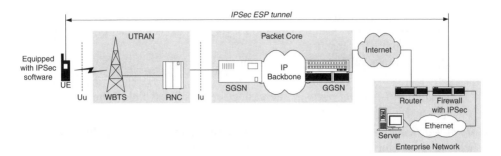

Figure 5.53 GRPS VPN solution

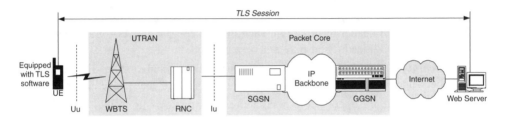

Figure 5.54 TLS with GPRS

not usually the client. After keys are exchanged all the traffic is protected using strong encryption (usually RC4). This configuration is shown in Figure 5.54.

5.10   INTERNET PROTOCOL VERSION 6 (IPv6)

The TCP/IP protocol suite was originally designed over 25 years ago to connect the US government's vast array of computers. Today, as IP version 4, it is used as the Internet protocol to connect thousands of routers and millions of users around the world. This explosive growth was never expected or anticipated. The protocol has been constantly evolving with new protocols being introduced through the IETF's RFCs. For example, in the case of the UDP protocol, it is now used to transfer speech and video. However, although additional protocols have been introduced, the core protocols have not changed a great deal since their inception, as this would have a major impact on the thousands of routers and millions of users. One of the main reasons for the change to IP version 6, or IP next generation, as it was also known, is the insufficient number of available addresses in IP version 4. IP version 4 has a 4-byte addressing range giving a combination of $2^{32} = 4\,294\,967\,296$ unique addresses. On the surface this seems like plenty of addresses, but the address classifications are very inefficient. Organizations which require more than the 256 addresses available with class C have historically been given a class B address, which allows them to have 65 536 addresses, clearly far too many. Recent stop gaps such as DHCP and CIDR have allowed IP version 4 to continue for some time but this is not seen as a permanent solution but rather a way of delaying the inevitable. While not part of the UMTS specification, at the application layer, it is expected that devices should be able to support both IPv4 and IPv6.

The key capabilities of IPv6, in comparison to IPv4, are:

- *Expanded addressing and routing*: increasing the IP address field from 32 to 128 bits in length and incorporation of address hierarchy.
- *Simplified header format*: eliminating or making optional some of the IPv4 header fields to reduce the packet handling overhead. Even with the addresses, which are four times as long, the IPv6 header is only 40 octets in length, compared with 20 octets for IPv4. An example is fragmentation, which has been removed from the core header and placed in an extension header.
- *Extension headers and options*: IPv6 options are placed in separate headers located after the core IPv6 header information, such that processing at every intermediate stop between source and destination may not be required.
- *Authentication and privacy*: required support in all implementations of IPv6 to authenticate the sender of a packet and to encrypt the contents of that packet, as required.
- *Autoreconfiguration*: support from node address assignments up to the use of DHCP.
- *Incremental upgrade*: allowing existing IPv4 hosts to be upgraded at any time without a dependency on other hosts or routers being upgraded.
- *Low start-up costs*: little or no preparation work is needed in order to upgrade existing IPv4 systems to IPv6, or to deploy new IPv6 systems.
- *Quality of service capabilities*: a new capability is added to enable the labelling of packets belonging to particular traffic flows for which the sender has requested special handling, such as non-default QoS or real-time service.
- *Mobility support*: since the end system is identified with the end system identifier (ESI), in many ways IPv6 makes implementing mobile IP simpler.

5.10.1   The IPv6 header

The IPv6 header is 40 octets in length, with eight fields. This can be compared to IPv4, which has only 20 octets of header but 13 fields within it. The IPv6 header is shown in Figure 5.55 and its fields are described in Table 5.14. It can be seen that the following have been removed:

- *Header length*: since the header of IPv6 is a standard 40 bytes there is no need for a length indicator.
- *Fragment offset, identification, flags*: these have been moved to the extension headers.
- *Checksum*: this has been removed completely. Error checking is typically duplicated at other levels of the protocol stack. Requiring routers to perform this check has reduced performance on today's Internet. Also, many point-to-point connections are now fibre and this medium has a very low error rate.

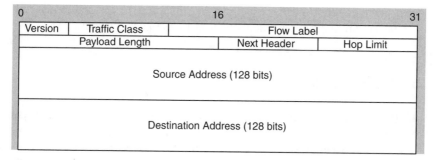

Figure 5.55   IPv6 header format

Table 5.14   IPv6 header fields

Header	Comment
Version	4-bit IP version number = 6
Traffic class	8-bit traffic class field
Flow label	20-bit flow label
Payload length	16-bit integer to indicate length of payload in bytes
Next header	8-bit selector. Identifies the type of header immediately following the IPv6 header
Hop limit	8-bit unsigned integer. Decremented by 1 by each node that forwards the packet
Source address	128-bit address of the originator of the packet
Destination address	128-bit address of the intended recipient of the packet (possibly not the ultimate recipient, if a routing header is present)

5.10.2   Traffic classes

The 8-bit traffic class field in the IPv6 header is available for use by originating nodes and/or forwarding routers to identify and distinguish between different classes or priorities of IPv6 packets. This is equivalent to the IPv4 type of service byte, which forms the 'differentiated service' for IP packets. The traffic class field in the IPv6 header is intended to allow similar functionality to be supported in IPv6. Both IPv4 and IPv6 support the RSVP protocol. The following general requirements apply to the traffic class field:

- The service interface to the IPv6 service within a node must provide a means for an upper-layer protocol to supply the value of the traffic class bits in packets originated by that upper-layer protocol. The default value must be zero for all 8 bits.

- Nodes that support a specific (experimental or eventual standard) use of some or all of the traffic class bits are permitted to change the value of those bits in packets that they originate, forward or receive, as required for that specific use. Nodes should ignore and leave unchanged any bits of the traffic class field for which they do not support a specific use.

- An upper-layer protocol must not assume that the value of the traffic class bits in a received packet are the same as the value sent by the packet's source.

5.10.3 Flow labels

The 20-bit flow label field in the IPv6 header may be used by a source to label sequences of packets for which it requests special handling by the IPv6 routers, such as non-default QoS or real-time service. Hosts or routers that do not support the functions of the flow label field are required to set the field to zero when originating a packet, pass the field on unchanged when forwarding a packet, and ignore the field when receiving a packet. There may be multiple active flows from a source to a destination, as well as traffic that is not associated with any flow. A flow is uniquely identified by the combination of a source address and a non-zero flow label. Packets that do not belong to a flow carry a flow label of zero. A flow label is assigned to a flow by the flow's source node. All packets belonging to the same flow must be sent with the same source address, destination address and flow label.

5.10.4 The payload length field

The 16-bit payload length field measures the length, given in octets, of the payload following the IPv6 header. Any extension headers present are considered part of the payload, i.e. included in the length count. Payloads greater than 65 535 are allowed, and these are called *jumbo payloads*. To indicate a jumbo payload, the value of the payload length is set to zero, and the actual payload length is carried in a jumbo payload hop-by-hop option.

5.10.5 The next header field

The 8-bit next header field identifies the header immediately following the IPv6 header. This field uses the same value as the IPv4 protocol field, as outlined in RFC 1700. Examples are shown in Table 5.15.

5.10.6 The hop limit

The 8-bit hop limit is decremented by one by each node that forwards the packet. If the hop limit equals zero, the packet is discarded and an error message is returned. This allows 256 routers to be traversed from source to destination. It actually replaces the TTL field in the IPv4 header, which was initially included in IPv4 to give an indication of the number of seconds that the packet had been alive. The field was never used for this; it in fact also counted the hops (intermediate nodes) but unfortunately the name stuck.

Table 5.15   Example IPv6 next header fields

Value	Header
0	Hop-by-hop options
1	ICMPv4
4	IP in IP (encapsulation)
6	TCP
17	UDP
43	Routing
50	Encapsulating security payload
51	Authentication
58	ICMPv6
59	None (no next header)
60	Destination options

5.10.7   The source address

The source address is a 128-bit field that identifies the originator of the packet.

5.10.8   The destination address

The destination address field is a 128-bit field that identifies the intended recipient of the packet. Although this is usually the ultimate recipient, if the routing header is present this may not be the case.

The IPv6 design simplified the existing IPv4 header by placing many of the existing fields in optional headers. In this way, typical packet transfer is not complicated by intermediate routers having to check all the fields. Where they are needed, the more complex conditions are still provided for via the extension header with the obvious overhead. An IPv6 packet, which consists of an IPv6 packet plus its payload, may consist of zero, one, or more extension headers. Figure 5.56 shows an example of the next header field. The original next header first indicates the value 00. This means that the first extension is a hop-by-hop extension. The hop-by-hop extension has a value of 43, indicating that the next extension is a routing extension. This routing extension has 06 in its next header field to indicate that the next header is the TCP header. Each of the extension headers is an integer multiple of 8 octets long to maintain alignment for subsequent headers. The hop-by-hop options header carries information that must be examined and processed by every node along a packet's delivery path, including the destination node. As a result, the hop-by-hop options header, when present, must immediately follow the IPv6 header. The other extension headers are not examined or processed by any node along a packet's delivery path, until the packet reaches its intended destination(s). When processed, the operation is performed in the order in which the headers appear in the packet.

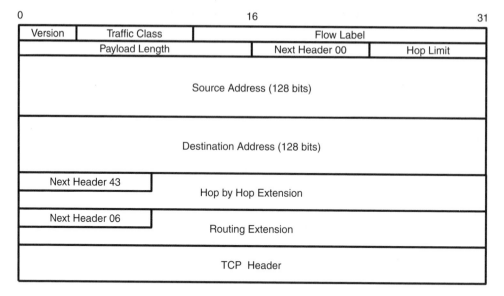

Figure 5.56   Example header extensions

5.10.9   IPv6 address representation

IPv4 address are typically represented in dotted decimal notation. As such, a 32-bit address is divided into four 8-bit sections separated by periods. Each section is represented by a decimal number between 0 and 255, for example 152.226.51.126.

This would be a cumbersome way of representing IPv6 addresses since they are 128 bits long, so a different method of representation is required. This new method is specified in the addressing architecture document, RFC 2373. The preferred method of representation is:

$$x : x : x : x : x : x : x : x$$

where each x represents 16 bits, and each of those 16-bit sections is defined in hexadecimal. For example, an IPv6 address could be of the form:

$$DEFC : A9BE : 1236 : DE89 : D7FE : 4535 : 908A : 4DEF$$

Note that each of the 16-bit sections is separated by colons, and that four hexadecimal numbers are used to represent each 16-bit section. If any sections contain leading zeros, then those zeros can be omitted. For example:

$$DEC5 : 0000 : 0000 : 0000 : 0009 : 0600 : 3EDC : AB41$$

may be simplified to:

$$DEC5 : 0 : 0 : 0 : 9 : 600 : 3EDC : AB41$$

If long strings of zeros appear in an address, a double colon ':::' may be used to indicate multiple groups of 16 bits of zeros, which further simplifies the example shown above:

DEC5 :: 9 : 600 : 3EDC : AB41

The use of the double colon can only appear once in an address, although it may be used to compress either the leading or trailing zeros in an address. For example, a loopback address of:

0 : 0 : 0 : 0 : 0 : 0 : 0 : 1

could be simplified as:

:: 1

5.10.10   The transition from IPv4 to IPv6

Figure 5.57, shows how IPv4 and IPv6 can coexist on an Ethernet network. The type field identifies which protocol is in the payload, with 0x0800 for IPv4 and 0x86DD for IPv6.

The benefits derived from a new protocol must also be balanced by the costs associated with making a transition from the existing systems. These logistical and technical issues have been addressed in RFC 1933, 'Transition Mechanisms for IPv6 Host and Routers'.

The developers of IPv6 recognized that not all systems would upgrade from IPv4 to IPv6 in the immediate future, and that for some systems, that upgrade may not be for years. To complicate matters, most internetworks are heterogeneous systems, with various routers, hosts etc. manufactured by different vendors. If such a multivendor system were to be upgraded at one time, IPv6 capabilities would be required on all of the individual elements before the large project could be attempted. Another (much larger) issue is the worldwide Internet, which operates across 24 different time zones, Upgrading this system in a single process would be even more difficult.

Given the above constraints, it therefore becomes necessary to develop strategies for IPv4 and IPv6 to coexist, until such time as IPv6 becomes the preferred option. At the time of writing, two mechanisms for this coexistence have been proposed: a dual IP layer and IPv6 over IPv4 tunnelling. These two alternatives are discussed in the following sections.

5.10.11   Dual IP layer

The simplest mechanism for IPv4 and IPv6 coexistence is for both of the protocol stacks to be implemented on the same station. The station, which could be a host or a router,

Dest Add	Source Add	0800	IPv4 packet	FRC
Dest Add	Source Add	86DD	IPv6 packet	FRC

Figure 5.57   Example of Ethernet carrying IP

is referred to as an IPv6/IPv4 node. The IPv6/IPv4 node has the capability to send and receive both IPv4 and IPv6 packets, and can therefore interoperate with an IPv4 station using IPv4 packets and with an IPv6 station using IPv6 packets. The IPv6/IPv4 node would be configured with addresses that support both protocols, and those addresses might or might not be related to each other. Figure 5.58 illustrates a station with both IPv4 and IPv6.

5.10.12 Tunnelling

Tunnelling is a process whereby information from one protocol is encapsulated inside the frame or packet of another architecture, thus enabling the original data to be carried over that second architecture. The tunnelling scenarios for IPv6/IPv4 are designed to enable an existing IPv4 infrastructure to carry IPv6 packets by encapsulating the IPv6 information inside IPv4 packets.

Figure 5.59 illustrates how two stations that use only the IPv6 protocol can communicate via an IPv4 network such as the current Internet. The IPv4 packets contain both an IPv4 header and IPv6 header, as well as all of the upper-layer information, such as the TCP header, application data, etc. The tunnelling process involves three distinct steps: encapsulation, decapsulation and tunnel management. At the encapsulating node (or tunnel entry point), the IPv4 header is created, and the encapsulated packet is transmitted. At the decapsulating node (or tunnel exit point), the IPv4 header is removed and the IPv6 packet is processed. In Figure 5.59, the encapsulation and decapsulation is accomplished by the routers, which will have the capability of processing both IPv4 and IPv6.

RFC 1933 defines four possible tunnel configurations that could be established between routers and hosts:

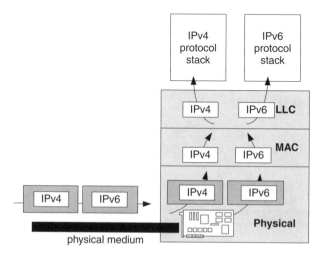

Figure 5.58 IPv4 and IPv6 support

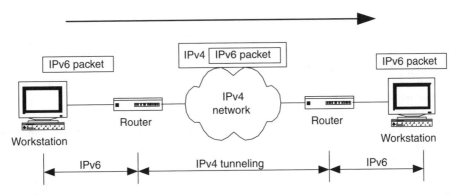

Figure 5.59   IPv6 tunnelling

- *Router-to-router*: IPv6/IPv4 routers that are separated by an IPv4 infrastructure tunnel IPv6 packets between themselves. In this case, the tunnel would span one segment of the packet's end-to-end path.

- *Host-to-router*: an IPv6/IPv4 host tunnels IPv6 packets to an IPv6/IPv4 router that is reachable via an IPv4 infrastructure. In this case, the tunnel would span the first segment of the packet's end-to-end path.

- *Host-to-host*: IPv6/IPv4 hosts that are interconnected by an IPv4 infrastructure can tunnel IPv6 packets across the IPv4 infrastructure. In this case, the tunnel spans the packet's entire end-to-end path.

- *Router-to-host*: IPv6/IPv4 routers can tunnel IPv6 packets to an IPv6/IPv4 host which is the final destination. In this case, the tunnel would span only the final segment of the packet's end-to-end path.

5.11   SERIAL LINE IP (SLIP) AND POINT-TO-POINT PROTOCOL (PPP)

SLIP (RFC 1055) is an extremely simple protocol which can be used to encapsulate IP packets on a serial line (or other point-to-point connection). SLIP defines two special character values, END (192 decimal) and ESC (219) decimal. Figure 5.60 shows the SLIP packet. Each IP packet is sent as is and terminated with an END character. If the END character is found inside the packet, two characters are sent instead: ESC END. If ESC is found then ESC ESC is sent.

		END
IP header	Payload	192

Figure 5.60   SLIP packet format

Note that SLIP is only an encapsulation scheme and does not allow the automatic allocation of IP addresses or other configuration information. It also does not support the negotiation of such things as header compression. For these reasons it has been mainly replaced by PPP.

Point-to-point protocol (PPP) is a datalink protocol that allow access to the Internet and other networks using TCP/IP. It is a multi-protocol transport system which can also transport IPX, AppleTalk and other protocols simultaneously over a single connection. PPP negotiates configuration parameters at the beginning of your connection and these details are transparent to the user. Further information can be found in RFC 1548.

PPP contains three main components:

1. A derivative of the high-level datalink control (HDLC) protocol for encapsulating datagrams over serial links.

2. A link control protocol (LCP) to establish, configure and test the datalink connection.

3. A group of network control protocols (NCPs) for establishing different network layer protocols.

To establish communication over a point-to-point link, PPP first sends out LCP frames to configure and optionally test the link. After the link has been negotiated and configured there is an optional authentication phase, where one or both parties have to prove their identity. Finally, PPP sends out NCP packets to negotiate and configure the actual network layer protocols to be used over the link. Once this is complete then the link can be used to transfer the actual data packets of the particular protocol setup. Figure 5.61, shows the PPP packet format, each of the fields in the packet is described below.

5.11.1   LCP link establishment

The first procedure required when connecting using PPP is link establishment. Figure 5.62 Table 5.16 shows the format of an LCP packet. The code determines the type of packet and these types are describe in Table 5.18. The basic procedure for opening a connection is as follows.

The initiator sends a configure-request. This can contain a number of requested link options, for example the maximum frame size or the authentication protocol for the link. The responder will respond with one of the following:

• Configure-ack, in which case the options have been accepted and the link is open.

• Configure-nak is used to reject the options requested by the sender but allows for some renegotiation of the options listed.

• Configure-reject is used to reject the options requested by the sender.

Flag 8 bits	Address 8 bits	Control 8 bits	Protocol 16 bits	Information variable	FCS 16 bits

Figure 5.61   Point-to-point protocol

Table 5.16   Summary of PPP fields

Field	Description
Flag	A single byte that indicates the beginning or end of the frame; it always consists of 01111110
Address	A single byte that is always 11111111 (FF$_h$), since PPP does not assign individual layer 2 addresses
Control	A single byte that always contains the sequence 00000011, which means that the data will be unsequenced frames
Protocol	Identifies the datagram that is being transferred. Protocols starting with a value of 0–3 indicate the network layer being transported. A field starting with 8–b indicates an associated NCP. Values of c–f indicate link layer control protocols such as LCP. A number of example protocols and their associated values are listed in Table 5.17.
Data	Zero or more bytes that contain the datagram for the protocol specified, i.e. if the protocol carried is IP, then the user data will be encapsulated within this packet. The default maximum size for the information field is 1500 bytes
FCS	Frame check sequence for error detection.

Table 5.17   PPP protocol field

Value	Protocol name
0021	Internet Protocol
002b	Novell IPX
002d	Van Jacobson compressed TCP/IP
002f	Van Jacobson uncompressed TCP/IP
8021	Internet Protocol control protocol
802b	Novell IPX control protocol
c021	Link control protocol
c023	Password authentication protocol
c223	Challenge-handshake authentication protocol

flag	address	control	protocol	code	identifier	length	data	
01111110	11111111	3	c021	8-bits	8-bits	16-bits	variable	FCS

Figure 5.62   PPP LCP packet

To close the link down, the terminate-request is sent followed by a response of the terminate-Ack. Code-reject is sent if the responder does not understand the code number. This could happen if different versions of PPP are used on each end of a link. Protocol-reject is sent if one of the PPP peers has received a PPP packet with an unknown or unauthorized PPP protocol field type. Echo-request and echo-reply are used to test the link, as is discard-request.

Table 5.18 LCP packet codes

Code	Name	Description
1	Configure-request	Requests establishment of link, can include request for options
2	Configure-ack	Confirms establishment of link
3	Configure-nak	Rejects option values requested
4	Configure-reject	Rejects options outright
5	Terminate-request	Request to close connection
6	Terminate-ack	Confirms closing of connection
7	Code-reject	Unrecognized code value
8	Protocol-reject	Packet with unsupported or un-negotiated PPP protocol field sent
9	Echo-request	Link test
10	Echo-reply	Link test
11	Discard-request	For test purposes; this packet is discarded silently

5.11.2 PPP authentication

PPP also offers two methods for automating logins: password authentication protocol (PAP) and challenge-handshake authentication protocol (CHAP). PAP is an insecure method of authentication since it sends the user's login name and password unencrypted across the channel. CHAP, on the other hand, is more secure since the login name and password are encrypted before being sent across the channel.

CHAP, as described in RFC 1999, is an authentication protocol for use with PPP. Four messages types are supported.

- *challenge*: contains the random number;
- *response*: contains hash of random number plus secret;
- *success*: indicates that the user was authenticated successfully;
- *fail*: indicates authentication failed.

Each CHAP message is encapsulated within a PPP frame with a protocol ID of c223. The basic operation of the protocol is shown in Figure 5.63. In this case user B is requesting authentication from user A. User B sends a challenge to user A consisting of a long random number. User A computes the response using a hash function and the shared secret and sends this back to the user B. If user B sees that the hash value is correct, user A is authenticated and success is indicated, otherwise a failure message is returned.

This protocol has a number of important features. Without the original secret the hash value cannot be computed correctly. As long as the random number is long and random enough, the challenge will nearly always be different. This makes it very difficult for a hacker to use recorded responses to fool the authenticating party. Also the password is not sent over the network (only the hashed value), protecting the shared secret from being snooped.

RFC 1999 allows for various hash algorithms to be used but only requires support for MD5. The algorithm type is negotiated using PPP protocol negotiation. The length of the response values for MD5 is 16 bytes. The responder will concatenate the identifier,

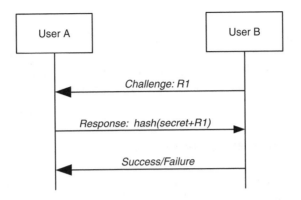

Figure 5.63   CHAP

shared secret and challenge and perform a hash function over the result. The hash value is sent back in the response. Finally, the success or failure packets are used to indicate the result of the authentication. The name on the challenge and response packets indicates the identity of the system transmitting the packet. For a client wanting authentication from a network access server, the name will indicate the client's username. The message in the success and failure packets can be displayed to the user when their request is accepted or denied.

5.11.3   Network control protocol (NCP) for IP

To enable IP transport over the PPP connection the NCP protocol for IP must be used. The NCP protocol for IP is called PPP Internet Protocol control protocol or IPCP (see RFC 1332).

IPCP allows the configuration of IP options such as the IP host address and header compression. The format of the packet is shown in Figure 5.64.

Note that this is the same format as the LCP packet except the protocol code has been changed. The codes have the same meaning as LCP with only code numbers 1–7 being supported; other values are to be ignored. It is also similar in operation in that options are requested and the responder either accepts or rejects them. There are two options supported by IPCP.

IP address

A peer can request the responder to give it an IP address by sending an IP address option with the IP address set to all zeroes

flag	address	control	protocol	code	identifier	length	data	
01111110	11111111	3	8021	8-bits	8-bits	16-bits	variable	FCS

Figure 5.64   IPCP packet format

Header compression

A peer can request the responder to provide Van Jacobson TCP/IP header compression (RFC 1144).

A client requesting an IP connection will send a configure-request. The network access server will respond with either a configure-ack, configure-nak or configure-reject.

5.11.4   IP packet encapsulation

The encapsulation options for IP within PPP are shown in Figure 5.65. At the top is a standard IP datagram with a protocol field of 0021. In the middle is a TCP/IP packet with a Van Jacobson compressed header and at the bottom a Van Jacobson uncompressed TCP/IP header. In the uncompressed case the IP protocol field of the IP header indicates the slot to be updated.

5.11.5   PPP in 3G

PPP has been introduced as an access protocol in UMTS, allowing the transparent transportation of protocols such as AppleTalk and IPX. Figure 5.66, shows how PPP is used in this context.

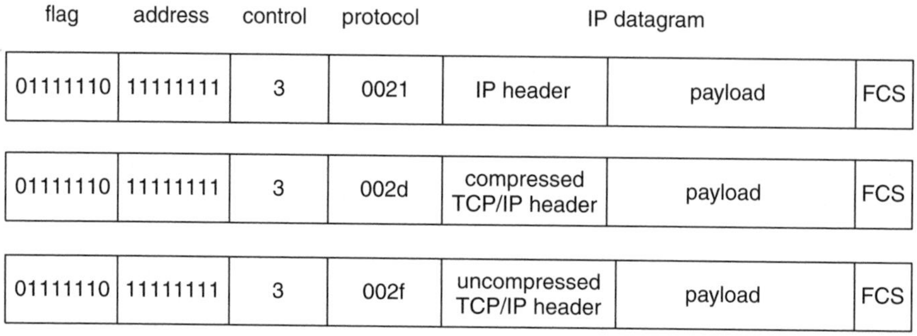

flag	address	control	protocol	IP datagram		
01111110	11111111	3	0021	IP header	payload	FCS
01111110	11111111	3	002d	compressed TCP/IP header	payload	FCS
01111110	11111111	3	002f	uncompressed TCP/IP header	payload	FCS

Figure 5.65   PPP IP encapsulation

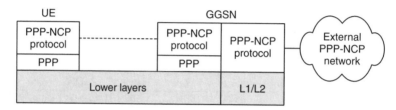

Figure 5.66   PPP protocol tunnelling

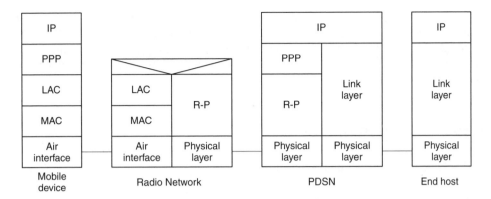

Figure 5.67 PPP for CDMA2000

PPP and IPCP are also used to provide IP access for CDMA2000 systems. The configuration is shown in Figure 5.67. PPP is used to provide transport between the mobile station and the interworking function. Since PPP is being used any network protocol (e.g. AppleTalk) can be transported as long as it is supported by both the interworking function (IWF) and the mobile station (MS).

5.12 RADIUS ACCOUNTING, AUTHORIZATION AND AUTHENTICATION (AAA)

The remote authentication dial-in user service (RADIUS), defined by RFC 2865, is a protocol developed to carry out authorization, authentication and accounting functions (AAA). The protocol allows a network access server to access a centrally located shared authentication and accounting server. Figure 5.68 shows an example RADIUS configuration for GPRS/UMTS. Here the GGSN is acting as a network access server and is therefore a RADIUS client, communicating with the RADIUS server on behalf of the mobile user. Managing a central database of authentication information in this way is the simplest and most secure method of allowing authenticated access for a dispersed set of access lines. RADIUS is used in CDMA2000 to provide connections between the packet data serving node (PDSN) and the central authentication and accounting servers.

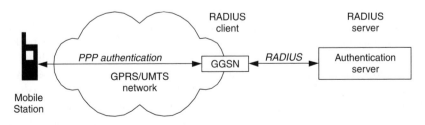

Figure 5.68 RADIUS configuration

Each network access server acts as a RADIUS client, requesting services from the central authentication (or accounting) server which acts as the RADIUS server. RADIUS is not used to transport authentication data end-to-end, the actual authentication mechanism between the network access client and the access server will be some other protocol such as PPP authentication using CHAP.

5.12.1   RADIUS functions

RADIUS provides the following functions:

- *Authentication*: determination of identity of a user via the use of a shared secret.
- *Authorization*: allowing or rejecting user access to the network based on their profile and the current security policy.
- *Host configuration*: providing configuration data for a user connecting to the NAS, for example providing an IP address for the host to use.
- *Accounting*: controlling usage statistics for accounting purposes (RFC 2866).

The user connecting to the network access server (NAS) can request a range of services via the RADIUS architecture, for example a PPP connection or login connection to a particular host using telnet.

5.12.2   RADIUS authentication and configuration

When the NAS has a client requesting access to the network, it constructs an access-request message. This message will contain the username and password and also information about what type of service is being requested. The username will typically be in the format of a network access identifier. This looks very much like an email address, i.e. user@host. An example would be bobsmith@orbitage.com. The password in the request is obscured by only sending a hash dependant of the password instead of the password itself. The NAS (RADIUS client) then sends the access-request to the authentication server (RADIUS server).

The RADIUS server will look up the user's name within its database and retrieve the user's password. It will then use the password to check the authenticity of the request. If the identity of the user is confirmed and the server determines they should have access (authorization check) an access-accept message is sent back to the RADIUS client. If any of the conditions for access are not met, an access-reject message is sent back. If the username is not present in the database the access request will be silently discarded.

It is also possible for the server to generate an access-challenge message in response to the original request. In this case the radius client (NAS) is expected to re-submit the original access request using the shared secret to generate the hash as well as state information found in the access-challenge message. This technique is more secure than the standard access procedure.

The configuration data for the RADIUS client is contained with the access-accept message. Possible configuration values are:

- IP address to be allocated to the host;
- compression technique for link;
- remote host address for login.

5.12.3   RADIUS accounting

RADIUS accounting is covered by RFC 2866. The NAS acts as a client to the RADIUS accounting server and sends information to it regarding the usage statistics of clients.

Two types of message are supported, accounting-request and accounting-response. The accounting-request is sent from client to server; the server replies with the accounting response. Each type of accounting request is distinguished by the acct-status-type attribute. Possible values for this field are;:

- *start*: start accounting for a user session;
- *stop*: stop accounting for a user session;
- *on*: turns on accounting (for example on NAS start up);
- *off*: turns off accounting for this user (on scheduled NAS shutdown).

When a client connects to the NAS, a start accounting-request message is sent to the accounting server. This contains a unique session ID attribute, username and an identifier (or IP address) for the NAS. This information allows the accounting server to build a unique record for each session.

At the end of the session a stop accounting-request is sent. This contains attributes giving the user's usage statistics for example the number of bytes sent/received. It will also contain the same session information as found in the start accounting-request, so the server can update the correct charging record.

5.13   DIAMETER AAA

The AAA working group of the IETF focuses on authentication, authorization and accounting functions for network access. The group identified the following requirements for AAA. 3GPP recommends the use of DIAMETER to provide accounting functions in the IP multimedia subsystem (IMS; see Chapter 9).

- Support for IPv6
- Backwards compatibility with RADIUS
- Explicit proxy support
- Lightweight security model.

A number of RFCs have been produced by the AAA working group. The core functionality is provided by a protocol called DIAMETER (www.ietf.org/internet-drafts/draft-ietf-aaa-diameter-17.txt). This is a base protocol which is extendable to produce a range of AAA applications. Table 5.19 shows some of the AAA documents released to date.

The DIAMETER base protocol does not provide full AAA functionality on its own and is used in conjunction with the application protocols. For example, a NAS will be expected to support the DIAMETER base protocol as well as the DIAMETER network access application.

The base protocol supports session management and the transfer of attribute value pairs (AVPs) (between peers, these are explained in the following section). It also provides a base set of commands to handle simple accounting transactions. DIAMETER supports enhanced reliability using the dynamic peer discovery. A domain will be configured with both primary and backup DIAMETER servers.

5.13.1 Attribute value pairs (AVPs)

DIAMETER and DIAMETER application protocols move data between peers using AVPs. Each AVP consists of a data structure containing an attribute descriptor and its value. Figure 5.69 shows the basic AVP format.

The AVP code in conjunction with the vendor ID identifies each attribute uniquely. The use of the vendor ID is optional and is indicated by the setting of the V bit to 1. This allows a particular vendor to develop their own specific application by defining their own set of AVPs. The M or mandatory bit is set for AVPs that must be supported for a given request. If a peer receives a DIAMETER request containing an AVP with the M bit set to 1 and it does not understand the AVP, it must then reject the request. The P or privacy bit indicates that the AVP should be protected on the link using encryption end-to-end.

Table 5.19 AAA documents

Name	Functionality
Diameter base protocol	Peer-to-peer session management, accounting, fail over and proxy Defines support for TLS and IPSec
Diameter network access server application	AAA service for NAS Support for PPP CHAP and RADIUS/DIAMETER interworking
Diameter Mobile IPv4 application	AAA service for mobile IPv4
AAA transport profile	Use of TCP and SCTP for AAA
DIAMETER EAP application	Use of DIAMETER to provide extensible authentication protocol (EAP) to PPP users
DIAMETER cryptographic message syntax application	Transport of X.509 certificate between DIAMETER peers

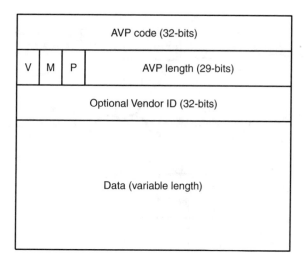

Figure 5.69 AVP format

5.14 MOBILE IP

Since many users on the Internet move from one location to another, it would be advantageous for them to be able to access their home network resources and services on the move. This is becoming more important as more users connect via their PDAs or laptops. Ideally a user would like to connect to any network access point within a customer premises or even use a mobile connection and be connected transparently to their home network via the Internet.

The problem here is with the Internet addressing itself. Recall that an Internet address consists of 4 bytes, e.g. 152.226.23.45. The 152.226 in this example identifies the home network of a user; all packets addressed to the user machine will be directed by the Internet routers to this network. If the user resides on this network then there is no problem. However, if the user moves to another network, e.g. 145.67, then the packets will never reach him or her. Using DHCP, a user can attach to a new network and be given a new IP address but this does not solve the problem, since it allows access to this visited network but not the user's home network.

How does the system work in practice? As illustrated in Figure 5.70, the user's home network must have a *home agent* set up and the visited network must have a *foreign agent* set up. Once the computer is attached to the visited network, it will contact the foreign agent by soliciting an advertisement (or just wait until the foreign agent advertises itself). In its advertisement the foreign agent will provide a list of care-of addresses that can be used by the mobile node while it is residing at the foreign network.

5.14.1 Mobile IP routing

The care-of address is used as follows. Each mobile IP user has an associated care-of address which is registered with its home agent. This is referred to as the home user's

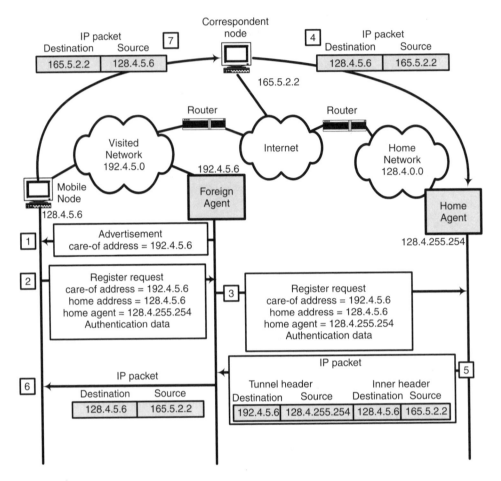

Figure 5.70   Example of mobile IP

address binding. When packets arrive for the station, the home agent will intercept them and then divert them to the foreign agent using the mobile node's care-of address. Each packet received by the mobile node is tunnelled between the home and foreign agents by encapsulating it in an outer IP header. This allows the original packet to remain unmodified and the process of mobility to be transparent to the end hosts. When the foreign agent receives the tunnelled packet, it removes the outer header and forwards the contents to the mobile user directly. In the reverse direction, the packets can be sent directly to the source and do not have to traverse the mobile user's home network.

Looking at the call scenario in Figure 5.70 in some detail, the following steps are performed.

1. The mobile host receives an advertisement containing the care-of address 192.4.5.6.
2. The mobile node sends a registration request containing the care-of address, home agent's address (128.4.255.254), mobile node's home address (128.4.5.6) and

authentication data to the foreign agent. An expiry time value is also included (not shown) which indicates how long the binding is valid for.

3. The foreign agent forwards the registration request to the home agent. The home agent then uses this information to create a binding between the node's home address (128.4.5.6) and the care-of address (192.4.5.6).

4. The packet sent from the correspondent node to 128.4.5.6 is intercepted by the home agent.

5. The home agent tunnels the packet using the care-of address registered for 128.4.5.6.

6. The foreign agent receives the tunnelled packet, removes the outer header then forwards it to the mobile node.

7. The mobile node sends a reply back to the correspondent node directly.

5.14.2   Mobile IP security

Since mobile IP registration requests are used to alter the routing of IP packets, rogue registrations (i.e. from unauthorized users) could be used to facilitate a denial of service attack on the network. For this reason mobile IP registration messages contain an authentication field. This field is generated via the use of a secret which is shared between the mobile user and its home agent. When a registration arrives at the home agent, the authentication field is checked; if it is found to be invalid, the request is ignored.

5.14.3   Route reverse tunnelling

From Figure 5.70, it is clear that the packet sent directly from the mobile node to the correspondent node has an illegal source address. The network prefix for this packet is 128.4, that of its home network. However, now it is residing on a network with prefix 192.4.5, and hence there is a mismatch. It is common for security devices (e.g. firewalls) to filter out packets which have illegal IP source addresses. This is to protect the network from becoming a source of certain types of denial of service attack. For the example given all packets sent directly in the return direction would be filtered. To get round this problem, a scheme called route reverse tunnelling has been developed. In this mechanism packets sent in the reverse direction from the mobile user to the correspondent address are tunnelled between the foreign and home agent back to the home network. The home agent removes the tunnel header before forwarding the packets to their final destination.

5.14.4   Route optimization

One can see from Figure 5.70, that the route packets take from the correspondent node to the mobile node is non-optimal. This is because the correspondent node is unaware of the mobile node's care-of address. Ideally, there should be a mechanism which allows the mobile node to optimize the route. To achieve this a scheme has been proposed which allows the mobile node to update a correspondent node directly with its care-of address binding. This is illustrated in Figure 5.71.

1. The first packet sent from the correspondent's address is via the home agent, which forwards it to the mobile node. Of course, to do this the mobile user must have still registered with the home agent.
2. The mobile node then sends a binding update to the correspondent node, containing its care-of and home address binding.
3. The correspondent node then sends packets directly to the user's care-of address.

The use of route optimization is difficult, however, due to problems of authentication. The binding update must be authenticated (to guard against denial of service attack). If the updates are unauthenticated, an attacker could send a spoof binding, fooling the correspondent node into sending packets to the wrong destination. Authentication between the mobile node and the home agent is relatively simple since they can both be configured with a shared secret. Authentication between the correspondent node and the mobile node is a lot more difficult, since the correspondent can be any node on the Internet. Obviously configuring a different shared secret between the mobile node and all other nodes on the Internet would be totally impractical. The problem of security with route optimization is still work in progress for the IETF.

5.14.5 Mobile IP for IPv6

For IPv6, the use of the foreign agent is dropped and all care-of addresses are co-located at the mobile host. Three messages are supported: binding update, binding acknowledgment and binding request.

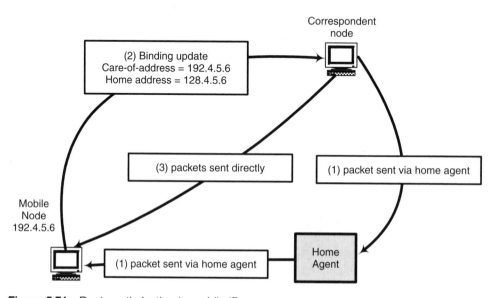

Figure 5.71 Route optimization in mobile IP

The binding update serves the same purpose as the registration request but does not contain an authentication field. This is because authentication for IPv6 is supported using IPSec as a standard header extension (i.e. AH or ESP). Binding updates are sent to the user's home agent, and they can also be sent to a correspondent node directly (to achieve route optimization). However, the authentication problem for route optimization still exists. For this reason, a mobile node is only allowed to send binding updates to correspondents with which they can develop an IPSec security association. The binding acknowledgement is sent in reply to a binding update, to ensure reliability.

Finally, the binding request is sent by a correspondent to request a new update; for example, it may have a care-of address listed for the mobile node which is about to expire. The mobile node is expected to reply with a new binding update.

In summary, mobile IP with IPv6 is simpler and more scalable than with IPv4. It uses the inherent security mechanisms provide with IPv6 (i.e. IPSec) and provides support for route optimization as standard.

5.14.6   Foreign agent handover and mobile IP

When the mobile station is on the move using a cellular radio service, as it moves from one IP subnet to another, a new foreign agent must be contacted and the connection to the home agent must be re-established. However, packets may still be being delivered to the old foreign agent, which does not know the new foreign agent's care-of address. A hand off (handover) system is required to ensure a smooth transition with no packet loss.

To help guard against packet loss one possibility is to have a transition time when the home agent supports registrations to both foreign agents at the same time (see Figure 5.72). The ability to support multiple bindings at the home agent is supported via the use of the S (simultaneous) flag in the registration request. If the flag is set to 1 this instructs the

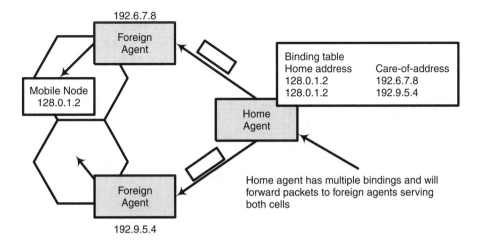

Figure 5.72   Mobile IP cell handover

home agent to add the new binding but leave all existing bindings in place. The procedure to achieve smooth handover will be as follows:

1. Contact the new foreign agent before the cell handover has taken place to obtain a new care-of address.
2. Register the new care-of address with the home agent with the S bit set. Now the home agent will deliver packets to both cells.
3. After a certain amount of time to allow the node to handover to the new cell, the mobile node re-registers the new care-of address but this time with the S bit set to 0 to remove the old binding.

5.14.7   Mobile IP for CDMA2000

Mobile IP is supported in CDMA2000. The foreign agent functionality is sited at the PDSN and supports the following functions:

- routing to mobile station from home agent;
- route reverse tunnelling back to home agent;
- hand off between PDSNs without involving home network;
- establishment of IPSec security association with home agent;
- sending agent advertisements to its served mobile stations;
- dynamic home address assignment;
- AAA service for visiting users.

Mobile IP provides similar functionality in CDMA2000 as the GTP within UMTS. Both support the roaming of the user within the radio access network without having to change IP address. The difference is that mobile IP allows the user to roam beyond the cellular domain and still remain contactable. The GTP tunnel, on the other hand, only exists within the operator's network and therefore cannot be used to provide mobile routing when the user attaches using a non-GPRS connection method.

5.14.8   Mobile IP for UMTS

TS23.923 is a feasibility study on the use of mobile IP within UMTS to provide tunnelling and mobility management. It describes two architectures. The first is an overlay of mobile IP on the current GPRS network to provide truly ubiquitous mobility, in which a user can be reached whether they are connected via GPRS or other means. This is a conventional mobile IP architecture and it is assumed that the foreign agent functionality is placed within the GGSN. In this case the GGSN will send out a foreign agent advertisement on receipt of a PDP context request. Since not all GGSN may support IP mobility the choice of this service would be decided by the APN of the original request.

The second proposal is a wholesale replacement of GTP with mobile IP. In this the SGSN and GGSN are combined into one unit called the Internet GPRS support node (IGSN). The IGSN would act as a foreign agent and provide a tunnel to transfer packets between itself and the user's home agent. Note that TS 23.923 is only a feasibility study and is not a mandatory requirement for UMTS.

5.15 SUMMARY

In this chapter the application of IP to 3GPP R99 UMTS networks is explored. The IP suite of protocols is examined as well as their application to the UMTS environment, in particular to the GPRS core network. The need for QoS is justified as well as a number of mechanisms on how it can be implemented, including the contrasting DiffServ and IntServ approaches. Also, security for IP is covered in two forms, session based and connectionless, and in particular the TLS and IPSec protocols were described in some detail. The use of PPP and CHAP to provide a configured and authenticated IP point-to-point service is examined, as is its application to CDMA2000. Finally, the AAA protocols are examined, and their use within IP, paying particular attention to RADIUS (for CDMA2000) and DIAMETER (for UMTS).

FURTHER READING

RFC 0114: File Transfer Protocol. A. K. Bhushan. Apr-10-1971.

RFC 0495: Telnet Protocol specifications. A. M. McKenzie. May-01-1973.

RFC 0760: DoD standard Internet Protocol. J. Postel. Jan-01-1980.

RFC 0768: User Datagram Protocol. J. Postel. Aug-28-1980.

RFC 0791: Internet Protocol. J. Postel. Sep-01-1981.

RFC 0793: Transmission Control Protocol. J. Postel. Sep-01-1981.

RFC 0799: Internet name domains. D. L. Mills. Sep-01-1981.

RFC 0826: Ethernet Address Resolution Protocol: Or converting network protocol addresses to 48.bit Ethernet address for transmission on Ethernet hardware. D. C. Plummer. Nov-01-1982.

RFC 0903: Reverse Address Resolution Protocol. R. Finlayson, T. Mann, J. C. Mogul, M. Theimer. Jun-01-1984.

RFC 1034: Domain names – concepts and facilities. P. V. Mockapetris. Nov-01-1987.

RFC 1035: Domain names – implementation and specification. P. V. Mockapetris. Nov-01-1987.

RFC 1055: Nonstandard for transmission of IP datagrams over serial lines: SLIP. J. L. Romkey. Jun-01-1988.

RFC 1058: Routing Information Protocol. C. L. Hedrick. Jun-01-1988.

RFC 1144: Compressing TCP/IP headers for low-speed serial links. V. Jacobson. Feb-01-1990.

RFC 1320: The MD4 Message-Digest Algorithm. R. Rivest. April 1992.

RFC 1321: The MD5 Message-Digest Algorithm. R. Rivest. April 1992.

RFC 1332: The PPP Internet Protocol Control Protocol (IPCP). G. McGregor. May 1992.

RFC 1518: An Architecture for IP Address Allocation with CIDR. Y. Rekhter, T. Li. September 1993.

RFC 1519: Classless Inter-Domain Routing (CIDR): an Address Assignment and Aggregation Strategy. V. Fuller, T. Li, J. Yu, K. Varadhan, September.

RFC 1548: The Point-to-Point Protocol (PPP). W. Simpson. December 1993.

RFC 1633: Integrated Services in the Internet Architecture: an Overview. R. Braden, D. Clark, S. Shenker. June 1994.

RFC 1752: The Recommendation for the IP Next Generation Protocol. S. Bradner, A. Mankin. January 1995.

RFC 1771: A Border Gateway Protocol 4 (BGP-4). Y. Rekhter, T. Li. March

RFC 1809: Using the Flow Label Field in IPv6. C. Partridge. June 1995.

RFC 1881: IPv6 Address Allocation Management. IAB, IESG. December 1995.

RFC 1918: Address Allocation for Private Internets. Y. Rekhter, B. Moskowitz, D. Karrenberg, G. J. de Groot, E. Lear. February 1996.

RFC 1933: Transition Mechanisms for IPv6 Hosts and Routers. R. Gilligan, E. Nordmark. April 1996.

RFC 1939: Post Office Protocol – Version 3. J. Myers, M. Rose. May 1996.

RFC 1999: Request for Comments Summary RFC Numbers 1900–1999. J. Elliott. January 1997.

RFC 2001: TCP Slow Start, Congestion Avoidance, Fast Retransmit, and Fast Recovery Algorithms. W. Stevens. January 1997.

RFC 2002: IP Mobility Support. C. Perkins, Ed. October 1996.

RFC 2060: Internet Message Access Protocol – Version 4rev1. M. Crispin. December 1996.

RFC 2131: Dynamic Host Configuration Protocol. R. Droms. March 1997.

RFC 2139: RADIUS Accounting. C. Rigney. April 1997.

RFC 2205: Resource ReSerVation Protocol (RSVP) – Version 1 Functional Specification. R. Braden, Ed., L. Zhang, S. Berson, S. Herzog, S. Jamin. September 1997.

RFC 2210: The Use of RSVP with IETF Integrated Services. J. Wroclawski. September 1997.

RFC 2211: Specification of the Controlled-Load Network Element Service. J. Wroclawski. September 1997.

RFC 2212: Specification of Guaranteed Quality of Service. S. Shenker, C. Partridge, R. Guerin. September 1997.

RFC 2246: The TLS Protocol Version 1.0. T. Dierks, C. Allen. January 1999.

RFC 2309: Recommendations on Queue Management and Congestion Avoidance in the Internet. B. Braden, D. Clark, J. Crowcroft, B. Davie, S. Deering, D. Estrin, S. Floyd, V. Jacobson, G. Minshall, C. Partridge, L. Peterson, K. Ramakrishnan, S. Shenker, J. Wroclawski, L. Zhang. April 1998.

RFC 2328: OSPF Version 2. J. Moy. April 1998.

RFC 2338: Virtual Router Redundancy Protocol. S. Knight, D. Weaver, D. Whipple, R. Hinden, D. Mitzel, P. Hunt, P. Higginson, M. Shand, A. Lindem. April 1998.

RFC 2401: Security Architecture for the Internet Protocol. S. Kent, R. Atkinson. November 1998.

RFC 2402: IP Authentication Header. S. Kent, R. Atkinson. November 1998.

RFC 2409: The Internet Key Exchange (IKE). D. Harkins, D. Carrel. November 1998.

RFC 2412: The OAKLEY Key Determination Protocol. H. Orman. November 1998.

RFC 2474: Definition of the Differentiated Services Field (DS Field) in the IPv4 and IPv6 Headers. K. Nichols, S. Blake, F. Baker, D. Black. December 1998.

RFC 2475: An Architecture for Differentiated Service. S. Blake, D. Black, M. Carlson, E. Davies, Z. Wang, W. Weiss. December 1998.

RFC 2507: IP Header Compression. M. Degermark, B. Nordgren, S. Pink. February 1999.

RFC 2510: Internet X.509 Public Key Infrastructure Certificate Management Protocols. C. Adams, S. Farrell. March 1999.

RFC 2511: Internet X.509 Certificate Request Message Format. M. Myers, C. Adams, D. Solo, D. Kemp. March 1999.

RFC 2581: TCP Congestion Control. M. Allman, V. Paxson, W. Stevens. April 1999.

RFC 2582: The NewReno Modification to TCP's Fast Recovery Algorithm. S. Floyd, T. Henderson. April 1999.

RFC 2597: Assured Forwarding PHB Group. J. Heinanen, F. Baker, W. Weiss, J. Wroclawski. June 1999.

RFC 2598: An Expedited Forwarding PHB. V. Jacobson, K. Nichols, K. Poduri. June 1999.

RFC 2616: Hypertext Transfer Protocol – HTTP/1.1. R. Fielding, J. Gettys, J. Mogul, H. Frystyk, L. Masinter, P. Leach, T. Berners-Lee. June 1999.

RFC 2821: Simple Mail Transfer Protocol. J. Klensin, Ed. April 2001.

RFC 2865: Remote Authentication Dial In User Service (RADIUS). C. Rigney, S. Willens, A. Rubens, W. Simpson. June 2000.

RFC 2866: RADIUS Accounting. C. Rigney. June 2000.

RFC 3022: Traditional IP Network Address Translator (Traditional NAT). P. Srisuresh, K. Egevang. January 2001.

RFC 3024: Reverse Tunneling for Mobile IP, revised. G. Montenegro, Ed. January 2001.

RFC 3095: RObust Header Compression (ROHC): Framework and four profiles: RTP, UDP, ESP, and uncompressed. C. Bormann, C. Burmeister, M. Degermark, H. Fukushima, H. Hannu, L.-E. Jonsson, R. Hakenberg, T. Koren, K. Le, Z. Liu, A. Martensson, A. Miyazaki, K. Svanbro, T. Wiebke, T. Yoshimura, H. Zheng. July 2001.

RFC 3096: Requirements for robust IP/UDP/RTP header compression. M. Degermark, Ed. July 2001.

RFC 3135: Performance Enhancing Proxies Intended to Mitigate Link-Related Degradations. J. Border, M. Kojo, J. Griner, G. Montenegro, Z. Shelby. June 2001.

3GPP TS23.002: Network Architecture.

3GPP TS23.101: General UMTS Architecture.

3GPP TS23.107: Quality of Service (QoS) concept and architecture.

3GPP TS23.207: End to end quality of service concept and architecture.

3GPP TS23.922: Architecture for an All IP network.

3GPP TS23.923: Combined GSM and Mobile IP mobility handling in UMTS IP CN.

3GPP TS24.008: Mobile radio interface Layer 3 specification; Core network protocols; Stage 3.

3GPP TS25.323: Packet Data Convergence Protocol (PDCP) specification.

3GPP TS29.060: General Packet Radio Service (GPRS); GPRS Tunnelling Protocol (GTP) across the Gn and Gp interface.

draft-ietf-aaa-diameter-09.txt: Diameter Base Protocol, P. Calhoun, j. Arkko, 03/04/2002.

draft-ietf-mobileip-optim-11.txt: Route Optimization in Mobile IP, C. Perkins, D. Johnson, 09/06/2001.

draft-ietf-mobileip-ipv6-15.txt: Mobility Support in IPv6, C. Perkins, D. Johnson, 11/21/2001.

S. Floyd, V. Jacobson (1993) 'Random early detection gateways for congestion avoidance', *IEEE/ACM Transactions on Networking*, **1**(4), 397–413.

A list of the current versions of the specifications can be found at http://www.3gpp.org/specs/web-table_specs-with-titles-and-latest-versions.htm, and the 3GPP ftp site for the individual specification documents is http://www.3gpp.org/ftp/Specs/latest/

6

Universal Mobile Telecommunications System

6.1 UMTS NETWORK ARCHITECTURE

Development is now well on the road towards third generation (3G), where the network will support all traffic types—voice, video and data—and we should see an eventual explosion in the services available on the mobile device. The driving technology for this is the Internet protocol (IP). Many cellular operators are now at a position referred to as 2.5G, with the deployment of the general packet radio service (GPRS), which introduces an IP backbone into the mobile core network. Figure 6.1 shows an overview of the key components in a GPRS network, and how it fits into the existing global system for mobile communications (GSM) infrastructure.

The interface between the serving GPRS support node (SGSN) and the gateway GPRS support node (GGSN) is known as the Gn interface and uses the GPRS tunnelling protocol (GTP, as discussed in Chapter 4). The primary reason for the introduction of this infrastructure is to offer connections to external packet networks, such as the Internet or a corporate intranet.

This brings the IP protocol into the network as a transport between the SGSN and GGSN. This allows data services such as email or web browsing on the mobile device, with users being charged based on volume of data rather than time connected.

The first deployment of the universal mobile telecommunications system (UMTS) is the release 99 (R99) architecture, shown in Figure 6.2.

In this network, the major change is in the radio access network (RAN) with the introduction of code division multiple access (CDMA) technology for the air interface, referred to as wideband CDMA (WCDMA), and asynchronous transfer mode (ATM) as a transport in the transmission part. These changes have been introduced principally to support the transport of voice, video and data services on the same network. The core

Convergence Technologies for 3G Networks: IP, UMTS, EGPRS and ATM   J. Bannister, P. Mather and S. Coope
© 2004 John Wiley & Sons, Ltd   ISBN: 0-470-86091-X

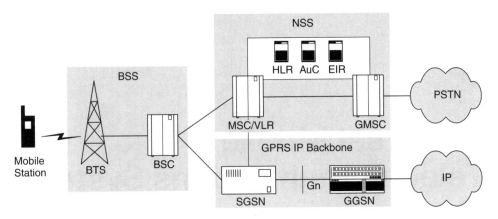

Figure 6.1 GPRS general architecture. Note that there are connections from the MSC and SGSN to the home location, visitor location and equipment identity registers (HLR/VLR/EIR)

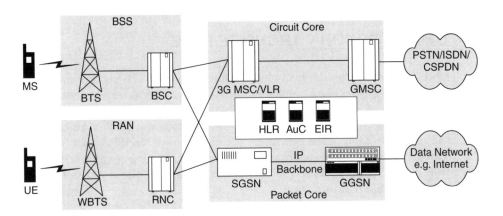

Figure 6.2 UMTS release 99 network

network remains relatively unchanged, with primarily software upgrades. However, the IP pushes further into the network, with the radio network controller (RNC) now transferring data with the 3G SGSN using IP.

The roles of the key new components in the network are described in the following subsections.

6.1.1 WCDMA base station (WBTS)

The Third Generation Partnership Project (3GPP) specifications refer to the base station as a Node B. However, it is more common to see this referred to as a WBTS, BTS or even BS. Throughout this book, the BTS notation will be used. Officially, a Node B is a network entity that serves a single cell. However, sectorized sites are much more efficient and economical so a commercial outdoor BTS would generally be expected to support multiple cells across the full spectrum of the required operating frequency. A typical BTS configuration is support of up to six sectors, with two carriers per sector.

The BTS is the termination point between the air interface and the transmission network of the RAN. It is therefore required to support both WCDMA and ATM, connecting through a plesiochronous or synchronous digital hierarchy (PDH or SDH) interface. The BTS must provide all the necessary signal processing functions to support the WCDMA air interface and this is where most of the complexity arises. In addition, the provision of interfaces to microwave PDH or SDH radio solutions is also desirable. Some solutions offer ATM cross-connection equipment, and ATM circuit emulation services to support combined transport of 2G and 3G traffic. It is also common that some manufacturers have multipurpose BTS solutions, which support multiple technologies on the one hardware platform, such as transceivers for GSM, enhanced data rates for global evolution (EDGE) and WCDMA. A BTS solution should also provide antenna diversity in both the uplink and the downlink.

6.1.2   Radio network controller (RNC)

The RNC is the heart of the new access network. All decisions of the network operation are made here, and at its centre is a high-speed packet switch to support a reasonable throughput of traffic. An RNC is responsible for control of all the BTSs that are connected to it, and maintains the link to the packet and circuit core network, that is the mobile switching centre (MSC) and the SGSN. It also needs to be capable of supporting interconnections to other RNCs, a new feature of UMTS. Most of the decision-making process is software based, so a high processing capacity is required. This chapter deals with much of the functionality of the RNC, such as radio resource management (RRM).

6.1.3   3G mobile switching centre (3G MSC)

For UMTS R99, the changes to the core network side are minimal, and these should be mostly in the form of software upgrades to support the new access network. The role played by the 3G MSC is exactly the same here as in GSM. However, a 2G MSC is a narrowband device and connects to the access network via the A interface. The traffic is expected to be in 64 kbps, and for voice this should be 64 kbps pulse code modulation (PCM). The RAN, on the other hand, presents the circuit core network with an interface which is transporting speech across ATM, and uses the adaptive multirate (AMR; refer to Section 6.13), which codes speech to a range from 4.75 kbps to 12.2 kbps. Therefore, an interworking function (IWF) is needed between the RAN and the MSC. The role of the IWF is twofold: first, for user traffic it is responsible for transcoding of speech to and from 64 kbps PCM. If the traffic is circuit switched data, then it is responsible for transferring to and from the narrowband time division multiplexing (TDM) time slots. Second, for control information, it is responsible for mapping between the MSC signalling messages and the signalling messages to the RAN (RANAP protocol, see later). The combination of this IWF and a 2G MSC is considered a 3G MSC. For many manufacturers, this IWF is a separate functional unit, which enables them to retain the existing hardware of a GSM network. It is generally known as a media gateway (MGW) as this hardware platform can be reused in subsequent UMTS releases to switch traffic between TDM, ATM and IP technologies.

6.2 NETWORK EVOLUTION

The next evolution step is the release 4 (R4) architecture (Figure 6.3). Here, the GSM core is replaced with an IP network infrastructure based around voice over IP (VoIP) technology.

The MSC evolves into two separate components: an MGW and an MSC server (MSS). This essentially breaks apart the roles of connection and connection control. An MSS can handle multiple MGWs, making the network more scalable.

Since there are now a number of IP clouds in the 3G network, it makes sense to merge these together into one IP or IP/ATM backbone (it is likely both options will be available to operators.) This extends IP right across the whole network, all the way to the BTS. This is referred to as the all-IP network, or the release 5 (R5) architecture, as shown in Figure 6.4. The HLR/VLR/EIR are generalized and referred to as the HLR subsystem (HSS).

Now the last remnants of traditional telecommunications switching are removed, leaving a network operating completely on the IP protocol, and generalized for the transport of many service types. Real-time services are supported through the introduction of a new network domain, the IP multimedia subsystem (IMS). The architecture of R4 and R5 is discussed further in Chapters 8 and 9.

Currently the 3GPP are working on release 6, which purports to cover all aspects not addressed in frozen releases. Some call UMTS release 6 4G and it includes such issues as interworking of hotspot radio access technologies such as wireless LAN.

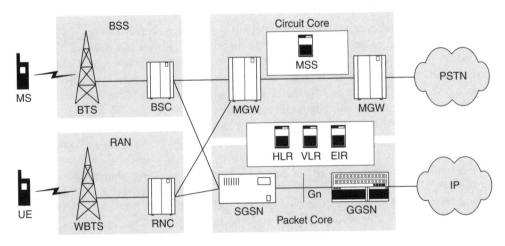

Figure 6.3 UMTS release 4 architecture

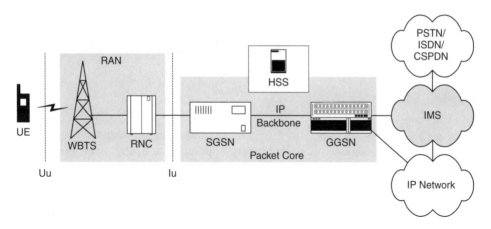

Figure 6.4   UMTS release 5 architecture

6.3   UMTS FDD AND TDD

Like any CDMA system, UMTS needs a wide frequency band in which to operate to effectively spread signals. The defining characteristic of the system is the chip rate, where a chip is the width of one symbol of the CDMA code. UMTS uses a chip rate of 3.84 Mchips/s and this converts to a required spectrum carrier of 5 MHz wide. Since this is wider than the 1.25 MHz needed for the existing cdmaOne system, the UMTS air interface is termed *wideband* CDMA.

There are actually two radio technologies under the UMTS umbrella: UMTS FDD and TDD. FDD stands for frequency division duplex, and, like GSM, separates traffic in the uplink and downlink by placing them at different frequency channels. Therefore an operator must have a pair of frequencies allocated to allow it to run a network, hence the term 'paired spectrum'. TDD or time division duplex requires only one frequency channel, and uplink and downlink traffic are separated by sending them at different times. The ITU-T spectrum usage, as shown in Figure 6.5, for FDD is 1920–1980 MHz for uplink traffic, and 2110–2170 MHz for downlink. The minimum allocation an operator needs is two paired 5 MHz channels, one for uplink and one for downlink, at a separation of 190 MHz. However, to provide comprehensive coverage and services, it is recommended that an operator be given three channels. Considering the spectrum allocation, there are 12 paired channels available, and many countries have now completed the licensing process for this spectrum, allocating between two and four channels per licence. This has tended to work out a costly process for operators, since the regulatory authorities in some countries, notably in Europe, have auctioned these licences to the highest bidder. This has resulted in spectrum fees as high as tens of billions of dollars in some countries.

The TDD system, which needs only one 5 MHz band in which to operate, is often referred to as unpaired spectrum. The differences between UMTS FDD and TDD are only evident at the lower layers, particularly on the radio interface. At higher layers, the bulk of the operation of the two systems is the same. As the name suggests, the TDD system separates uplink and downlink traffic by placing them in different time slots. As

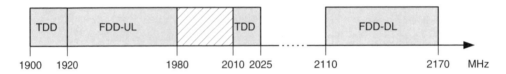

Figure 6.5   UMTS frequency allocation

will be seen later, UMTS uses a 10 ms frame structure which is divided into 15 equal time slots. TDD can allocate these to be either uplink or downlink, with one or more breakpoints between the two in a frame defined. In this way, it is well suited to packet traffic, since this allows great flexibility in dynamically dimensioning for asymmetry in traffic flow.

The TDD system should not really be considered as an independent network, but rather as a supplement for an FDD system to provide hotspot coverage at higher data rates. It is rather unsuitable for large-scale deployment due to interference between sites, since a BTS may be trying to detect a weak signal from a user equipment (UE), which is blocked out by a relatively strong signal at the same frequency from a nearby BTS. TDD is ideal for indoor coverage over small areas.

Since FDD is the main access technology being developed currently, the explanations presented here will focus purely on this system.

6.4   UMTS BEARER MODEL

The procedures of a mobile device connecting to a UMTS network can be split into two areas: the access stratum (AS) and the non-access stratum (NAS). The AS involves all the layers and subsystems that offer general services to the NAS. In UMTS, the AS consists of all of the elements in the RAN, including the underlying ATM transport network, and the various mechanisms such as those to provide reliable information exchange. All of the NAS functions are those between the mobile device and the core network, for example mobility management. Figure 6.6 shows the architecture model. The AS interacts with the NAS through the use of service access points (SAPs).

The UMTS terrestrial radio access network (UTRAN) provides this separation of NAS and AS functions, and allows for AS functions to be fully controlled and implemented within the UTRAN. The two major UTRAN interfaces are the Uu, which is the interface between the mobile device, or UE, and the UTRAN, and the Iu, which is the interface between the UTRAN and the core network. Both of these interfaces can be divided into control and user planes, each with appropriate protocol functions.

A bearer service is a link between two points, which is defined by a certain set of characteristics. In the case of UMTS, the bearer service is delivered using radio access bearers (RABs).

A RAB is defined as the service that the AS (i.e. UTRAN) provides to the NAS for transfer of user data between the UE and core network. A RAB can consist of a number of subflows, which are data streams to the core network within the RAB that

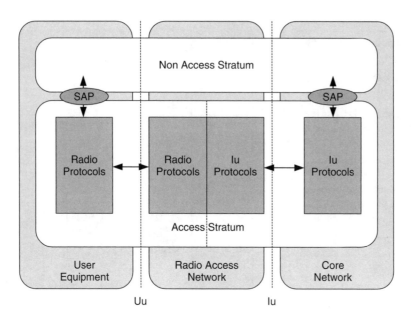

Figure 6.6   UMTS general architecture

have different quality of service (QoS) characteristics, such as different reliabilities. A common example of this is that different classes of bits with different bit error rates can be realized as different RAB subflows. RAB subflows are established and released at the time the RAB is established and released, and are delivered together over the same transport bearer.

A radio link is defined as a logical association between a single UE and a single UTRAN access point, such as an RNC. It is physically comprised of one or more radio bearers and should not be confused with a RAB.

Looking within the UTRAN, the general architecture model is as shown in Figure 6.7(a). Now shown are the Node B or BTS and RNC components, and their respective internal interfaces. The UTRAN is subdivided into blocks referred to as radio network subsystems (RNS), where each RNS consists of one controlling RNC (CRNC) and all the BTSs under its control. Unique to UMTS is the interface between RNSs, the Iur interface, which plays a key role in handover procedures. The interface between the BTS and the RNC is the Iub interface.

All of the 'I' interfaces – Iu, Iur and Iub – currently[1] use ATM as a transport layer. In the context of ATM, the BTS is seen as a host accessing an ATM network, within which the RNC is an ATM switch. Therefore, the Iub is a user-to-network interface (UNI), whereas the Iu and Iur interfaces are considered to be network-to-network interfaces (NNI), as illustrated in Figure 6.7(b).

This distinction is because the BTS to RNC link is a point-to-point connection in that a BTS or RNC will only communicate with the RNC or BTS directly connected to it, and will not require communication beyond that element to another network element.

---

[1]UMTS release 5 provisions for the use of IP as a RAN transport protocol.

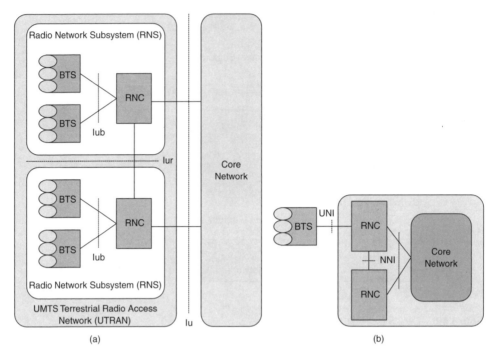

Figure 6.7 UTRAN architecture

For each user connection to the core network, there is only one RNC, which maintains the link between the UE and core network domain, as highlighted in Figure 6.8. This RNC is referred to as the serving RNC (SRNC). That SRNC plus the BTSs under its control is then referred to as the SRNS. This is a logical definition with reference to that UE only. In an RNS, the RNC that controls a BTS is known as the controlling RNC (CRNC). This is with reference to the BTS, cells under its control and all the common and shared channels within.

As the UE moves, it may perform a soft or hard handover to another cell. In the case of a soft handover, the SRNC will activate the new connection to the new BTS. Should the new BTS be under the control of another RNC, the SRNC will also alert this new RNC to activate a connection along the Iur interface. The UE now has two links, one directly to the SRNC, and the second through the new RNC along the Iur interface. In this case, this new RNC is logically referred to as a drift RNC (DRNC) (Figure 6.8). It is not involved in any processing of the call and merely relays it to the SRNC for connection to the core. In summary, SRNC and DRNC are usually associated with the UE and the CRNC is associated with the BTS. Since these are logical functions it is normal practice that a single RNC is capable of dealing with all these functions.

A situation may arise where a UE is connected to a BTS for which the SRNC is not the CRNC for that BTS. In that situation, the network may invoke the serving RNC relocation procedure to move the core network connection. This process is described in Section 6.19.

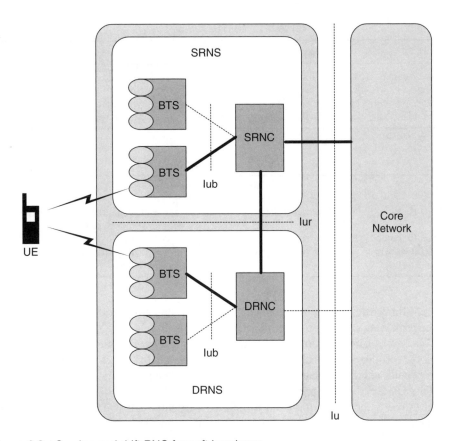

Figure 6.8 Serving and drift RNC for soft handover

6.5 UMTS QOS CLASSES

QoS has been defined by the ITU-T as 'the collective effect of service performance, which determines the degree of satisfaction of a user of a service'. QoS is associated with the user experience; the user is not concerned with how a service is provided but only whether or not they are satisfied with that service. So, from a user's point of view the QoS is a subjective matter and if the network does not perform adequately, the user may decide not to use a particular service or look around for another mobile network operator offering the same service with a better QoS. From the mobile network operator's point of view, this QoS requires technical analysis where the operator has to overcome certain implementation challenges within a cost constraint. 3G is the convergence of circuit switched networks such as GSM and the IP packet networks. A brief description detailing the state of QoS in these two different types of network now follows.

For UMTS, four different QoS classes are defined as well as an expected standard set of bit rates. These are summarized in Tables 6.1 and 6.2.

Table 6.1 UMTS QoS classes

Name	Delay	Buffering	Mode	Bit rate
Conversational	Minimal fixed	None	Symmetric	Guaranteed
Streaming	Minimal variable	Allowed	Asymmetric	Guaranteed
Interactive	Moderate variable	Allowed	Asymmetric	Not guaranteed
Background	Large variable	Allowed	Asymmetric	Not guaranteed

Table 6.2 UMTS bit rates

Bit rate	Comment	Coverage type
144 kbps (basic)	Peak rate for packet transfer	Rural/suburban, fast moving vehicles, outdoor
384 kbps (extended)	Peak rate for packet transfer	Urban, moving vehicles, outdoor
2 Mbps (hotspot)	Peak rate for packet transfer	Urban centre, walking speeds, indoor

For real-time traffic, only the conversational and streaming classes will be relevant since the remaining classes provide little QoS in terms of delay and bandwidth provision.

GSM networks have traditionally measured QoS in a network's busy hour call blocking probability, call drop rates, call setup delay and voice quality. Cost constraints limit the number of cells and the amount of transmission links to the base station controllers (BSCs) and BTSs. Once the call is registered, the user and mobile device are authenticated and checked to see if they are authorized to make a particular call. Once these checks have been completed successfully then the call goes through. This is not instantaneous and may take 5 or 10 seconds before the ringing tone is heard. Once the call goes through, the parties have a time slot for themselves over the air interface, a TDM slot between the BTS and the transcoder and rate adaptation unit (TRAU) and usually a 64 kbps link through the MSC and across the external network. There is little need for buffering in the network. If a user service on the mobile device wishes to send data more quickly then it is buffered in the mobile device. Thus there is little jitter and delays are comparatively constant. The resources were reserved at call setup and are used only by the called and calling parties.

The 3G network has a variety of services to offer, each of which requires its own distinct QoS provisions. A multimedia call may consist of voice, video and whiteboard connections, each of which requires its own QoS.

Within the scope of 3G, QoS is a requirement for many end-to-end communications. It can be seen in Figure 6.9, that for this end-to-end QoS, the QoS has to be guaranteed across a number of different areas. For the voice call, the PSTN or ISDN will guarantee a certain quality of service since at call setup the resources are provided and set aside for this call for the duration. However, it must also be noted that this call setup does require time, especially for an international call to a mobile device. This can have an adverse effect on the user perception of the quality of a call. One aspect of QoS which is a simple concept is that of delay. Once the call is established over the ISDN, the delay will be almost constant at approximately 20 ms, which is virtually unnoticeable by the calling

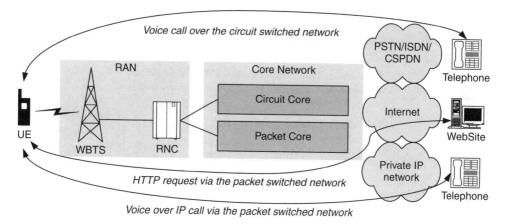

Figure 6.9 QoS considerations

parties. The end-to-end delay, however, takes into consideration:

Delay over ISDN + Delay over core network + Delay over RAN + Delay over air

+ Processing time in mobile device, and others + Buffering = Total delay

Although on the surface, delay is one of the more simple aspects of QoS to quantify, it is important to also take into consideration retransmissions (if data is retransmitted). The air interface works at the speed of light and thus on initial inspection seems to have little effect on the overall delay. However, this is where most of the interference takes place and there may be a number of retransmissions. How long should one wait for a packet? If the overall delay of the connection is more than 300 ms then for a voice call, this will most likely be noticed by the user.

Examining an IP network, it can be seen that the delay for browsing a web page is not so important. A user will already tolerate delays of the order of 10 seconds or more for a web page over a current dial-up connection. With the higher data rates available with UTMS, it is anticipated that a page could be downloaded in about 4 seconds. However, the design goal is to reduce this to 1 or 2 seconds.

QoS is provided in the UMTS network via bearer services. Figure 6.10 shows how the end-to-end service depends on the services below it. The UTRAN consists of the BTS and the RNC, the core network (CN) edge device will be the 3G MSC for circuit switched and SGSN for packet switched and the CN gateway will be the gateway-MSC for circuit switched and the GGSN for the packet switched network.

It can be seen for end-to-end QoS, there is a requirement that both the UMTS network and external bearer services provide QoS. In fact, if a user is working on a laptop which is connected to the mobile device then QoS is also needed between these devices (known as the local bearer service).

The RAB requires that there is a guarantee of service over the WCDMA air interface and also across the UTRAN. The air interface QoS for a particular mobile device is related to the power it transmits compared to the surrounding noise. The RNC provides the bearer control over the UTRAN.

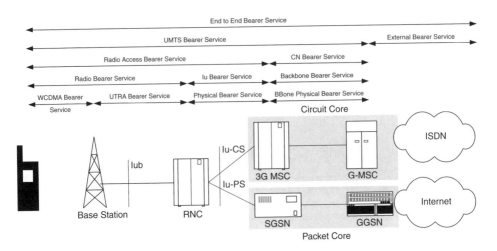

Figure 6.10   End-to-end QoS

6.6   UTRAN CHANNELS

UMTS channels at the UE are divided into a three-layer hierarchy: logical layer, transport layer and physical layer. A logical channel is a stream of information that is dedicated to the transfer of a particular type of information over the radio interface. A transport channel is how logical channels are transported between the UE and RNC. A physical channel is the actual channel across the air interface, defined by a WCDMA code and frequency. Figure 6.11 shows the layered model

As an analogy, consider that the post office offers the logical channel of conveyance of letters and parcels from one location to another. As transport, air mail, surface mail,

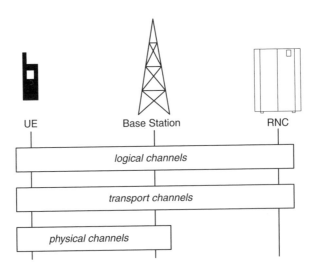

Figure 6.11   UMTS channel structure

registered etc. are offered, and the physical channel could be the particular vehicle in which the item is carried. The logical channel does not need to know that an Airbus A340 cargo plane is used in this case; it only needs to know that it will be physically transported and that the transport will meet the required QoS (e.g. a flight is used to minimize delay).

Figures 6.12 and 6.13, show the various channels available for logical, transport and physical in the downlink and uplink directions, and how the channels are mapped. The mapping of logical channels to transport channels is performed by the media access control (MAC) layer, discussed later. For clarity there are a number of channels which have only physical layer significance that are not included in this diagram. These channels, such as the access indication channel (AICH), will be discussed in reference to the channels below, as and when required.

6.6.1 Logical channels

- *Broadcast control channel (BCCH)*: this is a downlink channel which carries all the general system information that a UE needs to communicate with the network.
- *Paging control channel (PCCH)*: this is a downlink channel which carries paging information from the network to inform the user that there is a communications request.

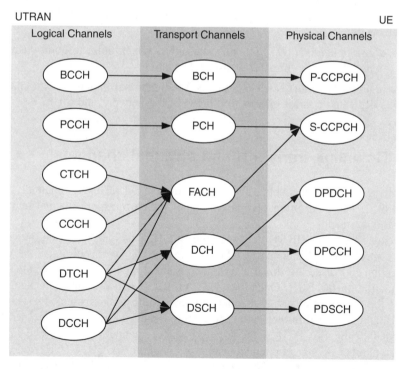

Figure 6.12 Downlink mapping of logical, transport and physical channels

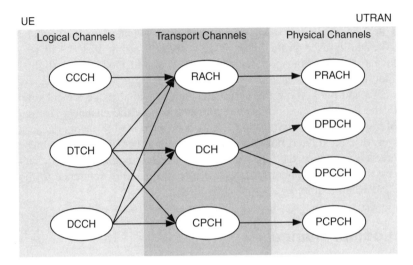

Figure 6.13 Uplink mapping of logical, transport and physical channels

- *Common traffic channel (CTCH)*: this is a downlink channel which carries dedicated user information to a group of specified UEs.
- *Common control channel (CCCH)*: this is a channel, both uplink and downlink, which carries control information from the network to UEs that do not have any dedicated channels.
- *Dedicated traffic channel (DTCH)*: this is a bidirectional, point-to-point channel, dedicated to one UE, for the transfer of user information.
- *Dedicated control channel (DCCH)*: this is a bidirectional, point-to-point channel that transmits dedicated control information between the network and a UE.

6.6.2 Downlink transport and physical channels

The following transport channels offer information transfer of logical channels as shown in Figure 6.12. There are often several options of which transport channel may be used to perform this transfer.

The broadcast channel (BCH) transports the BCCH. This is then carried in the primary common control physical channel (PCCPCH). The physical layer coding, i.e. the WCDMA channelization code used, for this is a system constant as all UEs need to be able to decode the information on the BCH. Once decoded, the BCCH will contain system information indicating the coding of all other channels present. The paging channel (PCH) transports the PCCH, and is carried in the secondary common control physical channel (SCCPCH). Also carried in the SCCPCH is the forward access channel (FACH). The FACH can transport a number of logical channels carrying common (CCCH, CTCH) and dedicated (DCCH, DTCH) control and traffic. Dedicated traffic and control can also be transported on a dedicated channel (DCH) or on a downlink shared channel (DSCH), where multiple users can be statistically multiplexed on the one transport channel. As will be seen, it is the RNC that decides, based on the user requirements, whether the user data is transported on a DCH or on a common or shared channel (FACH, DSCH). For example, a user may be allocated a low data rate DCH, and then simultaneously allocated larger, variable-rate transport on the DSCH. A more detailed example of this is presented in Section 6.15. At the physical layer, the DCH is carried in two physical channels (which are combined into one channel, time multiplexed together), separated into dedicated physical data channel (DPDCH) and dedicated physical control channel (DPCCH). The DPCCH contains physical layer control, such as the format of the data and power control information. The DSCH is carried in the physical downlink shared channel (PDSCH).

6.6.3   Uplink transport and physical channels

In the uplink, as shown in Figure 6.13, the random access channel (RACH) is present for the transport of dedicated traffic as well as common and dedicated control information. It is on the RACH that the UE makes its initial access to the network. The RACH is carried in the physical RACH (PRACH). The dedicated channel (DCH) is the same as the downlink; however at the physical layer, it is actually carried in two separate physical channels rather than multiplexed into a single channel as is the case in the downlink. The common packet channel (CPCH) is another channel for data traffic, but designed specifically for transport of packet data. It is carried in the physical common packet channel (PCPCH). The CPCH transport is associated with downlink transport on the FACH, where a user would receive data on the FACH, but be permitted to transmit on the CPCH. Since it is a common channel it requires a sophisticated control mechanism to ensure that two UEs do not try to access the channel at the same time. This mechanism introduces a number of other channels which only have relevance at the physical layer.

6.7   RADIO RESOURCE MANAGEMENT (RRM)

RRM is responsible for making sure that the resources of the UTRAN, and particularly of the air interface, are used efficiently. Its target is to maximize performance to provide all users with coverage and quality, regardless of which service they wish to use. RRM can be broken into two categories, those that are network based, and therefore performed for a whole cell, and those that are connection based, which are performed per connection. The RRM functions are summarized in Table 6.3 and a brief explanation of each is presented. The actual operation of the algorithm is implementation dependent. RRM is implemented in software and handled at the RNC.

6.7.1   Admission control

Admission control is used to provide resources for guaranteed, real-time services, such as voice calls. When a user requests resources, the admission control algorithm will verify

Table 6.3 Radio resource management

Category	RRM function
Network based	Admission control
	Packet scheduler
	Load control
Connection based	Handover control
	Power control

that it can meet the request by checking the loading of the cell and the available WCDMA and transmission resources. Based on this, if resources are available, the RNC will allocate and reserve them, and then update the load control and admit the user. It is at this point that the RNC will decide what type of channel to allocate a user. Should the resources not be available, the RNC will either reject the request or make a counter offer of resources that are available. The admission control is designed to meet a planned load level, placed a reasonable safety margin below the maximum resources available. A typical figure would be 50–70% planned load.

The decision process for admission control is referred to as call admission control (CAC), and in the UTRAN it is performed by the RNC. There are also admission control procedures performed in the core network. This role is executed in the CS domain by the MSC, and in the PS domain by both the SGSN and GGSN. Before a call is established, the network decides whether there are enough resources available for this connection. If not, either the call is queued, offered a lower quality of service or rejected. The RNC has to consider the effect on other users within the cell and possibly within adjacent cells if this call proceeds.

This is one of the most technically demanding aspects of implementing a soft capacity system. With GSM, CAC over the air interface is rather simple: a quick check to see if there is a time slot available; if so, admit the call. There are further checks for authentication and authorization, but these are separate issues to CAC. There are a number of performance measurements over the air interface which may be taken into consideration when considering the admission of another subscriber or modifying the data rate of an existing subscriber. These measurements can include, but are not limited to, the following:

- received signal strength of the current and surrounding cells;
- estimated bit error rate of the current and surrounding cells;
- received interference level;
- total downlink transmission power per cell.

Within the WCDMA system, each user shares the same 5 MHz spectrum at the same time as other users, and users are separated by a code. As more and more users are permitted connections in a cell, they cause interference for the other users in the cell and also for other users in adjacent cells since the frequency reuse is 1, meaning that adjacent cells share the same frequency. A user sending at a high data rate introduces a higher quantity of wideband noise than a user who is simply making a voice call. The network therefore has to know beforehand what data rate a user requires and estimate how much noise this will generate in the cell. If this noise value is too high it will cause too much interference for other users in the cell and their call quality will be reduced. WCDMA works on a 10 ms frame and it is therefore possible for users to modify their current connection data rates and QoS frequently, placing enormous strain on the system. It is also important for the CAC algorithm to take into consideration the load on surrounding cells. This is to ensure that users who are on the move from cell to cell do not lose their connection.

6.7.2 Packet scheduler

The packet scheduler uses the resources not currently allocated to schedule non-real-time (NRT) data. Since NRT data does not place any demands on the network in terms of delay requirements, this traffic only needs to be scheduled in terms of relative priority. Should real-time traffic need more resources, the NRT data will be rescheduled. Figure 6.14, illustrates the concept. Real-time traffic is considered as non-controllable load since the network must guarantee the resources and thus has no control over these. NRT traffic, however, offers such a compromise; it can expand and contract to fit the available resources and hence is termed controllable load.

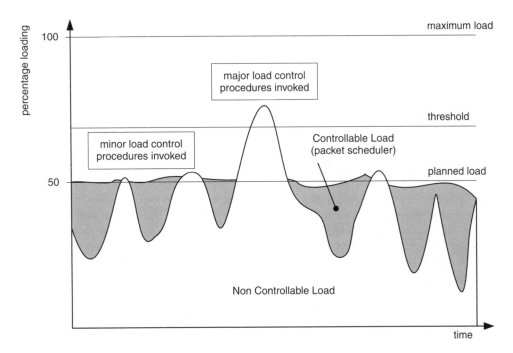

Figure 6.14 Load control in UMTS

6.7.3   Load control

Load control constantly monitors the air interface to maintain a clear picture of interference levels in the cells. It interacts with the admission control and packet scheduler to try to keep the loading below some target threshold level. Should the threshold be exceeded, then procedures will be invoked to bring it back down below again. These are regarded as emergency procedures and would include extreme actions such as suspension of NRT bearers or dropping of real-time bearers.

There are a number of strategies for admission and load control to enable a new user to enter a cell:

- The current mobile devices could increase their power but this causes more interference for other users. The mobile device is limited in the amount of power it can transmit due to the battery ampere-hour.
- The new mobile device could be granted a lower data rate than requested.
- Other voice call users could be shifted over to GSM (for example) if there is coverage.
- Soft handover could be performed for high data rate users. This would reduce the average transmitted power and hence reduce interference.
- Users on the fringe of the cell could be asked to move to other cells.
- Calls could be dropped in some controlled way.
- Use outer loop power control to reduce uplink signal to interference ratio (SIR).
- Decrease bit rate of speech calls by use of adaptive multirate (AMR) CODECS.
- Users who have been allocated dedicated channels but who have a low utilization can be moved to the shared or common channels.

By its nature, the air interface is prone to interference. The WCDMA system is designed to work within these constraints by constantly monitoring system loading.

6.7.4   Handover control

The RNC is in charge of management of handovers. In UMTS, two categories of handover exist: soft handover and hard handover. Soft handover is when a new radio link is initiated before the old link is released. It is unique to CDMA technology and possible since adjoining cells can operate at the same frequency. This provides for a better connection quality. In the context of the UTRAN, the handovers summarized in Tables 6.4 and 6.5 are available.

Table 6.4   Soft handover types

Handover type	Explanation
Softer handover	Handover between cells under the control of the same BTS
Intra-RNC soft handover	Handover between BTSs under the control of the same RNC
Inter-RNC soft handover	Handover between BTSs under the control of different RNCs

Table 6.5 Hard handover types

Handover type	Explanation
Intra-frequency hard handover	Handover between BTSs under the control of different RNCs where the Iur interface is unavailable
Inter-frequency hard handover	Handover between cells or BTSs operating at different frequencies
Inter-system hard handover	Handover between UMTS and another cellular technology. Initially, the focus is on support for UMTS/GSM handover
SRNC relocation	Moving of the SRNC, as detailed later

6.7.4.1 Soft handover

Soft handover is a mechanism which enables the mobile device to communicate with a number of base stations at the same time. This is only possible with the CDMA system since adjacent base stations will be working on the same carrier frequencies. When a mobile device transmits, it transmits in an omnidirectional area, which means that much of the signal will actually be in the wrong direction, away from the receiving antenna. It is interesting to note that transmission power from mobile devices is considered to be greater for non-voice services. This is since typically the voice user will place the device close to the head when making a call, and thus a large percentage of the radiated signal is absorbed. This is not generally the case for other services, where the device is held away from the body. Calculations for loss generally factor in a 3 dB 'head loss' figure for voice calls.

Soft handover allows the UE to communicate through another cell to maximize the utilization of this signal. This scenario is not possible with GSM since adjacent cells transmit and receive on different frequencies. In such a case, the mobile device would have to send the data on two separate frequencies and this defeats the whole purpose of soft handover. With reference to the soft handover, the cells that a UE can detect are broken into three categories (Table 6.6).

Figure 6.15 shows a mobile device in soft handover. In this example the mobile device is connected to three separate base stations and thus there are three separate connections over the Iub interface. The lower base station is controlled by a different RNC to the BTSs at the top of the diagram. This is referred to as a *drift RNC* (DRNC) and it simply passes the information for this mobile device across to the serving RNC. There is only one connection through to the core network and this is via the SRNC. All of the cells in the figure must be using the same frequency as each other for this system to function.

Table 6.6 Cell sets

Cell set	Description
Active set	Current set of cells in which the UE has an active connection and is sending and receiving information (i.e. >1 for soft handover)
Monitored set	A set of cells, not in the active set, that the UE has been instructed by the RNC to monitor as possible handover candidates
Detected set	All other cells that the UE has detected. Reporting of measurements only occurs for intra-frequency measurements

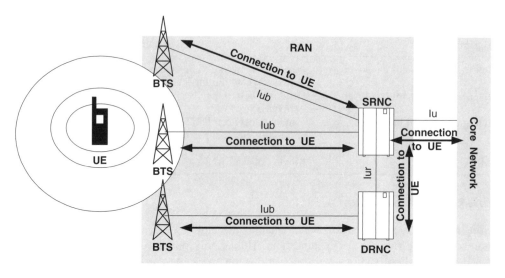

Figure 6.15   Soft handover

In the uplink the different base stations receive exactly the same information from the mobile device and these are combined at the SRNC in a procedure referred to as *macrodiversity*. Only one stream of information is passed across the Iu interface from the RNC towards the CN. The RNC takes the information streams from the Iub and Iur interfaces and simply ensures that the stream passed on conforms to the QoS requirements. The soft handover mechanism can generate a lot of traffic over the RAN Iub and Iur interfaces and this increased data transfer is weighed against efficiencies that can be made over the air interface. During the macrodiversity process at the SRNC, the different data streams are checked on a per-transport block basis for block error rate (BLER) or bit error rate (BER) against a reference target value. The best-quality block in each case is forwarded to the core network. Since the RNC has multiple streams from which to work, the overall data quality is increased. If this increase pushes the quality in excess of the required target to meet the QoS, this indicates that it is possible to reduce the transmit power of the mobile device, thus reducing the interference caused to other mobile subscribers in the cells, and increasing cell capacity.

In the downlink, a number of base stations may transmit exactly the same information to the mobile device. Again, this is used to reduce the downlink power over the air interface and the mobile device will internally perform a diversity process. The base station that is to be added to the active set needs to know the existence of the mobile device. Thus the RNC needs to pass it the scrambling code for the mobile device, as well as other connection parameters such as coding schemes, number of parallel code channels etc. Although the same wideband frequency is used by adjacent cells, the cells are physically identified by their different scrambling codes. The mobile device constantly monitors the common pilot channel (CPICH) of cells in its surrounding area for power levels. This information can then be used to assess whether or not the cell should be added to the active set. The mobile device passes this information in the form of measurement reports to the SRNC.

If a single base station is controlling a number of cells, typically referred to as sectors, each of which is utilizing the same frequency, it is possible for soft handover to occur within these cells. This form of soft handover is termed *softer handover*. In the downlink direction, there is no difference; however, in the uplink, the diversity operation is performed at the BTS, rather than the RNC. This diversity is an RF level procedure, where the combination is performed by Maximal Ratio Combining (MRC) as was explained in Section 2.8. For softer handover, however, the UE will need to take into account the different scrambling codes used by the two branches. A mobile device can simultaneously be in soft and softer handover.

Figure 6.16 shows a simplified graph of when a mobile device will perform a soft handover. It is assumed in the example that the mobile device is currently connected to cell-1 and is physically moving over timeout of this cell and into another cell, cell-2. The mobile device constantly monitors the cells in the surrounding area and passes the measurement reports of the CPICH signal strength to the SRNC. It can be seen that the measurement reports for cell-1 will indicate that the signal is getting weaker and for cell-2 it is getting stronger. When the reports from cell-2 indicate to the SRNC that the signal strength of the CPICH-2 is within a certain threshold of the stronger cell-1, a timer is set. If the CPICH-2 continues to be within the threshold of cell-1 for a predetermined amount of time then cell-2 will be added to the active set via an *active set update* message from the RNC to the mobile device. The timer is required to reduce the amount of exchanges, since if the subscriber decides to turn around and walk back towards cell-1 and CPICH-2 falls outside the threshold before the timer reaches its predetermined value then there will not be an update. During the time the mobile device is connected to both cells, the power it transmits at can be reduced, reducing interference for other users in both cells and also saving the battery life of the mobile device. As the subscriber moves further

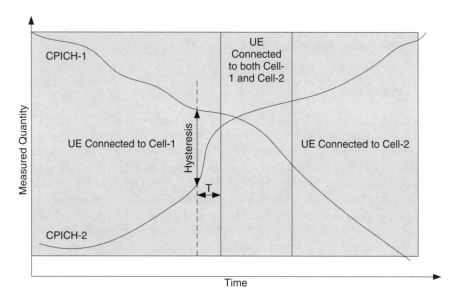

Figure 6.16   Example soft handover procedure

away from cell-1, eventually it will be dropped, and, again, there will be a threshold value and a timer. A mobile device may be connected to more than two cells during soft handover. The hysteresis in the diagram indicates that the two connections are being used simultaneously, and that the decision to add or remove a link is made independently based on the measurements recorded.

The advantages presented by soft handover are only really useful if the two (or more) signals are fairly equal in power, typically less than 3 dB difference.

6.7.5 Power control

Power control in a WCDMA system is crucial to its successful operation. This is because each handset transmits on the same frequency and at the same time as other handsets. Each of the handsets therefore generates interference, raising the overall noise level in the cell, and the base station has to be able to distinguish a particular user out of this interference. If a single mobile device is transmitting with too much power, or is physically closer to the BTS, this may drown out the other UEs. Conversely, if a UE is transmitting with too little power, or is physically further away, the base station will never hear it. This is commonly referred to as the *near–far problem*. There are two main concerns regarding power control: distance from the base station and fast fading. Within the WCDMA system three types of power regulation are used, open loop, inner loop and outer loop power control. Each of these mechanisms will be dealt with in turn in the following subsections. The key goals of power control are to:

- provide each UE with sufficient quality regardless of the link condition or distance from the BTS;
- compensate for channel degradation such as fast fading and attenuation;
- optimize power consumption and hence battery life in the UE.

6.7.5.1 Open loop

This is used when a handset first enters the cell. The handset will monitor the CPICH of the cell and will take received power level measurements of this channel. This information will be used when setting its own power level. The power radiated by the base station will reduce as the distance from the tower increases. Simply having a received power level is not enough information for the mobile device to set its own transmit power. This is because it does not know at this stage at what power level the CPICH was transmitted. The base station sends out power control information on the broadcast channel, which includes an indication of the power level at which the CPICH is being transmitted. The UE can now determine how much power has been lost over the air interface and thus has an indication of its own distance from the base station. The handset can now calculate what power to transmit at; if it has received a weak signal it will transmit a strong signal since it assumes that it must be a long distance from the base station. Conversely, if it receives a strong signal it assumes that it is near to the base station and can thus send a weaker signal.

The actual mechanism is that the UE will listen to the broadcast channel. From this, it will find out the following parameters:

- CPICH downlink transmit power
- uplink interference
- constant value.

The UE will then measure the received power of the pilot (CPICH_RSCP). The initial power used is then:

$$\text{Initial power} = \text{CPICH downlink transmit power} - \text{CPICH\_RSCP}$$

$$+ \text{UL interference} + \text{Constant value}$$

The constant value really provides a correction figure for better approximation of an appropriate start power. It should be noted that this is a rather approximate mechanism since the base station will be transmitting in a different frequency band to what the handset will use and the power loss over the air interface may be significantly different for each band.

Figure 6.17 gives an example of how two different mobile devices, depending on their distances from the base station, will send an access request on the RACH at different power levels. This will reduce interference to other mobile devices. The diagram shows separate BCCH and CPICH for the two different mobile devices; in reality only one of each of these is broadcast and all the mobile devices receive it.

6.7.5.2   Inner loop

Inner loop power control feedback is a form of closed loop power control. Information is sent in every slot, i.e. 1500 times per second. This can be compared with IS-95, which has feedback 800 times a second, and GSM, where it is only carried out approximately

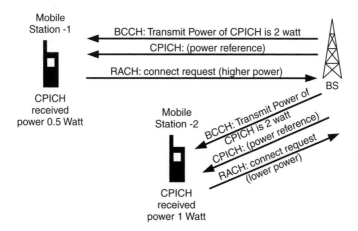

Figure 6.17   Example of open loop power control

twice per second. This type of power control is required because the open loop system can only give a rough estimate and is not accurate enough to deal with problems such as fast fading. To control the power level on the uplink, the base station performs frequent signal to interference ratio (SIR) measurements on the received signals from each of the mobiles. This value is compared to a target SIR value and if the power from the mobile station is deemed to be too high or low it will be told to decrease or increase the power accordingly. Since this task is executed 1500 times per second, it is much faster than power control problems, such as fast fading, that may occur, and hence can compensate for these. This fast power control is very effective for slow to moderate movement speeds of the mobile device. However, benefits decrease as the speed of the mobile increases. This also deals with the near–far problem, where signals from mobile devices which are far from the base station will suffer greater attenuation. The object of the fast power control is that the signal from each mobile should arrive at the BTS at its target SIR value. The same type of power control is used on the downlink. When communicating with mobile devices that are on the edge of the cell the base station may *marginally* increase the power it sends. This is required since these particular mobiles may suffer from increased other-cell interference.

6.7.5.3 *Outer loop*

As noted, inner loop power control measures the power from the mobile device and compares it to a set SIR target. This target value is set and adjusted by the outer loop power control within the RNC. This value will change over time but does not need to be adjusted at the same high frequency. The target value is actually derived from a target BER or BLER that the service is expected to meet. Some errors with the data received from the mobile device are expected. If there are no errors, then the UE is assumed to be transmitting at too high a power, with the consequences of causing interference and reducing the battery life of the device. To implement this method of power control the mobile device will compute a checksum before sending any data. Once received, a new checksum is computed on the data and this is compared to the one sent by the mobile device. The BTS will also measure the quality of the received data in terms of BER. If too many frames are being received with errors, or frames have too high a BER, then the power can be increased. The target set-point is not static: it does change over time. This is required so that the cell can be more efficiently utilized.

6.8 WCDMA PHYSICAL LAYER

The physical layer is structured into radio frames, each of 10 ms duration, and a radio frame is divided up into 15 slots, as shown in Figure 6.18. Blocks of data are transferred across the air interface in each radio frame, and the data rate at which information is sent may change with radio frame granularity. Within a slot, control functions such as power control take place.

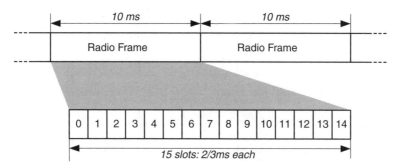

Figure 6.18 UMTS radio frame structure

6.8.1 Physical layer procedures

Figure 6.19 shows a synopsis of the procedures that must be performed at the physical layer for transfer of data across the Uu air interface. In this diagram, two transport channels are processed and then combined into a single channel, referred to as a *coded composite transport channel* (CCTrCH). Since this multiplexing of channels is done at the physical layer, this allows one user to have a number of simultaneous transport channels, each with their own QoS profile. This is used, for example, when a subscriber is making a phone call (logical DTCH) and has a control signalling channel at the same time (logical DCCH).

The physical layer must ensure that data is transferred reliably across the air, so the processes of cyclic redundancy check (CRC) attachment, coding and interleaving are designed to provide this. The composite channel is spread with the appropriate codes and then modulated for transmission. Each of these steps is now explained in further detail.

6.8.2 Data protection

UMTS provides for three basic layers of protection on the data: the CRC to detect for errors, convolution coding for forward error correction (FEC) and interleaving to distribute burst errors throughout the data. All of these general principles were described in Chapter 2. The CRC size can be 0, 8, 12, 16 or 24 bits, and the actual size to be used for a particular transport channel is defined when the channel is established by higher-layer signalling. At the receiver the CRC result is passed upward to higher layers and erroneous data is not always discarded since this corrupted data may still be of use in some applications.

For FEC, the options are to provide no coding, convolution coding or turbo coding. Again, the decision of which to use is made at channel establishment. Before passing through the coder, all of the transport blocks that have been sent in the defined transmission time interval (TTI, explained shortly) are concatenated together serially and coding is performed on this block. Should the resulting block be too large for the coder, it is segmented into appropriate sizes. The choice of which coding scheme to use for the

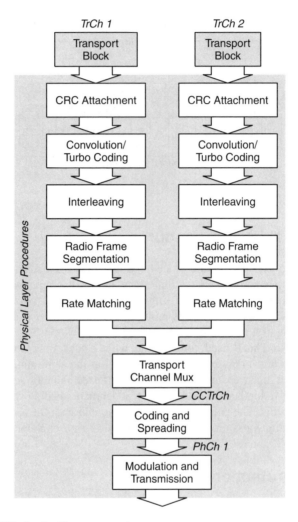

Figure 6.19 UMTS physical layer procedures

Table 6.7 Transport channel coding schemes

Transport channel	Coding	Rate
BCH		
PCH	Convolution coding	$\frac{1}{2}$
RACH		
DCH, FACH, DSCH	Convolution coding	$\frac{1}{3}$, $\frac{1}{2}$
or CPCH	Turbo coding	$\frac{1}{3}$
	No coding	n/a

channel is dependent on the level of error protection required. Table 6.7 shows the coding schemes available for different transport channels.

The resultant block of data then undergoes an interleaving process, as was described in Chapter 2.

6.8.3   Radio frame segmentation and rate matching

Depending on the data to be transmitted it will be split or concatenated into a number of blocks, which may consist of more than one 10 ms radio frame. These blocks are passed between the physical and the upper layers at certain time intervals. These blocks are referred to as transport blocks (TB). For example, a speech call may pass a transport block to the physical layer every 20 ms. This timing is defined when a connection is established and is known as the transmission time interval (TTI). For the previous example, the TTI is 20 ms. Where the TTI for the transport channel is more than the 10 ms radio frame size, the block must now be segmented into 10 ms radio frames. Rate matching is implemented to ensure that the coded transport block size is then brought to the same size as is expected by the spreading process. That is, the bit rate after the rate matching should equate to the channel bit rate. For example, at a spreading factor (SF) of 128, in the uplink the bit rate is 30 kbps. The rate matching process must bring the number of bits in each radio frame to 300. The rate matching achieves this by either bit repetition, if less than the required number, or by bit puncturing of some of the coding, should the size be larger. In Figure 6.20, it is assumed that a transport block of 244 bits is passed to the physical layer every 20 ms. For this block, the required error protection is an 8-bit CRC plus 1/2 rate convolution coding. At the physical layer, the CRC is attached and this block passed through the convolution coder. This results in a block of 504 bits. This is now rate matched by bit repetition to become 600 bits, and is then segmented into two 300-bit blocks. These are then spread with a code of SF = 128. Please note that this is a simplified form of the process and is indicative only.

If there is more than one transport channel, the rate matching process will consider all the blocks to be concatenated into a CCTrCH, and rate match appropriately such that the resulting CCTrCH is of an appropriate bit length.

It may seem a waste of valuable resources to perform such bit addition. However, the repetition of bits does make the data more robust, thus requiring less transmission power.

6.8.4   Spreading

The spreading process takes the CCTrCH in each radio frame, spreads it to the system chip rate of 3.84 Mcps and transmits it across the air at the same frequency and time as all the other common channels and currently active users. Each of these must be separated from each other. To provide this signal separation, two types of code are used in the WCDMA system. These are *scrambling* and *channelization codes*.

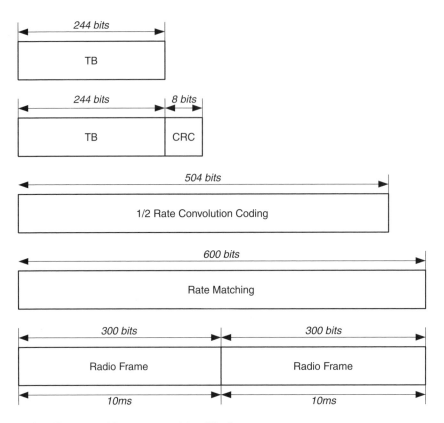

Figure 6.20 Rate matching process (simplified)

Scrambling code

In the downlink the scrambling code (or long code) is used to identify a particular cell. Each cell generally uses a single primary code. For greater capacity more than one scrambling code may be used in a cell. This scrambling code is not used in any neighbouring cells and identifies the cell within the vicinity. The codes can be reused in another geographical area as long as it is not within neighbouring cells. In the uplink the scrambling code is used to identify the particular user since each mobile device will have its own allocated scrambling code. This code allocation for the mobile device is performed by the RNC.

The use of a code to spread a signal over a wide bandwidth should result in a signal which without knowing the code is only interpreted as noise. In fact, the scrambling codes are Gold codes, as described in Chapter 2, where the codes are generated using two 25-stage shift registers. Both the BTS and the UE are required to have this mechanism. Since these codes are near orthogonal, if the receiver does not know the scrambling code used for transmission then the data transmitted will simply be received as extra noise. In the 3G WCDMA context, many signals are sent out from different mobile devices at the same time, i.e. a cell contains many mobile devices which will transmit simultaneously, over the same wide frequency band, to the base station. The base station needs to know how to distinguish each of these transmissions from the next and therefore each of the mobile devices needs to use a different scrambling code. Each of these coded signals introduces additional noise into the system and this noise will reduce the capacity of the cell. The higher the data rate, the more noise they will introduce.

In the downlink, the shift register can generate $2^{18} - 1$ codes, truncated to a length of one 10 ms radio frame, so that each code is 38 400 chips long. However, only the first 8192 codes are used, and these 8192 codes are broken down into 512 groups each consisting of a primary code and 15 secondary codes. Each of these will have a separate channelization code tree available. The limiting of codes to 512 groups is to reduce the number of codes that the mobile device needs to check through when it first enters a cell and thus speed up the cell search procedure. Currently in most UMTS implementations, only the primary code is used. In the uplink, there are 2^{24} possible codes that may be used. However, there is also an option to use simpler short codes which are only 256 chips long. These may be used to simplify the design of multi-user receivers at the BTS.

Channelization code

Since there is only one scrambling code in a cell in the downlink, user data and system information still need to be separated. This is the job of the channelization code. Each channel will represent a different mobile device or different control channel such as the BCCH. In the uplink the different DPDCHs and the DPCCH for a single mobile device are identified by their different channelization codes. These codes are also known as Walsh codes or orthogonal codes, as explained in Chapter 2.

It can be seen from Figure 6.21 that a number of different data streams (these may be user data or control information) each have their own channelization code. Channels with different data rates will have different length channelization codes, ranging from the lowest data rates at $SF = 256$ to the highest at $SF = 4$. These channels are combined in the adder and mixed with a single scrambling code. In the downlink the scrambling

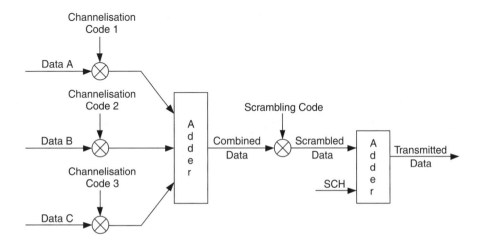

Figure 6.21 Downlink code usage in WCDMA

code does not include coverage of the synchronization channel (SCH), which has its own system-wide code.

For high data rates, a single subscriber can use a maximum of three out of four of the possible codes at SF = 4. In this case, the data is split and transferred over all three codes in parallel. The fourth code may not be used in its entirety since some codes are reserved for transmitting common channel information.

The usage of the different codes is summarized in Table 6.8 for both the uplink and downlink.

Figure 6.22 illustrates how the different codes are used within the WCDMA system. It is usual that cells 1 and 2 will be transmitting and receiving on the same frequency. BS1 transmits a single signal which includes the channelization code for user 1 (Ccu1) and the channelization code for user 2 (Ccu2), both of which are unified under a single scrambling code (Csbs1). In the downlink for cell 2 the same situation occurs with the exception that the scrambling code (Csbs2) must be different to that of cell 1. However, the same channelization codes which are used in cell 1 can also be used in cell 2. In the uplink each of the mobile devices is identified by its scrambling code (Csu1..4). The mobile devices can all use the same channelization codes since these are concealed by the unique scrambling code.

Table 6.8 Code usage table

	Channelization code	Scrambling code
Uplink	Separation of physical data and control channels from one user	Separation of users
Downlink	Separation of dedicated user channels	Separation of cells
Code length	Variable according to the required data rate	Fixed at one radio frame (10 ms or 38 400 chips)

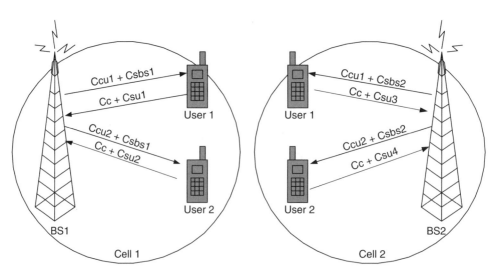

Figure 6.22 Example code usage

6.8.5   Modulation and transmission

UMTS defines the use of quadrature phase shift keying (QPSK) modulation for the air interface. With a QPSK modulation scheme, the complex signal that results from the spreading function is split by a serial to parallel converter into a real and an imaginary branch, each of which is multiplied with an oscillator signal. However, the imaginary branch is 90° out of phase with the real branch. When summed, the resulting signal can have four possible phase angles, each of which represents two data bits. Figure 6.23 illustrates the general principle.

QPSK modulation is specified for use in both the uplink and the downlink; however, the use of the QPSK modulation scheme does present some difficulties in the uplink. Consider that the amplifier is at a maximum output power and needs to change its signal by 180°. This consumes a considerable amount of power in the amplifier to retain the linearity of the signal, particularly across such a wide frequency band, and most of this power ends up wasted as heat. This is not so difficult a problem at the BTS, but is quite impractical at the UE, where cost, power consumption, battery life and heat dissipation are all significant issues. A common solution to this problem is to use offset quadrative phase shift keying (QPSK) in the uplink instead. With offset QPSK, there is a delay introduced into the imaginary branch to offset the phase shifting of this branch relative to the real branch. The result is that when a 180° phase shift is required, the shift is performed in two steps of 90°.

QPSK modulation provides a one-to-one relationship between the bit rate of an unmodulated signal and the symbol rate after modulation. In practice, this means that a 3.84 Mcps spread signal entering the modulator will emerge as a 3.84 MHz signal.

In the course of the modulation process, pulse shaping is also performed. WCDMA uses a root-raised cosine filter with a roll-off of 0.22. A modulated signal with this roll-off, plus the provision of a guard band between neighbouring frequencies, equates to the 5 MHz of spectrum allocated per WCDMA carrier (Figure 6.24). For frequencies licensed to the same operator, there can be less than 5 MHz spacing between carriers. However, the centre frequency must lie on a 200 kHz raster. This means that if, for example, the

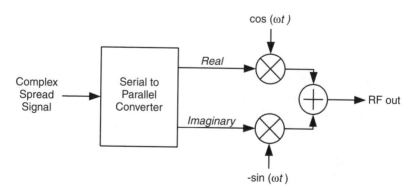

Figure 6.23   QPSK modulation

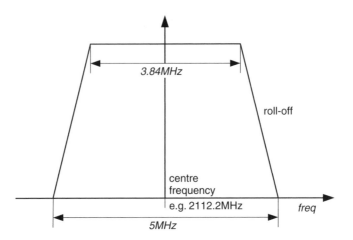

Figure 6.24 5 MHz frequency band

operator had been allocated 2010–2015 MHz downlink channel, the centre frequency could be 2012.4 MHz or 2012.6 MHz.

6.8.6 Common channels

The primary common control physical channel (P-CCPCH) is the channel that carries the system broadcast information. Since all users must be able to decode this channel, it is given a fixed channelization code of SF = 256, which is a system constant. The broadcast information is transmitted across two radio frames, and then repeats.

For the secondary CCPCH (S-CCPCH), it is offset from the P-CCPCH by a multiple of 256 chips. This carries the FACH and PCH channels. The SF and code used for this channel are variable, and the UE will receive the configuration information for this channel on the broadcast channel. For user paging, WCDMA uses a second channel known as the paging indication channel (PICH). This channel is offset from the S-CCPCH by $\tau_{PICH} = 7680$ chips. Figure 6.25 shows the relative timings, and the arrow indicates the S-CCPCH frame associated with the PICH frame.

The PICH is a fixed rate SF = 256 channel of 30 kbps. This means that each radio frame contains 300 bits. The first 288 bits of these are paging indicators, and transmission is switched off for the last 12 bits. These 288 bits contain N paging indicators representing different paging groups, where N can be 18, 36, 72 or 144. This means, for example, if there are 72 paging indicators, then each indicator is 4 bits (Figure 6.26).

When a UE connects to the network, it can calculate which paging group it belongs to, based on its international mobile subscriber identity (IMSI) number. It will then monitor the PICH, and if its paging group indicators are 1, it will decode the associated frame of the S-CCPCH for specific paging information. Since the UE knows the period when it needs to check the PICH, it can use discontinuous reception (DRX) in the interim, thus saving battery power.

The remaining common physical channels are discussed in Section 6.9.

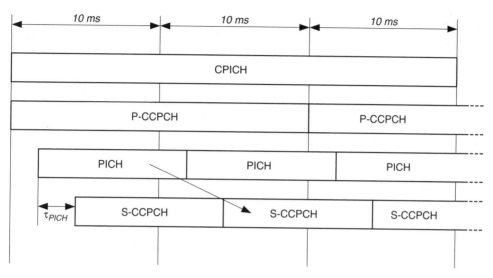

Figure 6.25   Common physical channel timings

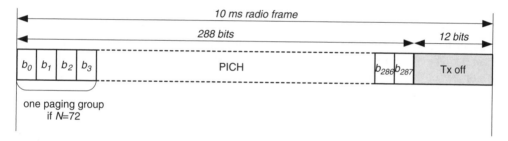

Figure 6.26   PICH structure

6.8.7   Dedicated physical channels

Dedicated channels (DCH) are transported at the physical layer using the dedicated physical data channel (DPDCH). This physical layer channel contains one or more transport channels, as formed by a CCTrCH. This physical layer channel has a control channel associated with it, the dedicated physical control channel (DPCCH), which contains relevant information with regard to the physical transmission of data. The structure of both is shown in Figure 6.27.

In the uplink these two channels (DPDCH and DPCCH) are separated using different channelization codes. They are, however, in phase/quadrative (I/Q) multiplexed together. There may be a number of DPDCHs on each radio link associated with one DPCCH. An example of this would be where higher data rates are achieved using multiple codes, and hence multiple DPDCHs. The DPDCH can have a variety of SFs associated with it and this is determined by the actual data rate required by the subscriber in each particular

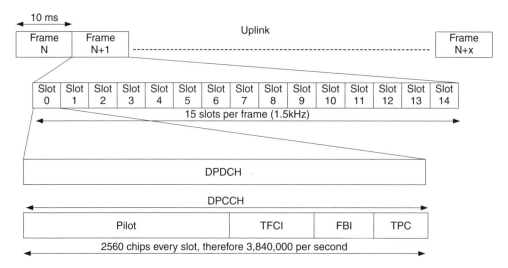

Figure 6.27 Uplink radio frame format

connection. However, in the uplink, the DPCCH always uses the lowest data rate, SF = 256. This equates to a channel symbol rate of 15 ksps, or 10 bits per slot.

The DPCCH is used to carry control information generated at layer 1. This control information consists of a number of different fields, some of which may or may not be present. The number of chips for each of these fields is not fixed and can vary. There are a number of defined slot formats, and the one being used is configured by higher layers. For example, slot format 0 is defined as shown in Figure 6.28.

The fields are defined as follows:

- The *pilot* bits are predefined, and are therefore known by both the UE and the BTS. These bits are used for channel estimation in a similar manner to the use of a training sequence in a GSM burst.
- The *transmit power control* (TPC) is used to request increased or decreased power on the downlink channels. This is actually only one bit, where 0 indicates that a power down is required, and 1 a power up. If there is more than one bit allocated for TPC in the DPCCH, then the bits are repeated. For the example configuration of slot 0, a power up command would be represented as TPC = 11. Repetition is required since the air interface is notoriously bad and the TPC is required to be received successfully while keeping the power level as low as possible.

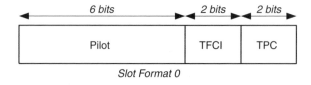

Slot Format 0

Figure 6.28 DPCCH slot format 0

- The *transport format combination indicator (TFCI)* is used to inform the BTS of the format of the specific transport channels mapped into the current DPDCH. Transport formats are defined by higher layers during radio link establishment. The TFCI field is not required for fixed-rate services, and is only used for variable-rate services where there are a number of possible rates.

- The feedback indicator (FBI) bits are used to support diversity techniques in the BTS which require feedback from the mobile device. These are explained in Section 6.11.

The coding of the TFCI bits into slots is performed as follows. One TFCI is needed to describe the format of the DPDCH in each radio frame. If it is less than 10 bits long, the TFCI is padded out to 10 bits long by adding zeroes at the most significant bit (MSB), and then coded using a Reed–Muller code to provide error protection. This generates a 32-bit coded TFCI. In the above example of slot format 0, the first 30 bits of this are transferred across each of the 15 slots of the radio frame, starting from the lowest index bit. This means slot 0 will contain bits 0 and 1 of the coded TFCI, etc. The most significant two bits are ignored.

In the downlink the two channels (DPDCH and DPCCH) are combined using TDM and transmitted under the same channelization code. This combined channel is referred to as the downlink dedicated physical channel (downlink DPCH) (see Figure 6.29). As with the uplink, the DPCCH is used to carry control information generated at layer 1. The control information consists of the same fields as the uplink with the exception of the FBI field. This is not required in the downlink since transmit diversity is not used on the mobile device. In the uplink, the DPDCH and DPCCH were I/Q multiplexed together. However in the downlink instead, the single downlink DPCH goes through a serial to parallel conversion and the odd and even bits are then I/Q multiplexed. This results in a doubling of the channel bit rate.

Once again, the format of the slot to be used is defined by higher-layer signalling at channel establishment. An example of slot format 9 is shown in Figure 6.30. For this

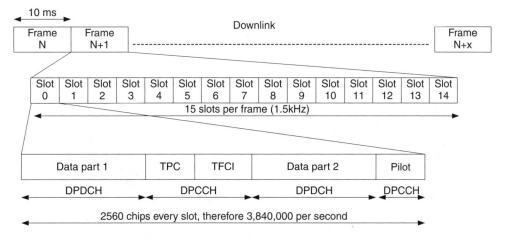

Figure 6.29   Downlink radio frame format

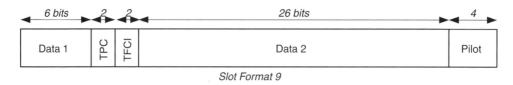

Slot Format 9

Figure 6.30   Downlink DDPCH slot format 9

format, SF = 128 is used, which provides a symbol rate of 30 ksps. However, because of the I/Q modulation used on the downlink, this equates to a channel bit rate of 60 kbps. Therefore there are 40 bits available per slot.

The reasoning behind time multiplexing the DPDCH and DPCCH in the downlink and I/Q multiplexing in the uplink is to avoid uplink interference. If the two uplink channels were time multiplexed together, then there would be situations where there was no data transfer, but the control channel still contained power control information. In this case, since power control is performed in each slot, this would generate a 1.5 kHz signal, which is located in the centre of the telephony band and causes unacceptable interference. Therefore the I/Q multiplexing method is employed. This 1.5 kHz signal does not present a problem in the downlink since there are many channels transmitting simultaneously.

6.9   INITIAL CONNECTION TO NETWORK

After the user switches on their mobile device, they must establish a connection to the network, principally to perform a location update to advise the network of their existence and location. Before a user can contact the core network to do this, the UE must establish a signalling connection to the RNC as described in Section 6.16. However, prior to that there are a number of steps that the device must go through, as summarized in Figure 6.31.

6.9.1   Synchronization procedures

As mentioned previously, within the WCDMA cell each user shares the same wide frequency band of 5 MHz. The frequency reuse within adjacent cells is 1, which means that the adjacent cells may also be using the same frequency range provided that they are within the same system, i.e. belonging to the same mobile operator. When a user switches on their mobile device it needs to connect to one of the base stations in the coverage area. Since according to the ITU-T frequency allocation for 3G, the maximum available bandwidth is 60 MHz, that means that there are at most 12 channels in use in a given country. Therefore the frequency search procedures are fairly trivial.

To ascertain the necessary parameters for connection to the network, the UE must access the BCCH, where all of this information is transmitted. Before the station can decode information on the BCCH, it must first synchronize with the system. It does this with the aid of the synchronization channel (SCH). The SCH is actually made up of two

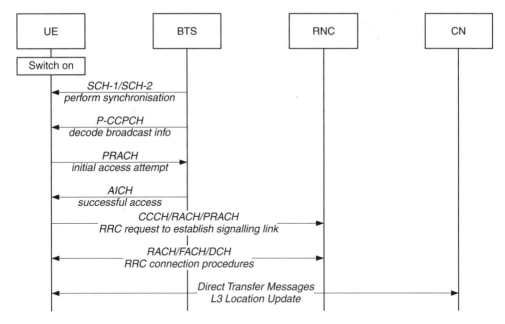

Figure 6.31   Initial network connection

separate channels known as the primary SCH and secondary SCH. Connecting to the BTS is known as 'camping on the cell'. The BCCH is carried in the P-CCPCH, which has a universally fixed channelization code and is always coded using the primary scrambling code for the cell. Therefore to 'camp on a cell' the UE has to initially determine the downlink scrambling code for that cell, and it also has to synchronize with the frame timing of the cell. There are three steps to the above:

1. slot synchronization
2. frame synchronization
3. scrambling code identification.

   The structure of the SCH and P-CCPCH is shown in Figure 6.32. The P-CCPCH is not transmitted during the first 256 chips as the SCH is sent during this period.

6.9.2   Slot synchronization

The SCH consists of a primary SCH (P-SCH or SCH-1) and secondary SCH (S-SCH or SCH-2). To obtain slot synchronization the UE listens to the P-SCH for a *synchronization code*, which consists of 256 chips and is referred to as the primary synchronization code (PSC). It is a system constant common to all BTSs and is easily identified (Figure 6.33).

   A matched filter is used, with the PSC as input, across the 5 Mhz frequency band. The output of the filter will be a series of peaks where there is a match between the PSC

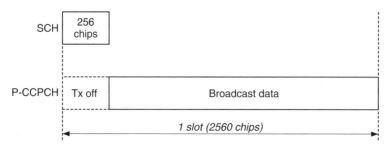

Figure 6.32 Multiplexing of physical layer SCH and P-CCPCH

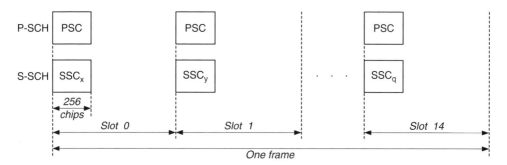

Figure 6.33 Format of synchronization channels

the UE is expecting and the P-SCH signal being transmitted. These peaks in the filter's output, as illustrated in Figure 6.34, are therefore the start points of each slot, thus slot synchronization has been achieved.

6.9.3 Frame synchronization

To obtain frame synchronization, the UE needs to listen to the S-SCH. Unlike the P-SCH, which continuously sends the same 256 chips at the beginning of every slot, the S-SCH rotates through a predefined sequence of different chip groups, known as the secondary synchronization codes (SSC). These sequences not only indicate the frame synchronization but are also used to provide the UE with a guide for determining the cell-specific scrambling code. Table 6.9 shows the first two groups (0 and 1); there are

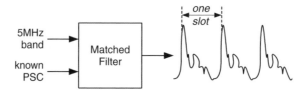

Figure 6.34 Output of matched filter

Table 6.9 SSC codes

Group	Slot number														
	0	1	2	3	4	5	6	7	8	9	10	11	12	13	14
Group 0	1	1	2	8	9	10	15	8	10	16	2	7	15	7	16
Group 1	1	1	5	16	7	3	14	16	3	10	5	12	14	12	10
⋮															
Group 63	9	12	10	15	13	14	9	14	15	11	11	13	12	16	10

64 groups in total. Each UE will have this table at its disposal. Using the unique codes, it is possible for the UE to obtain frame synchronization. It also completes the first step of finding the cell-specific scrambling code.

There is a defined set of possible 256-chip sequences that are used to represent each of the 64 groups, and all sets of sequences are unique with respect not only to each other, but also for every possible offset. As an example of how this mechanism works, consider that there are 15 slots, and that instead of a 256-chip sequence, rather a letter is transmitted in each slot, for example the word SYNCHRONIZATION. Then regardless of which slot the UE starts decoding the S-SCH, it can calculate where the first slot occurs. For example, consider the UE begins at slot 5, and then reads for 15 slots. It will then have the sequence: HRONIZATIONSYNC. It will then compare this to the stored table and therefore know where the first slot is positioned. Instead of just using one word, there are 64 groups of different letter sequences defined, for example.

Slot:	0	1	2	3	4	5	6	7	8	9	10	11	12	13	14
Group 0:	S	Y	N	C	H	R	O	N	I	Z	A	T	I	O	N
Group 1:	U	E	M	O	B	I	L	E	D	E	V	I	C	E	S
⋮															
Group 63:	M	A	N	C	H	E	S	T	E	R	U	T	D	F	C

Again, regardless of starting slot, the position of slot 1 can be found and the sequence of letters that is yielded is not equal to any other of the 64 letter sequences, at any slot position. Hence since the first slot is known, frame synchronization is achieved.

In a multi-sectored BTS site, a timing delay defined by the variable T_cell may be used as an offset between the SCH channels in different sectors. This avoids overlapping and interference of the SCH channels. T_cell is of resolution 256 chips and can have a value of 0–9.

6.9.4 Scrambling code identification

After frame synchronization, the UE also then knows which of the 64 words is being used in this cell. There are 512 possible primary scrambling codes used in the system, and these are broken up into 64 groups, each containing 8 codes. Therefore the code group has narrowed the scrambling code down to 1 of a possible 8. The UE will now look for the strongest correlation across the CPICH. The code which gives the best fit is assumed to be the cell-specific primary scrambling code. It is now possible for the UE to decode the BCCH and thus the cell- and operator-specific information can be read.

6.9.5 Random access procedure

It is now possible for the UE to receive and decode general information broadcast on the downlink from the BTS. However, at some stage it will need to transmit on the uplink to the BTS. Two questions then arise:

1. How does the cell know the scrambling code of the UE?
2. From where does the UE obtain its scrambling code?

Each UE in the cell must use a separate scrambling code. Although there are millions of scrambling codes available, it is impractical to provide each UE in the world with a unique code and place this in the SIM card. The scrambling codes used by the UE are actually allocated by the RNC. However, since the UE has not begun to communicate, what scrambling code should it now use? To overcome this problem, the UE must initially use a scrambling code that the BTS understands for the initial request for connection. The request is made via the RACH and thus this channel must use a code that the BTS understands. The BTS will broadcast a preamble scrambling code that should be used, as well as a number of signatures, which can be used on the RACH uplink channel by the UE. The signatures are 16-chip sequences, and there is a maximum of 16 possible signatures defined. The UE selects one of these signatures at random and uses it to code its request message to the BTS.

When this initial access is made, clearly there is a possibility of a collision with other UE requests. By having a number of signatures, with one chosen randomly, this is minimized somewhat; however, this alone is not enough. The RACH channel for initial access consists of a number of *access slots*, based on a slotted ALOHA mechanism. In addition to the signatures, the UE will also randomly pick one of these slots, further reducing the probability of collision. There are 15 possible access slots across two frames, spaced 5120 chips apart, as shown in Figure 6.35.

The initial access is in the form of a 4096-chip preamble, which consists of the selected signature repeated 256 times. The preamble is sent repeatedly until the network acknowledges that it has heard it. It does this on the acquisition indication channel (AICH) associated with the RACH, which will echo back the 4096 chips to the UE. The broadcast channel provides the UE with a power level for this initial access and a maximum time, within which the access will be acknowledged if the network has heard the UE. If there is no indication on the AICH within this time, the UE will power up, and send the preamble again. Once an indication has been received on the AICH, the UE will then send a 10 or 20 ms message part, which will typically be an RRC message requesting a connection. The process is illustrated in Figure 6.36.

It is also possible for the RACH resources to be divided into a number of different access service classes (ASC). This allows the resources to be given different priorities

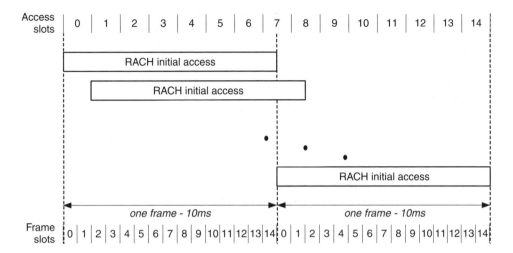

Figure 6.35 RACH initial access

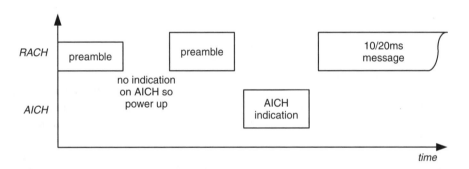

Figure 6.36 RACH initial access procedure

for different types of access. This would enable, for example, emergency call access to be given a higher priority level.

6.10 COMPRESSED MODE

For handover measurements, the rake receiver has the necessary resources to simultaneously make measurements of neighbouring cells, while the user is on an active connection. However, it can only perform this for cells that are at the same frequency as the current connection. For inter-frequency measurements, the UE must temporarily cease transmission and reception to allow the measurements to be made, but the connection must be maintained. Another alternative is for the UE to implement a second receiver for such measurements, but this is somewhat impractical. UMTS allows for this pause in transmission for inter-frequency measurement through compressed mode. This is used for performing measurements of both cells within UMTS that are at a different frequency, as well as inter-system measurements, for example to acquire GSM system parameters.

Compressed mode opens up periodic gaps in the data flow during which such measurements can be made. The RNC will instruct the UE and the BTS on the mechanism for providing a gap, and the position of the gap. As the name suggests, compressed mode takes the information to be sent within a 10 ms radio frame and compresses the time it takes to send. There are three methods that can be used to perform this compression:

1. *Puncturing*: rate matching is used to further puncture the data to achieve the necessary transmission gap.

2. *Reduction of SF*: the spreading factor is reduced by a factor of two during a compressed frame. This has the effect of doubling the data rate. However, since the amount of data has not changed, it halves the time in which it is sent, opening up a gap.

3. *Scheduling*: higher layers can permit only some of the defined transport formats to be used when compression is required, thus generating a gap. Clearly this method is only appropriate for variable-rate services.

Of these three, all are available for downlink compression, but method one, puncturing, is not permitted in the uplink. The largest allowed gap that can be created is seven slots in one radio frame.

Figure 6.37 shows an example of compressed mode. When the transmission gap is opened up, if SF reduction is used, it will result in an increase in transmission power, as illustrated. The gap can span two radio frames as long as there are at least 8 out of 15 slots in each frame containing transmitted data.

Compressed mode can be defined to be used in downlink, uplink or both. When used in both uplink and downlink simultaneously, the transmission gap must be coordinated to occur simultaneously in both directions.

For each of the physical channels, it was seen that there are predefined slot formats to describe the size of the DPDCH data and DPCCH control bits. For each defined format, there are associated formats for use in compressed mode. Two formats are defined for each: format A is to be used where the puncturing method is employed, and format B for SF reduction. As an example, consider the downlink slot format 9 that was shown previously (see Figure 6.30). This is expected to be transmitted in all 15 slots within a

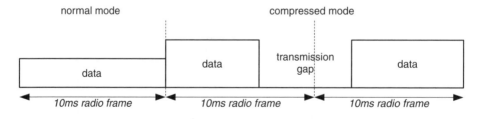

Figure 6.37 Compressed mode

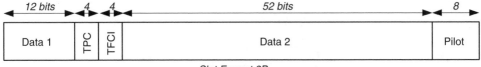

Slot Format 9B

Figure 6.38   Downlink compressed mode slot format 9B

radio frame. If compressed mode using SF reduction was indicated, slot format 9B would then be used (Figure 6.38), but only sent in 8–14 slots of the frame.

Slot format 9 uses SF = 128, channel bit rate = 60 kbps, whereas 9B uses SF = 64, channel bit rate = 120 kbps, and the respective fields within the slots are exactly doubled.

6.11   DOWNLINK TRANSMIT DIVERSITY TECHNIQUES

Both open loop and closed loop transmitter diversity techniques can be used at the BTS to improve the quality of the radio interface. The open loop modes are space time transmit diversity (STTD), time switched transmit diversity (TSTD) and site selection diversity transmit (SSDT). The simultaneous use of these techniques on the same physical channels is not allowed. There are also two modes of operation in the closed loop system. All of these are described below.

6.11.1   Space time transmit diversity (STTD)

Figure 6.39(a) illustrates how STTD functions. The symbols to be transmitted are grouped into blocks of four, b0–b3. Antenna 1 transmits these symbols in the same order and in the same polarity as they arrive. Antenna 2 modifies the order prior to transmission and also modifies the polarity of symbols 1 and 2. This method of open loop transmit diversity can be used on all channels except for SCH. The UE will then regenerate the original signal, now having two sources to use. Support for this open loop method of transmit diversity is mandatory in the mobile device and optional within UTRAN.

6.11.2   Time switched transmit diversity (TSTD)

TSTD is only used on the synchronization channel (SCH), and functions by alternating which of the two antennas will transmit the SCH. If TSTD is utilized, one antenna (antenna 1) will transmit both the primary and secondary SCH in even slots and the other antenna (antenna 2) will transmit both the primary and secondary SCH in odd slots. When an antenna is not transmitting the SCH channels, its transmitter will be switched off, thus reducing interference. In a situation where TSTD is not used, a single antenna will transmit the primary and secondary SCH on both the even and odd slots. Although TSTD support is mandatory in the mobile device, it is again optional to use this method within UTRAN.

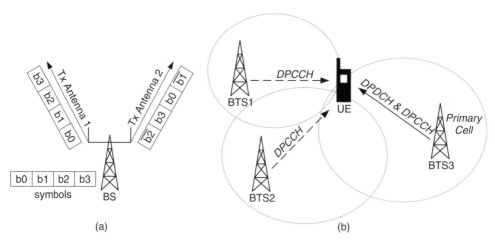

Figure 6.39 (a) STTD and (b) SSDT

6.11.3 Site selection diversity transmit (SSDT)

SSDT is a form of power control that can be used on the downlink while a mobile device is in the soft handover mode of operation. The UE will select the best cell in its active set on the basis of the quality of the CPICH channel. This is then deemed to be the primary cell, and is used to transmit the DPDCH. All of the other cells in the active set will now only transmit the DPCCH, so as to enable the mobile device to perform measurements and maintain synchronization. Each of the cells is given a temporary identifier with the primary cell referred to as the *primary id*. This identifier is transmitted on the uplink DPCCH to each BTS in the active set. The cell that has been selected as the primary cell will then begin transmitting at a sufficient power to maintain the desired SIR target. The primary cell can be periodically changed by the UE. This period is set by the UTRAN and can be 5, 10 or 20 ms. To enable SSDT requires each BTS involved to support this mode of operation, and its use is specified during the channel establishment phase. Simulations have demonstrated that SSDT is particularly suited to mobile subscribers who are moving at slow speed, where capacity gains of up to 50% can be achieved (TS25.922 Appendix D). As the speed of the user increases, the efficiencies diminish due to the limited frequency of update of the primary cell identifier.

Figure 6.39(b) illustrates the concept of SSDT. In the diagram, BTS3 is the primary cell and as such transmits both the data channel (DPDCH) and the control channel (DPCCH). The other two base stations simply transmit the control channel. On the uplink, the mobile device will transmit both the data and control channels, which will be received by all of the base stations and fed to the SRNC.

6.11.4 Closed loop mode transmit diversity

Figure 6.40 shows a general architecture for the closed loop mode. The DPCH channel is fed to both antennas for transmission on the downlink. There are actually two modes

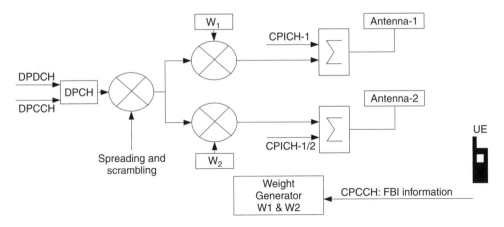

Figure 6.40   Closed loop transmit diversity

of closed loop, mode 1 where different CPICH channels are used and mode 2 where the same CPICH is fed to both of the summation units. The mobile device uses the CPICH it receives to estimate the adjustments that should be made to the transmitted signal and this information is fed back to the BTS. The value of the weight factors (w_1 and w_2) are determined through use of this feedback via the FBI bits from the mobile device on the DPCCH. Antenna-1 is used as a reference and in mode 1, the signal transmitted from antenna-2 is phase shifted from that of antenna-1. In mode 2 the signal transmitted from antenna-2 is phase shifted and also modified in amplitude.

6.12   RADIO INTERFACE PROTOCOL ARCHITECTURE

The radio interface is a layered structure, which forms the access stratum connection between the mobile devices and the RNC. It has three layers and roughly follows the OSI model, consisting of physical, datalink and network layers. The physical layer, shaded in Figure 6.41, consists of WCDMA and ATM. The datalink layer is comprised of a number of sublayers: the media access control (MAC), radio link control (RLC), packet data convergence protocol (PDCP) and the broadcast/multicast control (BMC) (see Figure 6.41). The PDCP layer is shown only half-way across the protocol stack. This is because it is only used for packet data such as IP, between the UE and the SRNC. The PDCP provides mechanisms for upper-layer header compression.

The network layer, layer three, is not shown but consists of the NAS signalling[2] between the UE and the core network.

The user plane is complemented by a control plane, shown in Figure 6.42. As can be seen, the control plane also uses the MAC and RLC layers. However, it has an additional protocol, considered to reside at the lower part of layer three, known as the radio resource

---

[2]NAS signalling consists of mobility management (MM), connection management (CM) and session management (SM).

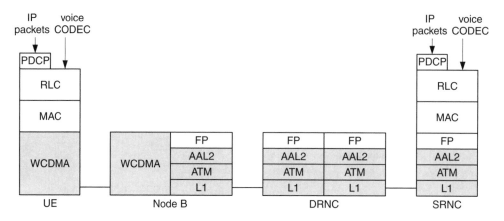

Figure 6.41 Radio network user plane protocol stack

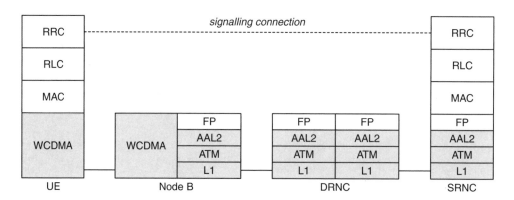

Figure 6.42 Radio network control plane protocol stack

control (RRC) protocol. RRC is the signalling protocol operating between the user and the RNC, and is responsible for controlling the lower RLC, MAC and physical layers. At the MAC layer, the signalling control and user data streams are combined together to present a transport channel to the physical layer. Notice that the protocol stack is very similar to the previous user plane stack and that the transport of this control information to the SRNC is also carried over the frame protocol (FP) and ATM adaptation layer 2 (AAL2).

For UMTS, the physical layer is considered in two stages: across the air interface, Uu, it consists of the WCDMA mechanisms, and across the Iub interface, it consists of the ATM transport and its respective underlying physical layers, such as SDH/fibre, etc. Figure 6.43, shows this architectural model. Across this transport network, an FP is added in between the ATM and upper layer service data unit (SDU). The purpose of this is to carry additional information, such as quality achieved on the air interface and user identification on common channels.

The following subsections will assume that data is passing down through the protocol stacks and will discuss each in that order. Layer 2 consists of the MAC, RLC, PDCP and

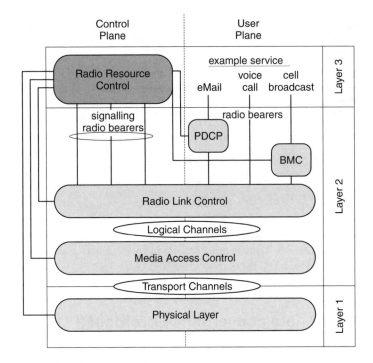

Figure 6.43   Radio interface protocols

BMC protocols, and is responsible for the mapping of data onto layer 1 across the transport channels. To understand this mapping, some general concepts about data transport must be defined. All transport channels are considered to be unidirectional and a UE can have simultaneously one or more transport channels in the downlink, and in the uplink.

6.12.1   Broadcast/multicast control (BMC)

The BMC protocol is used for scheduling and transmission of cell broadcast information to the UE. It is only used in the downlink direction, using the CTCH logical channel transported over the FACH. For R99, only the broadcast function is supported, with multicasting to be introduced in later releases. It is designed to send text broadcast messages over a certain geographical area, containing relevant information. Its service is similar in concept to the teletext system for television. Messages are limited to 1200 bytes, and currently SMS cell broadcast is the only supported service. There is work in progress at the 3GPP to expand the scope of this service to deliver richer content such as multimedia. The introduction of multicasting would also enhance the service considerably, opening up the possibility of provision of services such as selective advertising to a certain group of subscribers in a particular area.

The protocol stack for cell broadcast is shown in Figure 6.44 below. An IP-based cell broadcast centre (CBC) is located in the operator network. This is connected to the

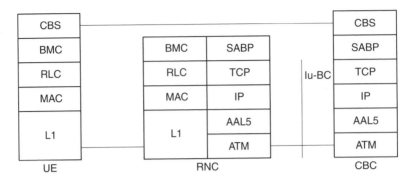

Figure 6.44   Cell broadcast service

RNC via the Iu-BC and broadcast messages are transported by the service area broadcast protocol (SABP). At the RNC, these are transferred to the BMC protocol and distributed on the FACH to all cells in the broadcast area.

6.12.2   Packet data convergence protocol (PDCP)

The PDCP maps the network-level protocol such as IP (v4 and v6) or PPP onto the underlying network. It is only used on the packet switched network and only for user data. The main role of PDCP is to support header compression of upper-layer headers, for example TCP/IP and RTP/UDP/IP compression. Headers associated with IP are only relevant at the UE and RNC, and therefore only add overhead across the Uu and Iub/Iur interfaces. Consider a TCP/IP packet; the header accounts for 40 bytes. Sequences of IP packets tend to have a considerable amount of static information in headers. By way of example, consider sending a file, which is broken up into many segments. Each segment will have a header containing the same IP source and destination address. Header compression takes advantage of this to substantially reduce the headers. A TCP/IP header can typically be reduced to 3–4 bytes. Currently, 3GPP defines only one header compression technique, which is 'IP Header Compression' as defined in RFC 2507. Each packet switched RAB is associated with its own PDCP.

6.12.3   Radio link control (RLC)

The RLC layer provides three types of data transfer to the higher layers.

Transparent data transfer

This exchanges packets with the higher layers without adding any protocol information. The encryption of user data for transparent mode is performed at the MAC layer. The functions provided by transparent mode, and the protocol data unit (PDU) format, are

Mode	Functions provided
Transparent (TR)	Segmentation and reassembly
	User data transfer
	Discard of errored SDU

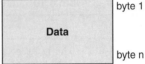

Figure 6.45   RLC transparent mode PDU

shown in Figure 6.45. Since no overhead is added, the format of the PDU is very simple, consisting only of the data from the upper layer.

For all types, error detection is by means of the physical layer CRC check, details of which are passed by the physical layer to the MAC and then on to the RLC.

Unacknowledged data transfer

This passes packets onward without ensuring delivery; packets are passed without acknowledgement. The RLC can segment data into an appropriate size for transmission, and reassemble it at the far end. Should the data be too short, it can be padded or concatenated to a valid length. The functions offered and the format of the unacknowledged mode PDU are shown in Figure 6.46. It contains a 7-bit sequence number to allow the segments to be reassembled. An extension bit (E) of 1 indicates that the next field will be a length indicator (LI). If $E = 0$, then data follows.

The LI points to where the SDU received by the RLC layer from the upper layer finishes within this PDU. It can be either 7 or 15 bits long, depending on the maximum length of the PDU, as defined in the upper-layer signalling. If the RLC size is greater than 125 bytes, the 15-bit LI is used. Generally, the RLC PDU size will be the transport block size (see later) less the MAC layer overhead. The LI field has some reserved values which are used to inform the receiver of such occurrences as the first part of an SDU, the last part of an SDU and padding. These are needed for the segmentation, concatenation and reassembly functions. For the 7-bit LI, these are shown in Table 6.10.

By way of example, consider that an RLC SDU needed to be segmented into three RLC PDUs. They would be arranged as shown in Figure 6.47. The first PDU indicates in the LI that the payload is the start of an SDU. The second has no LI as it is a continuation of the data, and the third has two LI fields. The first shows where the SDU finishes in

Mode	Functions provided
Unacknowledged Mode (UM)	Segmentation and reassembly
	User data transfer
	Discard of errored SDU
	Concatenation
	Padding
	Ciphering
	Sequence number checking

SEQ No.	E	byte 1
LI	E	
⋮		
LI	E	
Data		
PAD		byte n

Figure 6.46   RLC unacknowledged mode data PDU

Table 6.10 Length indicator field for an RLC UM PDU

Length indicator	Description
0x00	The last received PDU did not have a length indicator but contained the last part of an SDU
0x7C	The data in the payload is the start of an SDU
0x7F	The remainder of the PDU is padding

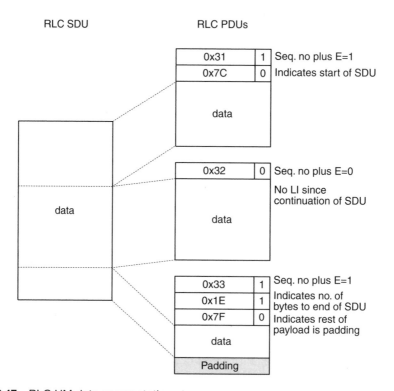

Figure 6.47 RLC UM data segmentation

the payload (in this case 0x1E bytes down), and the second indicates that the rest of the payload is padding. This would be as opposed to the first part of the next SDU for concatenation.

Acknowledged data transfer

This passes packets onward and guarantees delivery of the packets. It ensures that duplicate packets are discarded and that packets in error are retransmitted using an automatic repeat request (ARQ) mechanism. If delivery is not possible then an error message is send to the sender to notify it that there was a problem. Both in-sequence and out-of-sequence delivery modes are supported, since in many cases the upper-layer protocols are capable of restoring the correct order of the packets if required. It should be noted that there is a single RLC connection for each radio bearer. The functions provided by acknowledged

Mode	Functions provided
Acknowledged Mode (AM)	Segmentation and reassembly
	User data transfer
	Discard of errored SDU
	Concatenation
	Padding
	Ciphering
	In sequence PDU delivery
	Duplicate detection
	Error and flow control

D/C	SEQ No.		byte 1
SEQ No.		P	HE	byte 2
LI		E	
⋮			
LI		E	
Data			
PAD/STATUS			byte n

Figure 6.48   RLC acknowledged mode PDU

mode and the format of the data PDU are shown in Figure 6.48. There are a number of other PDU formats since this mode has control functions associated with it. Here, the sequence number is 12 bits and is used for retransmission control as well as reassembly. The D/C bit indicates whether this is a data (1) or control (0) PDU. The P bit indicates polling and is used to request a status report (if 1) from the receiver. The HE is a 2-bit field which indicates what follows. Currently only two values are defined to indicate data (0) or a length field (1).

The data PDU may have a status PDU piggybacked to the end of the frame to report the status between the two entities. Alternatively, a separate status PDU may be send. The format of the status PDU is shown in Figure 6.49.

The PDU type field is 3 bits in length and indicates that this is either a *status, reset* or *reset acknowledge* PDU. The other five values are currently reserved. The *reset* procedure is used to reset the RLC peer entities. During this procedure the hyperframe numbers (HFN) in UTRAN and the UE are synchronized. The status field (SUFI) contains status information and is indicated with a PDU type of 000. There are currently eight different types defined. The format is a type–length–field structure, where the length and field are dependent on the type. The possible types are also listed in Figure 6.49. The windowing is for flow control. Perhaps the most common status is the acknowledgement, Figure 6.50, which acknowledges receipt of all PDUs with sequence number less than *last sequence number* (LSN).

D/C	PDU type	SUFI₁	byte 1
SUFI₁			byte 2
⋮			
SUFIₖ			
PAD			byte n

PDU type value	Description
0	No more data
1	Window size
2	Acknowledgement
3	List
4	Bitmap
5	Relative list
6	More receiving window
7	More receiving window ack
8-15	Not currently used

Figure 6.49   RLC status PDU

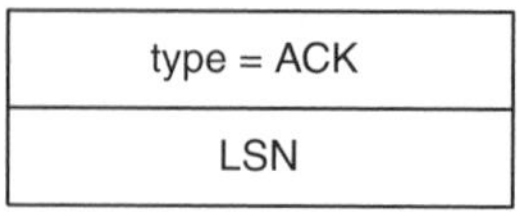

type = ACK
LSN

Figure 6.50   RLC acknowledgement

6.12.4   Media access control (MAC)

The MAC layer is responsible for mapping the logical user data and control channels to the transport channels. It ensures that the transport format for a data source provides for efficient usage of the transport channels. This is particularly important for 'bursty' data flows. The MAC can also prioritize data flows on the common and shared transport channels. When a mobile device is receiving on a common downlink channel or when it is sending on the RACH it needs to be identified uniquely within the cell. This indentification is also performed at the MAC layer. Multiplexing and demultiplexing of upper-layer packets on both the common and dedicated channels is also performed at this layer. If a higher data rate than is assigned on a common channel is required by the mobile device, the MAC layer can switch over to a dedicated channel if requested to do so by the RRC.

The MAC layer functions can be summarized as follows:

- mapping between logical and transport channels;
- selecting the transport format for each channel;
- QoS provisioning and scheduling;
- UE identification on common transport channels;
- multiplexing/demultiplexing of PDUs from upper layers onto both common and dedicated transport channels;
- encryption of RLC transparent mode data.

The architecture of the MAC layer is broken into three parts: the MAC-d, which handles dedicated data channels, the MAC-c/sh, which handles the common control and shared channels, and the MAC-b which deals with the broadcast channel. For each cell instance, there will only be one MAC-b and one MAC-c/sh; however, there will be one MAC-d for each user connection. The architecture is presented in Figure 6.51. The MAC-d to MAC-c/sh interface will often be internal to the RNC, but may be connected across the Iur interface when the SRNC of the user is not the CRNC of the common channel. This is described later. Note on the diagram that in some instances, the BCCH information can also be sent on the FACH.

As an example of data flow through the MAC layer, consider a dedicated logical traffic channel (DTCH), in the downlink direction. This channel is mapped at the MAC layer onto the appropriate transport channel, as directed by the RRC protocol. This mapping can be any one of the three options shown in Figure 6.52.

For a voice call, it makes sense to map this logical channel to a dedicated transport channel, DCH, since the nature of the application will have little change over a relatively long duration (of the order of minutes). However, many data applications, for example

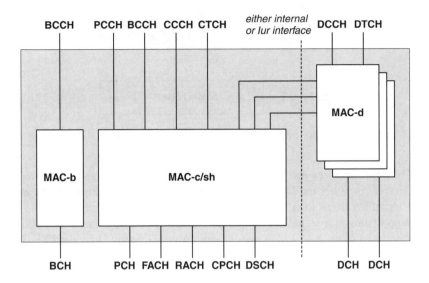

Figure 6.51   MAC architecture

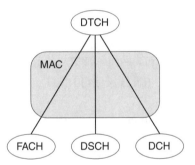

Figure 6.52   Logical to transport channel mapping

short emails or DNS queries, only send small amounts of information infrequently. There-fore, these can be mapped into either the common forward access channel (FACH) or the downlink shared channel (DSCH), where they share channel bandwidth with other users, allowing for statistical multiplexing. The RNC will monitor the bandwidth requirements and volume of data of the logical channel, and should it exceed some threshold, move it to a dedicated transport channel. Conversely, should the rate fall below a threshold, it can be moved from a dedicated to a common/shared channel.

The MAC layer adds a protocol overhead, and the type of overhead added depends on the logical to transport channel mapping.

For a DTCH/DCCH that is mapped to a DCH, there are two possible situations. The first is where there is no multiplexing of channels on the MAC, i.e. only one logical channel in the transport channel. Here there is no MAC header present (Figure 6.53(a)). The second is where multiplexing is present, and in this case, the channel is identified using a C/T fields (Figure 6.53(b)).

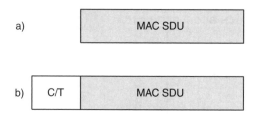

Figure 6.53   MAC format for dedicated transport channels

The C/T field identifies the logical channel number (1–15) where there are multiple logical channels in either the DCH, FACH or RACH. Channel multiplexing is used when there are a number of radio bearers defined to be carried on the same transport channel. An example of this is a signalling connection which defines a number of radio bearers, using different RLC modes. Which one is being used is identified by the C/T field.

When data is on the FACH/RACH, or on the DSCH, the specific UE needs to be identified. A target channel type field (TCTF) field is included, to identify the channel type, i.e. that this is a DTCH/DCCH. The UE ID type and UE ID fields are in the header, and optionally the C/T field, should multiplexing within a single UE be available (Figure 6.54).

- *UE ID type*: indicates whether the following UE ID is a U-RNTI (32 bits) or a C-RNTI (16 bits), both of which are explained in Section 6.16.2.
- *UE ID*: this will be either a U-RNTI or a C-RNTI as identified by UE ID type.

The various control channels, if transported on their own transport channel (e.g. PCCH mapped to PCH) do not require any header. However, in the case of logical channels mapped to the RACH/FACH, a TCTF header field is required, as shown in Figure 6.55.

- *TCTF*: this provides an identification of which logical channel type (e.g CCCH) is carried in the payload. It is required on the FACH and RACH channels. Note that the number of bits is dependent on the channel. The values currently designated are as shown in Figure 6.56.

TCTF	UE ID type	UE ID	C/T	MAC SDU

Figure 6.54   MAC format for data on common/shared channels

TCTF	MAC SDU

Figure 6.55   MAC format for control channels on FACH/RACH

FACH		RACH	
TCTF	**Channel**	**TCTF**	**Channel**
00	BCCH	00	CCCH
01000000	CCCH	01	DCCH/DTCH
10000000	CTCH		
11	DCCH/DTCH		

Figure 6.56   TCTF values for the FACH/RACH

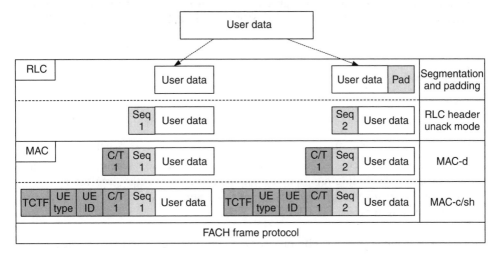

Figure 6.57   Data passing through RLC and MAC layer

Figure 6.57 shows an example of a segment of data passing through the MAC and RLC protocol stack. In this example, the RLC unacknowledged mode is used, and the DTCH/DCCH is mapped to the FACH. At the RLC layer, the data is segmented and a sequence number and length field are added to it. At the MAC layer, the UE is identified on the FACH using its U-RNTI or C-RNTI (see Section 6.16.2). The TCTF will have the value 11, which indicates that this is a DTCH (or DCCH) rather than a BCCH, CCCH or CTCH.

6.12.5   MAC and physical layer interaction

The following are definitions applied to data that is transferred between the MAC and physical layers. They describe the format and rate of the data.

- *Transport block (TB)*: this is the basic unit exchanged between the physical layer and the MAC. At the physical layer, a CRC may be appended to each TB.

- *Transport Block Set (TBS)*: a group of TBs, passed between the physical layer and the MAC during a given time interval and using the same transport channel.
- *Transport block size*: the number of bits in a given TB. All TBs in a TBS must have the same size.
- *Transport block set size*: the number of bits in a TBS. It is equal to transport block set × transport block size.
- *Transmission time interval (TTI)*: the time period during which a TBS is transferred across the radio interface. The MAC layer will deliver one TBS, i.e. one or more TBs, to the physical layer every TTI. The TTI is always a multiple number of radio frames (one radio frame is 10 ms.) An example of the MAC/physical layer exchange for two dedicated channels is shown in Figure 6.58. It can be seen that the data rate can be varied in every TTI by changing either the size of each TB or the number of TBs that are sent.

Transport format

The transport format describes the structure of the TBS that the physical layer will deliver in a TTI. It is broken into two parts, the dynamic part and the semi-static part:

- *Dynamic part*: this consists of the transport block size and the transport block set size.
- *Semi-static part*: this consists of the TTI, the error protection scheme (turbo, convolution or none, the coding rate, and the rate matching parameter), and the size of the CRC. This is set up via the upper layer RRC signalling and thus to modify requires further signalling between the UE and the RNC.

Consider that there are four TBs of size 336 bits, then the transport format could be defined as:

Dynamic part: 336 bits, 1344 bits
Semi-static part: TTI = 20 ms, turbo coding, 1/3 rate, rate match = 1, CRC = 16 bits

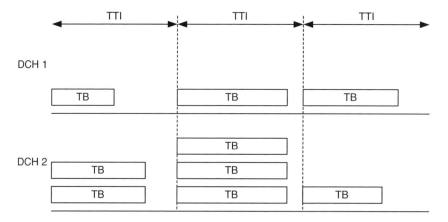

Figure 6.58 Example of transport blocks

Transport format set

The transport format set (TFS) describes all transport formats that are possible on a channel. This is to support variable data rates, since the rate can change every TTI. For a TFS, the semi-static part will be the same for all formats. For example, a TFS could be described as:

Dynamic part: (336 bits, 336 bits), (336 bits, 672 bits), (336 bits, 1008 bits),
                     (336 bits, 1344 bits)

Semi-static part: TTI = 20 ms, turbo coding, 1/3 rate, rate match = 1,
                     CRC = 16 bits

Here, the TB size remains the same for each TTI, but the number of blocks can vary between one and four. Consider that this uses RLC acknowledged mode with no MAC layer multiplexing, this accounts for 16 bits of overhead, meaning the data is actually 320 bits. This then corresponds to data rates of 16 kbps (320/20 ms), 32 kbps, 48 kbps and 64 kbps, and to transfer the data would take 80 ms (4 × TTI), 60 ms, 40 ms and 20 ms, respectively. This allows the data rate to vary every TTI without any need for any signalling between the UE and the network.

Transport format combination

The physical layer may take one single transport channel and place it into a physical channel, or it may multiplex several transport channels together. Each transport channel will have a TFS associated with it. However, if several are being multiplexed together, at a given instance, not all may be present. The transport format combination (TFC) is an authorized combination of transport formats that may be passed to the physical layer. This means that not all permutations from the TFS of the transport channels may be allowed. This could be used, for example, to prevent all multiplexed channels using their maximum rate simultaneously. The resulting channel consisting of several multiplexed transport channels is referred to as a *coded composite transport channel* (CCTrCH).

Transport format combination set

A transport format combination set (TFCS) is defined as a set of TFCs on a CCTrCH. It is the MAC layer that will decide which TFC from the TFCS to pass to the physical layer. However, the assignment of the TFCS is performed during bearer establishment signalling procedures, as described in Section 6.16.7, prior to transfer of data.

Transport format indicator

The transport format indicator (TFI) is a label for a particular transport format in a transport format set. Every time a transport block is passed between the MAC and the physical layer, a TFI for that transport block is also exchanged.

Transport format combination indicator

The physical layer may receive several transmission blocks that it will multiplex into a CCTrCH. Each block submitted by the MAC will also be associated with a TFI. The

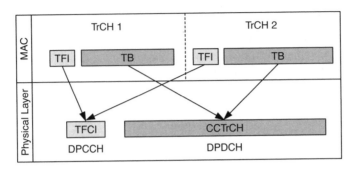

Figure 6.59 Transport format combination

physical layer will then generate a transport format combination indicator (TFCI) to indicate this current transport format combination. The TFCI is included in the physical control channel associated with the data to inform the receiver of the transport format in the current transmission.

Figure 6.59 shows an example of two transport channels (TrCH) being passed to the physical layer for transmission on a single dedicated data channel (DPDCH).

The TFI/TFCI part is optional and does not need to be transported if blind transport format detection (BTFD) is employed, where the receiver can work out the transport format using other information such as CRC, DPDCH/DPCCH power ratio, etc.

As an example of how all the definitions fit together, consider that a user is making a phone call using AMR 12.2 kbps, while also browsing the web. For the web traffic, a maximum rate of 64 kbps has been allocated, with intermediate rates of 16 kbps and 32 kbps. The TFS for the channels could look as follows:

TFS for TrCH 1 (12.2 kbps AMR)
Dynamic part: 244 bits, 244 bits
Semi-static part: TTI = 20 ms, convolution coding, 1/3 rate, rate match = 1,
                        CRC = 12 bits

TFS for TrCH 2 (64 kbps data, with 16 kbps and 32 kbps rates)
Dynamic part: (320 bits, 320 bits), (320 bits 640 bits), (320 bits, 1280 bits)
Semi-static part: TTI = 20 ms, convolution coding, 1/3 rate, rate match = 1,
                        CRC = 16 bits

TFCS
Dynamic part:
Combination 1 – TrCH 1: (244, 244) TrCH 2: (320, 320)
Combination 2 – TrCH 1: (244, 244) TrCH 2: (320, 640)
Combination 3 – TrCH 1: (244, 244) TrCH 2: (320, 1280)
Semi-static part:
TrCH 1: TTI = 20 ms, convolution coding, 1/3 rate, rate match = 1, CRC = 12 bits
TrCH 2: TTI = 20 ms, convolution coding, 1/3 rate, rate match = 1, CRC = 16 bits

Figure 6.60 illustrates the rates and possible combinations for these two channels. A value of 0 kbps indicates that there is no data being sent on the channel (discontinuous transmission; DTX).

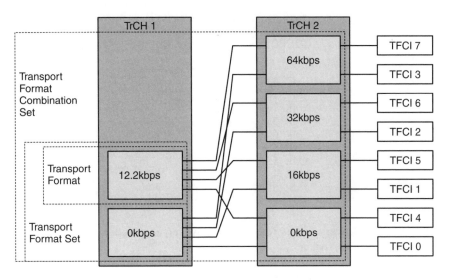

Figure 6.60 TFCS example

In practice, to provide a given user data rate requires the transport block size to include the RLC overhead, which is dependent on the chosen RLC mode of operation. Consider that the user application requires 64 kbps, and the RLC unacknowledged mode. If the TTI is 20 ms, then the application will send 320 bits each TTI. However, the RLC will add 2 bytes of overhead (7-bit sequence number + 7-bit length field + 2 × E bits). Therefore, the TB size will be 336 bits. Since 12.2 kbps speech is transferred in RLC transparent mode there is no additional overhead added for this channel.

6.13 ADAPTIVE MULTIRATE (AMR) CODEC

The previous example used the basic rate of 12.2 kbps for a voice call. It showed that this equated to a 244-bit transport block which is sent each TTI = 20 ms. However, due to the operation of the AMR CODEC in UMTS, the situation is somewhat more complex, but it serves as a useful example of the interoperation of the radio link protocols. The AMR speech CODEC is a multirate speech coding scheme consisting of eight source rates from 4.75 kbps to 12.2 kbps. Table 6.11 shows the rates.

These rates are CODECS from various sources; for example, the AMR_12.20 is the GSM Enhanced Full Rate (EFR) and the AMR_6.70 is the EFR CODEC as used in the PDC network in Japan. The AMR_SID is a silence descriptor rate for encoding of low-rate background noise and consists of 39 bits sent every TTI.

The rationale for support of a number of rates is that for traffic management, the source rate being used can be changed in the downlink when traffic on the air interface exceeds an acceptable load or when the link is exhibiting poor quality. In the uplink, it can also be changed because of loading considerations or to extend uplink coverage. In theory, the

Table 6.11 AMR CODEC rates

CODEC mode	Bit rate (kbps)
AMR_12.20	12.20
AMR_10.20	10.20
AMR_7.95	7.95
AMR_7.40	7.40
AMR_6.70	6.70
AMR_5.90	5.90
AMR_5.15	5.15
AMR_4.75	4.75
AMR_SID	1.80

rate can be changed every 20 ms (the TTI). However, in practice the adaptation would need to occur much less often.

The CODEC divides bits up into three different classes, classes A, B and C, according to their relative importance within the coding scheme. Class A bits are the most important and have a high level of error protection, including a CRC check. These bits, if corrupted, will result in audible degradation of speech quality. Erroneous speech frames are thus identified by the CRC check, and should have some error concealment techniques applied to them. For class B and C bits, their importance is less, with class C being less sensitive to errors than class B. These are afforded less error protection. Class B and C speech frames containing errors can be passed on. The voice quality will degrade with increased error content; however, the effects of errors are much less significant than for class A. The number of bits per class for each AMR rate is shown in Table 6.12. Note that not all rates have class C bits.

The CODECs are all based on the algebraic code excited linear prediction (ACELP) algorithm, which assumes that the vocal tract is a linear filter receiving inputs in the form of air from the lungs being vibrated at the vocal cords to produce speech. This means that the audio signal can be compressed by a high factor; however, the compression algorithm is optimized for voice.

It was seen in Chapter 2 that although in standard telephone networks, voice is allocated resources as a full-duplex application, in fact voice activity is closer to half-duplex, as

Table 6.12 AMR CODEC bit numbers

CODEC	Class A	Class B	Class C	Total
AMR_12.20	81	103	60	244
AMR_10.20	65	99	40	204
AMR_7.95	75	84	0	159
AMR_7.40	61	87	0	148
AMR_6.70	58	76	0	134
AMR_5.90	55	63	0	118
AMR_5.15	49	54	0	103
AMR_4.75	42	53	0	95
AMR_SID	39	0	0	39

Table 6.13 12.2 kbps transport format parameters

No. of TrChs		3	Comments
Transport block size	TrCh 1	0, 39, 81 bits	Class A
	TrCh 2	103 bits	Class B
	TrCh 3	60 bits	Class C
TFCS	1	1*81, 1*103, 1*60	12.2 kbps speech
	2	1*39, 0*103, 0*60	SID
	3	1*0, 0*103, 0*60	DTX
Error protection	TrCh 1	1/3 rate convolution + 12-bit CRC	
	TrCh 2	1/3 rate	
	TrCh 3	1/2 rate	
TTI		20 ms	

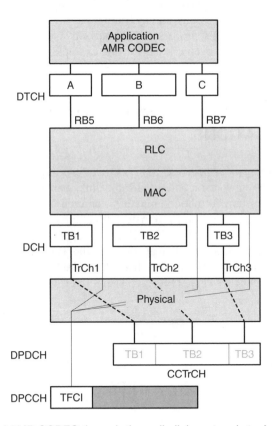

Figure 6.61 Flow of AMR CODEC through the radio link protocol stack

one user listens as the other speaks. In addition, normal speech is punctuated with silent
gaps. All the CODECs also work on silence suppression, where a voice activity detector
determines if the user is silent and whether speech frames should be sent or not. If the
system merely went dead when users ceased talking, it would be rather disconcerting for
the listener; therefore as a reassurance, the receiver side will introduce comfort noise. To

assist in the generation of this comfort noise, a Silence Descriptor (SID) frame at a low rate is sent. This is particularly applicable to environments where there is a relatively high level of background noise such as, for example, in a car, or if the user has a radio on.

Since the three classes of bits require different levels of error protection, this means that their QoS profiles are different and cannot therefore be combined at the MAC layer. This means that each requires a different transport channel (TrCh) which is combined at the physical layer into a CCTrCH. For simplicity consider only the 12.2 kbps rate, with the SID. The transport format information is as shown in Table 6.13.

There will be three radio bearers established, each using RLC transparent mode since the segmentation and reassembly function is not required. At the MAC layer, three different transport formats are defined for the different error protection schemes, with three transport blocks delivered to the physical layer every 20 ms. The physical layer performs the CRC attachment for class A bits, the convolution coding and the rate matching, and combines all these into a CCTrCH. This process is summarized in Figure 6.61. Note that radio bearers (RB) 5, 6 and 7 have been chosen for this example.

6.14 CALCULATED TRANSPORT FORMAT COMBINATIONS

When a connection is established, the RNC must inform both the BTS and the UE of the parameters for the new or reconfigured radio link. Part of this parameter set is the range of permitted transport format combinations, and the identifier of each. UMTS provides a simple, efficient mechanism for transfer of this information through the calculated transport format combination (CTFC). The CTFC is calculated as follows. Consider that there are I transport channels (TrCH) that make up the TFC, and that each of these has L tranport formats, where the TFI of the channel will be TFI $= 0, 1, 2, \ldots, L - 1$. For a given TFC, that is TFC $= (\text{TFI}_1, \text{TFI}_2, \ldots, \text{TFI}_I)$, the CTFC is calculated as:

$$\text{CTFC} = \sum_{i=1}^{I} \text{TFI}_i \bullet P_i$$

Where:

$$P_i = \prod_{j=0}^{i-1} L_j$$

And in this calculation, $L_0 = 1$.

The mathematics here may appear somewhat unwieldy; however, the following example illustrates their calculation. Consider that a user wishes to make a voice call and that the network establishes a radio link for this, plus an associated signalling channel. As discussed, the AMR CODEC generates class A, B and C bits, which each require their own TrCH, since the QoS profiles are different. Therefore, there are four channels established, three for the call and one for signalling.

	AMR			Signalling
	Class A	Class B	Class C	
	TrCH1	TrCH2	TrCH3	TrCH4
L	3	2	2	2
TFI				
0	0	0	0	0
1	81	103	60	148
2	39	-	-	-

Figure 6.62   AMR CODEC transport formats

The voice channels are as shown in Table 6.12, with the inclusion of the signalling bearer, which is a 3.4 kbps channel, with a TB size of 148 bits sent in a 20 ms TTI. The transport formats are as shown in Figure 6.62.

The permitted combinations are DTX, SID, 12.2 kbps, signalling, SID + signalling, 12.2 kbps + signalling, as shown in Figure 6.63.

The CTFCs are calculated as follows:

$$\text{TFC} = (\text{TFI}_{\text{TrCH1}} \times L_0) + (\text{TFI}_{\text{TrCH2}} \times L_0 \times L_{\text{TrCH1}})$$
$$+ (\text{TFI}_{\text{TrCH3}} \times L_0 \times L_{\text{TrCH1}} \times L_{\text{TrCH2}})$$
$$+ (\text{TFI}_{\text{TrCH4}} \times L_0 \times L_{\text{TrCH1}} \times L_{\text{TrCH2}} \times L_{\text{TrCH3}})$$

Ignoring L_0 since it is always 1, this yields the results shown in Table 6.14.

TFCI	$\text{TFI}_{\text{TrCH1}}$	$\text{TFI}_{\text{TrCH2}}$	$\text{TFI}_{\text{TrCH3}}$	$\text{TFI}_{\text{TrCH4}}$	Explanation
1	0	0	0	0	DTX
2	1	0	0	0	SID
3	2	1	1	0	12.2kbps
4	0	0	0	1	Signalling
5	1	0	0	1	SID+signalling
6	2	1	1	1	12.2kbps+signalling

Figure 6.63   Transport format combinations

Table 6.14 Calculation of CTFC

TFCI	Calculation	CTFC
1	$(0) + (0 \times 3) + (0 \times 3 \times 2) + (0 \times 3 \times 2 \times 2)$	0
2	$(1) + (0 \times 3) + (0 \times 3 \times 2) + (0 \times 3 \times 2 \times 2)$	1
3	$(2) + (1 \times 3) + (1 \times 3 \times 2) + (0 \times 3 \times 2 \times 2)$	11
4	$(0) + (0 \times 3) + (0 \times 3 \times 2) + (1 \times 3 \times 2 \times 2)$	12
5	$(1) + (0 \times 3) + (0 \times 3 \times 2) + (1 \times 3 \times 2 \times 2)$	13
6	$(2) + (1 \times 3) + (1 \times 3 \times 2) + (1 \times 3 \times 2 \times 2)$	23

This means that only these six numerical identifiers need to be sent, from which the table of transport formats can be constructed.

6.15 USE OF DSCH

The downlink shared channel is used in CELL-DCH mode see Section 6.16.1, and can provide extra variable data rate to a user. For variable rates, it can be more efficient to allocate many users to share the bandwidth of the DSCH. Consider the following example for a video streaming application (Sallent *et al.*, 2003). The service provides a basic rate of 32 kbps which is allocated on a dedicated channel, DSCH. In addition, an enhanced service at higher rate is allocated on the DSCH, which can vary between an additional 0–128 kbps. This allows the video rate to vary between a minimum of 32 kbps and a maximum of 140 kbps. This allocation can support a basic service delivery through the dedicated channel, and then supplement this on the DSCH when resources are available. However, the service can guarantee that at least 32 kbps will be provided.

For Release 5 of UMTS, the specification allows for use of advanced transmission techniques and the introduction of the 16-QAM (quadrature amplitue modulation) scheme to allow the DSCH to achieve data rates of up to around 10 Mbps. This scheme is referred to as high-speed downlink packet access (HSDPA).

6.16 RADIO RESOURCE CONTROL (RRC)

RRC is the radio network control protocol and provides the functions related to management and control of the radio network transmission resources such as the MAC, RLC and PDCP layers, etc. The primary purpose of RRC procedures is to establish, maintain and release radio resource connections between the mobile device and the network. Its functions include such things as handover procedures and cell reselection. RRC is the control protocol between the UE and the RNC. The major functions provided by RRC are listed in Table 6.15.

RRC must interact with each layer to provide control information, and in turn receive measurement feedback from the layers. Between the UE and UTRAN, RRC is responsible for the establishment of radio bearers for both transport of signalling and traffic, and

Table 6.15   RRC functions

Radio resource control functions
System information broadcast
Establishment, maintenance and release of RRC connection between UE and UTRAN
Assignment, reconfiguration and release of radio bearers
RRC connection mobility
Paging
Outer loop power control
Ciphering control
UE measurement and reporting

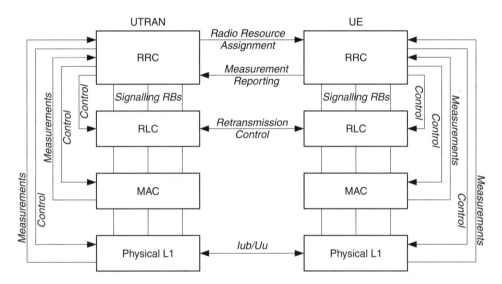

Figure 6.64   RRC control

the assignment of radio resources. These radio resources will include physical resources such as codes and frequencies, and transport resources such as transport block sizes and transport format sets. A conceptual model of RRC is shown in Figure 6.64.

This section describes the different RRC states that a UE can be in. It then explains how the UE is addressed in these different states. Subsections 1.16.3–1.16.11 describe various procedures that are frequently performed by RRC signalling. The RRC specification (TS25.331) is rather lengthy, running into many hundreds of pages, thus only a sample of all of the procedures are explained. These examples are in no particular order. However, an example of a *call life cycle* in Section 6.22 places them in context with each other, and the complementary signalling required at other levels as well. The procedures illustrated are:

- RRC connection
- Signalling radio bearers

- RRC security mode control
- RRC paging
- Radio bearer establishment
- Transfer of NAS messages
- Cell/URA update
- Measurement reporting
- Active set update.

6.16.1 RRC mobile states

At any given instant, the mobile device may or may not have a signalling connection. If no connection exists, the mobile device is referred to as being in idle mode, and must establish a connection before it can proceed with any further communication to the network. In idle mode, only the core network knows of the existence of the user, and this is to a location area (LA) or routing area (RA). In idle mode, the location of the mobile device is known only to the accuracy of an LA or RA and the UTRAN does not know anything about the UE. The UE will monitor the LA or RA and if it notices that it has changed, it must inform the core network. To do this, it must enter connected mode. The mobile device will enter the idle state either when the signalling connection is released or when is a failure of the radio resource connection.

When a *signalling* connection does exist, this is known as RRC connected mode. However, as has been seen, depending on the activity of the mobile device this connection can be transported either over dedicated or common/shared channels. If the mobile device has a lot of activity with the network, then its location may be tracked to the cell level. However, if the device sends data infrequently, then it may be tracked on a larger area consisting of a number of cells. In this latter case if a mobile device has to be located for a mobile terminated call, it is necessary to page the whole range of cells. However, this tracking over a larger area does minimize the amount of location update messages sent by the UE.

If a signalling connection does not exist then the mobile device will require to be paged across a location area (routing area) to receive a mobile terminated call. This page will instruct the mobile device to establish a dedicated signalling connection. If the mobile device is to be able to receive mobile terminated calls, it must have already registered with the core network so that paging (to indicate an incoming call, for example) can be directed to the correct geographical area.

In the connected state the location of the mobile device is known within the core network to the accuracy of the serving RNC (SRNC). The SRNC will then know the location of the mobile device to either the cell or the UTRAN registration area (URA) level. The connected state actually consists of four separate states known as CELL-DCH, CELL-FACH, CELL-PCH and URA-PCH.

CELL-DCH

In this state a dedicated physical channel is allocated for the mobile device in the uplink and downlink. In the uplink a dedicated transport channel, DCH, is used and in the downlink, this may be a DCH and/or DSCH. The location of the device is known by its SRNC to a cell level. The UE enters this state from idle mode through establishment of an RRC connection, and from CELL-FACH through the establishment of a dedicated channel. It may move to any other state from CELL-DCH by explicit signalling. The UE does not generally listen to the broadcast channel in this state and any major changes to system information will be indicated on the FACH channel.

CELL-FACH

No dedicated channel is allocated to the mobile device in this state, the RACH and FACH are used instead for both signalling and small amounts of user data. This is ideal for such services as SMS and multimedia messaging service (MMS). The mobile device can monitor the BCH and may also be asked to use the CPCH to transfer data in the uplink. The UE must constantly monitor the FACH in the downlink.

CELL-PCH

Again, the mobile device can be located to a specific cell but it cannot transfer data and must be paged by the network. If the mobile device moves cell and a cell reselection is required, the mobile device will change to the CELL-FACH state to perform the procedure and will then return to the CELL-PCH state. CELL-PCH allows the mobile device to use DRX. This allows the mobile device to 'sleep' and only wake up at specific times to monitor the paging indication channel (PICH). This enables the device to consume less power. A counter within the network can monitor the number of cell updates a mobile device makes and if a threshold is met the mobile device may be moved to URA-PCH mode to conserve its power and to reduce the amount of signalling. No uplink activity is possible in the CELL-PCH state. The dedicated control channel cannot be used in this state.

URA-PCH

In the URA-PCH state, the mobile device does not update its location every time it selects a new cell. The URA is a new area that has been introduced into the UMTS network to complement the LA and RA, which are still used. Unlike the LA and RA, which are controlled from the core network, the URA is controlled within UTRAN. The core network only knows to which RNC to send data; it is up to the UTRAN to locate a specific mobile device. Each cell in the UTRAN will belong to at least one URA and the mobile device will be allocated a URA to which it is attached. The mobile device will monitor the cell broadcast channel to see if the current cell broadcast is in the same URA as the one it has been allocated. If they are the same then there is no update; if they are different then the mobile device will enter the CELL-FACH state to perform a URA update, which is similar to a cell update. It should be noted that a cell may broadcast a number of URAs. By introducing this overlapping feature, the UTRAN can reduce the number of URA updates. Thus, two mobile devices may have different URAs assigned to them but be attached to the same cell. The dedicated control channel cannot be used in this state. The four states and the transitions between them are shown in Figure 6.65.

CELL-PCH and URA-PCH states are really designed for packet traffic. For a phone call, there is a very clearly defined start and end of the transaction, so returning to idle

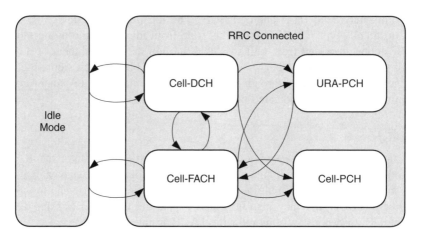

Figure 6.65 RRC Modes

mode following call hang-up could be seen as logical. However, this does mean that the subscriber will need to be paged for a mobile terminated call over the LA; if it had been transferred to Cell-PCH for a certain period of time it takes little in the way of resources and can be paged directly in a single cell. However, for packet traffic, this is not the case as it is impossible for the network to decide if a user has completed their data transfer. A typical example would be web traffic where there are periods in between information exchange where the user is reading the pages. It is useful to keep the UE 'connected' but release any resources it is utilizing by moving the UE to one of these two states.

A scenario may go as follows. The UE is transferring a reasonable amount of data, say email or file transfer, and the UTRAN allocates a dedicated channel, moving the UE to CELL-DCH. Subsequently, the RNC observes that the quantity of data being transferred has significantly reduced, and instructs the UE to release its dedicated connection and instead use the FACH/RACH or CPCH. The UE then moves into CELL-FACH. After a further period of inactivity, the RNC instructs the UE to move to CELL-PCH, where all resources are released; however, the RNC is still aware of the UE, as it is performing cell updates. The RNC will generally use an inactivity timer to decide when the UE should change states. However, this is part of the RRM strategy, which is the remit of the manufacturer. This state would be ideal for a stationary data user, since the burden of signalling is reduced to periodic cell updates between the UE and RNC, and DTX mode can be used. If the user then moves, these cell updates will happen more frequently, and the RNC may decide to reduce this further by moving the UE into the URA-PCH state. The precise mechanisms controlling how and when a UE moves states, and into which state, are governed by the radio resource policies in the RNC. However, these policies may be currently considered quite subjective, with the absence of an adequate working test environment in which to evaluate.

As previously stated, if the mobile device is in idle mode then it will still update the core network of its location to the LA or RA. However, if the mobile device is in connected mode, the UTRAN itself deals with the location of the device, removing some of the burden from the core network. The core network only needs to know the location of the device to the correct RNC; it is up to the RNC to locate the mobile device within a cell or number of cells. If a mobile device has a connection to the packet switched core network and another subscriber wishes to make a mobile terminated voice call to the same mobile device, the page request is directed to the correct RNC from the MSC/VLR. The RNC in this case will know exactly where the mobile device is since it is already in packet switched connected mode. The RNC will use the IMSI number of this device to coordinate the use of the existing connection for the page. If the circuit switched call is accepted, the UE will now have two connections, one from the RNC to the CS-CN and one to the PS-CN. There will, however, only be one instance of RRC between the mobile device and the RNC. The IMSI is only stored in the RNC for the duration of the RRC connection. It should be noted that if the mobile device is in CELL-PCH or URA-PCH then paging is done in a similar method to that of paging a mobile device in idle mode.

6.16.2 UTRAN UE identifiers

Throughout the UTRAN, once a user is RRC connected, a number of radio network temporary identities (RNTI) are used to identify the UE and are contained in signalling messages between the UE and UTRAN. There are four types of RNTI:

- *Serving RNC RNTI (S-RNTI)*: this is allocated to a UE by the SRNC when it establishes an RRC connection (see Section 6.16.3) to the network. It is unique within that SRNC. Should the RRC connection change, a new S-RNTI is allocated. It is used subsequently by the UE to identify itself to the SRNC, and by the SRNC to address the UE. It is also used by the DRNC to identify the UE to the SRNC when the Iur is being used.
- *Drift RNC RNTI (D-RNTI)*: this is allocated to a UE by the DRNC when the UE establishes a connection to it. The SRNC maintains a mapping between S-RNTIs and D-RNTIs. It is only used by the SRNC to identify the UE to the DRNC and is not used by the UE at all.
- *Cell RNTI (C-RNTI)*: this is allocated by the controlling RNC when the UE accesses a new cell, and is unique within that cell. It is used on common channels for the UE to identify itself to the CRNC, and also by the CRNC to address the user.
- *UTRAN RNTI (U-RNTI)*: this is allocated to a UE having an RRC connection and identifies the UE within the whole UTRAN. A U-RNTI consists of the RNC ID of its SRNC and its S-RNTI. The RNC ID is a unique identifier for the RNC within the UTRAN.

6.16.3 RRC connection

From idle mode, once the UE is ready to connect to the network, for example in response to a page or to perform a location update, it will send an *RRC connection request* message to the network. Since this is an initial access and the UE has no other connection to use, it will be sent on the RACH using the logical common control channel (CCCH). At the physical layer there is a chance for collision on this channel, since more than one mobile could attempt to access the network at the same time. The physical layer provides a number of mechanisms to reduce the collision probability and deal with collisions, should they occur.

The RNC will respond to an *RRC connection request* with an *RRC connection setup* message, again on the logical CCCH, which in turn is confirmed by the UE with an *RRC connection setup complete* message. Since the connection setup message informs the user of the connection that the RNC has established, the UE will then use this channel (DCCH) to reply to the RNC with the *complete* message. Figure 6.66, shows an example of the establishment of a dedicated signalling connection. Here, once established, the UE can send subsequent RRC messages over its dedicated transport channel. Note that RRC messages are transparent to the BTS.

The RNC will allocate resources and establish bearers for the signalling using the Node B application part (NBAP) and the access link control application protocol (ALCAP). This will establish a connection for the subscriber between the RNC and BTS. NBAP and ALCAP signalling is only required if a DCH is being set up; if the user is offered a signalling connection over the RACH/FACH then this is not required. This NBAP and ALCAP signalling is shown in Figure 6.66 as *signalling bearer establishment*. Once completed, it will respond to the UE with the *RRC connection setup* message.

The parameters of the *RRC connection request* are very straightforward. It uses the RLC transparent mode and contains the initial UE identity, which is the IMSI, TMSI or P-TMSI, and an establishment cause to indicate why this RRC connection is being made. It could indicate, for example, registration, for a location update, or originating interactive call, etc. It will also include some measurements made on the pilot channel, such as the received power level, to assist in power control procedures. A common pilot channel measurement is the *CPICH Ec/No*. This has a range of values from 0 to 47, which map to 0.5 dB measurement ranges of the CPICH Ec/Io. Table 6.16 shows these mappings. This information is used by the BTS to select a suitable power level on which to reply to the UE.

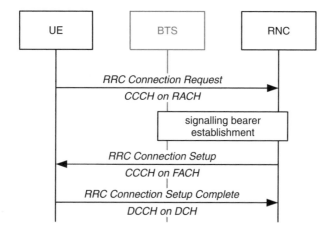

Figure 6.66 RRC connection establishment

Table 6.16   CPICH Ec/Io values

Value of CPICH Ec/No	Actual value range (dB) CPICH Ec/Io
0	CPICH Ec/Io < −24
1	−24 = CPICH Ec/Io < −23.5
2	−23.5 = CPICH Ec/Io < −23
⋮	⋮
48	−0.5 = CPICH Ec/Io < 0
49	0 = CPICH Ec/Io

Table 6.17   RRC connection setup elements

Information Element	Explanation
Initial UE identity	IMSI, TMSI or P-TMSI number
RRC transaction ID	Unique identifier for this transaction
New U-RNTI or C-RNTI	Allocation of a U-RNTI/C-RNTI to the user For a FACH connection, C-RNTI is allocated
RRC state indicator	The connected state, e.g. CELL-DCH for a dedicated channel
DRX cycle length coefficient	Used by the UE to calculate the frame numbers in which it should check the PICH for an incoming page
Signalling radio bearer information	Information on the signalling bearers being established (see Section 6.16.4)
UL/DL transport channel information	Information regarding the uplink and downlink transport channels, such as transport formats

For establishment of a dedicated channel, the *RRC connection setup* message will contain the key information elements listed in Table 6.17. Note that this connection is only through as far as the RNC. If and when a connection through to the core network is required, this must be established also.

6.16.4   Signalling radio bearers

The 3GPP defines four radio bearers for use as signalling bearers for RRC messages. Signalling radio bearer 0 (SRB0) is used for all messages sent on the CCCH. In the uplink direction, this uses RLC transparent mode and in the downlink unacknowledged mode. SRB1 is for messages sent on the DCCH using unacknowledged mode. SRB2 is for messages on the DCCH using acknowledged mode, and SRB3 is for transport of non-access stratum messages, such as mobility management messages, on the DCCH in acknowledged mode. Providing two separate acknowledged-mode signalling radio bearers allows for the prioritization of RAN-related RRC messages over those for the non-access stratum. There is an optional fifth bearer, SRB4, which is also an acknowledged-mode bearer for NAS signalling. This allows for two priorities of NAS signalling, with SRB3 deemed 'high priority' and SRB4 'low priority'. High priority is used for NAS signalling that uses service access point identifier 0 (SAPI 0), and low priority for signalling using SAPI 3, which is defined for SMS traffic. These different radio bearers are distinguished as different logical channels at the MAC layer using the C/T field.

For conformance, the specifications require that all UEs must be able to support standalone signalling radio bearers of 1.7 kbps, 3.4 kbps and 13.6 kbps. For signalling radio bearers associated with data radio bearers, for example voice and signalling, the rate is usually defined at a bit rate of 3.4 kbps; however, for lower rates such as AMR 4.75 kbps, the 1.7 kbps rate can be used. The 3.4 kbps channel is established at the transport layer with a transport block size of 148 bits and a TTI of 40 ms. It assumes an RLC/MAC overhead of 12 bits (RLC-UM: 8 bits, MAC C/T field: 4 bits). The RLC-AM signalling radio bearers will be slightly less than 3.4 kbps due to the extra overhead imposed by the RLC layer. It is the RNC that will decide, according to its admission control policy, which standalone signalling bearer rate the user is allocated. The *RRC connection setup complete* message provides the RNC with the core network domain it wishes to connect to, and the capabilities of the UE. This information element covers a whole range of UE capabilities such as for RLC, security and RF capabilities.

6.16.5 RRC security mode control

User authentication in UMTS differs from GSM in that the procedure validates not only the UE to the network, but also the network to the UE. During the NAS security authentication procedures, two keys are generated for ciphering (CK) and integrity checking (IK). The security mode procedure starts (or reconfigures) the security process (Figure 6.67). Subsequently, all signalling messages may be checked for integrity to verify that they have not been altered en route. The UE and RNC will discard any messages for which the integrity check fails.

The procedure for integrity checking is described in further detail in Section 6.21.

6.16.6 RRC paging

The RRC protocol is responsible for delivering paging messages to the UE. Two types of paging mode exist, and the method employed is dependent on the state of the UE:

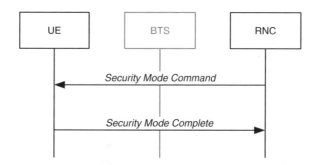

Figure 6.67 Security procedure

- *Paging type 1*: this is used where there is no longer a dedicated connection between the UE and the network. It is applicable to idle, CELL-PCH and URA-PCH modes. A paging type 1 message is transferred on the paging control channel (PCCH). In idle mode, the paging message will be sent to all cells in the location/routing area. In CELL-PCH and URA-PCH, the RNC knows the location of the user to a cell or URA, and can transmit the message on the PCCH just in that cell or URA.

- *Paging type 2*: this is used to transmit paging information to the UE when the RNC has an established connection to the network, i.e. the UE is in CELL-FACH or CELL-DCH mode. In this instant, the paging message is delivered along the FACH or DCH, respectively.

6.16.7 Radio bearer establishment

Another RRC procedure is the establishment of radio bearers. This is required if the UE wishes to send or receive data. If the UE already has a suitable bearer established, it may reconfigure that bearer for the new traffic. It is the job of the RNC to perform the radio bearer setup in response to Layer 3 signalling. Figure 6.68 shows the general procedure flow.

The contents of the radio bearer setup and setup complete are much the same as those outlined for the RRC connection and will define which RLC mode each bearer should use, and relevant parameters for this, the transport block sizes and formats, and the physical layer parameters. Prior to this establishment, the RNC will either set up a new radio link, or reconfigure the existing radio link to support this new bearer.

Table 6.18 shows some examples of the recommended bearers and bearer combinations expected to be supported in the system, with their relevant parameters. For this table, the transport format for the 3.4 kbps signalling bearer when included with a data bearer is the same as described for the standalone. Notice that for the last case of a simultaneous packet and circuit connection, the voice call does not add to the physical layer resource usage, and the same SF for UL/DL is used. For further details, please refer to TS34.108.

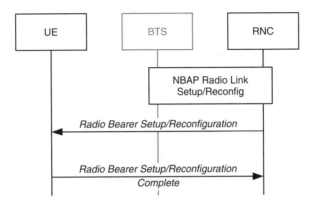

Figure 6.68 Radio bearer setup or reconfiguration

Table 6.18 Example data bearer configurations

Description	Bit rate (kbps)	Transport format	SF UL	SF DL
Standalone signalling bearers	1.7	TB size: 148 bits TTI: 80 ms 1/3 rate CC, 16-bit CRC	256	512
	3.4	TB size: 148 bits TTI: 40 ms 1/3 rate CC, 16-bit CRC	256	256
	13.6	TB size: 148 bits TTI: 10 ms 1/3 rate CC, 16-bit CRC	64	128
Speech + signalling	12.2 + 3.4	TB size: 244 bits (refer to Section 6.13) TTI: 20 ms 1/3 rate + 1/2 rate CC, 16-bit CRC + no CRC	64	128
Circuit switched non-transparent data + signalling	28.8 + 3.4	TB size: 1×576, 2×576 bits TTI: 40 ms TC, 16-bit CRC	32	64
Circuit switched transparent data + signalling	64 + 3.4	TB size: 1×320, 2×320, 4×320, 8×320 bits TTI: 40 ms TC, 16-bit CRC	16	32
Packet switched data + signalling	16 + 3.4	TB size: 1×336, 2×336 bits TTI: 40 ms TC, 16-bit CRC	32	64
Packet switched data + signalling	64 + 3.4	TB size: 1×336, 2×336, 3×336, 4×336 bits TTI: 20 ms TC, 16-bit CRC	16	32
Asymmetric packet switched data + signalling	64 UL/ 384 DL + 3.4	UL: TB size: 1×336, 2×336, 3×336, 4×336 bits TTI: 20 ms TC, 16-bit CRC DL: TB size: 1×336, 2×336, 4×336, 8×336, 12×336, 16×336, 20×336, 24×336 bits TTI: 20 ms TC, 16-bit CRC	16	8
Asymmetric packet switched data + signalling	64 UL/ 2048 DL + 3.4	UL: TB size: 1×336, 2×336, 3×336, 4×336 bits TTI: 20 ms TC, 16-bit CRC DL: TB size: 1×656, 2×656, 4×656, 8×656, 12×656, 16×656, 20×656, 24×656, 28×656, 32×656 bits TTI: 10 ms TC, 16-bit CRC	16	4×3 DPDCH

Table 6.18   (*continued*)

Description	Bit rate (kbps)	Transport format	SF UL	SF DL
Packet switched data + circuit switched voice call + signalling	64 + 12.2 + 3.4	As above for 'Speech + signalling' and 'Packet switched data + signalling'	16	32

CC, Convolution coding; TC, turbo coding

6.16.8   Transfer of NAS messages

RRC also offers procedures for transferring non-access stratum messages between the UE and the core network. For this, the direct transfer messages are used. There are three types of direct transfer message, as explained in Table 6.19. The direct transfer message transparently transports the NAS message, providing only an indication of which core network domain (CS or PS) is being used. The direct transfer message is transported to the core network across the Iu interface transparent to UTRAN by RANAP. For example, an *MM location update request* message from the UE will be transferred to the RNC using *RRC intial direct transfer*, and then from the RNC to the CN using the *RANAP initial ue Message*.

6.16.9   Cell/URA update

For a location or routing area update, the UE must establish a dedicated RRC connection (CELL-FACH/CELL-DCH), and transfer the signalling using the direct transfer message over a DCCH. However, for UTRAN mobility, there is no need to do this, but rather, in CELL-PCH/URA-PCH, the UE can send an update message across the CCCH to the RNC. The format of this is shown in Figure 6.69.

Table 6.19   Direct transfer messages

Message	Explanation	RANAP equivalent
Initial direct transfer	This is used to establish a signalling connection to the CN and transport the initial NAS message	Initial UE message
Uplink direct transfer	This is subsequently used to transfer all NAS messages to the CN	Direct transfer
Downlink direct transfer	This is used to transfer all NAS to the UE	Direct transfer

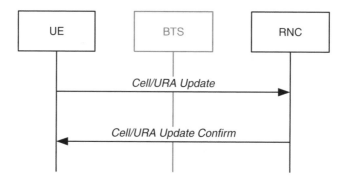

Figure 6.69 RRC cell/URA update

The UE will initiate a *cell update* procedure for the following reasons:

- uplink data transfer
- paging response
- radio link failure
- re-entry of service area
- cell reselection
- periodic cell update.

It will initiate a URA update in the following situations:

- URA reselection
- periodic URA update.

For a periodic update, the UE performs this based on the value of timer T305, the value of which is known from system information (SIB 1). Once this timer expires, the UE will perform a periodic cell/URA update. The default value of T305 is 30 minutes.

The cell/URA update message will contain the U-RNTI, the update cause, and some measurement results, such as the measured power of the CPICH. The RNC will reply with the *cell update confirm* message, which may also be used to allocate a new RNTI to the UE.

6.16.10 Measurement reporting

For the purposes of handover, the UE must make measurements across the Uu interface and pass these back to the RNC to allow it to make an informed decision about UE handover. This measurement reporting is controlled by the RNC using the *measurement*

Table 6.20 UE measurement criteria

Measurement	Description
Intra-frequency	Measurements on downlink physical channels in cells at the same frequency as the active set
Inter-frequency	Measurements on downlink physical channels in cells at frequencies that differ from the frequency of the active set
Inter-RAT	Measurements on downlink physical channels in cells belonging to another radio access technology (RAT) than UTRAN, e.g. GSM
Traffic volume	Measurements of uplink traffic volume
Quality	Measurements of downlink quality parameters of a channel, e.g. downlink transport block error rate
UE-internal	Measurements of UE transmission power and UE received signal level

Table 6.21 Cell sets

Cell set	Description
Active set	Current set of cells in which the UE has an active connection and is sending and receiving information (i.e. >1 for soft handover)
Monitored set	A set of cells, not in the active set, that the UE has been instructed by the RNC to monitor as possible handover candidates
Detected set	All other cells that the UE has detected. Reporting of measurements only occurs for intra-frequency measurements.

Table 6.22 Measurement control

Measurement	Description	Example
Command	Setup, modify or release a measurement	Release
Type	Description of what type of measurement the UE should make	Intra-frequency
Object	Description of what the UE should measure	Cell
Quantity	The quantity the UE should measure	CPICH Ec/N0
Reporting criteria	Indication of when the UE should report the measurement	500 ms

control command. The different types of measurement that the UE may be asked to make fall into the broad categories listed in Table 6.20.

The cells that a UE is monitoring are broken into three categories (Table 6.21).

The measurement control message will contain the elements listed in Table 6.22.

For each measurement that has been set up, the UE will pass back the measured quantity in a *measurement report* message according to the reporting criteria (Figure 6.70). The UE receives its initial measurement control information from the broadcast channel in system information blocks 11 and 12. This provides sufficient information and reporting criteria for the UE to perform measurements and file measurement reports. Any subsequent new

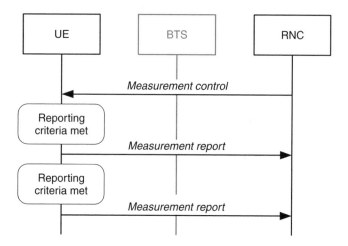

Figure 6.70 RRC measurement reporting

or updated measurement information can be sent to the UE by the RNC in a *measurement control* message.

6.16.11 Active set update

For soft handover (in CELL-DCH), the *active set update* is used by the RNC to instruct the UE to modify its active set (Figure 6.71).
    The possibilities are:

- add radio link
- delete radio link
- combined add and delete radio link.

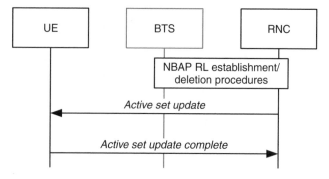

Figure 6.71 RRC active set update

Prior to the update, where a new radio link is being added, the RNC will have made the appropriate transport link preparations in the network. Upon receipt, the UE will update its active set and report this back to the RNC with an *active set update complete* message.

6.17 BROADCAST SYSTEM INFORMATION

System information is carried in the downlink BCCH channel, which is mapped to the BCH and optionally the FACH. The BCH is carried at the physical layer in the P-CCPCH. System information provides the UE with the necessary general parameters for both system and cell level to enable it to communicate with the network, such as details of the structure of the RACH and PRACH. The system information is split into a number of system information blocks (SIB), where similar information is contained in one SIB. The scheduling and frequency of occurrence of these SIBs is referenced in a master information block (MIB), which acts like an index to the SIBs contained in the broadcast channel. Optionally, up to two scheduling blocks (SB) may also be present, which also contain SIB scheduling information. This is implemented in a hierarchical manner, with the MIB referencing the SBs, which in turn reference the SIBs (Figure 6.72).

System information is considered an RRC function. However, much of the system information remains static, and thus to minimize the quantity of signalling traffic on the Iub interface, signalling information is sent to the BTS by the CRNC over NBAP when the cell is created or restarted. The BTS then transmits this information out on the P-CCPCH. Subsequently, the RNC will only update the BTS with changes to system information. In addition, there is some information that changes rapidly and is only available to the BTS, and it is the responsibility of the BTS to place this information on the BCCH. An example of this is uplink interference in the cell (SIB7).

The specifications define 18 different SIBs, with two of those (13 and 15) having a number of sub-blocks. For example, SIB15 also has SIB15.1, 15.2, etc. Not all of the SIBs are required to be present on the broadcast channel, and the choice of which ones are present is dependent on the equipment implementation. Some of the SIBs represent information about features that may not be currently implemented, and are therefore irrelevant. An example might be UE positioning information for location-based services. For all SIBs, with the exception of 15.2, 15.3 and 16, their content is static, and can only

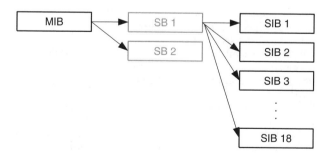

Figure 6.72 System information structure

change when directed by the RNC. For 15.2, 15.3 and 16, the information can vary in different occurrences of the blocks.

Table 6.23 lists the SIBs relevant to FDD, and the information each presents. Note that SIB14 and SIB17 are only for the TDD system, and are therefore omitted from the list.

Consider that the initial rollout of a simple UMTS network configuration can substantially narrow down the system information it must broadcast, as follows:

- Do not use different parameters for connected mode – eliminates SIB4, SIB6 and SIB12
- No common packet channel implemented (CPCH) – eliminates SIB8 and SIB9
- No Dynamic Resource Allocation Control (DRAC) – eliminates SIB10
- Only GSM–mobile application part (MAP) core network – eliminates SIB13–SIB13.4
- No UE positioning support – eliminates SIB15–SIB15.4

Table 6.23   UMTS system information blocks

Information Block	Description
MIB	SIB scheduling information
SB1	Optional SIB scheduling information
SB2	
SIB1	NAS system information as well as UE timers and counters to be used in idle mode and in connected mode
SIB2	URA identity
SIB3	Parameters for cell selection and reselection
SIB4	Parameters for cell selection and reselection to be used in connected mode
SIB5	Parameters for the configuration of the common physical channels in the cell
SIB6	Parameters for the configuration of the common and shared physical channels to be used in connected mode
SIB7	Fast-changing parameters uplink interference and dynamic persistence level
SIB8	Static CPCH information to be used in the cell
SIB9	CPCH information to be used in the cell
SIB10	Information to be used by UEs having their DCH controlled by a dynamic resource allocation control (DRAC) procedure
SIB11	Measurement control information to be used in the cell
SIB12	Measurement control information to be used in connected mode
SIB13	ANSI-41 system information
SIB13.1	ANSI-41 RAND information
SIB13.2	ANSI-41 user zone identification information
SIB13.3	ANSI-41 private neighbour list information
SIB13.4	ANSI-41 global service redirection information
SIB15	Information useful for UE-based or UE-assisted positioning methods
SIB15.1	Information useful for UE positioning differential GPS corrections
SIB15.2	Information useful for GPS navigation model
SIB15.3	Information useful for ionospheric delay, UTC offset and Almanac
SIB15.4	Information useful for OTDOA-based UE positioning method
SIB16	Radio bearer, transport channel and physical channel parameters to be stored by UE in idle and connected mode for use during handover to UTRAN
SIB18	PLMN identities of neighbouring cells to be considered in idle mode as well as in connected mode

Table 6.24 Typical SIB usage

Information block	Description
MIB	SIB scheduling information
SB1	Optional SIB scheduling information
SB2	
SIB1	NAS system information as well as UE timers and counters to be used in idle mode and in connected mode
SIB2	URA identity
SIB3	Parameters for cell selection and reselection
SIB5	Parameters for the configuration of the common physical channels in the cell
SIB7	Fast-changing parameters uplink interference and dynamic persistence level
SIB11	Measurement control information to be used in the cell

- No UE storage of channel parameters – eliminates SIB16
- No other public land mobile network (PLMN) identities – eliminates SIB18

This leaves only the SIBs shown in Table 6.24, with the details of each also presented.

6.17.1 Master information block (MIB)

The MIB contains some basic information about the network, and scheduling information for the SIBs that are present (optionally the SBs). The network information includes the supported PLMN types: GSM-MAP, ANSI-41 or both, and for GSM, the PLMN ID, which is the mobile country code (MCC) + mobile network code (MNC). The scheduling information provides for each SIB present, the SIB type (SIB1, SIB2, etc.), the number of segments present in the SIB (SEG COUNT, 1–16), where the first segment is located (SIB_POS), and how often it repeats on the channel (SIB_REP), referenced to the system frame number (SFN).

6.17.2 System information block 1

This contains the NAS information for the CS and PS domain, and the UE timers and constants. For idle mode, the parameters listed in Table 6.25 are defined.

Table 6.25 UE timers and constants in idle mode

Timer/ counter	Description	Default
T300	RRC connection request retransmission timer	1000 ms
N300	RRC connection request retransmission counter	3
T312	The timer for supervising successful establishment of a physical channel	1 s
N312	Maximum number of successive 'in sync' indications received from L1 during the establishment of a physical channel	1

Table 6.26 UE timers and constants in connected mode

Timer/counter	Description	Default
T302	Cell/URA update retransmission timer	4000 ms
N302	Cell/URA update retransmission counter	3
T305	Periodic cell update/URA update timer	30 min
T307	Timer for transition to idle mode, when T305 has expired, and the UE detects an 'out of service area'	30 s
T316	Started when the UE detects an 'out of service area' in URA-PCH/CELL-PCH state	180 s
T317	Started when the T316 expires and the UE detects an 'out of service area'	180 s

When the UE initiates an RRC connection request, it will start T300 and wait for a response from the network. If it does not receive an RRC connection setup message, it will try again, up to a total of N300 attempts. If it fails to receive a network response, it will return to idle mode. When a UE is establishing a physical channel, it will start T312. During the T312 time period, it should receive N312 'in sync' indications from L1, to consider that the physical channel establishment has been successful. If T312 expires before this, a failure has occurred.

For connected mode, there are several timers and counters that are relevant. Table 6.26 defines some of the more relevant ones.

The T302 timer is started when a UE cell/URA update message is sent; if there is no response from the network before it expires, then the UE will try again up to a total of N302 times. Should there still be no response, the UE will enter idle mode.

Should the UE be in CELL_FACH, CELL_PCH or URA_PCH, it will perform periodic cell/URA updates once T305 expires. If T305 expires and the UE detects it is out of coverage (out of service area), it will start T307 as a time period to wait for transition back to coverage. Should it expire, it will move to idle mode. This timer is to prevent cell/URA update failure as a result of momentary loss of coverage. In CELL_PCH or URA_PCH state, if the UE detects out of service area, it will start T316. If T316 expires, the UE checks for coverage. Should it be in service area, it will perform a cell/URA update with an indication that it was temporarily out of coverage. However, at expiry should it still be out of service, it will start T317 and try to do a cell/URA update anyway. If this does not succeed before T317 expires, it will move to idle mode, and release its signalling connection resources. These timers are to try to keep the UE connected again, even if there are temporary losses in coverage. An example might be a user entering a lift. From the default values, as long as the UE is not out of coverage for up to 6 minutes (by default), it will remain connected.

6.17.3 System information block 2

This contains the list of UTRAN registration area identities for the cell in which the block is transmitted. A cell can be a member of up to a maximum of eight URAs.

Table 6.27   SIB3 parameters

Parameter	Description
$S_{intrasearch}$	Threshold (in dB) for intra-frequency measurements and for the HCS measurement rules
$S_{intersearch}$	Threshold (in dB) for inter-frequency measurements and for the HCS measurement rules
$S_{searchHCS}$	This threshold is used in the measurement rules for cell reselection when HCS is used. It specifies the limit for cell selection receive level (Srxlev) in the serving cell below which the UE shall initiate measurements of all neighbouring cells of the serving cell
Qqualmin	Minimum required quality level in the cell in dB
Qrxlevmin	Minimum required RX level in the cell in dBm.
$Qhyst1_s$	Hysteresis value (Qhyst). It is used for cells if the quality measure for cell selection and reselection is set to CPICH RSCP
$Qhyst2_s$	Hysteresis value (Qhyst). It is used for cells if the quality measure for cell selection and reselection is set to CPICH Ec/No
$Treselection_s$	Cell reselection timer value
HCS_PRIO	HCS priority level (0–7) for serving cell and neighbouring cells
Qhcs	Quality threshold levels for applying prioritized hierarchical cell reselection
T_{CRmax}	Duration for evaluating allowed amount of cell reselection(s)
N_{CR}	Maximum number of cell reselections. Default 8
$T_{CRmaxHyst}$	Additional time period before the UE can revert to low-mobility measurements

HCS, Hierarchical cell structure: this is where there is an overlay of macro, micro and pico cells.

6.17.4   System information block 3

This contains the parameters for cell selection and reselection. It will indicate the cell ID, which is a unique identifier for the cell within the network, a list of information for cell selection and reselection, plus any cell access restrictions, for example, cell barred. The cell selection and reselection process is based on the quality of the cell, as measured on the pilot channel. The cell selection and reselection quality measure can be chosen either as the CPICH Ec/No or CPICH RSCP.[3]

The parameters cover the required quality levels and thresholds, in dB, for cell selection and reselection, and also define the maximum allowed uplink transmit power (in dBm). The important parameters are listed in Table 6.27.

6.17.5   System information block 5

This block contains the configuration parameters for the common physical channels in the cell (Table 6.28).

---

[3]The received signal code power (RSCP) is a measure of the received power of the pilot after despreading. The Ec/No is the energy per chip (Ec) of the pilot, divided by the power density in the frequency band (No). Ec/No is actually the RSCP/RSSI, where RSSI, the received signal strength indicator, is the wideband received power within the relevant channel bandwidth.

Table 6.28 SIBS parameters

Parameter	Description
PICH power offset	The power transmitted on the PICH minus the power of the primary CPICH
AICH power offset	The power level of the AICH, AP-AICH and CD/CA-ICH channels minus the power of the primary CPICH
Primary CCPCH info	Indication if transmit diversity is supported
PRACH system information list	Including physical layer information (available signatures, spreading factors, preamble scrambling code, available subchannels), RACH transport channel information (RACH TFS)
SCCPCH information	FACH/PCH information (FACH/PCH TFS), PICH information (channelization code, no. of PI per frame)

6.17.6   System information block 7

This block contains the fast-changing parameters uplink interference and dynamic persistence level in the cell, as follows:

Uplink interference – total uplink interference in dBm

6.17.7   System information block 11

This block contains measurement control information to be used in the cell. Again, it defines whether the CPICH RSCP or Ec/N0 should be used, and the measurement parameters for intra-frequency, inter-frequency, inter-RAT (radio access technology), traffic volume and UE internal measurements that should be made. This information is used by the UE to make measurements and when reporting criteria are fulfilled, feed the relevant measurement information back to the RNC during connections for handover evaluation.

6.18   FRAME PROTOCOLS

On the Iu, Iur and Iub interfaces, transmission blocks are encapsulated in a user plane frame protocol for transfer across the interfaces. It sits immediately above the transport layer, which on these interfaces is currently ATM. The frame protocol indicates to the destination the format of the transmission blocks held within. The format of this frame protocol is dependent on the interface, the nature of the data being transported, and also whether it is being transported on dedicated or common/shared channels.

6.18.1   Dedicated user data on the Iub/Iur interface

Once a radio bearer has been established between the UE and RNC for a dedicated user data channel (DCH), the frame protocol (FP) is used across the Iub and Iur interfaces to transfer this user data. The FP provides the following services to the user data:

- transfer of transport blocks across the Iub and Iur interfaces;

- transfer of outer loop power control information between the base station and the serving RNC;

- synchronization;

- transfer of transport format information from SRNC to base station for DSCH channel;

- transfer of radio interface parameters from SRNC to base station.

The structure of the FP is slightly different in the uplink and the downlink, and both are now examined.

6.18.1.1   Uplink data frame

The data frame contains the transport blocks that comprise the dedicated channel. A user may have more than one dedicated channel, so these multiple channels may be transferred in one user data frame. It also provides synchronization information and reliability by using frame numbers to check for correct sequencing. The format of the uplink frame is shown in Figure 6.73(a).

- *Header CRC (7 bits)*: provides error checking on the header.

- *Frame type (FT, 1 bit)*: indicates if this is a control (1) or data (0) frame.

- *Connection frame number (CFN, 8 bits)*: this acts like a sequence number, and is linked to the system frame number (SFN) used across the Uu interface at layer 1.

- *Transport format indicator (TFI, 5 bits)*: indicates the TFI for each of the data channels within the payload, extracted from the TFCI received at the physical layer.

- *Transport block (TB, variable)*: this is the actual data block being transported. Its size is variable and the format is as indicated by the TFI. Each transport block of each data channel is included in the payload. The transport blocks are padded out to a byte boundary.

- *Quality estimate (QE, 8 bits)*: this is calculated from the BER of either the transport channel or the physical channel. The choice is dependent on what was selected during the NBAP/RNSAP channel establishment procedure. The transport channel BER is the measured bit error rate on the DPDCH, across a transmission time interval (TTI). It provides a measure of how much error has been introduced to the data by traversing the air interface. By including such information, this provides the SRNC with a measure of the quality of the air interface, which can be used for both macrodiversity and outer loop power control. The physical BER measurement is performed on the DPCCH and provides a more general quality estimate of the air interface.

- *CRC indicator (CRCI, 1 bit)*: in addition to the QE, the uplink FP also includes an indicator for each TB of each DCH to inform the SRNC whether the CRC check on the data across the air interface was correct (0) or not (1). If there was no CRC check across the air, then the field is set to 0.

- *Payload CRC (16 bits)*: this is an optional CRC check on the payload contents.

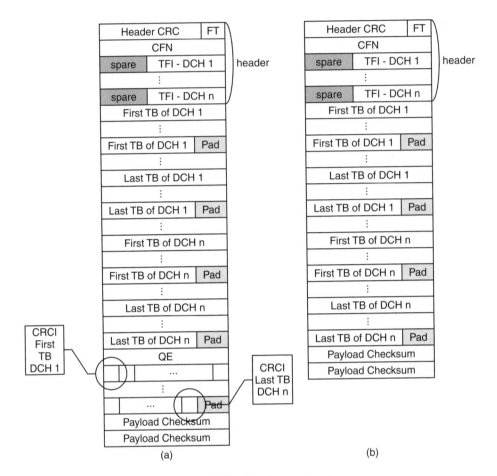

Figure 6.73   (a) Uplink and (b) downlink FP data frame structure

The purpose of the uplink frame protocol is to transport additional information that the SRNC needs for such tasks as handover evaluation, macrodiversity and channel quality estimates for outer loop power control. Figure 6.74, summarizes the role the BTS plays in forming the frame protocol. Across the air, the BTS receives a physical dedicated data channel and a control channel, DPDCH and DPCCH. From the control channel, it must pass the TFCI for each physical data channel. The BTS will extract the TFCI for the CCTrCH and decode it to the TFI for each transport block. From the data channel, it extracts each of the transport blocks to the frame protocol, checks the physical layer CRC (if present) and makes an evaluation of the transport/physical channel BER, if requested to do so. This information is also included in the frame protocol.

For uplink data transmission, there are two modes of operation defined: *normal mode* and *silent mode*. Which is used is decided by the SRNC at the time the transport channel is established. For *silent m*ode, the BTS will only transmit data frames to the RNC if it has received transport blocks over the air, and if it receives an indication that there are no transport blocks, it will not send a transmission. For normal mode, the BTS will always

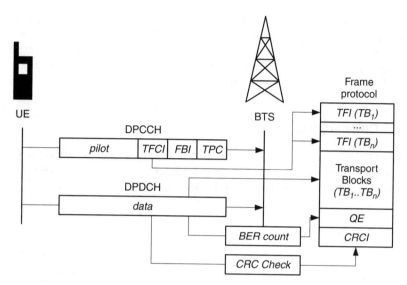

Figure 6.74   Role of uplink frame protocol

send an uplink data frame, even if it has not received a transport block over the air. In this case, it will send an empty data frame. This is to preserve a timing relationship between the BTS and RNC.

6.18.1.2   Macrodiversity

This mechanism enables a reduction in the required E_b/N_o when soft handover is used as compared to having a single radio link. For soft handover, macrodiversity is performed at the SRNC, which increases traffic over the Iub and possibly Iur (if a radio link is via a D-RNC). This therefore introduces a tradeoff between decreased interference over the air and increased traffic volume on the fixed network. This tradeoff has to be carefully managed and planned for, but with the air interface generally being a limiting factor in CDMA capacity, any reduction in interference over the air is seen as advantageous.

In a simple implementation a method of selection combining can be used at the SRNC to ensure that a signal with a successful CRC check is passed onto the core network whereas any frames with a CRC error are discarded. A more advanced system of recombining will make use of the QE that has been passed to the SRNC in the frame protocol. Using this in addition to the CRC check will enable the SRNC to pass data with errors to the core network from the radio path with the best quality, with reference to SIR, and hence BER.

Consider that the UE is in soft handover, and data from two active connections is being received by the SRNC. For each received TB, the SRNC can compare the CRC checks using the CRCI bits in the frame protocol. If both pass, then either TB can be passed to the CN. If one fails, then the TB which is correct can be passed. However, if both fail, then the SRNC can resort to the QE value, if available, and select the best quality TB to send.

Although this may not be useful when dealing with TCP/IP data packets, packets with errors can still be useful for voice traffic over both the CS-CN and the PS-CN. Simulations have shown that for soft handover to be effective there should be limited difference in the signal power from all links. It is suggested that this limit be around 3 dB but of course this depends on the actual implementation.

6.18.1.3 Downlink data frame

The downlink data frame, shown in Figure 6.73(b), is almost identical to the uplink frame aside from the absence of the QE and CRCI, since these are only relevant to uplink information extracted from the Uu interface performance.

6.18.1.4 Control frames

The control frame procedures transfer control information regarding the DCH from the SRNC to the base station. The general format of a control frame is illustrated in Figure 6.75.

The control frame type field can have the values shown, to indicate the nature of the control information contained within. Only a subset of the control procedures are discussed here. For further information, the reader is referred to TS25.427.

6.18.1.5 Synchronization

To ensure that the network is stable and that information is distributed with correct timing, synchronization is an important feature of the UMTS network. It is particularly important for the downlink delivery of information to the mobile device, to minimize the transmission delay and buffering time. The procedures for synchronization can be broken down into a number of important areas, as discussed in the following subsection.

Value	Control Frame Type
1	Outer loop power control
2	Timing adjustment
3	DL TrCH synchronisation
4	UL TrCH synchronisation
5	DSCH TFCI
6	DL node synchronisation
7	UL node synchronisation
8	Received timing deviation
9	Radio interface parameter update
10	Timing advance
11-255	Not currently used

Frame CRC	FT
Control Frame Type	
Control Information	
⋮	
Control Information	

header

Figure 6.75 Control frame general format

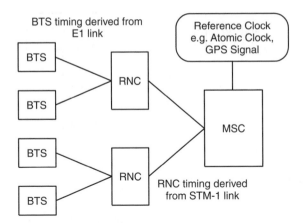

Figure 6.76 UMTS network synchronization

Network synchronization

This involves the distribution of a common reference clock to all the nodes throughout the system. This is performed in a hierarchical manner. An example is shown in Figure 6.76.

Node synchronization

This provides a measure of the timing differences between two nodes. For example, BTS-RNC node synchronization enables the RNC to know what are the time differences between itself and the BTSs connected to it. This is essential for correct delivery of information, especially in soft handover situations. Node synchronization is achieved by the RNC sending an FP control frame to the BTS(s) containing a sent time reference, T1. Upon receipt, the BTS will respond, echoing T1, and supplying T2, the time the control frame from the RNC was received, and T3, the time the BTS responded. When the RNC receives this reply it notes the time, T4. Now the RNC can simply calculate the round trip time (RTT) according to

$$RTT = (T2 - T1) + (T4 - T3)$$

This now allows the RNC to factor in the link delay, particularly in soft handover when communicating simultaneously with two or three BTSs with different RTTs.

Transport channel synchronization

The role of the transport channel synchronization procedure is twofold. First, it achieves or restores synchronization between nodes by establishing a common frame numbering between the two. Second, it acts as a keep-alive procedure across the Iub and Iur interfaces. Figure 6.77 shows a synchronization exchange between an SRNC and a base station.

First, the RNC sends a downlink synchronization message to the BTS. This message contains a connection frame number (CFN) to be used for the data transfer, the format is shown in Figure 6.78(a).

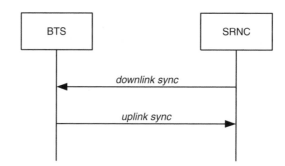

Figure 6.77 Iub synchronization procedure

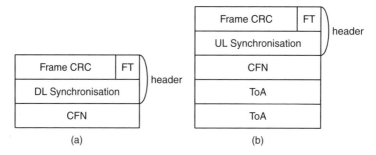

Figure 6.78 Downlink synchronization

Once the BTS has received this message, it immediately responds with an *uplink synchronization* message, which echoes the CFN received, and also contains a time of arrival (ToA) value (Figure 6.78(b)). This is an indication of the time difference between the arrival time of the downlink frame and the defined time of arrival window end point (ToAWE). During the NBAP establishment phase of the radio bearer, a timing window is defined during which a frame may arrive, so that a correct timing for downlink data transfer is established between the RNC and the BTS. This timing window is defined in terms of a time of arrival window startpoint (ToAWS) and a ToAWE.

Should a data frame arrive outside this window, either before or after, a timing adjustment procedure is invoked and a message sent to indicate the ToA of the frame. This ToA can be either a negative value, indicating a frame received after the ToAWE, or a positive value, indicating a frame received before the ToAWE. This allows the RNC to adjust its transfer time of data frames to keep it within the window. The ToA field is a 16-bit field that covers the range from -1280 ms to $+1279.875$ ms in steps of 0.125 ms.

Figure 6.79 shows a simplified form of this in practice. The receiving window is defined between ToAWS and ToAWE. The RNC sends a frame with CFN number 30 to arrive at the BTS within this window to provide the BTS enough time to process it and transmit it across the air interface to reach the UE at the expected time. Beyond the ToAWE is another value, the latest time of arrival (LToA). This is the last possible time that the BTS can receive the frame and still have enough time to handle it. Any frames received after LToA are discarded. The LToA is set at time *Tproc* before the transmit point, where *Tproc* is the BTS processing time. This figure will be naturally vendor dependent.

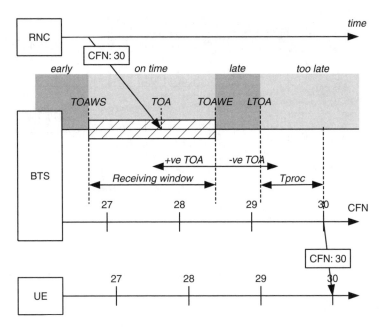

Figure 6.79   UTRAN transport channel synchronization

It should be noted that on the air interface, frames are numbered according to the cell SFN which is sent on the BCH. The CFN is not transmitted over the air but rather is mapped by layer 1 to the SFN of the first radio frame being transmitted. The SFN cycles every 4096 frames.

Figure 6.80 shows an example exchange of synchronization information between a SRNC and a BTS.

Radio interface synchronization

Radio interface synchronization is used to ensure that when the UE is receiving frames from several cells, it gets the correct frames synchronously so as to minimize the buffering requirements in the UE. During the establishment of a radio link, RRC signalling will inform the UE of when the frames it will receive in the downlink can be expected. At handover, the UE will measure the time difference between its existing dedicated physical channel and the SFN in the target cell. It will then report the timing deviation to the SRNC, which will in turn use this information to calculate appropriate timings for the new BTS, and deliver this information to the BTS.

6.18.1.6   Outer loop power control

As previously discussed, the outer loop power control is used by the SRNC to modify the SIR target used at the BTS for received user data. The SIR is modified to adjust the quality of the user signal (either reduce or increase) depending on the BER of the signal received. The uplink QE and CRCI can be used in the RNC power control algorithm to

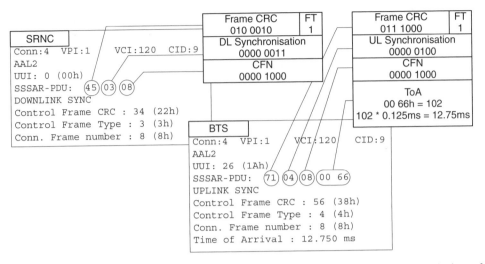

Figure 6.80 Trace example of synchronization procedure. Reproduced by permission of NetHawk Oyj

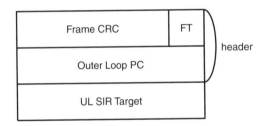

Figure 6.81 Outer loop power control frame

determine the required adjustment necessary to the SIR target. Once an outer loop power control message is received, the BTS will then update its SIR target for the user, and perform the necessary inner loop power control adjustments. The frame format is shown in Figure 6.81.

The UL SIR target value is an 8-bit field, encoding a value in dB in the range −8.2 dB to +17.3 dB in 0.1 dB steps. For example the value 98 implies a SIR target of +1.6 dB.

An example trace of outer loop power control is shown in Figure 6.82.

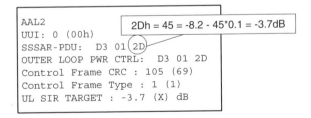

Figure 6.82 Sample trace of outer loop power control

6.18.2 User data on Iub common channels

The FP format for data transport on the common channels is slightly different to that of the dedicated channels. This covers transport on both the FACH/RACH and the shared data channels DSCH/CPCH. The differences are since the termination point of these channels is the controlling RNC (CRNC) and not the serving RNC (SRNC), although they may often be the same entity. Recall that these channels are handled by the MAC-c/sh layer at the CRNC, which separates the control and data, with the data (DTCH/DCCH) being passed to the MAC-d layer of the SRNC. If the MAC-c/sh and MAC-d are in separate network elements then data frames will be transferred over the Iur using the Iur FP, which is described later in this section. For this reason, extra information is required. There is no need for a quality estimate on uplink common channels since there will be no macrodiversity performed because there is no soft handover on common channels.

6.18.2.1 RACH/FACH transport

For the RACH, which is an uplink channel, the data frame is almost the same as shown in Figure 6.73, without the QE and CRCI fields but with the addition of a propagation delay field at the end of the header. This is an 8-bit field which indicates, in numbers of chips, the radio interface delay during the RACH access. Another difference is that the TFCI field in the channel will only be a transport format indicator (TFI) here since only one transport channel is permitted per frame. For the FACH, the additional header field is a transmit power level, which encodes a negative offset to the maximum power for this channel, indicating to the BTS the power level recommended for use for transmission of this transport block across the Uu interface. This is an 8-bit field, with a range of 0–25.5 dB in 0.1 dB steps. Figures 6.83 and 6.84 show examples of a RACH and a FACH exchange.

6.18.2.2 CPCH/DSCH transport

On the common packet channel, which is an uplink channel, the format is the same as the RACH frame. For the downlink shared channel, the header format is as shown in Figure 6.85.

The additional fields are defined as follows:

- *Code number (8 bits)*: indicates the code number of the physical downlink shared channel, since there may be more than one.
- *Spreading factor (SF, 3 bits)*: indicates which SF should be used on the physical DSCH. The mapping is as shown in Table 6.29.
- *Multicode information (MC info, 4 bits)*: indicates the number of parallel physical channel codes on which the data will be transported. For data rates higher than 384 kbps, a user must use more than one CDMA code in parallel, with the data distributed across these codes on the Uu interface. If more than one channel code is used, the SF of all codes is the same.

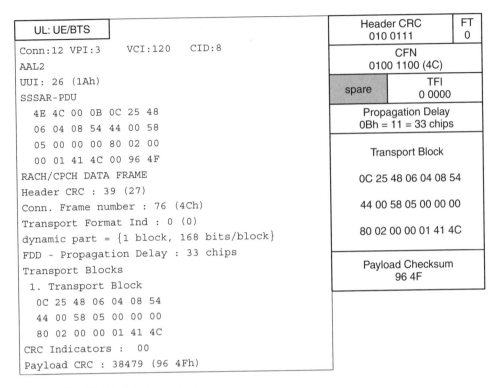

Figure 6.83 RACH data frame trace

6.18.2.3 Control frames

As with the dedicated channels, there are similar control procedures for the common channels, however not all procedures are applicable to each channel. Table 6.30 details the procedures, and the channels to which they apply.

The DSCH TFCI signalling procedure is sent every 10 ms on the TFCI2 transport bearer when there is DSCH data to be sent. It provides the BTS with information to allow it to generate the TFCIs to be transmitted on the DPCCH.

Table 6.29 SF mapping

SF value	SF on PDSCH
0	4
1	8
2	16
3	32
4	64
5	128
6	256

6.18.3 User data on Iur common channels

The FP across the Iur interface is a little more complex. It is required to provide the following services:

- transfer of MAC-c/sh SDUs from DRNC to SRNC for RACH and CPCH channels;
- transfer of MAC-c/sh SDUs from SRNC to DRNC for FACH and DSCH channels;
- flow control between MAC-c/sh and MAC-d.

The reason for the significant difference here is that the MAC-c/sh for common/shared channels is terminated at the CRNC whereas the MAC-d terminates at the SRNC. Recall that the CRNC controls the BTS, and thus the common channels, whereas the SRNC manages the UE. Most of the time, calls on common channels will have coincident CRNC and SRNC, as shown in Figure 6.86 for the RACH channel. The diagram for the FACH is identical except for the directions of the arrows.

```
 DL: CRNC

Conn:13 VPI:3      VCI:120    CID:9
AAL2
UUI: 0 (00h)
SSSAR-PDU
   30 72 02 CD 40 51 F8
   30 E7 12 A4 03 02 04
   2A 22 00 20 37 00 20
   40 96 06 91 40 52 A5
   C1 09 70 85 98 D4 28
   80 30 B8 95 0D 2E 28
   4B 8C 4C C6 63 28
FACH DATA FRAME
Header CRC : 24 (18h)
Conn. Frame number : 114 (72h)
Transport Format Ind : 2 (2h)
dynamic part = {2 blocks, 168 bits/block}
Transmit Power Level : 20.5 dB
Transport Blocks
1. Transport Block
   40 51 F8 30 E7 12 A4
   03 02 04 2A 22 00 20
   37 00 20 40 96 06 91
2. Transport Block
   40 52 A5 C1 09 70 85
   98 D4 28 80 30 B8 95
   0D 2E 28 4B 8C 4C C6
Payload CRC : 25384 (63 28h)
```

Header CRC 001 1000	FT 0
CFN 0111 0010 (72h)	
spare	TFI 0 0010
Tx Power Level CDh = 205 = 20.5dB	
First Transport Block	
40 51 F8 30 E7 12 A4	
03 02 04 2A 22 00 20	
37 00 20 40 96 06 91	
Second Transport Block	
40 52 A5 C1 09 70 85	
98 D4 28 80 30 B8 95	
0D 2E 28 4B 8C 4C C6	
Payload Checksum 63 28h	

Figure 6.84 FACH data frame trace

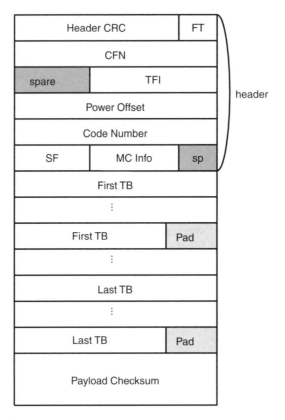

Header CRC	FT
CFN	
spare	TFI
Power Offset	
Code Number	
SF	MC Info	sp
First TB	
⋮	
First TB	Pad
⋮	
Last TB	
⋮	
Last TB	Pad
Payload Checksum	

header

Figure 6.85 DSCH format

Table 6.30 Common channel control procedures

Transport bearer	Control frames			
	Timing adjustment	DL transport channel sync	Node sync	DSCH TFCI signalling
RACH	N	N	N	N
FACH	Y	Y	Y	N
CPCH	N	N	N	N
DSCH	Y	Y	Y	N
TFCI2	Y	Y	Y	Y

However, there are some circumstances where a UE may be communicating with the core network through a SRNC where it is not physically connected to any BTS for which this is the CRNC. Such an example is where an inter-RNC hard handover has been performed but a SRNC relocation has not. Figure 6.87 shows the RACH protocol stack for this. In this particular situation information arriving from the UE on a common channel will be checked at one RNC, the CRNC, to see if the information is user data or common control messages. It will deal with common control messages itself but

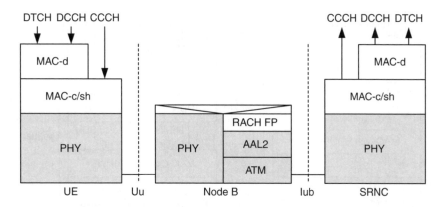

Figure 6.86 Protocol for RACH with coincident CRNC/SRNC

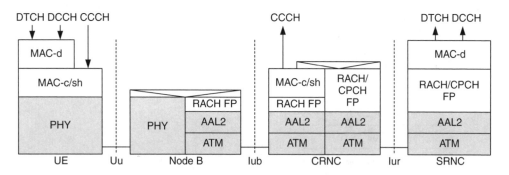

Figure 6.87 RACH protocol stack for separate CRNC and SRNC

user data (including user dedicated signalling) will be passed to another RNC, the SRNC via the Iur.

For this reason, a more detailed set of information needs to be sent in the FP to deal with identifying the CRNC and SRNC.

6.18.3.1 RACH/CPCH transport

The format of the RACH and CPCH data frames is the same, and is as shown in Figure 6.88(a). The additional fields are explained as follows:

- *Serving RNC radio network temporary identifier (SRNTI, 20 bits)*: as previously discussed, the SRNTI identifies the UE context in the SRNC.
- *MAC-c/sh SDU length (13 bits)*: identifies the length of each MAC-c/sh SDU in the payload.
- *Num of SDU (8 bits)*: indicates the number of MAC-c/sh SDUs in the frame payload.

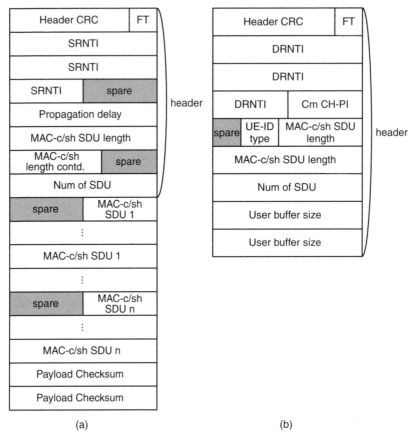

Figure 6.88 (a) RACH/CPCH and (b) FACH Iur frame format

With this additional information, the RACH/CPCH can carry multiple users across the Iur in the same logical channel.

6.18.3.2 FACH transport

The format for the FACH frame protocol is shown in Figure 6.88(b). Note that the diagram shows only the header, since the payload is identical in format to that of the RACH/CPCH frame protocol.

- *Drift RNC temporary network identifier (DRNTI, 20 bits)*: identifies the UE in the DRNC.
- *Common transport channel priority indicator (Cm CH-PI, 4 bits)*: used to identify the priority of this frame on the common channel. In this way, QoS can be provided on the common channels across the Iur.

- *UE-ID type (1 bit)*: indicates to the DRNC/CRNC which identifier should be used in the MAC-c/sh header. The options are either URNTI (0) or DRNTI (1).
- *User buffer size (16 bits)*: indicates the amount of buffer space a user has available.

6.18.3.3 DSCH transport

The DSCH frame protocol format is almost the same as that for the FACH except that it does not need to include the DRNTI field in the header. The header format is shown in Figure 6.89.

6.18.3.4 Control protocol

The frame control protocol provides flow control across the Iur interface for downlink common channels. The rationale for providing this only in the downlink is that it is designed to deal with limited buffering and processing capability in the mobile device, whereas network components are in a position to offer significantly higher levels of each. To perform flow control, the SRNC may send a capacity request for either the FACH or DSCH to the DRNC. This contains the user buffer space as an indication of the capacity requested on the common channel. The SRNC may send this command again should it not get a reply from the DRNC within a reasonable time. The DRNC will reply to the SRNC with a FACH flow control or DSCH capacity allocation. This request will contain a credit value, which is the number of MAC-c/sh SDUs that may be transferred, and for the DSCH, an interval during which the transfer may take place. A credit value of 0 blocks the SRNC from transfer of any more data and a maximum credit value of 255 indicates that the SRNC may send unlimited data.

6.18.4 User data on the Iu interface

The Iu user plane (UP) protocol is used to transport user data across the Iu interface between the RNC and either the circuit or packet core network. The UP protocol is defined

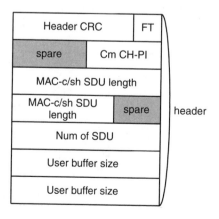

Figure 6.89 DSCH Iur frame protocol header

on a per-RAB basis, and does not refer to a particular domain in the core network. Each instance of the Iu UP is associated with a particular RAB, therefore if one mobile device has a number of RABs established, they will be carried in separate Iu UP instances. There are two modes of operation available:

- transparent mode (TrM)
- support mode for predefined SDU sizes (SMpSDU).

The decision as to which mode is used for a particular RAB is made at the time the RAB is established. The RAB is established using the RANAP protocol, as discussed in the next section. As its name suggests, the transparent mode is used where the data requires no extra facilities or support from the UP, and is to be carried transparently to/from the core network. It adds no overhead to the data. The transparent mode is typically used for transfer of GTP-U PDUs between the RNC and the packet core, and also for circuit switched data on the Iu-CS interface. The support mode is used when further information is required, and a frame protocol is added to the data to transfer this information. The support mode also provides a number of control procedures between communicating entities, such as rate control and error control. An example of the use of support mode is for the transfer of AMR speech PDUs.

6.18.4.1 *Transparent mode*

As mentioned, the transparent mode is very simple in structure as it adds no overhead to the data, and simply provides the following services:

- user data transfer;
- in-sequence delivery of user data, if required by the RAB being transferred, as defined at RAB establishment.

The frame format is shown in Figure 6.90. There is no restriction on the length, which is dependent on the nature of the user data being conveyed. The transparent mode does not carry a length field and this information will be supplied by a higher layer.

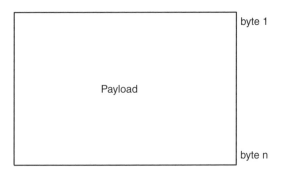

Figure 6.90   Transparent mode

6.18.4.2 Support mode for predefined SDU sizes

The support mode adds a frame protocol overhead to the data and also offers a number of control functions to provide the necessary support. The functions provided by support mode are:

- user data transfer, in-sequence if required
- initialization
- rate control
- time alignment
- error handling
- frame quality classification.

For user data transfer, there are two frame formats defined:

- data transfer with error detection (PDU type 0)
- data transfer with no error detection (PDU type 1).

The frame format for type 0 is shown in Figure 6.91(a) below. It has a fixed 4-byte header and a variable length payload, which is padded to a byte boundary. There is also a spare extension at the end (not shown) to allow for the addition of extra fields in future versions. It is not currently used.

The header fields are defined as follows:

- *PDU type (4 bits)*: this indicates the type of PDU, in this case, 0.
- *Frame number (4 bits)*: this acts as a sequence number and allows a destination to check the PDU order and check for lost frames.

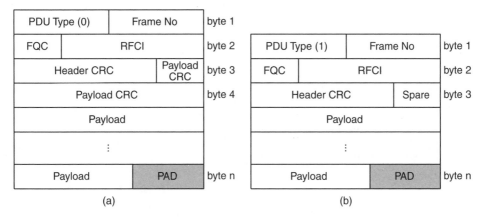

Figure 6.91 Support mode PDU type (a) 0 and (b) 1

- *Frame quality classification (FQC, 2 bits)*: this is used to indicate to the destination the state of frames received. This parameter is valid if the RAB has been established with the attribute 'Delivery of erroneous SDUs' set to YES. If this attribute has been set to NO, then SDUs that contain errors are discarded. The possible values for the FQC are listed in Table 6.31.

  In the uplink direction, the FQC chosen is dependent on whether the radio frame quality classification has been set to 'good' or 'bad', which will be assessed on the basis of whether the frame received at the RNC on the Iub interface has achieved its required QoS based on BER/BLER. If the RAB attribute 'Delivery of erroneous SDUs' is set to *no-error detection-consideration*, the FQC is irrelevant. If at least one of the subflows in the payload has 'Delivery of erroneous SDUs' set to YES or NO, Table 6.32 shows the actions of the RNC and how the FQC is chosen.

  In the downlink direction, the same rules apply for setting of the FQC, except that 'bad' is used instead of 'bad radio' since the quality classifier will be a CRC check.

  Considering an AMR voice call, the class A part will have the 'Delivery of erroneous SDUs' attribute set to YES, since actions need to be taken if this section contains errors; however the class B and C bits will be set to *no-error detection-consideration*.

- *RAB subflow combination indicator (RFCI, 6 bits)*: the RFCI is used to indicate to the destination the structure of the payload data. The payload may contain one or more RAB subflows. The definition of the allowed subflow combinations is defined by the initialization procedure, described shortly.

- *Header CRC (6 bits)*: this provides error protection for the frame protocol header.

- *Payload CRC (10 bits)*: this provides error protection for the frame protocol payload, including the padding section, if present.

The format of PDU type 1 is almost the same (Figure 6.91(b)). The only differences are that the PT field will be '1' to indicate PDU type 1, and the payload CRC is absent.

Table 6.31 FQC values

Value	Indication
0	Frame good
1	Frame bad
2	Frame bad due to radio
3	Not used

Table 6.32 FQC actions

Delivery of erroneous SDUs	Radio frame quality classification	Action
YES	Good	FQC set to 'good'
	Bad	FQC set to 'bad radio'
NO	Good	FQC set to 'good'
	Bad	Frame is discarded

6.18.5 Control procedures

To support the data transfer function, there are a number of control procedures defined. These procedures use a PDU type 14 (PT = 14), the general format of which is shown in Figure 6.92.

Several of the fields are the same as for the preceding PDU types; however, there are a few different fields:

- *Ack/nack (2 bits)*: the control procedures require a reliability mechanism to ensure correct operation. Therefore each control procedure exchange is acknowledged, and the acknowledgement indicates successful or unsuccessful operation, as shown in Figure 6.93. This field indicates whether the PDU type 14 is a control frame, a positive (ACK) or negative (NACK) acknowledge, as shown.
- *Frame number (2 bits)*: for control frames, the frame number is again to allow the destination to check for missing frames. It is also used in the acknowledgement mechanism, and the same frame number is used in the ack/nack as was used in the control frame being acknowledged.

PDU Type (14)	ACK/ NACK	Frame No.	byte 1
Iu UP Mode ver.	Procedure Indicator		byte 2
Header CRC		Payload CRC	byte 3
Payload CRC			byte 4
Procedure Data			
⋮			
Procedure Data			byte n

Figure 6.92 Control PDU general format

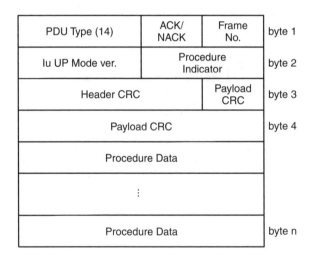

ACK/NACK value	Indication
0	Control frame
1	Positive ack
2	Negative ack
3	Not used

Procedure indicator	Procedure
0	Initialization
1	Rate control
2	Time alignment
3	Error event
4-15	Currently unused

Figure 6.93 Acknowledge procedure

- *Iu UP mode version (4 bits)*: this indicates the version of the UP mode being transferred. There can be up to 16 mode version available at the same time. This parameter is to allow new UP mode versions in the future. Currently the mode version is 1.

- *Procedure indicator (4 bits)*: this indicates which control procedure is contained within. Permitted values are shown in Figure 6.93.

6.18.5.1 Initialization procedure

For RABs that are using the support mode, prior to the transfer of user data, the destination must know the format of the RAB subflows contained within the UP frame. The initialization procedure originates at, and is controlled by, the SRNC. An RFCI is allocated by the SRNC to each subflow combination that may occur within the payload. For example, consider that a RAB is established to transport GSM enhanced full-rate speech (EFR, 12.2 kbps). In accordance with the specification, EFR consists of three RAB subflows with different QoS requirements and hence different BERs. Also, accompanied with EFR are two other rates, one for SID (silence descriptor) in silence suppression mode, where comfort noise is transferred, and one for DTX. The subflows are as shown in Figure 6.94.

The SRNC will allocate the RFCIs for each subflow combination, and this information is passed on during the initialization procedure (Figure 6.95).

RAB subflow combination		RFCI	RAB			Total
			subflow 1	subflow 2	subflow 3	
	set	0	81	103	60	244
		1	39	0	0	39
		2	0	0	0	0

Figure 6.94 Example RAB subflows

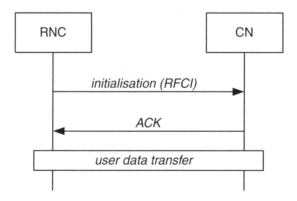

Figure 6.95 Initialization procedure

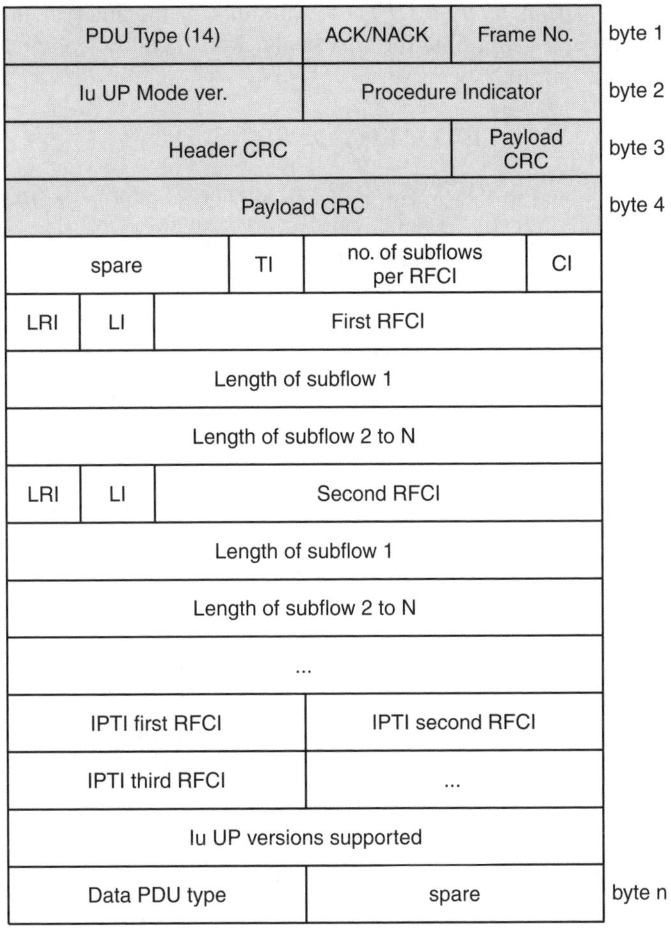

PDU Type (14)		ACK/NACK	Frame No.	byte 1
Iu UP Mode ver.		Procedure Indicator		byte 2
Header CRC			Payload CRC	byte 3
Payload CRC				byte 4
spare		TI	no. of subflows per RFCI	CI
LRI	LI	First RFCI		
Length of subflow 1				
Length of subflow 2 to N				
LRI	LI	Second RFCI		
Length of subflow 1				
Length of subflow 2 to N				
...				
IPTI first RFCI		IPTI second RFCI		
IPTI third RFCI		...		
Iu UP versions supported				
Data PDU type		spare		byte n

Figure 6.96 Initialization procedure frame format

The frame format of the initialization procedure is shown in Figure 6.96. It contains the RFCI and the format of each of the subflows.

The header follows the PDU type 14 format. The fields in the payload are defined as:

- *Timing information (TI, 1 bit)*: indicates if the frame contains timing information, contained in the optional IPTI fields. A 1 indicates that the IPTI field is present for each RFCI.
- *Chain indicator (CI, 1 bit)*: indicates whether this is the last control procedure frame related to a particular control procedure, 0 for last frame.
- *Last RFCI indicator (LRI, 1 bit)*: a value of 1 indicates that this is the last RFCI defined.
- *Length indicator (LI, 1 bit)*: a value of 0 indicates that the length values are 1 byte, whereas 1 indicates a 2-byte length field.

- *Inter PDU transmission interval (IPTI, 4 bits)*: this is the interval at which Iu UP PDUs can be sent at a given time for a particular RAB subflow combination (RFC). It is calculated according to:

$$\text{IPTI} = \text{RFC size/RFC bit rate}$$

If undefined, it is equal to the Iu Timing Interval (ITI), which is defined as:

$$\text{ITI} = \text{Max SDU size/Max bit rate}$$

- *Iu UP versions supported*: indicates which of the Iu UP versions are supported. Currently there is only one version defined.

- *Data PDU type (4 bits)*: indicates which PDU type (0 or 1) will be used for the data transfer.

Returning to the previous example, an initialization message for the RAB subflows would look as shown in Figure 6.97.

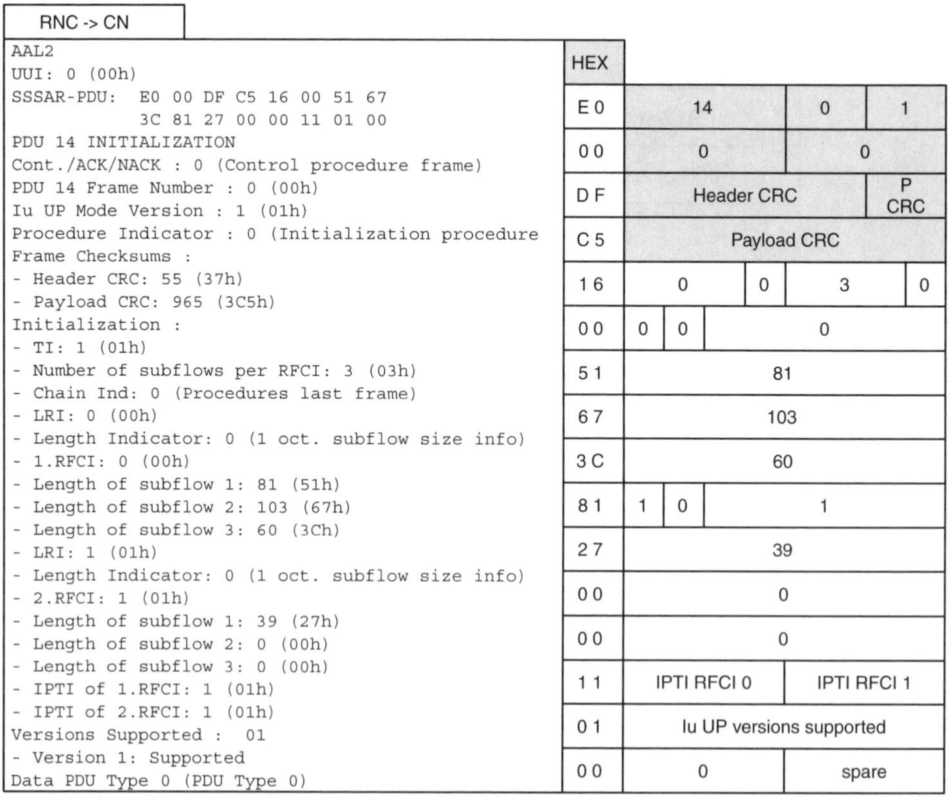

```
RNC -> CN

AAL2
UUI: 0 (00h)
SSSAR-PDU:  E0 00 DF C5 16 00 51 67
            3C 81 27 00 00 11 01 00
PDU 14 INITIALIZATION
Cont./ACK/NACK : 0 (Control procedure frame)
PDU 14 Frame Number : 0 (00h)
Iu UP Mode Version : 1 (01h)
Procedure Indicator : 0 (Initialization procedure
Frame Checksums :
- Header CRC: 55 (37h)
- Payload CRC: 965 (3C5h)
Initialization :
- TI: 1 (01h)
- Number of subflows per RFCI: 3 (03h)
- Chain Ind: 0 (Procedures last frame)
- LRI: 0 (00h)
- Length Indicator: 0 (1 oct. subflow size info)
- 1.RFCI: 0 (00h)
- Length of subflow 1: 81 (51h)
- Length of subflow 2: 103 (67h)
- Length of subflow 3: 60 (3Ch)
- LRI: 1 (01h)
- Length Indicator: 0 (1 oct. subflow size info)
- 2.RFCI: 1 (01h)
- Length of subflow 1: 39 (27h)
- Length of subflow 2: 0 (00h)
- Length of subflow 3: 0 (00h)
- IPTI of 1.RFCI: 1 (01h)
- IPTI of 2.RFCI: 1 (01h)
Versions Supported :  01
- Version 1: Supported
Data PDU Type 0 (PDU Type 0)
```

HEX			
E 0	14	0	1
0 0	0		0
D F	Header CRC		P CRC
C 5	Payload CRC		
1 6	0	0	3	0
0 0	0	0	0	
5 1	81		
6 7	103		
3 C	60		
8 1	1	0	1
2 7	39		
0 0	0		
0 0	0		
1 1	IPTI RFCI 0		IPTI RFCI 1
0 1	Iu UP versions supported		
0 0	0		spare

Figure 6.97 Example initialization PDU. Reproduced by permission of NetHawk Oyj

Table 6.33   Error causes

Error in CRC of frame header	Unexpected PDU type
Error in CRC of frame payload	Unexpected procedure
Unexpected frame number	Unexpected RFCI
Frame loss	Unexpected value
PDU type unknown	Initialization failure
Unknown procedure	Rate control failure
Unknown reserved value	Error event failure
Unknown field	Time alignment not supported
Frame too short	Requested time alignment not possible
Missing fields	Iu UP mode version not supported

6.18.5.2   Error handling

The error handling procedure is invoked in response to either error detection by another Iu UP procedure, such as receipt of a malformed or erroneous frame, or by request from an upper layer. The procedure contains an indication of the cause of the error, as listed in Table 6.33.

6.19   UMTS TERRESTRIAL RADIO ACCESS NETWORK (UTRAN)

Figure 6.98, shows the general protocol model for the interfaces within the UTRAN. It is divided into two horizontal layers, the radio network layer and the transport network layer. This separation is explicitly made so as to define clearly the role of each. The transport layer is a generic layer that offers bearer services to the radio network layer. This transport layer is currently implemented using ATM technology, but its separation and definition of primitives between the layer allows for the use of alternative transport technologies.

The radio network layer consists of a control plane and a user plane. The control plane is an application protocol, which defines the mechanisms used for establishing, releasing and managing bearers, as well as ancillary control procedures. The radio network user plane consists of the user data streams, and will be a UMTS transport channel plus a frame protocol for a given interface. It should be noted that 'user data' at this layer also encompasses control data for higher layers. As an example, on the Iub interface, the radio network user plane transports both RRC and NAS signalling messages.

The transport network layer offers the service of transport bearers for both signalling and data streams, where the signalling is the application protocol of the upper layer. These two bearer services are considered to be user plane at the transport network layer. In turn, the transport network layer needs its own control plane and protocols to establish, manage and release its own bearers. The generic name for this control plane component is the access link control application protocol (ALCAP). Again, in the context of UMTS, this is currently a protocol that is part of the ATM family, AAL2 signalling, which is discussed in Section 7.14.2.

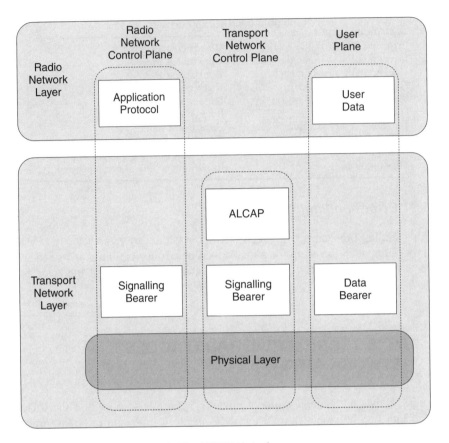

Figure 6.98 General protocol model for UTRAN interfaces

The ALCAP has its own signalling bearer to transport its control messages. A simple example of how these different protocol stacks may interact with each other is now listed:

1. A NAS message such as a *CM service request* would be sent from the higher layers of the user plane of the UE.
2. This would request that the network instruct the radio network control plane to set up a radio bearer.
3. The radio network control plane would then ask the transport network control plane to actually do this.

It would be possible to have only two planes laid out within the specification process, the user plane and the radio network control plane. However, although introducing a separate transport network control plane adds considerable complexity, it does enable flexibility of the underlying network, giving an operator the option of ATM or IP in later releases.

6.19.1   Iub interface

The Iub interface connects the BTS to the RNC. As previously mentioned, this is point-to-point in nature and is therefore a UNI interface. Its structure is shown in Figure 6.99.

The naming convention for the application parts is based on the node or subsystem to which they are connecting. So for the Iub, the interface connects to the BTS or Node B, so the application part is Node B application part (NBAP). The radio network user plane consists of the transport channels, for example, the FACH and RACH, DCH etc. In the transport network layer, the radio network user plane is offered the AAL2 adaptation layer for transport of user data. It has been seen that the user data across the Uu air interface has been fragmented into small segments by the RLC layer. The fragments continue across the Iub and the reassembly of these only takes place at the RNC, and not at the BTS. AAL2 is used to transport these small segments in its CPS-PDU, and allows them to be multiplexed together across a single virtual circuit. A channel ID is used to identify each individual AAL2 channel stream. For signalling transport, in this case NBAP, the only requirement is a bearer that can provide robust, in-sequence delivery of signalling

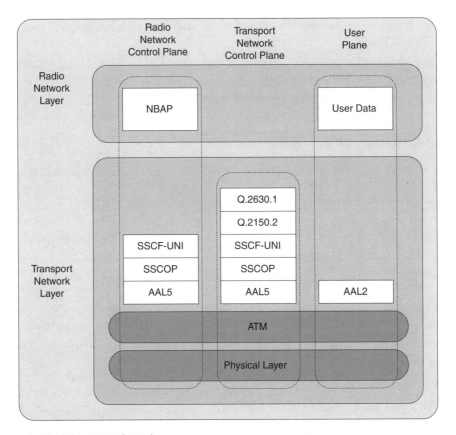

Figure 6.99   Iub protocol stack

messages. To meet this, the transport network layer provides the standard ATM signalling protocol stack, often referred to as the signalling AAL (SAAL). The details of this are described in Section 7.13.

6.19.2 Node B application part (NBAP)

The NBAP protocol provides all the control functionality required between a BTS and an RNC. Its procedures can be broken into two classifications: common and dedicated. Common NBAP procedures are defined as all general, non-UE specific, procedures, including those used to establish an initial context for the mobile user. Dedicated NBAP procedures relate to control of the UE once an initial context has been established. The key functions provided by NBAP are:

- Management of radio links – establishment, addition, reconfiguration and release of a radio link.
- Base station management – management of cell configuration and scheduling of broadcast information.
- Common channel management – control of the common transport channels such as the RACH and FACH.
- Measurement and supervision – reporting and control of measurement information to the RNC, such as power measurements.
- Fault management – reporting of general error situations.

6.19.2.1 Radio link establishment

As an example transaction, consider the initial connection request to establish a radio bearer. As shown in Figure 6.100, the RNC will issue a *radio link setup request* to the BTS. If successful, this will result in the following actions at the BTS:

- The BTS will reserve the necessary resources, as specified in the setup request.
- The BTS will begin reception on the link.
- The BTS will respond to the RNC with a *radio link setup response* message.

The key parameters involved in the setup message are listed in the table in Figure 6.100. The command passes the BTS only transport and physical layer parameters. Information about radio bearers carried by this link is only relevant to the UE/RNC, since the BTS does not deal with those layers of the protocol stack.

The allocation of uplink channelization code is not provided, but rather the minimum spreading factor that should be used. The decision of which code to use is made by the UE. The actual bearer for this will be initiated within the transport layer, at the request of the RNC, using the transport network layer signalling protocol, ALCAP, which here is AAL2 signalling. The NBAP response provides the AAL2 end system address, i.e. the

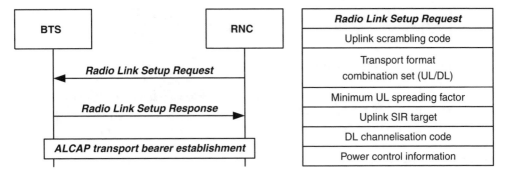

Figure 6.100   NBAP radio link setup procedure

AAL2 address of the BTS. The actual addressing scheme is implementation dependent, but would typically allocate an AAL2 address for each cell. It also provides a *binding ID*. This value is used as an explicit reference to a set of signalling instructions. The binding ID is generated by the BTS and sent to the RNC over NBAP. It is passed to and carried within the ALCAP signalling protocol as an identifier to link the NBAP and ALCAP signalling procedures for a given transaction. The RNC will then incorporate this value to be carried in its ALCAP signalling procedure.

6.19.2.2   Common channel establishment

An example of a common procedure, somewhat similar in format to the radio link setup, is the common channel setup, where the common channels such as the RACH and FACH are established. This procedure is normally done at cell setup, such as when a BTS reset is initiated. The procedure is shown in Figure 6.101.

As an example, consider the RNC is establishing the RACH, the key parameters shown will be included in the setup. Notice that the respective physical acquisition indication channel (AICH) is established at the same time. The BTS will respond with a *common transport channel setup response* to indicate the success or otherwise of the transaction.

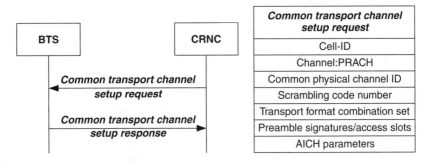

Figure 6.101   NBAP common channel setup

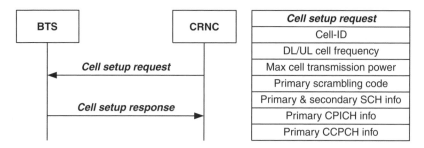

Figure 6.102 NBAP cell setup

6.19.2.3 Cell setup

Another common NBAP procedure is cell setup, which establishes a cell and the initial downlink channels for that cell. The procedure is shown in Figure 6.102.

The key elements of the cell setup are also shown. The RNC must allocate a cell-ID to uniquely identify the cell in the RNS, and also indicate the uplink and downlink frequency. This is passed as a UTRA absolute radio frequency channel number (UARFCN) value, calculated as follows:

$$UARFCN = \text{frequency (in MHz)} \times 5$$

For example, consider an operator has been allocated the frequency range 2110–2115 MHz in the downlink. The central point of this, aligned to a 200 kHz raster, is 2112.4 MHz. The UARFCN of this is $2112.4 \times 5 = 10\ 562$.

Also included is the primary scrambling code for the cell, and the maximum transmission power in the downlink for all the channels summed together. It is in the range 0–50 dBm (1 mW to 100 W). The physical channels established in the cell setup are those for synchronization (P-SCH and S-SCH), the primary pilot channel (CPICH) and the P-CCPCH, which carries the broadcast channel. The principal piece of information defined for each is the transmission power. For this, the absolute value of the pilot channel is defined, in dBm, and all other channels are defined as a dB offset to this value. If there are any secondary pilot channels, they are also defined here. The S-CCPCH, which contains the paging channel and the FACH, is defined using a common channel configuration message.

6.19.3 Iur interface

As was previously mentioned, the Iub interface is defined as a UNI due to its point-to-point nature. The Iur and Iu interfaces are defined as NNI, and therefore their structure, while adhering to the standard format, is inherently more complicated. Figure 6.103 shows the Iur interface protocol stack.

At the radio network layer, the format is much the same, but now with RNSAP as the application part. However, at the transport network layer, there are some distinct

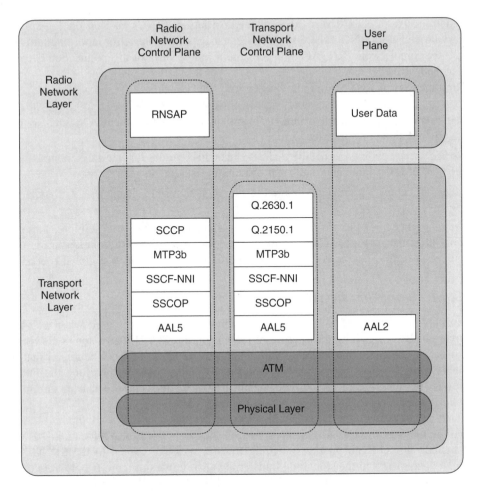

Figure 6.103   Iur protocol stack

additions. On top of SAAL is placed the MTP3b protocol. For the radio network signalling bearer, the SCCP layer is implemented, while for ALCAP, Q.2150.1 is used, as discussed in Chapter 7. The rationale for introducing these new layers is again to separate this networked (NNI) signalling bearer infrastructure from the underlying ATM technology. These layers are a broadband SS7 stack, as described later. One difference is that MTP3b is used, which is a broadband version of MTP3 that supports transport of larger message formats.

6.19.4   Radio network subsystem application part (RNSAP)

The Iur interface and RNSAP procedures are what enable Inter-RNC soft handover, where a mobile device is connected through more than one radio link, and the BTSs are under the control of different CRNCs. As explained earlier, the RNCs take on the logical role of serving and drift RNC, where the SRNC manages the connection and maintains the link to the core network. RNSAP is responsible for the control of this connection between DRNC and SRNC.

The functions of RNSAP may be broken into four key areas:

- *Basic mobility*: procedures which deal with mobility within the RAN, such as user paging and signalling transfer.
- *Dedicated*: procedures to handle dedicated channels on the Iur interface, such as management of radio links and dedicated channel measurement.
- *Common channel*: procedures related to common transport channels, e.g. RACH and FACH
- *Global*: procedures that do not relate to a particular user context. Therefore, these are procedures between peer CRNC entities, since the DRNC/SRNC relationship is only with respect to a specific context. Error reporting is an example of a global procedure.

6.19.4.1 Signalling transfer

An example of the use of RNSAP is in signalling transfer. This is used when a DRNC receives an RRC signalling message from the UE on the CCCH that is for its SRNC. In this instance, the signalling message will contain the U-RNTI of the UE. Recall that the U-RNTI consists of the user's S-RNTI and the ID of the SRNC. Therefore, the DRNC must pass the message to the correct serving RNC. For this purpose it uses the *uplink signalling transfer indication*, as shown in Figure 6.104.

If not already done, the DRNC will allocate the user a C-RNTI. If this is the first signalling message the DRNC has received from this UE, it will also allocate a D-RNTI, and include this D-RNTI in the signalling transfer message passed to the SRNC. The other key parameters in the signalling transfer are also shown in Figure 6.104. Note that it must contain the UTRAN cell-ID (RNC ID + cell ID) to uniquely identify this cell within UTRAN. The normal cell-ID only identifies the cell within an RNS. To reply to the UE in the downlink, the SRNC will send the downlink signalling transfer request, containing the signalling message to send to the UE, containing the cell-ID that was included in the uplink message. Upon receipt of this signalling message, the DRNC will transfer the RRC signalling to the UE on the CCCH in the cell identified by the received cell-ID.

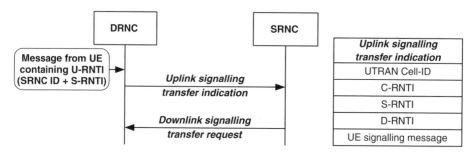

Figure 6.104 RNSAP signalling transfer

6.19.5   Iu interface

The structure of the Iu interface is split into two parts since the Iu connects to the circuit and packet domains of the core network. The Iu-CS, as shown in Figure 6.105, connects to the 3G MSC and is much the same as the Iur interface. Again, the transport and radio network control plane are built on top of an ATM and broadband SS7 signalling stack, with the RANAP protocol sitting on the SCCP layer to provide connection-oriented and connectionless services. As before, user data is transported using AAL2 and thus the AAL2 signalling stack is required for transport network control, i.e. establishment of user plane AAL2 connections. The user data is placed in the Iu frame protocol, as described previously.

To place this in context with the radio network protocol stack, consider the following diagrams, which illustrate the complete connection for control and data. Figure 6.106

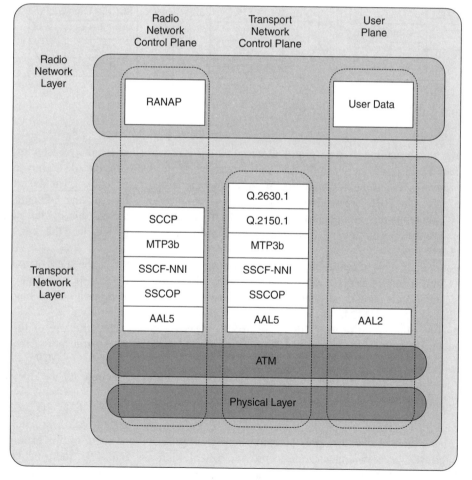

Figure 6.105   Iu-CS protocol stack

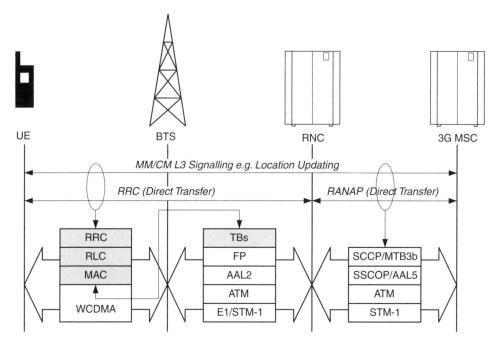

Figure 6.106 Signalling protocol stack

shows the signalling control protocol stack, which is established for the transfer of L3 signalling messages between the UE and core network, in this case, the MSC (however, the stack is the same towards the SGSN). The signalling connection is made up of two parts, the RRC connection and the RANAP connection. Between the UE and the RNC, although this is signalling, as far as the UTRAN is concerned, this is viewed as user data since it is coming from the mobile, and thus is transported over the standard AAL2 bearer.

For the transport of real-time voice data between the UE and CS-CN, the stack overview is as shown in Figure 6.107. Between the UE and RNC, the call is transported using the RLC transparent mode, and then to the MSC using the Iu user plane protocol in support mode, over AAL2/ATM.

For the connection to the SGSN, the Iu-PS interface is used. The structure of this is somewhat simpler than the other interfaces, with the absence of any transport network signalling (Figure 6.108). This is because the user data is transported over a GTP tunnel, using an IP over ATM connection. As will be seen in the next section, all the relevant information for this tunnel is sent in RANAP procedures.

Once again, considering an overview of this connection from the UE to the SGSN, Figure 6.109 shows the stacks. Since this is IP traffic to the packet core, the packets first have their headers compressed using the PDCP protocol, and then, in this example, use the RLC acknowledged mode to provide reliable delivery across the radio network. Subsequently, they are transported in a GTP tunnel and by IP over ATM towards the packet core.

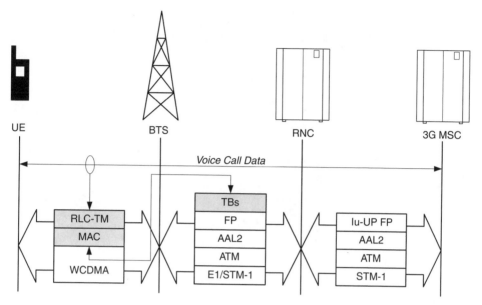

Figure 6.107   Voice call overview

6.19.6   Radio access network application part (RANAP)

The RANAP protocol provides the radio network signalling between the core network and the radio access network across the Iu interface. RANAP covers procedures for both the circuit and packet domains of the core network. The general services offered by RANAP are general control services, notification services and dedicated control services. The transport layer is expected to provide connectionless and connection-oriented services to support RANAP procedures, which it does through the broadband SS7 stack. RANAP expects the underlying signalling transport to provide in-sequence delivery of messages, and maintain a connection per active UE.

The key RANAP functions are as follows:

- *RAB management*: establishment, modification and release of radio access bearers.
- *Serving RNC relocation*: shifting of resources and functionality between RNCs.
- *Iu connection release*: release of all resources related to one Iu connection.
- *Iu load control*: adjusting of Iu loading and overload handling.
- *User paging*: offers the core network user paging service.
- *Transport of NAS information*: the transfer of NAS information, such as layer 3 signalling, between a user and the core network. An example would be the transfer of location/routing area update messages.
- *Security mode control*: transfer of security keys for ciphering and integrity protection to the radio access network.
- *General error reporting*: the reporting of general error situations.

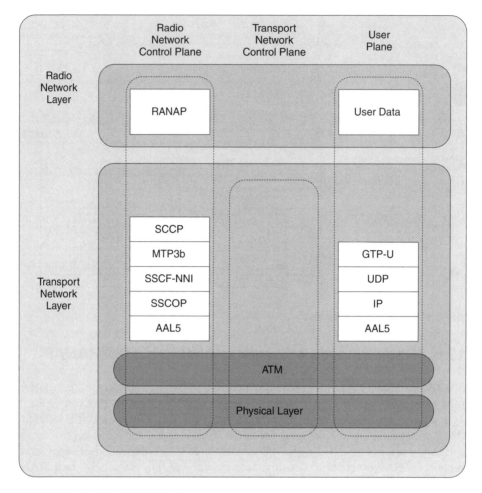

Figure 6.108 Iu-PS protocol stack

Table 6.34 highlights some of the principal RANAP procedures. For a complete definition of RANAP, the reader is referred to TS 25.413. This section will address one or two of the common ones in further detail.

6.19.6.1 Transfer of NAS messages

Two RANAP commands are provided for the transfer of NAS signalling messages. For both of these, the contents are not interpreted, but merely passed transparently across the Iu interface. The *initial UE message* is used to pass initial NAS messages from the UE to the CN. It is used when no signalling connection exists between the user and core network. An example of NAS messages that use this would be a location update request or a CM service request. Since this is an initial message, it not only transfers the NAS

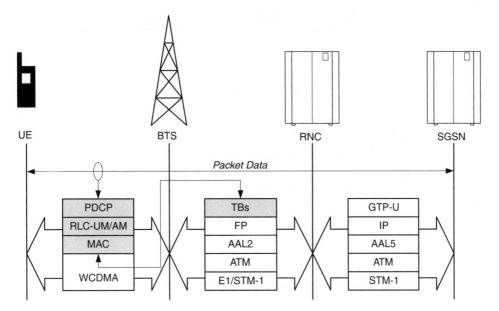

Figure 6.109   Packet connection overview

Table 6.34   RANAP procedures

Message name	UTRAN procedure
Initial UE message	NAS signalling connection establishment
Direct transfer	Uplink direct transfer
(of NAS messages)	Downlink direct transfer
Radio access bearer	Radio access bearer establishment
assignment	Radio access bearer release
	Radio access bearer modification
Iu release	RRC connection release
	UTRAN to GSM/BSS handover
Paging	Paging for a UE in RRC idle mode
	Paging for a UE in RRC connected mode
Relocation	Hard handover with switching in the CN
	SRNS relocation
	UTRAN to GSM/BSS handover
	GSM/BSS to UTRAN handover
Security	Security mode procedures

message, but also contains some extra information to support this first connection. The additional information is shown in Table 6.35.

Subsequent NAS messages between a UE and CN domain, in either direction, are transported over the *direct transfer* message. This is much simpler and just transfers the NAS message. However, in the downlink direction, the CN will specify the SAPI used for this message (either SAPI 0 or SAPI 3), indicating a priority level. This will tell the RNC which signalling bearer to use for RRC transport.

Table 6.35   Initial UE message contents

Initial UE message
Core network domain indicator (CS or PS domain)
Location area ID (LAI)
Routing area (RAC) (conditional on PS domain)
Service area ID (SAI)
NAS message
Iu signalling connection identifier (a unique identifier allocated by the RNC)
Global RNC-ID

A service area identifier (SAI) is a subdivision of a location area. The SAI is defined as follows:

$$SAI = LAI + SAC \text{ (service area code)}$$

where an $LAI = PLMN \text{ ID} + LAC$.

6.19.6.2   RAB establishment, modification and release

When a user wishes to establish a call or connect to the network, either self-initiated or as a response to a page, a RAB must be set up between the RNC and the CN to transport that traffic. The *RAB assignment request* is used by the core network to establish a radio access bearer. The procedure may request one or more RABs to be established, so there may be a number of RAB assignment responses (Figure 6.110). The RAB assignment request is the message that initiates the establishment of a radio link and associated radio bearers on the Iub/Uu interfaces. The request contains all the necessary QoS parameters for the RNC to determine the requisite resource allocation for the radio link, and the requirements on the radio link to meet the QoS.

The assignment request will contain the elements listed in Table 6.36.

Note that for a connection to the PS core, the RAB assignment request contains both the IP address and GTP tunnel end point identifier (TEID) of the SGSN, and therefore the

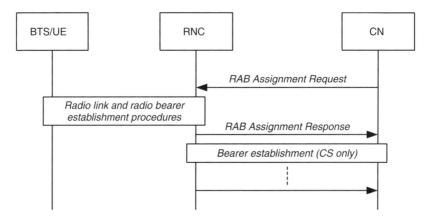

Figure 6.110   RAB assignment procedure

Table 6.36   RAB assignment request elements

Information element		Description
RAB ID		Unique identifier for the RAB across the Iu interface
RAB parameters		See Table 6.37
User plane info	Mode	Indicates either transparent or support for predefined SDUs
	Versions	Indicates the UP version, currently 1
Transport layer info	Address	Although not interpreted by RANAP, in practice for the CS-CN this will be the AAL2 ATM address of the CN element, and for PS-CN, the IP address
	Association	This will either be a binding ID for use with AAL2 signalling for CS-CN connection, or a GTP TEID for the PS-CN

Iu-PS does not require any ALCAP signalling since this is enough information to create the GTP tunnel.

The RAB parameters are listed in Table 6.37.

For the residual bit error rate, the delivery of erroneous SDUs parameter is specified. Its values can be:

- *Yes*: error detection is used and SDUs with errors are delivered

- *No*: error detection is used and SDUs with errors are discarded

- *No-error-detection-consideration*: SDUs are delivered, and no error detection is performed.

The use of this for frame classification is described in Section 6.18.4.2.

Table 6.37   RAB parameters

Parameter	Description
Traffic class	This will be one of conversational, streaming, interactive and background
RAB asymmetry indicator	Indicates whether this is symmetric or asymmetric, bidirectional or unidirectional (and whether uplink or downlink)
Maximum bit rate	A maximum rate for this RAB in bits/second in the range 1 bps to 16 Mbps
Guaranteed bit rate	For conversational and streaming classes, a minimum guaranteed bit rate is specified
Delivery order	Indicates if the SDUs need to be delivered in order or not
Maximum SDU size	Indicates how big, in bits, an SDU can be
SDU parameters	See Table 6.38
Transfer delay	For conversational and streaming classes, indicates the maximum tolerable transfer delay (in ms)
Traffic handling priority	For the interactive class, specifies a priority for the SDU. The priority range is 1 to 14 with 1 being highest priority. A value of 15 indicates no priority
Source statistics descriptor	For conversational and streaming classes, indicates whether the RAB contains *speech* or *unknown* traffic

Table 6.38 SDU parameters

Parameter	Description
SDU error ratio	The fraction of SDUs that may be either lost or contain errors. It is expressed in terms of a mantissa and exponent where the rate is given by mantissa $\times 10^{-\text{exponent}}$
Residual bit error rate[a]	Indicates the tolerated BER for each subflow, expressed as above. It also provides an indication of how errors should be dealt with, as described in the text
SDU format information[a]	Provides details of the size, in bits, of each subflow combination possible

[a]Expressed for each subflow.

As an example, Figure 6.111 shows an abridged trace sample of a RANAP RAB assignment request from the CS-CN to the RNC. In this example, a circuit switched call is being established and a RAB for AMR 12.2 kbps (max bit rate: 12 200). For 12.2 kbps rate, the SDU size is 244 bits, and there are three subflows. The RAB has a drop rate of 7×10^{-3} SDUs (i.e. BLER), and the subflows have the format shown in Table 6.39. The message also specifies a 20-byte NSAP address (see Section 7.13), which identifies the core network element originating the signalling message. This and the binding ID are used in the AAL2 signalling to establish the actual transport layer connection.

On receipt of an assignment request for a circuit switched connection, the RNC will invoke the transport layer signalling (AAL2 signalling) to establish the AAL2 transport bearer for the RAB. In the case of a packet connection, all the necessary information is within the assignment request. To establish, the RNC will also apply its internal arbitration to determine if it can meet the request, and what it is in a position to allocate. Once successfully completed, the RNC will reply with a *RAB assignment response* message. This generally only contains the RAB ID to relate it to the request message.

The RAB assignment request is also used by the CN to modify and release a RAB. In the case of a release, the message will contain the RAB ID of the RAB to be released, and a cause for the release, such as UE release, or successful relocation. The RNC may also release a RAB using the *RAB release* message. Again, this will indicate the particular RAB to release and the cause.

If there are no longer any RABs established across the Iu interface for a particular user, the core may initiate the *Iu release* command. On receipt, the RNC will clear the resources allocated for this connection and reply with an *Iu release complete* (Figure 6.112). Only the CN can release the Iu connection. However, the RNC can request this release through the *Iu release request*. If the CN agrees, it will then invoke the *Iu release*.

6.19.6.3   *RANAP paging message*

When the core network wishes to page the user, it will send a *paging* message on the Iu interface to the RNC covering the user's location/routing area. If the location/routing area covers more than one RNC, the core will repeat the paging messages to the relevant RNCs. The paging message will contain at least the CN domain identifier (CS or PS) and the permanent NAS UE identity, which is the IMSI (or TMSI/P-TMSI) number. The RNC

```
RANAP-PDU
RAB-AssignmentRequest
  protocolIEs
    RAB-SetupOrModifyList
        - rAB-ID: '00000001'B
      rAB-Parameters
        - trafficClass: conversational
        - rAB-AsymmetryIndicator: symmetric-bidirectional
      maxBitrate
        - MaxBitrate: 12200
      guaranteedBitRate
        - GuaranteedBitrate: 12200
        - deliveryOrder: delivery-order-requested
        - maxSDU-Size: 244
      sDU-Parameters
        sDU-ErrorRatio
          - mantissa: 7
          - exponent: 3
        residualBitErrorRatio
          - mantissa: 1
          - exponent: 6
        - deliveryOfErroneousSDU: yes
        sDU-FormatInformationParameters
          - subflowSDU-Size: 81
          - subflowSDU-Size: 39
        residualBitErrorRatio
          - mantissa: 1
          - exponent: 3
          - deliveryOfErroneousSDU: no-error-detection-consideration
        sDU-FormatInformationParameters
          - subflowSDU-Size: 103
          - subflowSDU-Size: 0
        residualBitErrorRatio
          - mantissa: 5
          - exponent: 3
          - deliveryOfErroneousSDU: no-error-detection-consideration
        sDU-FormatInformationParameters
          - subflowSDU-Size: 60
          - subflowSDU-Size: 0
        - transferDelay: 80
        - sourceStatisticsDescriptor: speech
      userPlaneInformation
        - userPlaneMode: support-mode-for-predefined-SDU-sizes
        - uP-ModeVersions: '0000000000000001'B
      transportLayerInformation
        - transportLayerAddress:
            '45 00 00 00 00
             00 65 96 8F 00
             00 00 00 00 00
             00 00 00 00 00' H
      iuTransportAssociation
        - bindingID: '0100954F'H
```

Figure 6.111   RANAP assignment request trace. Reproduced by permission of NetHawk Oyj

Table 6.39   RAB subflows

	Subflow 1	Subflow 2	Subflow 3	Rate (kbps)
No. of bits	81	103	60	12.2
	39	0	0	1.8
BER	10^{-6}	10^{-3}	5×10^{-3}	

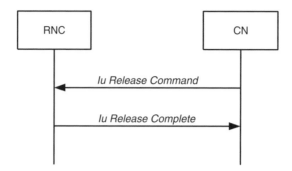

Figure 6.112   Iu release

will decide how to page the user across the UTRAN depending on the RRC connection status of the UE.

6.19.6.4   *Relocation*

The relocation procedures are invoked when it is necessary to perform an SRNC relocation, which is moving the UE connection to the core network from one RNC to another as a result of user movement. This situation arises when a user is no longer connected to any BTSs that are under the control of its SRNC. The relocation procedure for this is shown in Figure 6.113.

First, the RNC informs the CN that a relocation is necessary using the *relocation required* message. This will contain identifiers for the source (SRNC) and target RNCs. On receipt, the CN will send a *relocation request*, which is much the same as a *RAB setup request*, to the target RNC, which will then perform transport layer setup and respond with the *relocation request acknowledge*. The CN will then send a *relocation command* to the source RNC to release its RABs. The source and target RNC will use RNSAP signalling between them to commit this relocation, and, once successful, the target will respond with the *relocation complete* message. The CN will then call the *Iu release* procedure as described above.

This relocation procedure is also used for inter-system handover from UTRAN to GSM. In this case, the *relocation required* message will contain GSM-related parameters.

6.19.6.5   *RANAP security control*

Another important feature is the security mode procedures. In this, security information is passed from the core network to the RNC to protect both data and signalling messages across the UTRAN. The CN issues a *security mode command*, which contains ciphering, and integrity information. This information is a list of available algorithms being used for ciphering and integrity, plus the keys for each that have been generated during the UE-CN authentication procedure. The RNC will choose an algorithm to use for each, and inform the CN of this choice using the *security mode complete* message (Figure 6.114). Subsequent data and signalling messages in UTRAN are now protected.

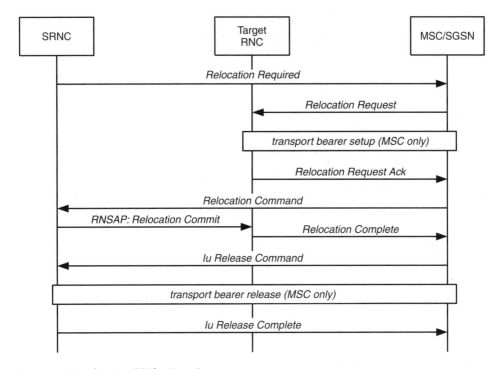

Figure 6.113 Serving RNC relocation

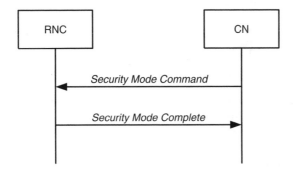

Figure 6.114 Security procedure

6.19.7 Broadband SS7

For the Iu and Iur interfaces, a broadband SS7 stack is introduced to support the control transport, as shown in Figure 6.115. The components of this are explained briefly.

- *Message transfer part level 3 broadband (MTP3b)*: this layer allows signalling messages to be transported over a complex network, i.e. there does not need to be any direct connection between signalling points (network elements). This layer deals with

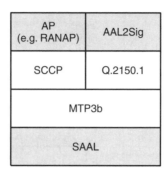

Figure 6.115 Use of broadband SS7

message routing, re-routing when a link fails and congestion control. It has many similarities with the IP layer on the Internet. By insertion of this SS7 stack, it decouples the ATM layer from the UMTS protocols, which provides the flexibility that the ATM can be removed and replaced with any transport network that can support the same functionality.

- *Signalling connection control part (SCCP)*: this sits on top of the message transfer part and provides both connectionless and connection-oriented network services. In UMTS, the connectionless mode is used, since it sits on top of the SAAL, which already provides the connection-oriented service (see Chapter 7). General control and notification messages use connectionless, and dedicated control uses connection-oriented. The application part (AP) sitting on top of SCCP could be RANAP or RNSAP. Together with MTP it is referred to as the *network service part* (NSP). For dedicated messages, SCCP is used to support transfer of signalling messages between the CN and the RNC, providing the service of establishing a unique RANAP or RNSAP signalling connection per active user.

Signalling points can have a number of attached applications operating simultaneously, SCCP introduces the subsystem number (SSN) to ensure that the correct application is accessed. In this case, the SSN will identify RANAP or RNSAP. This layer can be seen as analogous to TCP in the Internet suite of protocols. For RANAP, SCCP can also be used to provide address translation capabilities, known as global title translation (GTT). However, use of this is implementation dependent.

Consider the initial establishment of a signalling connection with the RANAP initial UE message. The SCCP message used to transport this is the *connection request* (CR) message (Figure 6.116). The CR contains a local reference which uniquely identifies the signalling connection being created. The CN will cite this number as a destination local reference in subsequent messages relating to this connection. The CR also contains calling and called party addresses, which identify the RANAP subsystem. The CN replies with a *connection confirm* (CC) message, which provides another reference number for associated messages in the opposite direction. Should the connection have failed or been refused, the *connection refused* (CREF) would have been sent. Now the connection is established, and the following RANAP signalling messages sent will contain

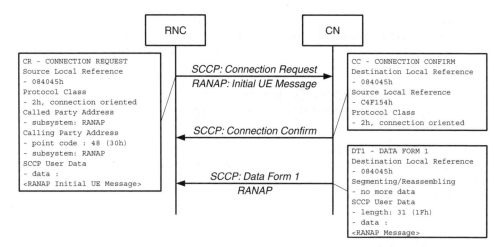

Figure 6.116   SCCP connection establishment. Reproduced by permission of NetHawk Oyj

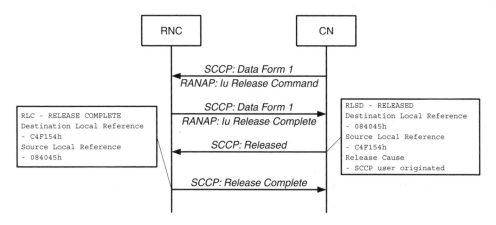

Figure 6.117   SCCP connection release. Reproduced by permission of NetHawk Oyj

the connection reference at the SCCP layer to link the message to a specific UE. These RANAP/RNSAP signalling messages are transported in the data form 1 format, which is an unacknowledged connectionless service used to transparently transport the user data between two nodes.

Once the CN has completed a signalling transaction, it will issue an *Iu release* message. Once the RNC has acknowledged this, SCCP will then release the SCCP connection (Figure 6.117). The *released* message specifies the reference in both directions to identify which connection to release, which is in turn acknowledged by the *release complete* message.

- *Q.2150.1*: this is a signalling converter, which adapts the AAL2 signalling messages to be transported by MTP3b. For the Iub interface, Q.2150.2 is used, which is designed for an UNI interface.

For the Iu-PS domain, in R99, the specifications allow for the use of IP to transport signalling in addition to data. This is covered in more detail in Section 8.9.

6.20   MOBILITY MANAGEMENT FOR PACKET SWITCHED OPERATION

The UTRAN connection to the packet core network domain uses the Iu-PS interface for operation. The operation of the packet core consists of three states, referred to as the packet mobility management (PMM) states. These Iu states are PMM-detached, PMM-idle and PMM-connected, as described below.

6.20.1   PMM-detached

In the detached state there is no communication between the mobile device and the SGSN, and the SGSN does not hold a valid location for the device. In order to establish a MM context, the mobile device has to perform a *GPRS attach* procedure. This is usually executed by the subscriber switching on the mobile device. The packet switched signalling connection consists of two separate parts, the RRC connection and the Iu connection.

6.20.2   PMM-idle

In PMM-idle state, the location of the mobile device is known in the SGSN to the accuracy of a routing area. Paging is required from the network if the mobile device is to be reached. The mobile device will monitor its routing area and will perform a routing area update if its routing area changes.

6.20.3   PMM-connected

In the PMM-connected state, the location of the mobile device is known in the SGSN to the serving RNC. This is different to the 2G GPRS system, where the SGSN keeps track of the location of the mobile device to the precise cell. It is the responsibility of the serving RNC to keep track of the location of the mobile device to the cell or URA. The mobile device will, however, inform the SGSN if its routing area changes. While the mobile device is in PMM-connected state, a packet switched signalling connection is established between the mobile device and the SGSN. If this connection is lost then the mobile device will revert to the PMM-idle mode.

Figure 6.118 shows how the mobile device may change between the three different states. The mobile device may have a packet data protocol (PDP) connection in either PMM-idle or PMM-connected states. The mobile device can move directly from the PMM-detached state to the PMM-connected state by issuing the GPRS attach request

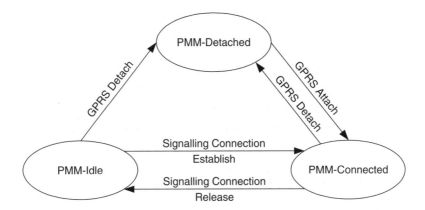

Figure 6.118 Mobility management

message. The mobile device may move from PMM-connected to PMM-detached when the GPRS detach command has been issued, when an RA update is rejected by the SGSN or if a GPRS attach request has been rejected by the SGSN. The mobile device will move from PMM-idle to PMM-detached by an implicit GPRS detach. This would occur if the battery expires or the SIM card is removed from the device. The mobile device will move from PMM-idle to PMM-connected when a signalling connection is established. This would occur, for example, if the mobile device wished to transfer some data. After a certain period of inactivity the mobile device would revert back to PMM-idle from the PMM-connected state.

6.21 UMTS SECURITY ARCHITECTURE

Security is an integral part of UMTS to provide a robust, secure framework on which a subscriber can rely. UMTS security has evolved from principles first used in GSM. However, there have been a number of key enhancements compared with GSM security, to address perceived weaknesses in the GSM security architecture. Also, UMTS integrates security mechanisms to reflect the nature of traffic that the network is intended to transport, and here borrows heavily from the experience with security on the Internet. The UMTS system has a major advantage over other networks such as the Internet since the issue of key exchange is solved through the issuing of a user services identity module (USIM) card containing the key, which is also stored in the home environment HLR/AuC. However, the keys used in the system all use a standard key length of 128 bits, which is assumed to be currently unbreakable by brute force. A brute force attack is where the attacker exhaustively tries all the possible keys for a match. Statistically, the probability of a match will occur after half the keys have been tried. A key length of 128 bits generates enough possible keys (2^{128}) that it would take an unrealistic length of time to crack. The addition of one extra bit to the key length doubles the number of keys available. Of course, attacks can be made in other ways, by looking for weaknesses in the algorithms used. Any secure system should provide for the following framework for user protection:

- *Privacy*: ensures that only authorized individuals can read the information.
- *Authentication*: proves the identity of the message sender.
- *Data integrity*: ensures data cannot be altered without detection.
- *Non-repudiation*: undeniably proves that a message was sent, usually including when it was sent.

From GSM, the following services are retained:

- user authentication
- radio interface encryption
- user identity confidentiality
- independent hardware security module (SIM).

However, the following areas have been shown to be weak in GSM:

- no network authentication, hence attack by a 'false BTS' is not prevented;
- keys and authentication data are not protected either within or between networks;
- the encrypted relationship is only between the mobile device and the BTS, therefore data is in plain text beyond the BTS. Often, this data is sent across microwave links between the BTS and the BSC;
- GSM security algorithms have already been broken;
- there is no provision of a data integrity mechanism;
- the security mechanisms are fixed and do not have the flexibility to be upgradable.

The UMTS architecture addresses all of these issues. The basic services provided are detailed in the succeeding sections. UMTS does not specify particular algorithms that must be used, but rather includes a default chosen set of tried and trusted algorithms within a framework that allows for other algorithms to be introduced. The algorithms included within the specifications are the f8 ciphering and f9 integrity algorithms, both based on the Kasumi block cipher algorithm. Kasumi also forms the basis of the A5/3 encryption algorithm, which has recently been introduced retrospectively to GSM. Unlike the security algorithms used in GSM, which were developed in secret, the Kasumi algorithm is well defined and has thus been exposed to the rigours of testing among the broader community. The Kasumi algorithm, is a variant of the MISTY algorithm, which was developed by Mitsubishi in 1995.

User protection in the mobile device is provided on the USIM. A user can protect access to the USIM by using a user-defined password, which must be entered at power-on to activate the card. The USIM is housed in a UMTS integrated circuit card (UICC), as is done currently with GSM. This card, which contains the user master key, is designed as a tamper-proof module which prevents direct access to data stored within.

6.21.1 User identity confidentiality

In UMTS this relates not only to the identity of the user, that is the IMSI, but now also to the location of the user. The method to provide the former is to use temporary identities, such as the TMSI/P-TMSI, as much as possible. Obviously this is not available the first time a user accesses the network; however, subsequently the temporary identity will be used. In addition, once the security relationship has been established, the network may periodically reallocate this temporary identity so as to avoid use of the same identity for a long period of time. Any signalling messages that could potentially reveal the identity of the user are encrypted.

6.21.2 Authentication

UMTS provides for the mutual authentication of both user and network. This process ensures that not only is the network ensured that only authorized users can access, but also that users validate that the network to which they are connecting is authorized by the user's home network. This may not seem initially such a major point, since in a GSM context, the economy of scale is not present for a would-be snooper to acquire a BTS to eavesdrop on user conversations. However, for the 3G framework, many different radio access technologies can be integrated into the system. This means that in the future, a wireless LAN access point could be providing user access to their services.

The authentication procedure flows as shown in Figure 6.119. Once a new user attempts to access the network, for example to perform a location update, the VLR/SGSN will send an authentication request to the HLR/AuC, identifying the user by its IMSI. The HLR will then use the user key, K, to generate a set of n authentication vectors, which it returns to the VLR/SGSN. An authentication vector consists of the following components:

- random number, RAND
- expected response, XRES
- cipher key, CK
- integrity key, IK
- authentication token, AUTN.

The generation of this vector is explained below. The rationale for generating multiple vectors is that subsequent user authentication requests can be handled at the VLR/SGSN without recourse to the home environment. The VLR/SGSN can take the next vector from the set it has stored.

Now the VLR/SGSN sends an authentication request to the user, containing just the RAND and AUTN components. The user USIM uses the AUTN value to validate that the

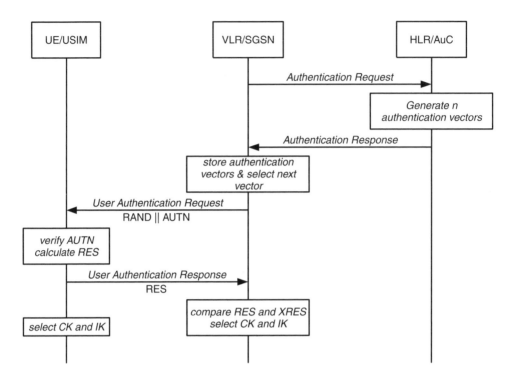

Figure 6.119 UMTS authentication procedure

network is trusted, and calculates its result using the RAND and its key, K. The USIM also uses the AUTN to generate the keys CK and IK. This result, RES, is sent back to the VLR/SGSN, which compares it to the expected result, XRES, and, if matching, thus validates the user on the network. Both the USIM and the VLR/SGSN then distribute the generated keys to the relevant units that will be performing the encryption and integrity functions. In the case of the VLR/SGSN, this is the SRNC.

The mechanism used in the HLR/AuC to generate the authentication vectors is as shown in Figure 6.120. First, the HLR/AuC generates a sequence number, SQN, and a random number, RAND. The sequence number will increment for each new authentication vector. These are then used with the user key, K, authentication algorithms (f1, f2) and key generation algorithms (f3–f5) to generate the authentication vector fields. The authentication token, AUTN, is:

$$AUTN = SQN \oplus AK\|AMF\|MAC$$

where $\|$ means concatenation, and each whole vector, AV, is:

$$AV = RAND\|XRES\|CK\|IK\|AUTN$$

The use of f5 is to protect the sequence number by using an anonymity key, AK. If this level of protection is not required, then f5 = 0. The authentication and key management field, AMF, can be used to provide services such as support for multiple security

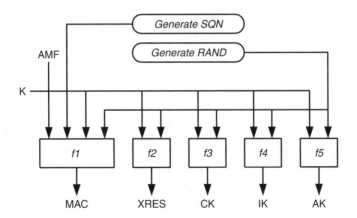

Figure 6.120   Authentication vector generation

algorithms. Upon receipt of the RAND∥AUTN field, the USIM uses a similar process with the same key and algorithms to extract the relevant fields.

6.21.3   Security mode establishment

The entire procedure to establish a secure relationship between a user and the network is shown in Figure 6.121. Section 6.16.3 showed how an initial signalling connection is established, and it was seen that the UE responds to the *RRC connection setup* message with an *RRC connection setup complete* message. This message contains all the UE capabilities and includes the supported algorithms for both integrity protection (UMTS integrity algorithm, UIA) and encryption (UMTS encryption algorithm, UEA). Also included is a value called START, which is a 20-bit field stored in the USIM and passed to the mobile device at power on. This start value is used by the RNC as the most significant bits of a counter for subsequent messages, referred to as a hyperframe number (HFN). The reason this is used is that the standard sequence numbers, for example those used by RRC and RLC, are not long enough to prevent replays, where old sequence numbers are reused at some point later in time.

The initial L3 message, which could be, for example, a *location update request* or a *GPRS attach request*, prompts the authentication procedure as described above.

After the authentication procedure has completed, the VLR/SGSN makes a decision of which integrity and encryption algorithms to use (UIA/UEA) in order of preference, and issues this information to the SRNC, along with the respective keys, IK and CK, in the RANAP *security mode command*. The SRNC compares the selected UIA/UEAs to those that the UE is capable of, and selects the highest priority match. The UE is then informed of this choice through the RRC *security mode command*, which also contains a random value, FRESH, generated by the SRNC. Attached to this message is the message authentication code for integrity (MAC-I) field, which enables the UE to validate that the message has come from a trusted source, and has not been altered en route. The UE verifies acceptance of the security mode with the RRC *security mode complete* and

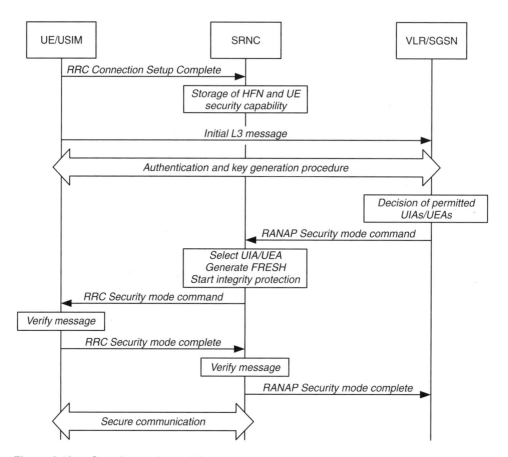

Figure 6.121   Security mode establishment procedure

the entire procedure completes with the transfer of the RANAP *security mode complete* message to the core network. There now exists a secure relationship between the UE and RNC, where all transfers of both control and data can be both integrity checked and encrypted.

The procedure for generating an integrity check authentication field for a message is shown in Figure 6.122. The message is passed in to the f9 algorithm with the integrity key (IK).

On receipt of a signalling message, the message authentication field (MAC-I) can be checked to verify it came from the correct user, and has not been tampered with.

The other components of the integrity check are as follows:

- *Count-I*: a 32-bit counter, which is incremented for each new signalling messages, so message sequences can be checked. The format of count-I is as shown in Figure 6.123. The RRC HFN is initialized by placing the START value in the most significant 20 bits, and padding the remaining 8 bits with 0. The lower 4 bits are the sequence numbers

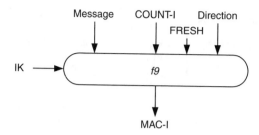

Figure 6.122   Message integrity check

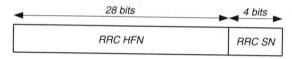

Figure 6.123   Composition of count-I

for each RRC signalling message. Once the RRC SN cycles, the HFN number is incremented.

- *FRESH*: a random number generated by the RNC and passed to the UE in the security mode command. The FRESH value is valid for the duration of the signalling connection.

- *Direction*: indicates if this signalling message is from UE to RNC (0) or from RNC to UE (1). This protects against use of the downlink authentication field to attempt to validate an uplink message.

The UE or RNC uses the same mechanism on any received messages to generate an expected message authentication code for integrity (XMAC-I), which is then compared to the received MAC-I to validate the message. Any messages received that fail the integrity check will be rejected.

6.21.4   Confidentiality

As was seen, encryption of data carried in a given radio bearer is performed at two possible points in the radio link protocol stack. For RLC acknowledged and unacknowledged modes, it is performed at the RLC layer, and for RLC transparent mode, at the MAC layer. UMTS performs encryption using a stream cipher, where the encryption algorithm generates a keystream which is added bit per bit to the plain text to generate the cipher text. Confidentiality protection may be applied to both data and signalling messages. The operation of the algorithm is shown in Figure 6.124.

The fields in the algorithm are:

- *Count-C*: this is a 32-bit ciphering counter, similar to the Count-I for the integrity check. However, for encryption there are three different formats depending on which

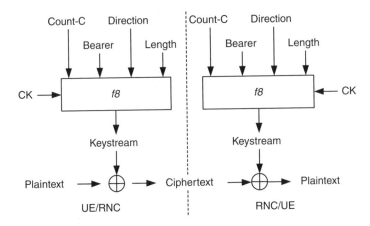

Figure 6.124 UMTS encryption algorithm

RLC mode is being used. All formats use the START value to initialize the most significant bits of the HFN, with the remaining bits filled with 0. For RLC transparent mode, the MAC-d HFN is 24 bits and the lower 8 bits are taken from the connection frame number (CFN) (see Figure 6.125(a)). For RLC unacknowledged and acknowledged mode, the RLC HFN is 25 bits and 20 bits, with the lower bits comprised of the RLC sequence number, 7 bits for unacknowledged mode and 12 bits for acknowledged mode (Figure 6.125(b) and (c)).

- *Bearer*: this is a 5-bit identifier of the radio bearer being encrypted, so as to avoid identical keystreams for different radio bearers from the same user.
- *Direction*: again, this is a 1-bit field which indicates if this signalling message is from UE to RNC (0) or from RNC to UE (1).
- *Length*: since data is presented to the RLC layer in blocks, the generated keystream must match the data block length. This is a 16-bit field which indicates the required length of the keystream block to be generated.

The counters provide an anti-replay service, where a message cannot be captured and reused by a potential intruder at a later time. However, the communication is still vulnerable

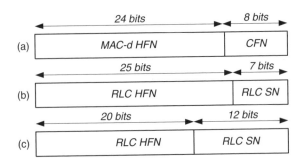

Figure 6.125 Composition of Count-C

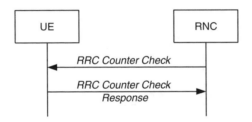

Figure 6.126 Counter check procedure

to a 'man-in-the-middle' attack, where an intruder imposes itself on a legitimate connection, posing as either the UE or network. To minimize the risk from this, the RNC can at any stage request the UE to inform it of its current counter value, to verify that the UE has received the same amount of data during the RRC connection as the RNC has sent. If the counter at the UE does not match the one at the RNC, the RNC may release the RRC connection. The procedure uses RRC signalling, as shown in Figure 6.126. The *counter check* message contains the 25 most significant bits of the Count-C for each radio bearer. If all is well, then the UE sends a *counter check response* to acknowledge this. If there were differences, then the UE will respond with the different counter values it has. This procedure is available for RLC-UM and RLC-AM modes only.

6.22 UMTS CALL LIFE CYCLE

The previous sections have detailed the necessary signalling protocols throughout the network to establish, maintain and release connections between the UE and the core network. It is useful to combine these in describing the life cycle of a UMTS call. Consider that a user arrives in a cell and switches on their mobile device. Section 6.9 detailed the initial synchronization procedures required prior to establishing a signalling connection. Following these procedures, the following sequence of events is assumed:

1. User performs a location update to the circuit core.
2. User receives a phone call.
3. During call progress, the user moves to another cell.
4. After hanging up, the user proceeds to check their email.

The basic steps to fulfil these four tasks are as shown in Figure 6.127.

The following subsections go through these four signalling procedures as a means of demonstrating how all of the signalling protocols tie together.

6.22.1 Signalling connection establishment

This is the first signalling process a UE will perform when it wishes to connect to the network. This is referred to as establishing an RRC connection, and in this example, the establishment of a dedicated transport channel for signalling is shown (see Figure 6.128).

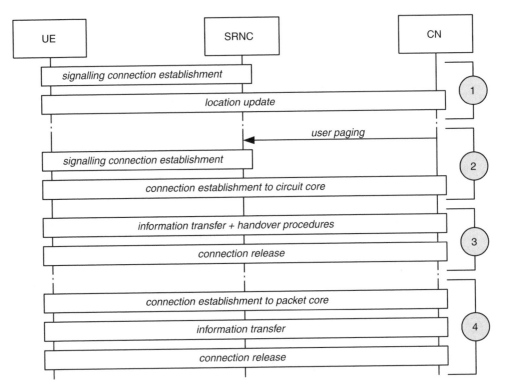

Figure 6.127   Basic UMTS call life cycle

1. The UE initiates the establishment of an RRC connection by sending the *RRC connection request* message. This message is sent on the common control channel, transported on the RACH/PRACH (Figure 6.129). The parameters included in this initial connection are the initial UE identity, typically the IMSI or TMSI/P-TMSI, and the establishment cause, which indicates why the UE is requesting a connection. The cause here is registration, since it will perform a location update next.

2. This request is carried transparently by the BTS. The SRNC receives the request, and will perform an admission control procedure to decide whether the request should be serviced, and if so, which type of transport channel to establish. In this case, the SRNC has chosen to use a DCH for this RRC connection. It will also allocate a RNTI and the required radio resources for the RRC connection. To establish the bearer for this connection, the SRNC will issue the NBAP *radio link setup request* message to the BTS. The parameters included will be the cell ID, transport format set, transport format combination set, frequency channel, uplink scrambling code and downlink channelization code to be used, and power control information.

3. The BTS will allocate the requested resources, and is therefore ready to begin uplink physical reception. It will then respond to the SRNC with the NBAP *radio link setup response* message. The parameters in this will include the transport layer

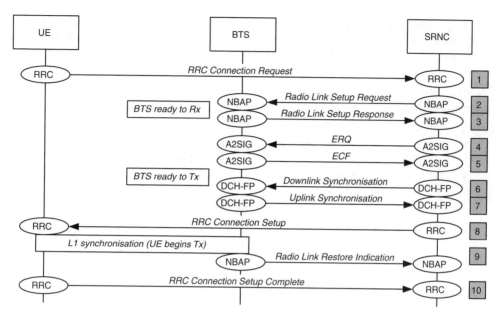

Figure 6.128   RRC connection establishment

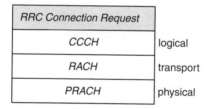

Figure 6.129   RRC connection request layers

addressing information (AAL2 address, AAL2 binding identity), which are required for the Iub data transport bearer (AAL2 connection) setup. This is only being established for signalling at this point and not for data transfer.

4. Once the BTS has responded, the SRNC then uses AAL2 signalling (ALCAP) to establish an Iub data transport bearer. It will use the AAL2 address to contact the BTS transport layer entity, and this *establish request* (ERQ) will contain the AAL2 binding identity to tie the Iub data transport bearer to the DCH.

5. The BTS will acknowledge the setup request with the *establish confirm* (ECF) command.

6. The RNC will send the *downlink synchronization* message, which contains the connection frame number.

7. The BTS responds with the connection frame number, and timing information through the *uplink synchronization*. The BTS will start downlink transmission at this point.

8. The SRNC will send the *RRC connection setup* message to the UE on the CCCH. This will include the following parameters: initial UE identity, allocated U-RNTI, capability update requirement, transport format set, transport format combination set, frequency, DL scrambling code (FDD only), and power control information.

9. Once the BTS has detected the UE at the physical layer (*L1 synchronization*) it will report this to the SRNC with the *radio link restore indication* message.

10. The UE responds to the SRNC with the *RRC connection setup complete* message, but now it can use the DCCH that has been established. The parameters here are UE capability information, such as integrity and ciphering information. There is now a signalling connection between the UE and SRNC but data transfer cannot take place just yet.

Note that in the following sections, some of the involved procedures described above, such as the *L1 synchronization* and the *downlink/uplink synchronization*, do still take place but are omitted from the descriptions for simplicity.

The decision to allocate a user a dedicated signalling connection as shown above is made in the RNC through its admission control policy. Alternatively, as illustrated in Figure 6.130, the RRC connection could be established on the RACH/FACH common transport channels.

1. As before, the UE initiates an *RRC connection request* on the CCCH/RACH/PRACH. Notice that this is a much simpler procedure with NBAP and ALCAP signalling not being required. This is because the CCCH channels will be set up already when the cell was first established or reset.

2. After performing admission control, the SRNC decides to use the FACH/RACH for this RRC connection, and allocates the UE both a U-RNTI and a C-RNTI identifier. The *RRC connection setup* message is sent on CCCH, including the same parameters as before with the addition of the C-RNTI. The C-RNTI is required since the CCCH is now being used for signalling as well as data transfer; the signalling will be destined for the controlling RNC of this BTS. At this point the CRNC MAC-c/sh will be able to determine if this is signalling or user data; the user data will be forwarded to the MAC-d of the SRNC, which may or may not be in the same physical RNC.

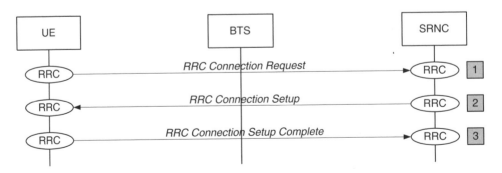

Figure 6.130   RRC connection on FACH/RACH

3. The UE responds with the *RRC connection setup complete* message on a DCCH logical channel, which is mapped to the RACH transport channel.

6.22.2   Location updating

Once the connection to the circuit core has been established, the UE proceeds to perform a location update (Figure 6.131).

1. The *L3 location updating request* is carried to the 3G MSC using the RRC *initial direct transfer* message and the RANAP and *initial UE message*. As was discussed in Section 6.19.7, at the lower SCCP layer, this initial UE message will establish a RANAP unique signalling link between the SRNC and the 3G MSC. Since this is carried over AAL5 there is no need to set up an AAL2 connection as was required over the Iub. The UE will identify itself in the *location updating request* by its IMSI or TMSI.

2. Then the core network will invoke the authentication procedure, during which there is mutual authentication between the UE and network, keys for encryption and integrity protection are generated, and suitable algorithms are selected. Please refer to Section 6.21 for full details of these procedures.

3. After the authentication procedure has successfully completed, the 3G MSC sends the UE a *location updating accept* message.

4. This now completes the transaction, so the connection is no longer required. The 3G MSC thus sends the *Iu release* message to release this connection, and the RNC replies with the *Iu release complete*.

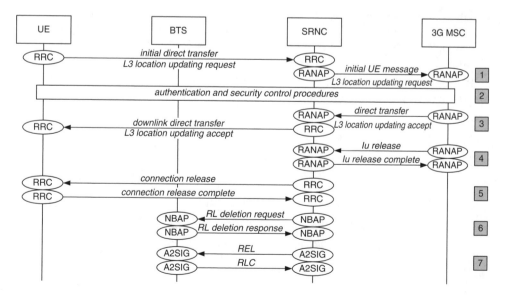

Figure 6.131   UMTS location update

5. In turn, the RRC connection may also be released, and the UE may return to idle mode. The layer 3 signalling actually has a follow-on parameter, which may be set by the UE if it is going to send another message, such as make a phone call, immediately after the location updating request. In this situation the RRC connection may not be released. This, however, is not the case in this particular scenario.

6. The RNC then instructs the BTS to delete the radio link.

7. The AAL2 transport channel for that radio link is also deleted.

6.22.3 Paging

As previously discussed, there are two types of paging used in UMTS. For type 1, the UE has no connection to the network so needs to be paged on a paging channel. For type 2, the UE has a connection on the FACH or DCH and can be paged along that connection.

Paging type 1

This is the paging for a UE with no FACH/DCH connection. This applies to a UE in RRC idle mode and RRC connected mode for both CELL-PCH and URA-PCH states. In this example, shown in Figure 6.132, the paging is performed for a UE in RRC idle mode. The core network domain could be the CS or PS domain, and in idle mode, the RAN knows nothing about the user. The UE is only known to a CN location, either an LA for CS or an RA for PS. Therefore, the paging message must be distributed across that whole LA/RA to reach the user. The example shows that the area spans two RNCs, so the paging message must be sent to both.

1. The CN initiates the paging of a UE across an LA, which spans two RNCs. The paging is performed using the RANAP paging message and contains the following parameters: CN domain indicator (CS or PS), permanent NAS UE identity (e.g. IMSI, TMSI or P-TMSI) and the paging cause.

2. Paging of the UE is then performed by the RNCs. This is done by sending a paging message, type 1, across the Iub interface using the Iub common channel frame

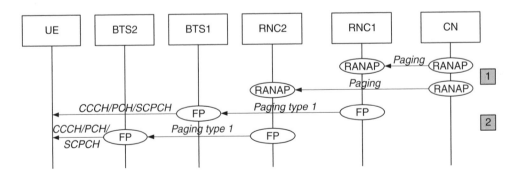

Figure 6.132 Paging in idle mode

protocol. This is then carried on the air interface on the PCH, carried in the physical secondary CPCH. The page is sent across all the cells in the LA (only two are shown in the diagram). The UE will answer the page in whichever cell it is currently in, and will then reinitiate an RRC connection.

This procedure described for RRC idle mode also applies to RRC connected mode for CELL-PCH and URA-PCH states. The difference here is that the RNC will only page on the PCH across either one cell or one URA. Also, in these states, the paging message will identify the user using the U-RNTI, rather than the IMSI/TMSI/P-TMSI.

Paging type 2

Paging type 2 is used when an RRC connection has already been established for this particular mobile device, i.e. it has an RRC connection (DCH or FACH). In this case, the mobile device is paged directly over the established RRC signalling link rather than the paging channel. The SRNC determines that an RRC has already been established for this mobile device since the RRC connection is associated with the IMSI within the RNC. The following example shows how paging is performed for a UE in RRC connected mode (CELL-DCH and CELL-FACH states) when the UTRAN coordinates the paging request with the existing RRC connection using DCCH (Figure 6.133). The scenario set out in the section will not use paging type 2 since the RRC connection was released after the location update; however, it is presented here for a comparison.

1. As before, the CN initiates the paging of a UE via the RANAP *paging* message. The parameters are CN domain indicator, IMSI/TMSI/P-TMSI and the paging cause.
2. The SRNC sends the *RRC paging type 2* message. This is on the established logical DCCH, which can be transported over a DCH, the FACH or DSCH.

6.22.4   Connection establishment: circuit core

Once the UE has detected the paging message, it will answer the call and reinitiate a connection to the network (Figure 6.134). It should be noted that although there is no signalling connection to the network, the CN still retains information about the UE.

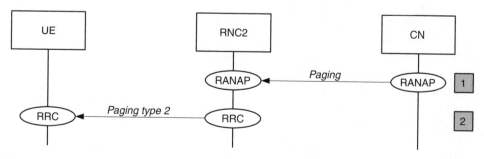

Figure 6.133   UE paging using dedicated channel

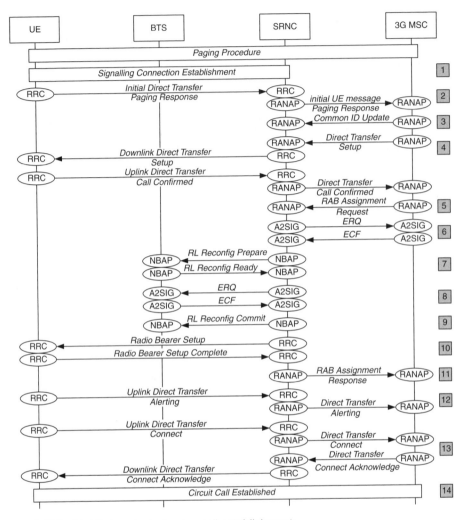

Figure 6.134 Mobile terminated voice call establishment

1. After the user is paged, once again a signalling connection must be established. The procedure is the same as described above for that prior to a location update. All that will be different is the establishment cause. In this case, the cause is *terminating conversational call.*

2. The UE then sends an initial connection management (CM) service request *paging response* to the circuit core. This message will indicate the IMSI/TMSI of the UE. Recall that all UMTS has introduced is a new access layer, so for R99, the basic structure and operation of the core networks are unchanged. These L3 procedures are the same as would be seen for a GSM call establishment, and it is the underlying access bearer establishment where the changes are significant. Please refer to Chapter 3 on GSM for more details. The paging response will be followed by authentication and security procedures as previously described.

3. Now, the core network informs the RNC of the permanent NAS UE identity of the user, which is the IMSI, through the *common ID update* message. This is so as to tie the RRC connection for this UE to its permanent identity for such procedures as paging coordination.

4. The L3 procedure for call establishment continue with the call control (CC) *setup* and *call confirmed* messages. These are Q.931 basic call control messages. For further details, refer to Chapter 3.

5. The core network now requests the SRNC to set up a radio access bearer using the *RAB assignment request*, which will contain all the QoS information for the call. It is this message that will result in both Iub and Iu bearers being established as described in the following steps. In this case, since a voice call is being made, there will be three RAB subflows set up for the different classes of AMR CODEC bits, classes A, B and C. The RAB assignment request also contains the AAL2 address and binding ID for coordination with AAL2 signalling over the Iu interface.

6. An AAL2 bearer is established across the Iu interface for this call.

7. On the UTRAN side, a bearer must now be set up for the call. The first stage of this is to inform the BTS. Since a radio link which is being used for signalling already exists between the UE and RNC, the procedure is to reconfigure this link to now carry data as well. Since timing is critical for a voice call, it is necessary that there is a synchronized procedure for reconfiguring the link. This means that initially the BTS is passed all the new transport and physical layer information in a *radio link reconfiguration prepare* message, and responds with a *radio link reconfiguration ready* message. At this stage, it has not committed the link, since the AAL2 transport bearer is not ready. The *radio link reconfiguration ready* also contains the AAL2 address and binding ID of the new transport bearer to be established.

8. An AAL2 bearer is established across the Iub interface for this call.

9. Since the AAL2 bearer is ready, the RNC can now commit the reconfigured radio link, *radio link configuration commit*.

10. The RNC now needs to establish new radio bearers between it and the UE with the *radio bearer setup*. For a voice call, three RBs are set up for the three classes of AMR bits. Since the different classes require different QoS, each of these RBs have a separate transport channel, only being concatenated at the physical layer into a CCTrCH. Both the new/modified transport and physical layer parameters are passed in this message also. The UE responds with a *radio bearer setup complete*.

11. Now that the connection on the UTRAN side is complete, the SRNC will finalize the process with the RAB assignment response.

12. The Q.931 signalling continues with the *alerting* message from the UE, indicating that the phone is ringing.

13. Once the connection is made on the UE side, i.e. the user answers the phone, the *connect* message is sent, with the network responding with the *connect acknowledge*.

14. At this point, all the connections have been established and the call proceeds, with speech flowing across the bearers.

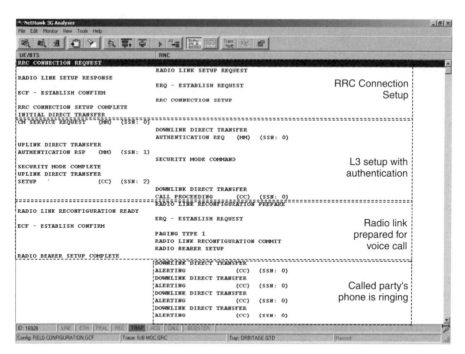

Figure 6.135   Sample Iub trace of part of a mobile originated call. Reproduced by permission of NetHawk Oyj

Figure 6.135, shows a sample trace file of part of a mobile originated call, captured on the Iub interface. First, there is an RRC connection setup procedure followed by the L3 CM service request and subsequent authentication procedures. Then the radio link is reconfigured for the new bearer, with the transport bearer established by AAL2 signalling ERQ/ECF. Finally in this sample, the alerting message is seen, indicating that the called party's phone is now ringing.

6.22.5   Handover control

During the course of the call, the user is free to move around. One of the key aspects of the WCDMA system is the ability to perform soft handovers, where the new connection is made before the existing connection is dropped. Soft handover also provides the advantage that diversity may be used to provide better quality signals while maintaining minimum power levels. Consider that during the call established in the last phase, the user moves towards an adjacent cell employing the same frequency and performs an intra-RNC soft handover.

1. Once the measurement criteria of the neighbouring cell is fulfilled, the UE sends a *measurement report* to the SRNC. A common criteria may be that the power of the CPICH pilot channel in the cell has exceeded a defined threshold, Figure 6.136.

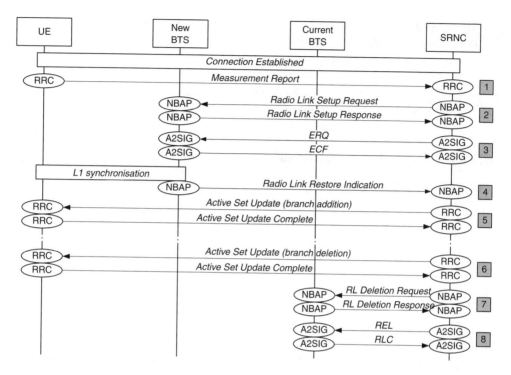

Figure 6.136   Intra-RNC soft handover

2. The SRNC performs an evaluation of the report and decides that the new cell should be added to the active set. It then initiates a *radio link setup procedure* with the new BTS. For a softer handover, this procedure passes the required parameters to the same BTS, but for the new cell.

3. An AAL2 bearer is established across the Iub interface for this new branch as described previously in this scenario. If this was a softer handover procedure, there would be no requirement for a new AAL2 bearer to be established, since the signals are combined at a single BTS so only one transport bearer is required.

4. Once the new BTS has achieved *L1 synchronization* with the UE, it sends a *radio link restore indication* message to the SRNC.

5. The SRNC informs the UE that the cell it was monitoring should now be placed in the active set with an *active set update* message, and the UE replies with an *active set update complete*. The UE is now in soft handover, and is using two active radio connections.

6. Sometime later as a result of another measurement report, the SRNC evaluates that the initial branch is now no longer contributing anything and should be removed from the active set. Once again, an *active set update* message is sent to inform the UE to remove the first link from the active set.

7. The RNC then requests the BTS to delete the radio link, *radio link deletion request/radio link deletion response*.

8. The RNC releases the AAL2 connection, *REL/RLC*.

If this were an inter-RNC soft handover, the same NBAP commands between the SRNC and the BTS would be seen also through RNSAP on the Iur interface.

6.22.6 Circuit call termination

After some time, the user hangs up the phone.

1. Once the user hangs up, the UE sends the *disconnect* request to the network (Figure 6.137).

2. The network responds with the *release* message, confirmed with the *release complete* message.

3. The core network then releases the Iu connection, *Iu release/Iu release complete*, which has the effect of closing the entire connection to the core, both the radio access bearers and the signalling connection.

4. The AAL2 transport bearer is also released.

5. The UTRAN side now deletes the RRC connection, the radio link and the AAL2 transport bearer, as previously described for the location update.

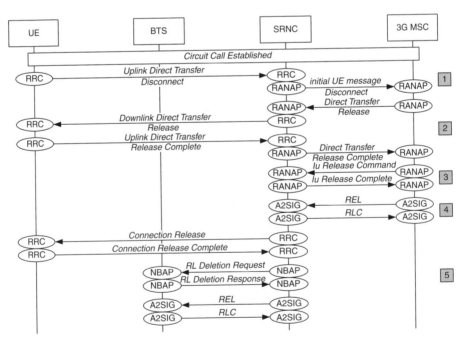

Figure 6.137 Circuit call release

6.22.7   Packet core connection

After the call, the user decides to check his email and must establish a packet core connection. Once again, the higher-layer signalling is the same as for GPRS, as described in Chapter 4. Another observation that should be made is that the mechanisms for establishing radio bearers are much the same as for a circuit call; only the QoS profile of the connection that is established differs.

1. Once again, the signalling connection is established (Figure 6.138).
2. The initial message from the UE is the *GPRS attach request*.
3. Once the authentication and security procedures have successfully completed, the SGSN will reply with a *GPRS attach accept*.
4. The *common ID update* is performed.
5. To transfer data, the UE must establish a PDP context. This is done through the *activate PDP context request*. There will also be a *create PDP context/create PDP context response* between the SGSN and the GGSN, during which the UE is allocated an IP address which is then bound to its IMSI.

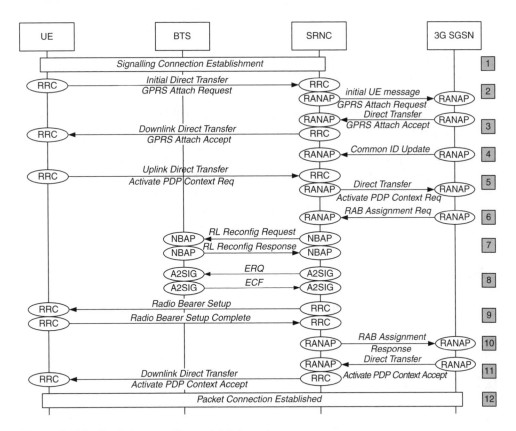

Figure 6.138   Packet connection establishment

6. The SGSN will now use the *RAB assignment request* to set up the bearer. Typically, one RAB is established, and the QoS of it at this point is dictated by the service profile for the user. For example, the user may have subscribed for a background service with a maximum data rate of 64 kbps. Notice that there is no AAL2 signalling over the Iu interface following since this is an IP connection. Instead, the *RAB assignment request* will contain the IP address of the SGSN and the GTP TEID.

7. Once again, the radio link is reconfigured for this data transfer. Notice that since this is data, time is not critical, so a *radio link reconfiguration request/response* is used, instead of the synchronized *radio link reconfiguration prepare* which was used for the circuit call.

8. An AAL2 transport bearer is also established.

9. Now, a new data radio bearer is established between the UE and the SRNC with the selected QoS profile. The QoS profile here can be downgraded by the SRNC from that requested by the SGSN, due perhaps to resource limitations within UTRAN.

10. Once completed, the SRNC replies to the SGSN with the *RAB assignment response*. This will contain the RNC IP address and the GTP TEID to complete the information about the other end of the tunnel.

11. Now the SRNC may inform the UE that the connection is ready with the *activate PDP context accept*.

12. The packet connection is now established, and the user is free to send and receive data through their permitted access points.

The nature of this packet connection is also different from the circuit side. With a phone call, there is a very clearly defined start and finish. However, with packet data, there may be long idle periods where the UE is 'connected' to the network, but is not transferring any data. This is the 'always connected' aspect of GPRS. What it actually means is that the SGSN holds the PDP context of the UE. In the above case, consider that the user has finished reading the email but remains connected to the network. After some time period of no data transfer, the SRNC will most likely request that the Iu connection be released using the *Iu release request* message, stating the reason as user inactivity. Once the UE sends or receives data, the UTRAN connection can be quickly re-established. For the packet core, a *paging* message indicates downlink packet data for the user.

Figure 6.139 shows a trace file captured on the Iu-PS interface, where R1 and R2 represent the RNC and SGSN, respectively. Here the GPRS attach and PDP context activation procedures are shown, followed by IP data transfer. The last exchange is a 'ping' on the interface.

6.23 CDMA2000

The original work for this system started out as TIA TR45.5 and has been further developed by 3GPP2 as an integrated part of the IMT2000 specification suite under

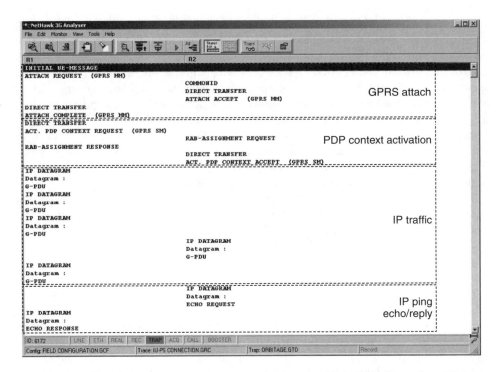

Figure 6.139 Sample Iu-PS trace packet connection establishment. Reproduced by permission of NetHawk Oyj

the IMT-MC (multicarrier). The CDMA2000 system leverages on the 2G IS-95 (commonly known as cdmaOne) system and does not require a major overhaul of the existing system. CDMA2000 is backward compatible with IS-95 and as such allows operators to upgrade the 2G system to 3G in stages. Upgrades are required to the BTS and BSC; as well as this a *packet data server network* (PDSN) is also required. Because of the backward compatibility, existing IS-95 subscribers, will be able to use their mobile devices on the new network and CDMA2000 devices will work on IS-95 networks, and handover between these will be seamless. Services such as those required for voice, data, SMS as well as over-the-air provisioning and activation are also supported by CDMA2000. There are a number of variations to the CDMA2000 system and these will be described in the following sections. CDMA2000 is designed to work with two spreading rates referred to as rates 1 and 3: rate 1 is 1.2288 Mcps and rate 3 is 3.6864 Mcps.

6.23.1 History of cellular in the USA

The original first generation (1G) telecommunication system established on the cellular concept was the advanced mobile phone system (AMPS) which was developed by Bell labs in 1947. This system was based on an analogue rather than a digital technology and was adopted by many countries throughout the world. In the early 1980s the AT&T/Bell

monopoly was broken up under the *modification of final judgement* (MFJ) and the AMPS technology was given to a number of regional Bell operating companies (RBOCs). To introduce competition, the frequency bands available in the US were separated into A and B frequency bands. The A bands were licensed to *nonwire* operators, i.e. new cellular operators who did not have any association with the local fixed-line carriers, and the B band was licensed to the *wireline* companies, which essentially comprised the local RBOCs. These frequencies were licensed for operation with both analogue and digital systems. In the early 1990s the US government auctioned six new nationwide licences, which were referred to as the Personal Communication Service (PCS) A–F band carriers (Figure 6.140) and resided within the 1900 MHz band. The PCS bands were licensed to be used by digital systems only such as TDMA, CDMA and GSM.

It can be seen from this diagram that these frequencies overlap with those that were set aside by the ITU for IMT2000 (1920–1980 MHz and 2110–2170 MHz). The spectral deployment is primarily within the cellular and PCS bands of North America: 824–849 MHz in the reverse link, 869–894 MHz in the forward link (the same as for South Korea) and 1850–1910 MHz reverse link and 1930–1990 MHz for the forward link. These are almost the same frequencies that are allocated in South Korea and Japan.

There are two digital access systems that have wide-scale availability in the US. These are TDMA and CDMA. Both of these systems work with the ANSI-41 (formerly IS-41) core network, which is also implemented in the AMPS system. However, the market there is no longer so clear cut as GSM is also now making inroads into the US in the 1900 MHz band. As discussed in Chapter 2, digital systems have a number of advantages over analogue systems and thus many of the AMPS networks in the US have been replaced by TDMA, CDMA or GSM systems. It should be noted that whereas GSM/UMTS refer to the *uplink* and *downlink*, in CDMA these are termed *reverse* and *forward* link, respectively.

This situation in the US may be set to change. At the time of writing, the US regulator, the Federal Communications Commission (FCC), has made a radical reform of its spectrum allocation policy, moving to a market system, where spectrum can be refarmed. This has been prompted by the perceived failure of the traditional approach of regulated designation, where some allocated frequencies are severely underutilized, but cannot be used by others. Other related developments are the increase in availability of unlicensed technologies such as WiFi, and the potential of software defined radio (SDR), where the air interface and frequency used is merely an application at the transceiver. This would allow the software to dynamically utilize frequency not currently occupied. The FCC has

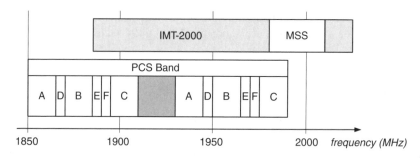

Figure 6.140   US PCS bands

defined interference levels as a real-time gauge of frequency usage, noting that interference rejection is already a key component of digital wireless communication systems.

6.23.2   The TDMA system

The TDMA system is a digital system which was initially introduced to work within the same cellular spectrum as the analogue AMPS system but to give increased efficiency. The first of these was the IS-54 standard, which was a dual-mode system working with the AMPS attributes but allowing user data to be sent over digital traffic channels. The introduction of digital signal processor (DSP) techniques made it three times as efficient as the old AMPS system since three subscribers could share the same 30 kHz frequency band simultaneously. Digital control channels were added in the IS-136 standard and this gave further efficiencies, extending mobile device battery life through the sleep facility and also introducing SMS. In October 1996 a revised version of IS-136 was published for use in the 1900 MHz PCS band.

6.23.3   The CDMA system

The original CDMA system, which is referred to as cdmaOne or IS-95, comprises a 1.25 MHz channel and uses a chip rate of 1.2288 Mcps. As well as voice, IS-95A will support data up to 9.6/14.4 kbps; for packet data transfer, IS-95B requires the addition of an Interworking Unit and can support data in excess of 64 kbps. It can be seen as a 2.5G solution similar to GPRS.

6.23.4   Evolution path

The path to higher speed data via the CDMA2000 route consists of three evolutionary steps, as illustrated in Figure 6.141.

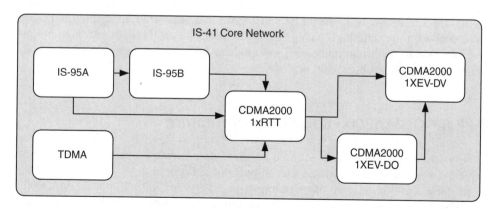

Figure 6.141   CDMA2000 network evolution

6.23.5 CDMA2000 1xRTT

The first step is referred to as 1x or 1xRTT (radio transmission technology), which allows for significantly higher voice capacity as well as enabling data rates up to 153 kbps in a 1.25 MHz carrier, with 307 kbps defined for later releases. Voice capacity is doubled. From the operator's perspective there is little effort required in upgrading from cdmaOne to a 1xRTT system, since much of the existing infrastructure simply requires slight modifications. It is available in both the 850 MHz and 1900 MHz bands.

6.23.6 CDMA2000 1xEV

The evolution of CDMA2000 beyond 1x is referred to as CDMA2000 1xEV. This actually consists of two phases: 1xEV-DO (data only, Phase 1) and 1xEV-DV (data and voice, Phase 2), both utilizing the standard 1.25 MHz carrier. In common with aspects of UMTS, 1xEV uses existing IP protocols extensively throughout the network to allow smooth connectivity to external data networks. 1xEV-DO provides a new data channel, similar in concept to the UMTS downlink shared channel (DSCH), which allows for downlink data rates up to a theoretical 2.4 Mbps peak best effort service, with 153 kbps in the uplink. 1xEV-DO networks are currently deployed extensively by operators in Korea. 1xEV-DV offers data rates of up to 3.09 Mbps and will support real-time applications such as video streaming.

6.23.7 CDMA2000 3xMC

There is also a multi-carrier (MC) mode, which has a carrier bandwidth of 3.75 MHz. In the forward link it actually consists of three consecutive 1.25 MHz channels, each with a chip rate of 1.2288 Mcps. In the reverse link the chip rate is 3.6864 Mcps, which aids in multipath reconstruction. There is also an alternative hybrid configuration whereby the forward link is three consecutive carriers and the reverse link is a single 1x carrier using the normal 1.2288 Mcps. It should be noted that the multi-carrier approach to CDMA is not restricted to 3x, and higher chip rates over wider bandwidths (6x, 9x 12x) are also possible. Due to spectrum limitations, and minimal system advantages, it is possible that this system will never be widely deployed.

6.23.8 CDMA2000 network architecture

This system is also compatible with the ANSI-41 core network and can be implemented using the existing frequency bands that an operator is licensed to use. It is therefore seen as a simple and logical way of migrating to a 3G system. Like IS-95, CDMA2000 is a synchronous system which requires the timing of these networks to be aligned within a high degree. They are also aligned with the IS-95 networks for reasons of system

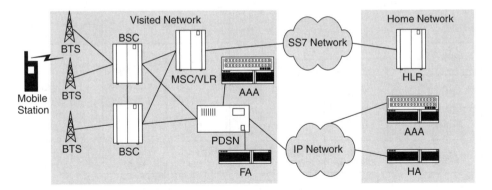

Figure 6.142   CDMA2000 example network

interoperability. This stringent timing is achieved through the use of the global positioning system (GPS). In the forward link, each carrier is identified through a scrambling code. Unlike UMTS, where each carrier has its own unique code, in CDMA2000, a single code is used throughout the system, and each carrier is identified by its offset (phase difference) from the reference code. In a similar fashion to UMTS on the reverse link, the scrambling code (long code) is used to identify a particular mobile device which has a dedicated channel. A simplified view of the CDMA2000 network architecture is illustrated in Figure 6.142.

It can be seen that the radio network consists of CDMA BTSs and BSCs to control them. The BSCs are connected to each other, enabling the soft handover mechanism to function. The link between the BTS and the BSC is ATM, and both AAL2 and AAL5 are used. The interface between the packet control function (PCF) located in the BSC and the PDSN is referred to as the R-P interface and is used to transfer both packet data and signalling messages.

The packet data serving node (PDSN) is used to connect to the external packet switched networks. In CDMA2000 the point-to-point protocol (PPP) is used between the mobile device and the PDSN to transfer both user data and signalling messages. This can be compared to both GPRS and UMTS. In a GPRS system logical link control (LLC) is used between the SGSN and mobile device, and the GTP protocol is implemented between the SGSN and GGSN. In UMTS an RRC connection is established between the RNC and mobile device and GTP tunnels are set up between the RNC and SGSN and between the SGSN and GGSN. A single PPP session is allowed between a mobile device and the PDSN and once established it is maintained while the mobile device is in the transmitting phase but also when the mobile device is in the *dormant state* (equivalent to the idle state in GSM/GPRS). The basic protocol stack for user data transfer is as shown in Figure 6.143. The link access control (LAC) and media access control (MAC) play similar roles to the UMTS RLC and MAC layers.

The functions of the PDSN are as follows:

- establish, maintain and terminate the PPP session with the subscriber;
- direct authentication, authorization and accounting (AAA) for the session to the AAA server;

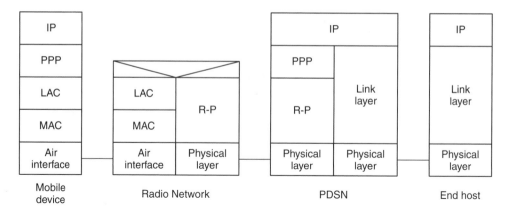

Figure 6.143   CDMA2000 protocol stack

- collect usage data to be relayed to AAA server;
- support IP and mobile IP services;
- maintain the logical link;
- route data to and from the external packet network.

More details of the application of AAA and PPP to CDMA2000 can be found in Chapter 5. The capacity of the PDSN can be measured in throughput or number of PPP sessions that can be served. For a small-scale network this may be 50 000 PPP sessions and for a larger implementation could be 400 000. The interfaces may be 100 Mbps Ethernet toward the external packet network and ATM 155 Mbps OC-3 towards the radio network. Since the PDSN is seen as a carrier class device redundancy and fail safe mechanisms are also required.

6.23.9   Simple IP and mobile IP

Before any transfer of IP datagrams between the mobile device and the PDSN, the PPP datalink must first be established. Once this is accomplished, the specification supports two methods of accessing external networks, simple IP and mobile IP:

- *Simple IP*: here mobile devices requiring a change in PDSN as they roam through the network will have their IP session terminated, since the new PDSN is required to assign a new IP address. The mobile device may be assigned a static IP address or may be assigned a dynamic address via the PDSN.
- *Mobile IP*: when a mobile device that has an established mobile IP session wishes to transfer from one PDSN to another, the mobile device may do this and maintain its originally assigned IP address. This is due to the added flexibility that the mobile IP mechanism introduces, and this is discussed in Chapter 5.

It is requirement that the PDSN supports the following packet data transfers from a single mobile device simultaneously:

- simple IPv4 and simple IPv6
- simple IPv4 and mobile IPv4
- mobile IPv4 and simple IPv6
- mobile IPv4, simple IPv6 and simple IPv4.

The mobile device may or may not support all of the above. IP and its related protocols are discussed in more detail in Chapter 5.

6.23.10   Mobility management

Mobility is achieved through a system of handoffs (handovers). When this is between PCFs which are connected to the same PDSN it is termed a *PCF to PCF handoff* whereas when the PCFs are connected to different PDSNs a *PDSN to PDSN handoff* is required.

- *PCF to PCF handoff*: the connection over the R-P interface changes and as such a new connection between the PDSN and the target RN needs to be established for each of the packet service instances. This type of handoff may occur when the mobile device is in either the active or dormant phase, and although the R-P connection changes, the mobile device will maintain the PPP connection and the IP address(es). Allowing the mobile device to maintain the PPP session even in the dormant state ensures that usage of the airlink is kept to a minimum, thus reducing signalling etc.
- *PDSN to PDSN handoff*: if the mobile device has a mobile IP session activated during the handoff then the IP address will be maintained even though the PDSN changes, otherwise the connection will be released. To do this the mobile device needs to re-register with its home agent (HA).

6.24   TIME DIVISION-SYNCHRONOUS CDMA (TD-SCDMA)

This system was jointly developed by the Chinese Academy of Telecommunications Technology (CATT) and Siemens, which have spent over $200 million on the standardization process. It was proposed by the China Wireless Telecommunications Standards (CWTS) group to the ITU in 1999 and was given its imprimatur as an official 3G standard under the ITU-TC grouping, along with UMTS-TDD. China has reserved more spectrum for this time division duplex (TDD) variant of the 3G system than to the FDD versions. In fact, whereas 60 MHz has been reserved for CDMA2000 and UMTS, 155 MHz has been reserved for TD-SCDMA.

It has also been adopted as part of the UMTS 3G R4 specification as UMTS-TDD LCR (low chip rate, TR25.834). In common with UMTS-TDD, TD-SCDMA does not require separate uplink and downlink bands (i.e. paired spectrum) and offers speeds from as low as 1.2 kbps up to 2 Mbps. Uplink and downlink traffic can be transferred in the same frame but in different time slots, and there can be up to 16 codes allocated per slot, enabling a number of simultaneous users. For asymmetric traffic such as web browsing more time slots can be devoted to downlink transfer than in the uplink. This allocation of time slots is dynamic and if a symmetric allocation is required, which is usually the case for a telephone call, then this will also be allocated the required resources. The minimum frequency band required for this system is 1.6 MHz and the chip rate is 1.28 Mcps. TD-SCDMA does not have a soft handover mechanism but has a system similar to GSM where the mobile devices are tightly synchronized to the network, and it is from here that the term 'synchronous' is derived. It is designed to work with a GSM core network in a similar way to WCDMA and can also use the UTRAN signalling stack when it is deployed as a complementary technology. The frame is 5 ms rather than 10 ms in WCDMA and is split into seven slots.

Although it is an official 3G standard, it is debatable whether TD-SCDMA will be deployed outside of China. However, the large population of China does indeed form a sufficient market to sustain the system.

6.25 SUMMARY

This chapter describes in considerable detail the architecture and operation of the UMTS network, with particular emphasis on the protocols and signalling procedures. This are enhanced by extensive use of trace captures to illustrate their operation. The UMTS logical, transport and physical channels are described, as are the PDCP, RLC and MAC layers which support them. Each of the control protocols – RRC, NBAP, RNSAP and RANAP – are explained, with reference to the key procedures for establishment, maintenance and release of radio access bearers. To assist in understanding the interoperation of all these protocols, a typical user scenario is discussed at length, where the user receives a phone call, and then proceeds to check their email. This is again augmented by reference to live network traces. Finally, there is a brief examination of the other CDMA-based IMT2000 technologies: CDMA2000 and TD-SCDMA.

FURTHER READING

O. Sallent, J. Perez-Romero, R. Agusti *et al.* (2003) 'Provisioning multimedia wireless networks for better QoS: RRM strategies for 3G W-CDMA.' *IEEE Communications Magazine* **41**(2), 100–107.
3GPP TS22.105: Services and service capabilities.
3GPP TS23.060: General Packet Radio Service (GPRS) Service description; Stage 2.
3GPP TS23.101: General UMTS Architecture.
3GPP TS23.110: UMTS Access Stratum Services and Functions.
3GPP TS23.121: Architecture Requirements for release 99.

3GPP TS23.207: End to end quality of service concept and architecture.

3GPP TS23.821: Architecture Principles for Release 2000.

3GPP TS24.008: Mobile radio interface Layer 3 specification; Core network protocols; Stage 3.

3GPP TS25.101: UE Radio transmission and reception (FDD).

3GPP TS25.133: Requirements for support of radio resource management (FDD).

3GPP TS25.141: Base station conformance testing (FDD).

3GPP TS25.201: Physical layer – general description.

3GPP TS25.211: Physical channels and mapping of transport channels onto physical channels (FDD).

3GPP TS25.212: Multiplexing and channel coding (FDD).

3GPP TS25.213: Spreading and modulation (FDD).

3GPP TS25.214: Physical layer procedures (FDD).

3GPP TS25.215: Physical layer; Measurements (FDD).

3GPP TS25.301: Radio Interface Protocol Architecture.

3GPP TS25.303: Interlayer procedures in Connected Mode.

3GPP TS25.321: Medium Access Control (MAC) protocol specification.

3GPP TS25.322: Radio Link Control (RLC) protocol specification.

3GPP TS25.323: Packet Data Convergence Protocol (PDCP) specification.

3GPP TS25.324: Broadcast/Multicast Control (BMC).

3GPP TS25.331: Radio Resource Control (RRC) protocol specification.

3GPP TS25.401: UTRAN Overall Description.

3GPP TS25.402: Synchronisation in UTRAN Stage 2.

3GPP TS25.410: UTRAN Iu Interface: General Aspects and Principles.

3GPP TS25.411: UTRAN Iu interface layer 1.

3GPP TS25.412: UTRAN Iu interface signalling transport.

3GPP TS25.413: UTRAN Iu interface RANAP signalling.

3GPP TS25.414: UTRAN Iu interface data transport & transport signalling.

3GPP TS25.415: UTRAN Iu interface user plane protocols.

3GPP TS25.419: UTRAN Iu-BC interface: Service Area Broadcast Protocol (SABP).

3GPP TS25.420: UTRAN Iur Interface: General Aspects and Principles.

3GPP TS25.421: UTRAN Iur interface Layer 1.

3GPP TS25.422: UTRAN Iur interface signalling transport.

3GPP TS25.423: UTRAN Iur interface RNSAP signalling.

3GPP TS25.424: UTRAN Iur interface data transport & transport signalling for CCH data streams.

3GPP TS25.425: UTRAN Iur interface user plane protocols for CCH data streams.

3GPP TS25.426: UTRAN Iur and Iub interface data transport & transport signalling for DCH data streams.

3GPP TS25.427: UTRAN Iur and Iub interface user plane protocols for DCH data streams.

3GPP TS25.430: UTRAN Iub Interface: General Aspects and Principles.

3GPP TS25.431: UTRAN Iub interface Layer 1.

3GPP TS25.432: UTRAN Iub interface: signalling transport.

3GPP TS25.433: UTRAN Iub interface NBAP signalling.

3GPP TS25.434: UTRAN Iub interface data transport & transport signalling for CCH data streams.

3GPP TS25.435: UTRAN Iub interface user plane protocols for CCH data streams.

3GPP TS25.853: Delay budget within the access stratum.

3GPP TS25.855: High Speed Downlink Packet Access (HSDPA); Overall UTRAN description.

3GPP TS25.931: UTRAN Functions, examples on signalling procedures.

3GPP TS26.071: AMR speech Codec; General description.

3GPP TS33.102: 3G security; Security architecture.

3GPP TS33.103: 3G security; Integration guidelines.

3GPP TS33.105: Cryptographic Algorithm requirements.

3GPP TS33.120: Security Objectives and Principles.

3GPP2 C.S0001-0: Introduction to cdma2000 Spread Spectrum Systems, Release 0.

A list of the current versions of the specifications can be found at http://www.3gpp.org/specs/web-table_specs-with-titles-and-latest-versions.htm, and the 3GPP ftp site for the individual specification documents is http://www.3gpp.org/ftp/Specs/latest/

7

UMTS Transmission Networks

7.1 INTRODUCTION TO RAN TRANSMISSION

Figure 7.1, shows again the basic structure of the universal mobile telecommunications system (UMTS) network. It is within the radio access network (RAN) that asynchronous transfer mode (ATM) is used to transport traffic, specifically, the Iu, Iub and Iur (not shown, but interconnects radio network controllers; RNCs) interfaces. The Uu interface is the air interface between the 3G terminal and the base station (BTS).

The UMTS core network consists of two domains of operation. The top domain is the circuit switched core network (CS-CN), which is essentially an upgraded version of the existing second-generation global system for mobile communications (GSM) core, dealing principally with voice traffic. The lower domain is the general packet radio service (GPRS) core, which introduces a packet switched backbone, based on the Internet protocol (IP). The GSM core then connects to other circuit switched networks, such as the PSTN, and the IP core connects to an external IP network, either the Internet or a private intranet.

The RAN must be able to connect to both domains, through the Iu interface. ATM is currently the only technology that can effectively connect to both, and the mechanisms are well established in the industry.

To link to the GSM core, the Iu interface is connected through an internetworking unit (IWU). This unit performs two functions: first, it interfaces the wideband switching and signalling of the ATM network to the 64 kbps time division multiplexing (TDM) circuits of the GSM 'A' interface; second, it performs voice transcoding functions between the two networks.

Convergence Technologies for 3G Networks: IP, UMTS, EGPRS and ATM   J. Bannister, P. Mather and S. Coope
© 2004 John Wiley & Sons, Ltd   ISBN: 0-470-86091-X

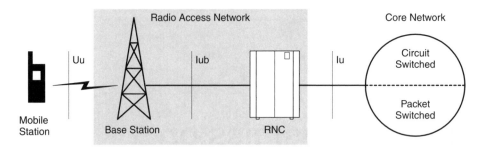

Figure 7.1   UMTS network structure

In addition, there are three major differences between the GSM and 3G cellular networks.

1. GSM is designed around a TDM structure, principally the 64 kbps connection. This is not the case with 3G, where wideband services up to a maximum of 2 Mbps are being offered.
2. Since GSM is intended for voice, all traffic is symmetric, that is, utilizing the same resources in the uplink and downlink. However, 3G applications such as video streaming or web browsing are typically asymmetric in nature and this should be reflected in the transport technology to maximize resource usage.
3. Voice traffic is always allocated a constant, small amount of resources in the GSM network, the *time slot*. However, both video and data traffic require support of variable bit rates, which can be provisioned for in ATM.

Another more practical consideration is that, as will be seen, ATM is independent of the underlying media or protocol, but generally is seen being transported over another technology, such as the synchronous digital hierarchy (SDH) or E1/T1. Since a cellular operator considering an upgrade to 3G will already have a network infrastructure, ATM provides the flexibility of working with that existing infrastructure.

7.2   INTRODUCTION TO ATM

The demands placed on modern networks have increased dramatically in recent years in terms of types of service and speed of operation. The following are some of the requirements:

- handling of different types of traffic on the same network (voice, video, data);
- provision of economically priced access to users;
- a reliable and flexible communications link.

ATM is now a widely used technology, which may best address all of these requirements. The different types of traffic pose vastly differing demands on a network (Table 7.1).

Table 7.1 Network demands of different traffic types

	Voice	Video		Data
		Real-time	Non-real-time	
Bandwidth	Small, constant	Variable	Variable	Variable
Error tolerance	High	Low	Low	Very low
Symmetry	Symmetric	Asymmetric	Asymmetric	Asymmetric
Example	Phone call	Video phone	Video on demand	Email

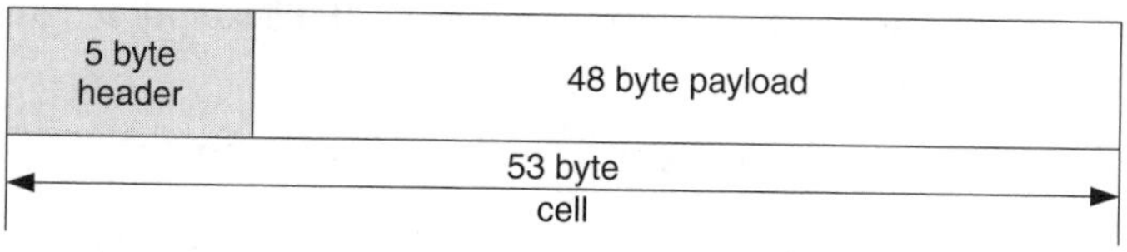

Figure 7.2 The ATM cell structure

Like UMTS, ATM is designed to support voice, video and data applications. To interface the wideband code division multiple access (WCDMA) air interface to this in the fixed-line network, ATM is the choice as the most appropriate technology to provide the backbone.

ATM is similar in many respects to Frame Relay. Just like Frame Relay, it assumes that the medium of transfer is of high quality and therefore does not protect the data from error. However, Frame Relay frames have a variable length, which introduces a variable delay. Hence, it is not well suited to sending voice and video. On the other hand, ATM uses a small, fixed packet size, called a *cell* – note that ATM is also referred to as 'cell relay'. An ATM cell is 53 bytes in size, consisting of a 5-byte header and 48 bytes of user data, as shown in Figure 7.2. Although ATM does not protect the data, it does include error checking in the form of a cyclic redundancy check (CRC), to look for single bit errors in the header, since even for a very low error medium such as fibre, most errors that do occur have been determined to be single bit. The data can be anything: voice or video packets, for example. But also, the data can be from some other protocol, split up into cells. LAN emulation, where Ethernet is sent over ATM, is a popular choice. In this case, the Ethernet frame can be up to 1518 bytes long and will be broken up into a number of cells for transportation. The different types of payload need to be handled differently and this is done by an ATM adaptation layer (AAL). The cells are transferred over a virtual circuit, which is prebuilt. This virtual circuit can be either switched, where it is built at connection time and torn down once finished, or permanent, where it is specified and built at subscription time.

The rationale of using a fixed size cell is that for real-time traffic, variation in delay is the most damaging. Therefore if the cells are fixed in length, this removes irregularities in the time cells are transmitted. However, the network must also keep the delay variation to a minimum, and this is guaranteed through negotiated quality of service (QoS), discussed later. Also, with a fixed length cell, intermediate devices, such as switches, need not examine the header to determine the length, as is the case with IP packets.

Being small, it takes a shorter time to fill the cells and transmit, minimizing the delay in 'packing' a cell, referred to as *packetization delay*. Consider Figure 7.3, where a 16 kbps

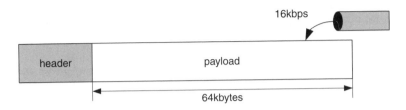

Figure 7.3 Packetization delay

voice channel is feeding into a large packet, such as an IP packet, of length 64 kbytes. It takes 32 seconds to fill this packet:

$$\text{Fill time} = (64 \times 10^3 \times 8)/16 \times 10^3 = 32 \text{ s}$$

This is an extreme example, since in practice IP packets are generally of the order of 1500 bytes, but is exaggerated to illustrate a point. Even at 1500 bytes, the packetization delay will be 750 ms. However, in comparison, an ATM cell will only take 24 ms to fill with voice data:

$$\text{Fill time} = (48 \times 8)/16 \times 10^3 = 24 \text{ ms}$$

The drawback is, of course, the relative size of the header when compared to the payload. With an IP packet, the basic header (IP + TCP) without extensions is 40 bytes, which becomes significant if the payload size were to be reduced to provide a minimal delay. ATM does not completely solve this problem, and will add nearly 10% of overhead to the traffic it carries. This is affectionately referred to as 'cell tax'.

ATM is a technology definition that is independent of physical medium; however, most ATM is carried over an optical fibre system, most notably SDH/SONET at 155 Mbps and 622 Mbps. One key advantage of ATM is that it is scalable in that it is easy to multiplex circuits together to provide faster circuits. ATM is termed 'asynchronous', since traffic can arrive at any time, and is not required to align to any framing or boundaries.

7.3 HISTORY AND STANDARDS

ATM refers to both a network technology and a set of standards. The standards are proposed by two standardizing bodies: the International Telecommunications Union – telecommunications standardization sector (ITU-T) and the ATM Forum. The latter organization is made up of many industry players and was established to hasten the rather lengthy standardization processes of the ITU-T and develop practical standards for ATM much quicker, particularly those concerning its application to smaller scale networks. The ATM Forum is more geared to the provision of interoperability standards among manufacturers of ATM products.

Its development essentially came out of the challenges facing the telephone companies to support a broad range of traffic types using multiple network types across a spectrum of circuit- and packet-switching technologies. The best solution seen was to come up with a single new network that would replace the telephone system and the existing networks with an integrated one supporting all types of information transfer, as mentioned above. In addition, the network should be able to offer a range of rates and be scalable up to high data rates. This new public network service conceptually led on from ISDN and hence was called broadband ISDN (B-ISDN), with ATM the underlying technology of B-ISDN. However, ATM is now also widely used as a backbone technology in private networks.

As a technology, ATM has developed from the operating principles of X.25 and Frame Relay in that it is based on the virtual circuit concept, where connections are established in advance of any data transfer, either by network management or through the operation of a signalling protocol. These technologies provide a statistically multiplexed system on top of a telephone infrastructure. Statistical multiplexing is more suitable to the transfer of data applications, since their behaviour and usage of bandwidth is unpredictable.

Initially the concept was that the introduction of this network would enable the phone companies to go head-on with the cable TV operators in the provision of such high bandwidth services as video on demand. However, the implementation of such a network is no mean feat. ATM is essentially designed to travel over fibre, but is capable of using category 5 and coaxial cable for short distances. The present telephone network uses fibre in the exchange backbone, but it is copper-based in the local loop. To implement ATM, all this would need to be replaced and upgraded with suitable physical media. In addition, the existing circuit switched equipment would also need to be replaced with ATM technology equipment.

A second major issue was that the telephone companies have many decades of experience with circuit switched technology, and relatively little with packet switched. Since ATM is based on the latter, all this accumulated experience would be thrown out, and a revolutionary shift made to a new technology. The compromise really has been that ATM has been relegated to the backbone, sitting on top of existing backbone telecoms systems (notably SDH/SONET), to provide higher bandwidth switched systems to support high-speed traffic transport, a lot of which is IP traffic. It is currently estimated that 40% of IP traffic travels over ATM. The initial idea of this ubiquitous ATM network has never come to fruition, but the concept is now being implemented through the use of IP.

As mentioned, the size of the ATM cell is small and fixed because one design consideration was to make it suitable for transmission of voice traffic, which is extremely sensitive to timing and delays. There is a tradeoff when considering a small packet. On the one hand, the smaller the cell, the lower the number of voice samples that can be sent in the cell, and hence the shorter the delay between speaking and hearing. On the other hand, as has been noted, the smaller the cell size, the more inefficient the transfer, since more header overhead is introduced. When a standard cell size was being decided by the ITU-T, it had to consider the views of two lobbies. The phone companies in the US wanted a cell size of 64 bytes, since their large area of operation meant that they had already installed much echo cancellation equipment to overcome delays. However, the European phone companies had not, so wanted a 32-byte cell. The compromise decided was 48 bytes.

7.3.1 Virtual circuits and virtual paths

ATM implements a virtual circuit type packet switched network. The basic unit of an ATM system is the virtual circuit, or virtual channel (VC). This is a connection from one

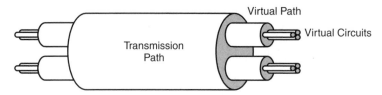

Figure 7.4 Virtual circuits and virtual paths

source to one destination. However, there is a provision in ATM allowing for multicasting. Virtual circuits are considered to carry information in one direction only, but for duplex operation two circuits are established at the same time. These two circuits are addressed as one, but the QoS properties of each, most commonly the data rate, may be different.

Between a source and a destination, a group of virtual circuits can be grouped together into a virtual path (VP). A virtual path provides the advantage that if a re-routing of circuits is required, then a re-routing of the virtual path automatically and transparently re-routes all the virtual circuits which it encapsulates. The concept is outlined in Figure 7.4.

The virtual circuits and paths can be one of three different types: permanent virtual circuits (PVC), soft PVC or switched virtual circuits (SVC). A PVC is established in advance, either by the network administrator or by arrangement with a carrier. Both the end points of the connection and the route through the network are predefined. This presents the problem that intermediate device failure results in failure of the entire PVC, unless the underlying infrastructure (e.g. SDH) can re-route below the ATM layer. It is similar to a leased line, and requires no setup phase. A soft PVC also predefines the end points of the connection; however, the route is not fixed and can be altered to deal with failure. In the early days of ATM, practical implementation of a soft PVC required establishing a network with equipment from a single vendor since the re-routing was proprietary. However, the ATM Forum has defined a routing protocol, the private network-to-network interface (PNNI), to deal with this in an open standard. PNNI is discussed in more detail later. An SVC is established when required, immediately prior to data transfer, and is set up by signalling. If there is failure somewhere in the network, then the SVC is broken and must be re-established.

7.4 THE ATM REFERENCE MODEL

To explore the application of ATM to a UMTS network, an understanding of the structure of the ATM protocol is required. In this section, the ATM reference model is discussed. Like most protocols, ATM can be split into a layered model. Each layer performs particular functions, but is self-contained, communicating with layers above and below through primitives. The point at which layers communicate and exchange primitives is referred to as a service access point (SAP). The block of data exchanged across a SAP, the contents of which are not altered, is known as a service data unit (SDU). This distinguishes it from a protocol data unit (PDU), which includes all the header and/or trailer information that may be added at that layer, i.e. the data plus protocol control information for the layer. A general model of this is shown in Figure 7.5.

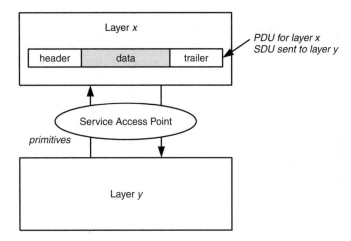

Figure 7.5   Layered model

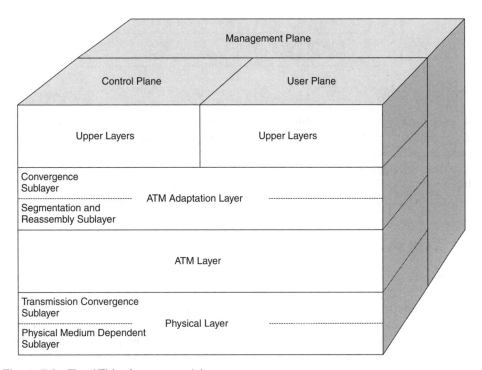

Figure 7.6   The ATM reference model

The reference model for ATM consists of three layers, plus the user layers operating on top of these. The three layers are the physical, ATM and AAL, as shown in Figure 7.6.

The ATM reference model is a three-dimensional one, with control, user and management planes. The role of each can be summarized thus. The user plane handles data transport, flow control, error correction and other user functions. The control plane handles

Layer		Function	
Higher Layers		**Higher Layer Functions**	
AAL	Convergence sublayer (CS)	Service specific (SSCS)	Layer Management
		Common part (CPCS)	
	SAR sublayer	Segmentation & reassembly	
ATM		Generic flow control Cell header generation Cell header extraction Cell VCI/VPI translation	
PHY	Transmission Convergence sublayer (TC)	Cell delineation Transmission frame generation & recovery	
	Physical medium dependent (PMD)	Bit timing physical medium	

Figure 7.7 ATM layer functions

connection management; for example, this is where the signalling protocols operate. The management plane handles resource management and interlayer coordination. The key operation and function of each of the layers is presented in Figure 7.7.

Note that although there are only three layers, both the AAL and PHY layers are further split into sublayers. The key roles of each is now explained in further detail, commencing at the physical medium and working up through the protocol stack.

7.5 THE PHYSICAL LAYER

The physical layer deals with interactions with the physical medium. However, ATM is designed to be independent of transmission medium and flexible in its use of the underlying infrastructure. ATM cells can travel by themselves on the medium, which is utilized for lower data rates on local connections, but more commonly the cells are packaged inside other carrier systems, for example ATM over plesiochronous digital hierarchy (PDH).

To deal with this, the physical layer is further divided into two sublayers. The lower sublayer is the *physical medium dependent* (PMD) sublayer, which interfaces to the physical medium. It is concerned with moving bits on and off the cable or physical layer protocol and handling timing. A different sublayer is used for different media or carriers. The upper sublayer is the transmission convergence (TC) sublayer. It is responsible for passing the cells to the PMD as a bitstream, and also for splitting an incoming bitstream up into cells. These two sublayers are now considered in more detail.

7.5.1 PMD sublayer

ATM specifications allow for travelling over media such as optical fibre, coaxial and twisted-pair cables, at a range of data rates. For example, an ATM to the desktop scheme travels at a rate of 25.6 Mbps over unshielded twisted-pair (UTP) category 3 cable in what is known as 'cell-stream', i.e. the cells are sent as-is and not framed within another protocol. Table 7.2, lists some of the principal carrier systems, the data rate and the medium/media that can be used.

7.5.1.1 Synchronous digital hierarchy (SDH)

Outside of the local area network (LAN), although ATM is independent of the underlying medium, it most commonly is implemented to run over SDH or SONET. SDH is an optical communications standard and was established by the Consultative Committee for International Telegraphy and Telephony (CCITT), now the ITU-T. It came about because so many telecommunications companies were running their own proprietary optical networks that interconnectivity was becoming a serious problem. A second standard, synchronous optical network (SONET), was developed by Bell Labs, USA, just prior to SDH. For the purposes of discussion, their differences are so minor that the two are normally discussed together.

SDH specifies the communications mechanism at the physical layer, i.e. on the fibre. Today, most long-distance telephone traffic runs over SDH, and because of the availability of SDH equipment, it is straightforward for companies to plug into the network. The SDH standard is expected to provide sufficient transport infrastructure for worldwide telecommunications for at least the next two or three decades.

The key aims of SDH may be summarized as follows:

- internetworking between different carriers;
- unification of world digital networks;
- multiplexing together of digital channels;
- operations, administration and maintenance support.

Table 7.2 ATM carrier schemes

Frame format	Data rate (Mbps)	Media
Cell-stream	25.6	UTP-3
Cell-stream	155.52	STP, MM fibre
STS-1	51.84	UTP-3
FDDI	100	MM fibre
STS-3c	155.52	UTP-5, coax pair
OC-3	155.52	SM fibre, MM fibre
STS-12	622.08	SM fibre, MM fibre
E1	2.048	UTP-3, coax pair
T1	1.544	UTP-3

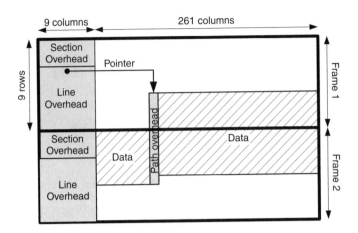

Figure 7.8 SDH frame structure

SDH is a TDM system where the total bandwidth of the fibre is considered to be one channel. It is based on the same ISDN principles, and is made up of the basic telephone network building block, the 64 kbps call. It is a synchronous system and therefore the bits are transmitted at precise intervals, controlled by a master clock. An SDH frame is a block of 2430 bytes of data, sent out every 125 µs, with empty frames being transmitted if there is no data to send. The choice of 125 µs is because all digital telephone systems use a sampling rate of 8000 frames/s. The basic transmission rate is therefore $8 \times 2430 \times 8000 = 155.52$ Mbps (equivalent to 2430 64 kbps channels). This rate is referred to as STM-1 (synchronous transmission module) and all other lines are multiples of this. Shown in Figure 7.8 are two SDH STM-1 frames containing data.

The first nine columns of the frame are overhead. This comprises three rows of section overhead, followed by six rows of line overhead. The first row of the line overhead contains a pointer to the location of the first full cell. This byte is the first in a column of path overhead, with the data area containing for example ATM cells. Note that as illustrated, the data can be placed anywhere within the payload area and in fact, span more than one area, allowing data such as ATM that is asynchronous to be transported efficiently. For example, if an ATM cell arrives even as an empty frame is being constructed, it is inserted in the current frame rather than waiting for the next. To provide higher speeds of operation, the basic SDH building block, the STM-1, is multiplexed, to provide higher-order STM signals, as shown in Figure 7.9.

Table 7.3 summarizes the SDH signals, the SONET equivalent, and the associated data rates. Note that ATM was originally designed to travel over 155 Mbps.

7.5.1.2 UMTS physical layer

In the context of UMTS, Table 7.4 presents the specified formats for the physical carriers that may be used.

For reasons of reuse of existing infrastructure, the Iub interface will commonly be implemented using PDH, perhaps in conjunction with inverse multiplexing for ATM, as

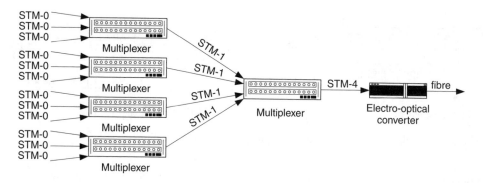

Figure 7.9   SDH signals

Table 7.3   SDH multiplexed signals

SDH	SONET		Data rate
	Electrical	Optical	(Mbps)
STM-0	STS-1	OC-1	51.84
STM-1	STS-3	OC-3	155.52
STM-4	STS-12	OC-12	622.08
STM-16	STS-48	OC-48	2488.32
STM-64	STS-192	OC-192	9953.28

Table 7.4   UMTS physical carriers

PDH		SDH/SONET	
Carrier scheme	Data rate (Mbps)	Carrier scheme	Data rate (Mbps)
E1	2	STM-0	51
T1	1.5	STS-1	51
J1	1.5	STM-1	155
E2	8	STS-3c	155
J2	6.3	STM-4	622
E3	34	STS-12c	622
T3	45		

described in Section 7.5.3. The Iu and Iur interfaces can utilize any of the above, but generally will require at least 155 Mbps to support the volume of traffic resulting from connection to multiple base stations.

When the cells are encapsulated in a protocol for transmission, the cells are placed in the payload of that protocol. An example is shown in Figure 7.10, where ATM cells are encapsulated in the structure of a 2 Mbps PDH line. The ATM cells are packed into time slots 1–15 and 17–31. Time slot 0 is used for framing and time slot 16 is reserved for signalling. Notice that there does not need to be an alignment between the ATM cell boundaries and the PDH medium frame delineation other than aligning to an octet

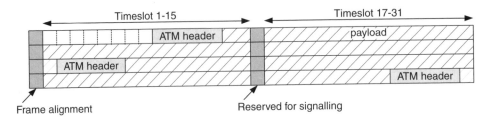

Figure 7.10   Framing of ATM cells in 2 Mbps PDH frame

boundary. When there are no ATM cells to send, idle cells will be inserted to maintain the cell rate. ATM cell delineation is performed by the header error control (HEC) as described shortly.

UTOPIA

In addition to the requirements of physical media for the major UMTS interfaces, ATM equipment needs internal connections, where systems or equipment cards are directly connected to each other. This type of situation arises where interface cards need to be connected to an ATM switching fabric. For this, the ATM Forum has created the UTOPIA standard. UTOPIA stands for Universal Test and Operations Physical Layer Interface for ATM. It is a specification that covers the physical layer of operation and outlines an open, common interface for the data and control connections between physical ATM devices. Control primitives such as timing and synchronization are defined within the specifications, and are controlled by a management entity. UTOPIA currently has four levels defined, providing for different data transfer rates and bus widths. The four levels are shown in Table 7.5.

The ATM cells are clocked across the bus between devices. UTOPIA forms an interface between the ATM layer and the physical layer, as illustrated in Figure 7.11.

As an example, in 8-bit mode, ATM cells are transferred across the interface as shown in Figure 7.12.

7.5.2 Transmission convergence (TC) sublayer

The role of the TC layer can be summarized as shown in Table 7.6.

Table 7.5   UTOPIA levels

UTOPIA level	Maximum data rate	Maximum bus width (bits)
1	155 Mbps	8
2	622 Mbps	16
3	2.4 Gpbs	32
4	10 Gbps	32

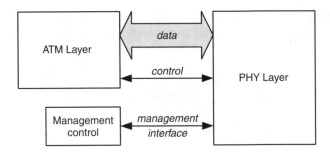

Figure 7.11   UTOPIA model

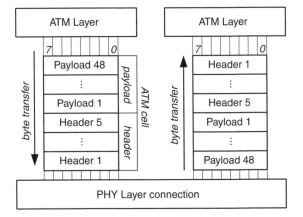

Figure 7.12   UTOPIA cell transfer

Table 7.6   Key TC functions

Transmission	Reception
Convert cells into bitstream	Convert bitstream into cells
Pack cells into transmission frames	Remove cells from transmission frames
Generate HEC	Check HEC
Insert idle cells	Remove idle cells

7.5.2.1   *Transmitting cells*

It has been seen that each cell contains a 5-byte header. One byte of this is a checksum which can detect errors and correct single-bit errors in the rest of the header. The checksum is calculated using cyclic redundancy checking with the polynomial $x^8 + x^2 + x + 1$. Often the header may contain mostly zeros, since low virtual path and channel numbers are allocated first. Therefore, the CRC has a constant 01010101 added to it. Since the CRC only checks for errors in the header and not the payload, it is referred to as the header error control (HEC). There is no provision at this layer for protection of the payload data since ATM assumes that the underlying medium provides reliability.

For an asynchronous medium or carrier, cells can be transmitted once they are ready. However, ATM also needs to support interfacing with physical carriers that are synchronous, such as the PDH schemes, resulting in a requirement for synchronous transmission. Since the cells then have a timing requirement, what should ATM do if there is no cell to send? In this case, the TC sublayer will generate idle cells to maintain the timing scheme. The format of an idle cell is presented in Section 7.6.

7.5.2.2 Receiving cells

For reasons of efficiency, the ATM cell has no boundaries defined and it is the job of the TC sublayer to decide where one cell ends and the next begins from the incoming bitstream. ATM is unusual in this respect as most protocols do define some mechanism for establishing where frames begin and end. To add framing information, such as the preamble used in Ethernet, would add a significant overhead to the cell, and detract from the benefits gained by using a small cell. It can be the case that the framing from the underlying carrier may indicate the frame boundaries, but this is not guaranteed.

ATM overcomes this problem by using the HEC to provide the framing information. Since, with carrier class equipment and reliable connections, the likelihood of a header error is remote (as shown above), the HEC can take on this second role. The header is 5 bytes long so if the cell is valid, the first 40 bits can be examined, with the right 8 bits used to check the remaining 32 bits, as shown in Figure 7.13.

If they match then the cell should be valid and the next 48 bytes will contain the data. This is implemented using a 40-bit shift register. The bits are shifted in and the HEC checked for validity. If it is not valid, all the bits are right-shifted and the next bit brought in until it is valid. Is there a problem? What are the chances of a random HEC being valid? The HEC only contains 8 bits, so therefore there is a probability of 1/256! At the speed of operation of ATM, this probability is too high. However, the probability of two random hits in succession is considerably lower, and for three, lower still.

Therefore, a mechanism must be introduced to check for a number of correct HECs before verifying that the incoming cells are synchronized and framed correctly. The scheme is implemented by a finite state machine. The details may be found in ITU-T I.432.1. Typically, it checks for 4–6 correct headers in a row before assuming it has correct framing.

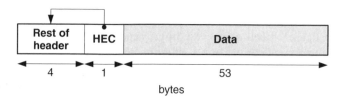

Figure 7.13 Header checking

7.5.3   Inverse multiplexing for ATM (IMA)

IMA is a standard defined for ATM which allows an ATM cell-stream to be inverse multiplexed and carried by multiple physical links, with the original stream retrieved at the other end of the links. It defines a new sublayer at the physical layer, the IMA sublayer. It may be used, for example, to transport an ATM stream over several E1/T1 lines, grouped to provide a suitable rate for the ATM, effectively creating a higher-bandwidth logical link, referred to as an IMA group. The inverse multiplexing process is performed on a cell-by-cell basis in a round-robin scheme. The concept is illustrated in Figure 7.14 below.

Also defined is a new type of operation, administration and maintenance (OAM) cell, which can either be an IMA control protocol (ICP) cell or a filler cell. The ICP cell contains information on the format in which the cells are being sent on the physical link, referred to as an IMA frame. This allows the receiver to adjust for differing delays on the physical links, since the frame format is known and timings can be established from the arrival times of ICP cells. The filler cells are used when there are no ATM cells to send. They are inserted to maintain timings and decouple the cell rate from that of the physical links. The receiver will remove filler cells when reconstructing the stream.

In summary, Figure 7.15, shows the physical layer with the various mechanisms discussed in their relative positions.

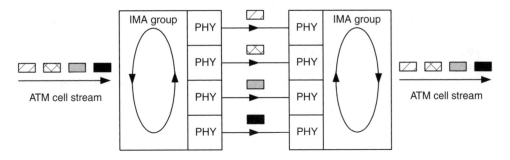

Figure 7.14   Inverse multiplexing for ATM

ATM Layer		
PHY Layer	IMA	UTOPIA
	TC Sublayer	
	PMD Sublayer	

Figure 7.15   The physical layer

7.6 THE ATM LAYER

The ATM layer provides a connection-oriented service and it is at this layer that the virtual circuit and path are established. However, ATM differs from most connection-oriented protocols in one major respect, as no acknowledgement is given for cells received, but it does guarantee that cells arrive in order provided that they travel on the same virtual circuit. Nevertheless, cells may be discarded should the network become congested.

At the ATM layer, two boundaries are considered, as shown in Figure 7.16. The first is that between a host and an ATM network, referred to as the user to network interface (UNI), and the second between two switches on the network, the network to network interface (NNI). The NNI defines the relationship between the two switches, regardless of whether they are on the same or different networks, or indeed lie on a boundary between a public and private ATM network. However, there are different protocols defined to deal with these different NNI relationships.

It has been seen that the ATM cell structure consists of 5 bytes of header and 48 bytes of data. The format of the header is slightly different for the UNI and the NNI, as shown in Figure 7.17.

As can be seen, the difference is in the generic flow control (GFC) field which is unused and overwritten once it reaches the first ATM switch with an extended virtual path identifier (VPI), since one switch can (and generally will) be connected to many hosts. The GFC is intended primarily for proprietary use between a user and a switch, allowing for the first switch to control the flow of information from a host. As an example, a switch may use the GFC to halt a host's cell flow. However, virtually all manufacturers ignore it and for UNI cells, the GFC field is filled with 0000. Note that it has no relevance to flow control within the network.

The VPI specifies the virtual path that should be used. For efficient routing arrangements, ATM switches may only store information about the path and not the individual circuits, meaning that a number of virtual circuits take up only one entry in the translation table of the switch.

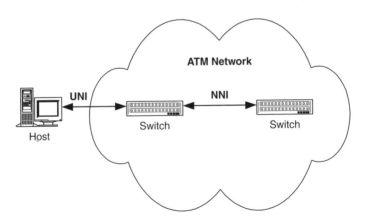

Figure 7.16 ATM interfaces

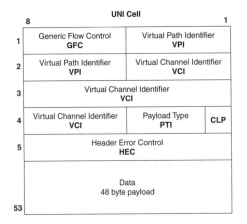
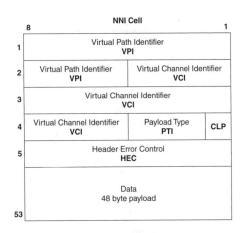

Figure 7.17 UNI and NNI headers

The 16-bit virtual circuit identifier (VCI) identifies a particular virtual circuit within a given virtual path. Since the user can specify an 8-bit VPI and a 16-bit VCI, this allows for the allocation of 256 paths, each containing up to 64 k circuits. In practice, the first 31 circuits are reserved for special purposes such as signalling.

Idle cells, which are inserted to maintain timing when synchronous communication is required, have the format shown in Figure 7.18.

The 3-bit payload type (PTI) field identifies the type of cell being transmitted. The most significant bit (msb) indicates whether it is a data cell (PTI msb = 0) or a management information cell (PTI msb = 1). The middle bit is used as a congestion indicator, when the msb is 0 (i.e. for data cells). If congestion is experienced, this bit is set to 1 by the switch so that the destination host is provided with an indication that there is congestion on the network. The destination can then feed this information back to the source, otherwise the source would be oblivious as to whether it is contributing to network congestion. In contrast to ATM, the TCP/IP protocol provides an acknowledgement of all data transmitted. The source can then use this acknowledgement mechanism to determine if the network is congested, and deal with this congestion using a sliding window implementation.

Finally, the least significant bit (lsb) of a data cell is used to indicate a type 0 or 1 cell. This is used by one of the adaptation layers, AAL5, and is described shortly. A summary of the values for the PTI field is presented in Table 7.7.

The cell loss priority (CLP) bit is used to identify the priority of a cell. Should the network become congested, the switches will first drop lower priority cells, CLP = 1.

The HEC checksum is as explained in the previous section.

header					
0000 0000	0000 0000	0000 0000	0000 0001	HEC	payload
byte 1	byte 2	byte 3	byte 4	byte 5	bytes 1-48

Figure 7.18 Idle cell format

Table 7.7 Payload type values

PTI	Significance
000	User data cell, no congestion, cell type 0
001	User data cell, no congestion, cell type 1
010	User data cell, congestion, cell type 0
011	User data cell, congestion, cell type 1
100	Maintenance between adjacent switches
101	Maintenance between source and destination switches
110	Resource management cell
111	Reserved

The payload section contains the data. However, note that all 48 bytes need not be data, since the higher layers will utilize some of this space to transfer their respective header information.

7.7 THE ATM ADAPTATION LAYER (AAL)

The AAL was conceived to provide a good interface between the different kinds of applications (voice, video and data) and the ATM network. Since the lower layers do not provide control functions, such as error and flow control, which are generally needed by applications, this layer is designed to bridge the gap. To define a suitable interface, the ITU-T considered that applications fell into three broad groupings as in Table 7.8.

This gives eight categories of service, of which four were chosen to support; the other four were considered not to be useful, or make logical sense. The four protocols defined for the services were termed AAL1 to AAL4, as shown in Table 7.9.

However, it was decided that since there was a lot of similarity between AAL3 and AAL4, they would be combined into one protocol, AAL3/4. Unfortunately, the computer

Table 7.8 Service categories

	Service	
1	Real-time	Non-real-time
2	Constant bit rate	Variable bit rate
3	Connection-oriented	Connectionless

Table 7.9 AAL service protocols

Mode	Connection-oriented				Connectionless			
Bit rate	Constant		Variable		Constant		Variable	
Timing	Real-	Non-real-time	Real-	Non-real-time	Real-	Non-real-time	Real-	Non-real-time
Protocol	AAL1		AAL2	AAL3				AAL4

Table 7.10 Revised AAL protocols

AAL protocol	Traffic	Example UMTS application
1	Constant bit rate	E1/T1 voice circuit emulation
2	Real-time variable bit rate traffic	Voice CODECS Packet video, data
3/4	Variable bit rate Best-effort data	Unused
5	Data	IP traffic to core network

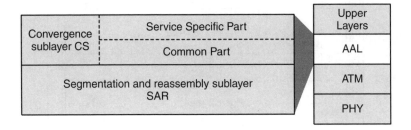

Figure 7.19 AAL sublayers

industry felt that none of the protocols was suited to their data transfer needs and so an extra protocol, AAL5, was implemented. The revised protocols are summarized in Table 7.10, with an introduction to their use within UMTS.

The AAL is split into two sublayers: the convergence sublayer (CS) and the segmentation and reassembly (SAR) sublayer, as shown in Figure 7.19.

The CS provides the interface to the applications for a particular adaptation layer. The lower part of the sublayer is common to all applications while the upper service-specific part is application specific. The CS accepts messages from applications and splits them up into units for transmission. A unit ranges from 44 to 48 bytes payload, and is dependent on the AAL used since some of the AAL protocols will add their own header information at the CS. When the CS receives cells, its role is to take the cells from the SAR sublayer and reconstruct the original message or data stream.

The SAR sublayer takes the cells and passes them to the ATM layer for transmission. Again, depending on the protocol, it may also add its own header information. As its name suggests, this sublayer also reassembles the cells received, and passes them to the CS. The overheads added by these sublayers are specific to the protocols, but generally cover such functions as message framing and error detection. The uses of the different AALs are shown in Figure 7.20.

7.7.1 AAL1

AAL1 is designed primarily for the transmission of traffic that requires a constant bit rate connection-oriented service. It provides the following services:

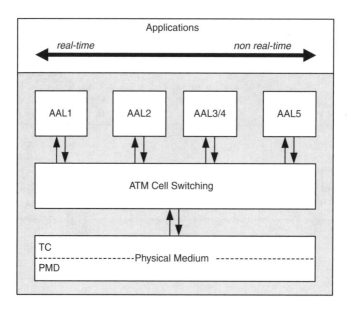

Figure 7.20 ATM adaptation layers

- transport of traffic with a constant data rate and guarantees maintenance of that data rate;
- exchange of timing information between communicating systems;
- if required, the transfer of structure information such as message boundaries;
- indication of errors.

The prime example of this is voice traffic. Since delay is of key concern, adding a lot of error checking would introduce too many delays so error detection and correction schemes are kept to a minimum, and rather mechanisms for detecting missing or misinserted cells are employed. Voice traffic is pretty robust and can handle errors in traffic.

Typically, AAL1 is used for transporting PDH connections, e.g. E1/T1, over an ATM network. The principal application of this in UMTS is for the combined transport of GSM and UMTS traffic on the same ATM backbone, thus simplifying the transport infrastructure. In many cases, this is advantageous, and cost-effective, for co-siting of GSM and UMTS BTSs.

The primary consideration for this protocol is that bits come from the application real-time, at a constant data rate, which must be maintained so that the application at the destination receives the data at the same rate with a minimum amount of delay. To minimize packetization delays, the header overhead added is kept as small as possible, hence the simplicity of the error detection functions. Since timing is crucial, empty cells will be transmitted, even if there is no traffic to send.

7.7.1.1 *Segmentation and reassembly (SAR) sublayer*

The function of the SAR sublayer is to provide a mapping between the CS and the ATM layer. At transmission, it receives a 47-byte block and a sequence number from the CS.

1 bit	3 bits	3 bits	1 bit	47 bytes
CSI	SN	SNP	P	Payload

Figure 7.21   AAL1 PDU format

It then adds a 1-byte header to make a 48-byte block, inserting the sequence number into this header. This header provides sequencing and header error protection. At the receiver end, the SAR accepts a 48-byte block from the ATM layer and extracts the header byte, leaving the remaining 47-byte block of data. It checks for header errors and then presents both the payload and the sequence number to the CS. The format of the SAR PDU is shown in Figure 7.21.

CSI is the convergence sublayer indicator provided by the CS, and is described below. SN is a 3-bit sequence number, used by the CS to check for missing or misinserted cells. It has a cycle of 8 (i.e. 0 to 7). The SNP is the sequence number protection field. It is again a 3-bit number, which is a CRC on the SN and CSI fields. It uses the polynomial $x^3 + x + 1$ to detect double-bit errors and correct single-bit errors. A further error check is provided by the P or parity bit, which uses even parity.

7.7.1.2   Convergence sublayer (CS)

During transmission, the CS takes data from an application and passes 47-byte blocks to the SAR sublayer, and at the receiver, takes 47-byte blocks and passes them to an application. It is at this layer that any variation in cell delay is handled to smooth the cells and maintain the constant bit rate required by the application. This is performed using a buffer. It may also deal with packetization delay, minimizing it by partially filling payloads. Since variation in delay is critical to real-time applications, methods to provide clock recovery are also provided.

The CS will process the sequence number received from the SAR and based on this, it can check for missing or misinserted cells. Should the application be video or high-quality audio, optional forward error correction (FEC) may be performed. However, as mentioned, the most common application is for PDH circuit emulation.

AAL1 can function in one of two modes of operation: structured and unstructured.

Structured data transfer (SDT)

Often, the data stream to be processed by the AAL is unbroken, or unstructured, meaning that there is no requirement for the ATM to provide any information with regard to framing, for example, an E1/T1 line. However, it can be structured, where boundaries between messages need to be preserved. An example of a structured application is the emulation of $n \times 64$ k circuits, such as support for fractional E1 circuits, where only some of the voice channels are being extracted for transport across the ATM network. Here the first byte of the 47-byte SAR payload is used as a pointer to indicate where the next message starts within the payload, measuring from the end of the pointer field. This would be another voice channel. The CSI bit of the SAR header is set to 1 to indicate that

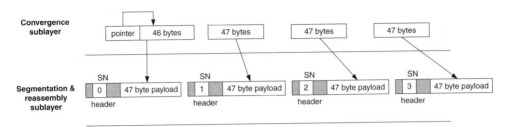

Figure 7.22 AAL1 structured data

this pointer is present. Only cells where the SN is an even number may contain pointers, so the pointer must be a number from 0 to 92 to cover two payloads, 46 bytes for this cell and 47 bytes for the next cell. The CSI in odd cells may be used for clock recovery, as described below for unstructured mode. Enabling data to start at any point in the cell through the use of this pointer means that message streams are not required to align to cell boundaries. The pointer occurs once and only once in every eight cells, since the 3-bit sequence number (SN) has a wraparound of 8. It may therefore be present only in cells with SN 0, 2, 4 and 6, occurring at the first possible chance. Note that the payload must align to an octet boundary. An example is shown in Figure 7.22.

If the block size is 1 byte long, for example, where only a single 64 kbps channel is being extracted, the pointer is not required. However, for larger block sizes it is needed to point to the start of the next full block in the payload. The block size equates to the number of channels being carried. For example, if three channels are used, then it will be 3 octets long. The block size being used is determined during the signalling establishment of the connection.

Unstructured mode

For non-structured cells, the pointer field is not used so the full 47 bytes are available for each cell. In even SN cells, the CSI has the default value of 0. However, for all cells with an odd SN, the CSI bits form a 4-bit number over a cycle of 8 cells, referred to as the residual timestamp (RTS), which encodes the difference between the clock of the sender and that of a common reference. This enables the receiver to synchronize to the sender. An example of the use of unstructured mode is the emulation of a full E1 circuit.

7.7.2 Circuit emulation service (CES)

The circuit emulation service application uses AAL1 to emulate a number of PDH circuits over the ATM network. It specifies mechanisms for transporting circuits with and without channel-associated signalling (CAS; see Section 3.7). As an example, consider that many operators of cellular networks will be looking at co-siting their new UMTS base stations with their existing 2G GSM base stations. If this is the case, it is useful to be able to use the same infrastructure for both 2G and 3G traffic (see Figure 7.23).

One option that is available is to transport the 2G traffic over the ATM connection that is being used to carry the 3G traffic to the RNC, via the 3G base station. To achieve

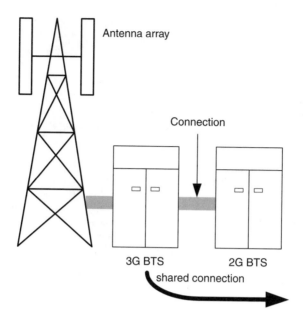

Antenna array

Connection

3G BTS   2G BTS

shared connection

Figure 7.23   2G/3G co-siting

this, the TDM circuit switched traffic of the GSM network, carried on E1/T1, is sent to the UMTS base station where it is carried over AAL1 using the circuit emulation service (CES). Both structured and unstructured mode can be used depending on whether full or fractional E1/T1 lines are being used by the 2G network. At the RNC/2G base station controller (BSC), some device, such as an ATM cross-connect, is used to separate the different traffic streams, and then the circuit emulation process is reversed, with the original E1/T1 stream being retrieved from AAL1 and forwarded to the 2G BSC. Note that the UMTS base station traffic is carried over the same ATM connection, but uses AAL2, which is discussed in the next section. An example of this is illustrated in Figure 7.24. This can be extremely cost-effective if the operator owns or has access to an existing ATM network that it wishes to use for both GSM and UMTS traffic. Consider that the links from base station to BSC/RNC are where the bulk of the infrastructure investment costs lie, since they are geographically separated, usually by a considerable distance. In the core network, much of the equipment may be located in the same building or even the same room, so cost of interconnection is trivial.

Consider the following use of structured mode. A GSM BTS consists of a three-sectored site with one TRX per sector. Each call requires 16 kbps of transmission, and with eight channels per TRX, this equates to 384 kbps of bandwidth ($8 \times 3 \times 16$ kbps). The GSM BTS is connected to the UMTS BTS using an E1 channel. Of the E1, only six channels are being used (6×64 kbps = 384 kbps). The CES function needs to extract these six channels and pack them into the ATM layer using AAL1. The block size will be six octets, using the first six channels from the E1 frame. This block is shown in Figure 7.25.

The six octets from each frame are extracted and placed in the AAL1 payload with the pointer indicating the start of the first block. Figure 7.26 illustrates this concept.

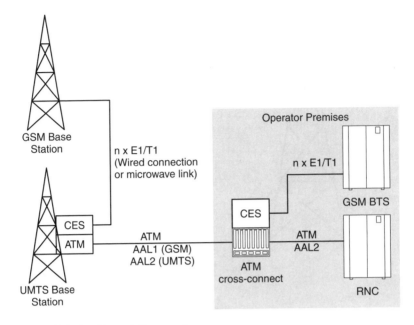

Figure 7.24 AAL1 circuit emulation service

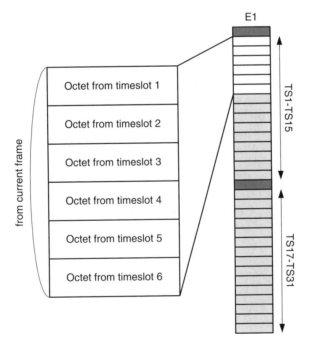

Figure 7.25 $n \times 64$ kbps block

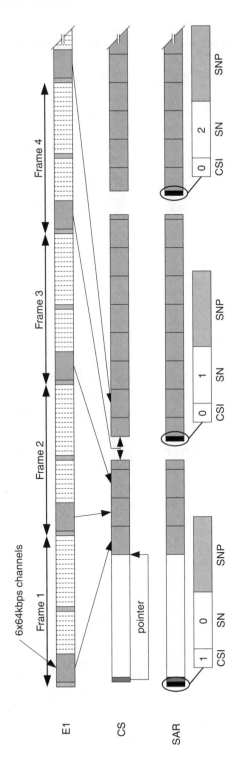

Figure 7.26   E1 encapsulation in ATM using AAL1

Note that the connection signalling establishes the required parameters such as mode (structured/unstructured), clock recovery mechanism, structured mode block size and required cell rate to support the constant bit rate.

For an $n \times 64$ kbps structured service, the cell rate is derived from the bit rate according to the following formula:

$$\text{Cell rate} = \text{Round-up} \ ((8000 \times N)/46.875)$$

where 8000 is the cycle time for the frame, N is the block size, and the 46.875 indicates the payload (i.e. 47 bytes but with 1 byte in 8 lost to the pointer).

In the E1 example, the cell rate would therefore be:

$$\text{Cell rate} = ((8000 \times 6)/46.875) = 1024 \text{ cells/s}$$

Limitations to AAL1

Even though AAL1 is designed for transport of real-time traffic, there are a number of limitations to its use in the context of meeting the requirements of real-time traffic in a modern network:

1. The AAL supports only one user over a virtual circuit. This requires a separate virtual circuit for each user call, necessitating a large amount of signalling or circuit setup.
2. Since cells are sent even if there is no traffic, bandwidth is wasted. If an operator is paying for bandwidth, this can be of great significance.
3. The AAL is designed for 64 k or $n \times 64$ k voice channels. This is not particularly suitable for the advanced CODECs used in cellular communications.
4. Currently, there is no mechanism for supporting these advanced CODECs which provide for compressed voice or voice with silence suppression, etc.

These limit the uses of AAL1 in the 3G network to those described above and render it unsuitable for the transport of 3G real-time user data.

7.7.3 AAL2

The AAL2 protocol was the last to be standardized, and the standard remained undefined for a number of years. It finally saw the light of day in late 1997 after a close working relationship between the ITU-T and the ATM Forum's working group on voice telephony over ATM (VToA). The standard has principal applications in trunking of narrowband services, such as compressed audio with silence suppression. In UMTS networks, AAL2 is the main transport of user data. The AAL2 adaptation layer is used to transport user traffic between the circuit switched core and RANs. Figure 7.27, shows the protocol stack for this traffic across both the Iub and Iu interfaces. Along the Iu interface, AAL2 carries only circuit switched traffic. However, along the Iub interface between the base station

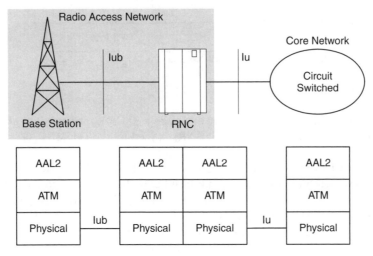

Figure 7.27   UMTS circuit switched user data transport

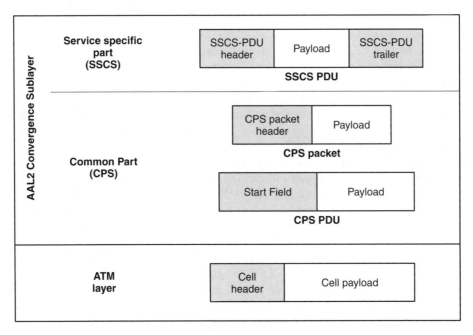

Figure 7.28   AAL2 structure

and the RNC, *all* traffic, voice, video and data, is carried by AAL2 and then segregated to/from the circuit- and packet switched cores at the RNC.

Unlike the other AALs, AAL2 has no SAR sublayer, but rather introduces a number of sublayers at the CS. The structure of AAL2 is shown in Figure 7.28.

The service-specific part of the CS (SSCS) sublayer is not part of the AAL definition, but rather may be specified by the application above the AAL. There can be multiple

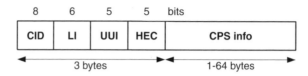

Figure 7.29 CPS packet format

SSCS layers defined, and indeed there need not be a SSCS layer present at all. However, the most commonly used SSCS is the SSSAR, defined by the ITU-T (I.366.1), described in Section 7.7.4.

Recall from Chapter 6 that at the BTS, transport blocks (TBs) will be encapsulated inside the frame protocol (FP) for the interface before being passed to the AAL2 layer.

The common part of the convergence sublayer (CPS) has two components, a CPS packet and a CPS PDU. The SSCS PDU is dependent on the particular service using AAL2 and specifics are defined in the standard for that service. The role of the CPS packet is to allow and identify a number of bidirectional AAL circuits, multiplexed over a *single* virtual circuit. This minimizes packetization delay, critical for voice and video applications, to reduce problems associated with echoing. The format of the CPS packet is shown in Figure 7.29.

The ability to multiplex several channels over a single virtual circuit is an extremely useful feature of AAL2, particularly in the context of 3G. The 8-bit channel ID (CID) field is used to identify the different AAL channels. There are up to 248 AAL channels allowed over a single VC, with CID 0–7 used for management functions or reserved for future uses. These are defined in Table 7.11.

Since a CID represents a bidirectional channel, the same CID is used in both directions. The length indicator (LI) gives the size, in bytes, of the payload, which is variable in length. The maximum length can be 45 bytes. Optionally it can be 64 bytes for ISDN compatibility. The user-to-user interface (UUI) is a means of identifying the particular SSCS layer being used, and to pass information to this layer. Values 0–27 are for different SSCS layers, 30–31 are for layer management and 28–29 are reserved for future use. The UUI field may be null if the application does not define an SSCS layer. Finally, the HEC provides header error control in the form of a CRC check over the rest of the header.

The CPS PDU fills its payload with 48 bytes worth of CPS packets, i.e. 47 plus the CPS packet header (Figure 7.30). Note that since the CPS packet is variable in length, there may be more than one packet in the CPS PDU payload, or indeed a CPS packet may span more than one PDU payload. The 1-byte start field (STF) consists of three components: the offset field (OSF) indicates the start of the next CPS packet header within the payload. This allows packets to span PDUs without wasting payload space, or requiring alignment to the PDU structure. A value of 47 indicates that no CPS packet start is present in the

Table 7.11 AAL2 CID designations

CID	Use
0	Unused
1–7	Management and future use
8–255	CPS user identification

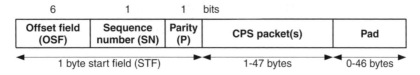

Figure 7.30   CPS PDU format

payload. The sequence number (SN) and parity (P) bits provide some error detection on the header. For instance, when there is no data received, the payload is padded out to fill the 48-byte ATM payload; this is to maintain real-time delivery.

Example of AAL2 application

In the context of UMTS, AAL2 provides a number of advantages in the transport of real-time applications. Considering voice CODECs, two major aspects have changed, and this is particularly evident in cellular systems:

1. CODECs used compress the voice to very small data rates in comparison to the legacy 64 kbps channels of the standard telephone network.
2. CODECs take advantage of silence detection and suppression mechanisms.

In GSM, for example, most networks now use the 12.2 kbps enhanced full rate (EFR) CODEC for voice. This enables a more efficient use of bandwidth across the air and BSS transmission network, where resources are 'expensive' either in quantity or monetary terms. However, GSM is still based on a circuit switched network where time slot resources are allocated for call duration. Therefore, even though the EFR CODEC ceases generating voice samples or moves to a low rate silence descriptor once the subscriber stops talking, the network cannot reuse that bandwidth. Its major advantage is in reduction of power consumption, and hence battery use, in the mobile device. However, AAL2/ATM allows this saving to be realized since the bandwidth during silence can be reused. Recall that a typical voice activity factor is about 50%, whereas standard circuit switched networks allocate resources on a full-duplex basis.

One of the original design features of ATM was that the payload size was chosen to provide a fixed, minimized packetization delay. With current coding schemes sampling voice at much lower data rates, filling a 48-byte payload is now considered too much delay.

In UMTS, the voice CODEC used is the adaptive multirate codec (AMR), which is really a suite of coding schemes. In the course of a call, the network can vary the coding scheme used as often as every 20 ms to optimize the use of resources. This means that if the packet/cell size is fixed, while the CODEC rate changes, a variable delay is introduced. AAL2 resolves this problem by implementing a small size, variable length packet. Each of the different data rates required by the CODEC generates a certain size of SDU for the AAL2 layer, so there can be a direct relationship between the coding scheme used, the delay requirements and the AAL2 packet size. Table 7.12 shows the AMR coding rates and data sizes (3GPP 26.102).

The different source rates in the AMR draw on many existing cellular coding rates. For example, 12.2 kbps is the GSM EFR coding scheme. The AMR SID is a low rate (1.8 kbps) used for transporting background noise.

Table 7.12 AMR CODEC rates

AMR source rate (kbps)	SDU Size (bits)
AMR SID	39
4.75	95
5.15	103
5.9	118
6.7	134
7.4	148
7.95	159
10.2	204
12.2	244

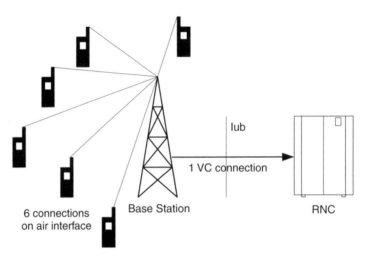

Figure 7.31 Six voice calls over a single virtual circuit

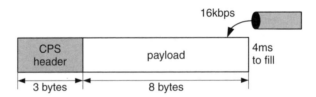

Figure 7.32 AAL2 interaction with voice channel

Consider the following example of six voice channels being transported from the base station to the RNC over one virtual circuit using AAL2, as shown in Figure 7.31.

For simplicity, in this example, each channel is 16 kbps adaptive differential pulse code modulation (ADPCM). Further consider that there is a requirement for a packetization delay of only 4 ms. A payload size of 8 bytes is chosen, since it will take 4 ms to fill the payload at 16 kbps, as illustrated in Figure 7.32. Note that ATM is only used over the transmission network, and not the air interface.

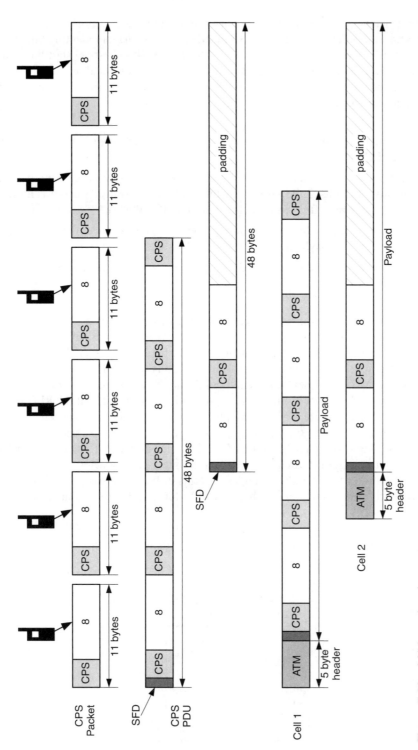

Figure 7.33 AAL2 overheads

The voice channels will pack into the ATM as shown in Figure 7.33. First, each channel fills a CPS packet, adding the 3-byte CPS header, with each channel identified by its own CID. Then, the CPS packets are packed into CPS PDUs with the 1-byte start field added, and padding where necessary. This is then inserted into ATM cells, adding the 5-byte ATM header. The padding may seem excessive, but delay requirements are of highest priority.

In summary, the advantages of using AAL2 connections are:

- AAL2 is particularly suitable for the transport of voice packets produced by advanced speech CODECs.
- AAL2 enables up to 248 channels to be multiplexed on a single virtual circuit. This is extremely advantageous if the virtual circuit is externally owned, since it can then be fully utilized. This may frequently be the case in a 3G network.
- The packetization delay introduced by filling a 48-byte ATM cell can further be reduced by using a small CPS payload.
- The delay can be kept fixed as the CODEC changes by allowing the size of the AAL2 packet to vary.

Typically in a UMTS environment, there will be many users, requiring different resources and data rates. All of these must be carried by AAL2 between the base station and the RNC on the Iub interface. At the user equipment (UE), all applications pass their data to the RLC/MAC layer where the data is formatted into TBs and then is forwarded to the physical layer for coding and radio frame segmentation. At the BTS, the data is brought back to TBs. If the application is sending IP packets, these will be segmented at the RLC layer to fit into an appropriate TB size. For example, a 32 kbps connection may generate a 336-bit TB every 10 ms (320 bits + 16 bit RLC overhead). The total SDU size used at the AAL2 layer will be the TB(s) size plus the overheads of the frame protocol. This protocol includes such overheads as transport formats and relevant air interface parameters.

The CODECs refer to voice data samples in terms of bits, whereas at the AAL2 layer, packet size is defined in multiples of bytes. The padding to byte boundaries is done by the frame protocol.

AAL2 presents no problem in the transport of these different sized TBs. Consider the following example of, again, six users, but this time with six different packet lengths. The user packets are as listed in Table 7.13.

Table 7.13 Example AAL2 user packet length

User	Packet length
User 1	45
User 2	15
User 3	22
User 4	8
User 5	24
User 6	40

As shown in Figure 7.34, the payload from user 1 is too large to fit into one CPS PDU and the last octet is placed in the second CPS PDU after the start of frame delimeter.

AAL2 connections may be established and released using the AAL2 signalling protocol, Q.2630, defined by the ITU-T. This signalling mechanism is outlined in Section 7.13. Note that AAL2 channels inherit the QoS of the virtual circuit in which they are carried and there is no standardized way to provide different QoS to individual CID streams.

7.7.4   Service-specific convergence sublayer (SSCS)

As was seen in Figure 7.28, the CS allows for an SSCS to provide any additional features to an application that are not directly supported in AAL2. Considering again a UMTS network, the Iub interface is required to transport traffic from multiple users, multiplexing them at the AAL2 layer. Although part of that traffic will be generated by small voice CODECS, some of it will consist of data traffic, such as IP packets. The payload for AAL2 is the frame protocol, and its size is dependent on the TB size and the number of TBs that may be sent simultaneously. Particularly for higher data rates, the size and number of TBs generated will create a frame protocol block that is larger than the AAL2 packet size of 45 bytes. Consider that a 384 kbps connection may generate a TB of 3840 bits (480 bytes) every 10 ms. Therefore, what is needed above the AAL2 is a layer that can provide segmentation and reassembly of these larger packets. The ITU-T has defined such a layer (I.366.1), referred to as the service-specific segmentation and reassembly convergence sublayer (SSSAR). It sits on top of the AAL2 CS layer and consists of three sublayers, as shown in Figure 7.35.

At a minimum, the service offered is merely segmentation and reassembly of large user data packets, and it is this function that is utilized in UMTS. This is performed by the SSSAR. The SSSAR will accept a packet of up to 64 kbytes in size (the maximum size of an IPv4 packet) from the upper layer, segment it and reassemble it at the far end. On a connection, there is no opportunity for cells to get out of order; therefore sequencing of the segmented portions is not necessary as this feature is inherited from the lower layers. What is required is a notification that segmentation and reassembly is being used, and an indication of the last segment received. This is performed using the UUI bits of the CPS packet header. Recall that a UUI value of 0–27 indicates the use of a SSCS layer. A UUI value of 27 is used to indicate that there is more data needed to complete the SSSAR SDU, while 0–26 indicate that the final segment has been received. Since there may be other SSCS layers implemented, normally a value of 26 is recommended to indicate the last piece. The format of the SSSAR PDU is shown in Figure 7.36.

It is usual that when segmentation occurs, all segments excluding the final one are the same length as determined by the maximum payload size of the CPS packet, i.e. 45 bytes. Consider a payload of 1200 bytes that needs to be segmented. The segmentation will consist of 26 segments of 45 bytes, with the CPS UUI field set to 27, and one segment of length 30, with the CPS UUI field set to 26.

In addition to the SSSAR, this SSCS also provides two further optional functions: SSTED and SSADT. The SSTED provides a mechanism to detect errors in the payload. It does this by adding a trailer of 8 bytes. The format of the trailer, shown in

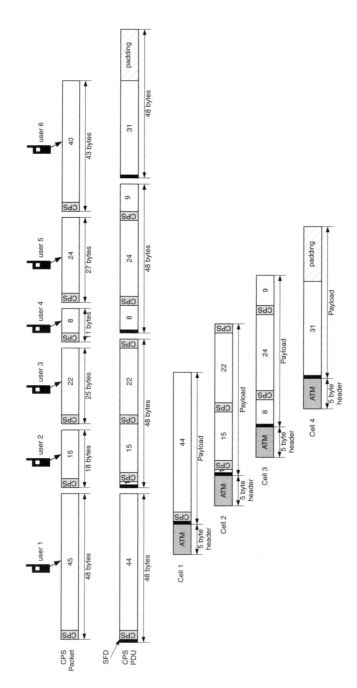

Figure 7.34 AAL2 with unequal payload

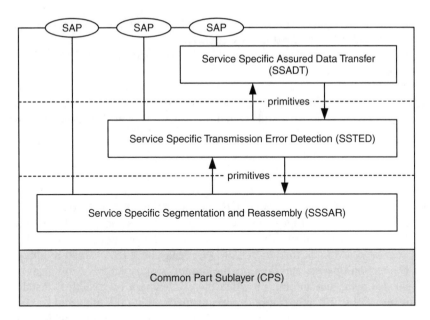

Figure 7.35 SAR service-specific convergence sublayer

Figure 7.36 SSSAR PDU

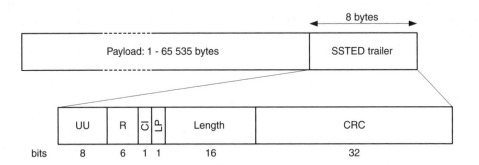

Figure 7.37 SSTED

Figure 7.37, is similar to that of AAL5, providing a 4-byte CRC check over the payload for error detection.

The UU (user to user) field is passed transparently to the user. The R (reserved) field is currently unused and filled with zeros. The CI (congestion indication) and LP (loss priority) are single-bit fields and come from the CI bit of the ATM layer PTI field and the

CLP bit of the ATM layer, respectively. The length field provides the number of octets in the payload, and the CRC field is for error checking.

These features are incorporated to decouple the AAL2 and associated SSCS layers from the underlying ATM network.

The SSSADT provides for assured delivery through acknowledgements and flow control. This mechanism is exactly the same as the service-specific connection-oriented protocol (SSCOP) as defined in the signalling stack. The intricacies of this protocol are discussed in Section 7.13.2.

7.7.5 AAL3/4

AAL3/4 is designed for data transfer and carries packets of up to 64 kbytes in size. It provides detection of corrupted, out-of-sequence and missing packets. An important feature of AAL3/4 is that it allows many sessions to be multiplexed on a single VCI, and then separated at the destination. Often, there may be a fee charged by the provider for each connection opened, and then for the time it is open. In this way, many sessions can be sent down the one circuit, provided there is enough bandwidth to handle all of them. This protocol is different from the previous two as both the sublayers add their own header and trailer to the payload. The CS adds 8 bytes to the payload, which can be up to 64 kbytes in length. The SAR then splits this up into 44-byte pieces to each of which it adds another 4 bytes of checksum and sequencing information.

AAL3/4 inserts a protocol overhead at both the CS and the SAR sublayer. The CS accepts messages from an application, which are in the range of 1–64 kbytes in length. The first task is to pad the message out to a multiple of 4 bytes before adding the header and trailer to facilitate segmentation. The format of the message is shown in Figure 7.38

The message is referred to by the ITU-T and ATM forum as the common part convergence sublayer protocol data unit (CPCS PDU).

The CPI field is the common part identifier, which indicates the message type and the counting units for the size and length fields. Currently there is only one type defined, which is for a byte as a unit, where CPI has a value of 0. The Btag and Etag fields are used to frame the beginning and end of the message. For a message they must be the same value, and are incremented for each new message sent, so that missing cells may be checked for. The BAsize (buffer allocation) field is an indication to the destination of how much buffer space to allocate for the message (2^{16} gives the 64 kbyte length). AL is an alignment field which is unused and filled with 0 s.

Once the CS has finished with adding its header and trailer, it passes the message to the SAR sublayer, which segments it into 44-byte units, adding its own overhead of 2 bytes at each end. The format of the SAR cell is as shown in Figure 7.39.

ST is the segment type and can have four possible values, shown in Table 7.14.

SN is a 4-bit sequence number, for detecting missing or misinserted cells.

1 byte	1 byte	2 bytes		0-3 bytes	1 byte	1 byte	2 bytes
CPI	Btag	BAsize	Payload: 1-64kbytes	Pad	AL	Etag	Length

Figure 7.38 AAL3/4 CS sublayer format

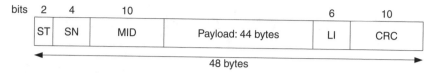

Figure 7.39   AAL3/4 SAR sublayer format

Table 7.14   Segment type values

Segment	Value	Significance
BOM	10	Beginning of message
COM	00	Continuation of message
EOM	01	End of message
SSM	11	Single segment message

Figure 7.40   Segmentation by AAL3/4

The CS may be working with a number of messages at one time, each belonging to a different session. Pieces of this message are passed to the SAR in an arbitrary fashion. The MID field is the multiplexing ID value and is used to keep track of the messages from different sessions. Each chunk of a given message will have the same MID value, allowing for the multiplexed messages to be segregated and reassembled at the destination. Since MID is 10 bits this allows for up to 1024 different sessions on one virtual channel. The length field is the number of bytes in the payload, since it can be less than 44 bytes for EOM and SSM messages, and the CRC is a cyclic redundancy check.

The segmentation of a message is illustrated in Figure 7.40.

It is generally considered that there is a problem with AAL3/4 in that it introduces a significant overhead to data, particularly for short messages, introducing a considerable amount of inefficiency into the transfer of data. The industry consensus was that this protocol was too cumbersome and so AAL5 was developed. AAL3/4 is virtually unused in practice.

7.7.6   AAL5

AAL5 is designed for the transport of large, non-real-time data, such as IP packets. It is the most widely used AAL, and in fact, its largest role is in the transport of IP traffic.

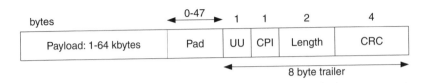

Figure 7.41   AAL5 PDU format

The AAL5 CS sublayer adds only 8 bytes to a payload which, again, can be up to 64 kbytes long. There is no overhead added at the SAR. Like AAL3/4, the application can pass messages of up to 64 kbytes in length. The CS adds a trailer to the message, as shown in Figure 7.41.

The Pad adds somewhere between 0 and 47 bytes to the data such that the entire length of the message, including the trailer, is a multiple of 48 bytes, so for a single-byte transmission, 48 bytes must be sent. The UU field allows for the transmission of one byte of user data by a higher protocol without it being examined by the AAL5 layer. The CPI is a common part indicator, similar to that for AAL3/4. Currently it is unused, and filled with 0s. The length field gives the true length of the data, excluding the padding, and the CRC is a standard 32-bit cyclic redundancy check.

The SAR adds no overhead to the message and merely segments it into 48-byte chunks. It also preserves message boundaries by informing the ATM layer to set the lsb in its PTI field to 1 to indicate the last 48-byte piece of a packet. Recall that this is cell type 1, as discussed in Section 7.6. AAL5 has a major advantage for data transfer over AAL3/4 in that only 8 bytes are added *per message* with no overhead added per cell. Note, however, that the message boundary mechanism is in direct conflict with the whole concept of layering, but substantially reduces the overhead. Due to its efficiency advantages, AAL5 is sometimes known as the *simple efficient adaptation layer* (SEAL).

As addressed later, AAL5 is also used for transport of signalling, management and routing protocols within ATM.

AAL5 is used across the UMTS Iu interface to connect to the GPRS packet switched core for the transport of user traffic. The stack is shown in Figure 7.42. Note that for the GPRS IP backbone, the traffic is carried over UDP, encapsulating the GPRS tunnelling protocol (GTP). Since signalling messages are generally quite large and need to be segmented for transport across ATM, AAL5 is used for the transport of all UTRAN signalling, for example the Node B application part (NBAP).

Although outside the remit of the specifications, a central part of a UMTS network is the provision of a network management system. For UMTS, just as was the case with GSM, network management is vendor specific. However, many manufacturers of UMTS equipment will utilize the ATM network to transport network management messages. AAL5 is used for this purpose.

7.7.7   Summary

Table 7.15, presents a comparison of the four protocols, with a summary of the role of the ATM layers and sublayers.

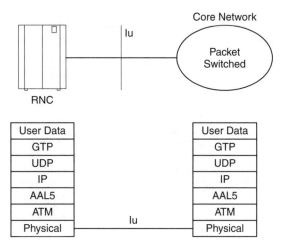

Figure 7.42 UMTS packet switched user data transport

Table 7.15 Comparison of AAL protocols

	Item	AAL1	AAL2	AAL3/4	AAL5
	Multiplexing	No	Yes	Yes	No
	Message framing	None	Pointer	Btag/Etag	Bit in PTI
	Buffer allocation	No	No	Yes	No
CS	Padding	0	0–45/64 Bytes	0–3 Bytes	0–47 Bytes
	Overhead	0	1 Byte[a]	8 Bytes	8 Bytes
	Checksum	None	5 Bits	None	32 Bits
SAR	Payload	46–47 Bytes	48 Bytes	44 Bytes	48 Bytes
	Overhead	1–2 Bytes	0	4 Bytes	0
	Checksum	None	None	10 Bits	None

[a]Note that it adds 1 byte to every 47 bytes of payload, but also adds 3 bytes overhead at the CPS packet sublayer to a variable length packet.

It is seen that AAL1 and AAL2 are designed to take small, real-time SDUs from an application, and, in the case of AAL2, variable length SDUs may be multiplexed together onto a single virtual circuit. In contrast, AAL3/4 and AAL5 are designed to facilitate the segmentation of larger messages in preparation for ATM transport.

7.8 TRAFFIC CLASSES

Thus far, data transport has been examined in terms of cells. The cell headers indicate the virtual path and channel that the cells should use, but do not provide any indication of the quality of service provided by them. This section addresses the issue of provision of QoS in an ATM network.

Applications and services using the network can be broadly broken into two types of service: guaranteed and best-effort. A guaranteed service provides a certain level of service

Table 7.16 ATM traffic classes

Class	Subclass	Description	Suggested layer	Example
Guaranteed	CBR	Constant bit rate	AAL1	Voice
	rt-VBR	Real-time variable bit rate	AAL2	Video conference
	nrt-VBR	Non-real-time variable bit rate	AAL2	Video streaming
Best effort	ABR	Available bit rate	AAL5	Web browsing
	GFR	Guaranteed frame rate	AAL5	Any IP traffic
	UBR	Unspecified bit rate	AAL5	File transfer

in terms of bandwidth, delay and probability of loss, provided the application stays within the rules of transfer. A best-effort service, as the name suggests, will only provide service if bandwidth is available and may drop packets should congestion occur. The two are analogous to travelling on an airline. If a reservation is made, it is a guaranteed service; however, if on standby, the passenger can go if the flight is not full, otherwise they will need to wait for an available flight. The airline makes a *best effort* to carry the passenger. Also like an airline, best-effort services are generally lower in cost than guaranteed.

Within these classifications, the ATM forum has defined traffic subclasses to form a standard of the service that are offered to customers. The classes are summarized in Table 7.16.

These traffic classes are as defined by the ATM Forum. The ITU-T defines the CBR class as deterministic bit rate (DBR) and the VBR classes as statistical bit rate (SBR). The traffic characteristics are the same. However, this section will focus on the definitions from the ATM Forum.

- *Constant bit rate (CBR)*: this is most similar to a telephone circuit and provides the same level of services as a physical medium such as copper or fibre. Traffic has a fixed bit rate with synchronous bit transmission. Typical applications are real-time voice and video. An application may transfer data at a rate lower than this fixed rate, so as to support such services as silence suppression. However, the service guarantees tight constraints on delay.

- *Variable bit rate (VBR)*: intended for services and applications that have variable bit rates, such as compressed video. Previously, video compression provided a constant bit rate, sacrificing quality when compressing. Current compression techniques, such as those offered by the MPEG schemes, maintain a fixed quality and instead provide a variable bit rate. Thus the bandwidth requirements will increase when there is a higher spatial and/or temporal resolution. VBR is further subdivided into two categories: real-time VBR (rt-VBR) and non-real-time (nrt-VBR). Non-real-time VBR is traffic that is still classified as VBR, but may be buffered at the receiver, and thus has much looser delay constraints. An example of this would be provision of a video streaming service.

- *Available bit rate (ABR)*: intended for traffic where the range of bandwidth requirements is known. Applications which are bursty can be supported here. This guarantees a zero loss as long as the application obeys the feedback from the network. In an ABR system, the network will ask the application layer to slow down if there is congestion through a feedback mechanism. Thus the ABR class is distinct from the other classes in that it is inherently a closed-loop system. With such a system, it is possible to define a minimum bandwidth that it will guarantee, and then a peak bandwidth that it will try to provide if and when it is needed. However, it will make no promises with regard to this. For example, the network may guarantee a minimum transfer rate of 2 Mbps, but with a peak of 5 Mbps. A typical application using this service would be web browsing.

- *Guaranteed frame rate (GFR)*: ABR traffic is inherently difficult to provision for as typically the peak data rate is not known or not relevant. Because of this, the ATM Forum defined a new type, GFR, to simplify the process. With GFR, a bandwidth level is defined such that traffic is guaranteed not to fall below this minimum bandwidth. The traffic may receive a performance above this bandwidth, but at a best-effort performance level. This service uses AAL5 and provides frame-level rather than cell-level guarantees. This means that in a situation of congestion, a whole AAL5 frame, containing, for example, an IP packet, will be discarded. This provides a much more suitable and efficient service for transport of other protocols.

- *Unspecified bit rate (UBR)*: provides no feedback or guarantees. The network will take all cells travelling UBR, and transfer if there is any capacity. However, in the case of congestion, they will be the first to be discarded and no information is sent back to the sender. This is suitable for sending IP packets, as they do not promise delivery. Typical applications would be email and file transfer.

Figure 7.43 shows how the different traffic types are placed within the bandwidth of the medium.

In UMTS, there is a defined set of end-to-end QoS classes, which must be met by any transport layer, including ATM. Table 7.17 shows the four service classes, their main characteristics and how they map to ATM.

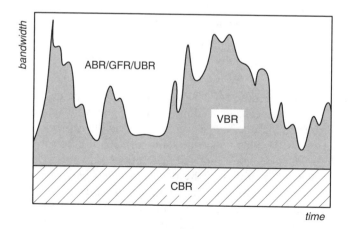

Figure 7.43 Traffic types

Table 7.17 UMTS service classes

Class	Conversational	Streaming	Interactive	Background
Delay	Fixed, small	Variable, small	Variable	Variable
Buffering	No	Yes	Yes	Yes
Symmetry	Symmetric	Asymmetric	Asymmetric	Asymmetric
Guaranteed	Yes	Yes	No	No
ATM class	CBR/rt-VBR	Nrt-VBR	ABR/GFR	ABR/UBR
Adaptation layer	AAL1/AAL2	AAL2	AAL5	AAL5
Example	Phone call	Video on demand	Web browsing	File transfer

7.9 TRAFFIC MANAGEMENT AND QUALITY OF SERVICE

Congestion occurs on networks for several reasons. One of the main ones is that data traffic is often bursty by nature, i.e. flow is not at a uniform rate. If traffic flow was smooth, then problems of congestion would be limited. To manage congestion, a traffic shaping policy is needed to force the cells to be transferred at a more predictable rate. When a user establishes a connection, it agrees to a traffic shaping for the transmission with the network. Provided the user sends data in a way that conforms to the agreed shaping, the network will play its part and promise to deliver all the cells. This reduces congestion on the network.

ATM offers a number of key functions to support management of traffic on the network, including:

- *Connection admission control (CAC)*: the network policy and actions during a connection setup which determine whether the connection should be admitted, rejected or have its parameters renegotiated.
- *Feedback control*: flow control mechanisms for the ABR class to maximize the bandwidth usage and efficiently share the available bandwidth among the users.
- *Usage parameter control (UPC)*: network policing functions to ensure that negotiated QoS parameters and constraints are adhered to.
- *Traffic shaping*: mechanisms implemented to ensure that QoS objectives are met by the network and that traffic conforms to agreed parameters.

Unlike technologies such as Ethernet, which is a best-effort, non-deterministic scheme, in an ATM network, the QoS mechanism allows a user to specify the type and level of service required during the transmission. In essence, the user and the network enter into a contract defining the service. ATM allows that the terms of such a contract may, and probably will, be defined for an asymmetric link and therefore be different in each direction.

The provision of QoS is dealt with by the ATM layer, based on the QoS class that has been defined for the incoming virtual circuit or virtual path. The QoS is established during the connection setup phase and relevant QoS parameters are passed through the signalling messages. These parameters will provide a traffic descriptor and related traffic parameters to characterize the QoS.

Before the connection can be admitted, each switch must not only figure out which output port can meet the service requirements, but also check if it can physically deliver the resources by examination of what has already been allocated. This is the principle of CAC, as illustrated in Figure 7.44.

To provide QoS across the network, each switch must also play a role to offer cells the service class defined for their connection. The key components of any QoS scheme are mechanisms to ensure that data passes through with a certain delay, delay variation (jitter) and loss characteristics. The method of implementation is through a *classify, queue and schedule* (CQS) architecture. Essentially, a number of *queues* corresponding to different service classes are established in the switch, and incoming traffic is *classified* and placed in the appropriate queue. A *scheduler* then takes data out of the queues, processing the queue with the highest priority, or level of service class, first, e.g. a voice call queue. Ideally, to guarantee QoS, a switch would maintain a separate queue for each virtual circuit that has been established. This mechanism is illustrated in Figure 7.45.

In a packet switched network environment, QoS offers only an approximation of the performance, based on the information available at the time of connection establishment. For some traffic types, such as voice, this will remain quite consistent for the duration of the connection, but for others, may vary from the initial situation and may be adversely affected by transient network events.

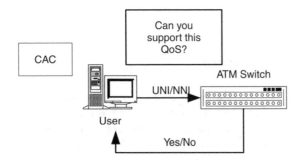

Figure 7.44   Call admission control

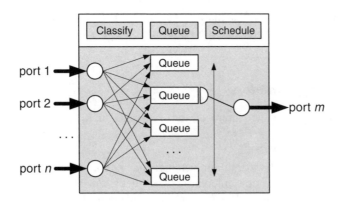

Figure 7.45   CQS architecture

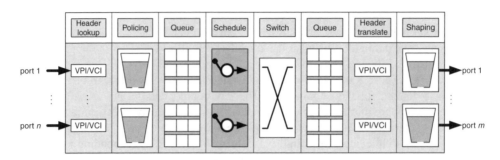

Figure 7.46 ATM switch CQS mechanism

CQS architecture is common in, and central to, ATM switches, but has not made a significant impact yet to the IP protocol due to its best-effort nature. An IP router generally has one queue and processes packets in a first in/first out (FIFO) order. However, CQS concepts will become increasingly important as IP introduces protocols to offer QoS, and a major driver is the emergence of IP switching technologies such as multi-protocol label switching (MPLS).

Enhancing Figure 7.45 to include more of the functionality of the system, Figure 7.46 shows the process cells are taken through as they pass through the switch. Each of these steps is discussed in more detail in the following sections.

7.9.1 Traffic descriptor

The traffic descriptor specifies a set of traffic parameters that classify the ATM service categories that were defined earlier. To specify the QoS, these parameters are negotiated between user and network, at connection establishment. The contract specifies the worst possible value of the parameter and the network then guarantees to at least meet this.

The traffic descriptor parameters split into two general categories. *Traffic parameters* pertain essentially to the speed and delay of transfer of traffic, and are what the user defines for the connection. The *QoS parameters* are tolerance levels that the user needs for the transfer. There are also some *non-negotiable network characteristics* which are not part of the traffic descriptor, but are rather measurements of the error performance of the ATM network.

7.9.1.1 Traffic parameters

These parameters govern the nature of the transmission in terms of speed and delay variation:

- *Peak cell rate (PCR)*: the maximum instantaneous rate at which the user will transmit. The inverse of this is the time interval between cells. For example, if the user defined the interval between cells as 10 μs, then the PCR would be 100 000 cells per second, i.e. 1/T. Figure 7.47 shows the cell interval.

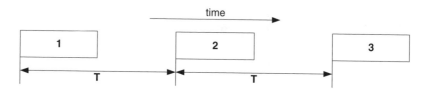

Figure 7.47 Interval between cells

- *Sustained cell rate (SCR)*: the average rate of cells measured over a long period. If, for example, real-time uncompressed video is considered (CBR) then the SCR and PCR will be the same, since there is no variation in rate of transmission. However, for bursty traffic, the SCR may be low but the PCR high.

- *Minimum cell rate (MCR)*: the minimum rate of cell transmission required by the user. For an ABR service, the bandwidth will be between MCR and PCR and will most likely vary a considerable amount between the two. However, it must never fall below the MCR. A value of MCR = 0 would be a UBR service.

- *Cell delay variation tolerance (CDVT)*: the level of variation in cell transmission times that the application can tolerate. It can be considered as an error margin, and defines an acceptable level of deviation in cell transmission times. For example, a user generating traffic at the PCR will need some margin of error as it is unlikely that they will be able to guarantee that each cell will be sent with exactly the same time interval. Conformance of this is measured in terms of the *peak-to-peak cell delay variation* (CDV).

- *Maximum burst size (MBS)*: it is expected that for variable rate applications, they will average at the SCR value, periodically bursting up to the MCR. The MBS defines the maximum number of cells that may be sent at the PCR.

7.9.1.2   QoS parameters

These characteristics are as measured at the destination.

- *Cell loss ratio (CLR)*: the percentage of cells that are not delivered to the destination. This can be due to lost cells resulting from congestion or errors, or cells that arrive much too late.

$$\text{CLR} = \frac{\text{No. of lost cells}}{\text{No. of transmitted cells}}$$

This parameter is only negotiable for guaranteed services.

- *Cell transfer delay (CTD)*: the average time taken for a cell to travel from source to destination. It will incorporate all delays experienced along the network such as queuing delays at switches en route, propagation delays, etc.

- *Cell delay variation (CDV)*: the measure of variance of CDT. If the value of this is high, it means that there will be buffering required for traffic that is sensitive to delay, for example, video.

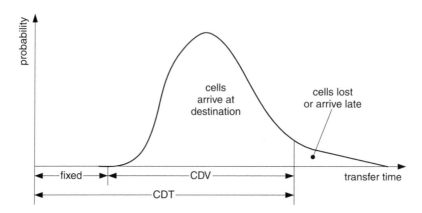

Figure 7.48 Probability of cell arrival times

Figure 7.48 shows a probability distribution for the arrival of cells at the destination. Normally, the value of CDV is picked so that the percentage of lost or late cells is very small. The fixed delay value at the start is made up of static delays such as propagation through the physical media, transmission system delays and processing delays at switches.

7.9.1.3 Non-negotiable network characteristics

In addition to these parameters, there are also some measured characteristics that relate to the reliability of the network, which cannot be negotiated:

- *Cell error ratio (CER)*: proportion of cells that are delivered with either single or multiple bit errors.
- *Severely-errored cell block ratio (SECBR)*: proportion of cells within a given block that have errors.
- *Cell misinsertion rate (CMR)*: number of cells ending up in the wrong place because of an undetected header error.

Table 7.18, presents a comparison of the QoS parameters across each of the traffic classes, stating which are specified and which are not. Note that for CLR, it can be defined separately for cells that have the cell loss priority (CLP) bit set or cleared.

As an example, a voice application will specify that it requires the CBR service class, and then supply the following QoS parameters: PCR, CLR, CTD and CDV.

Recall from the cell frame format that each cell has a 1-bit CLP field in the header, and that cells with CLP = 1 are lower in priority. A harsh policy in the switch would be just to discard a cell immediately a contract breach has been detected, particularly if there is no congestion on the network. A more humane solution is to lower the priority of the cell, by marking cells with CLP = 0 to CLP = 1. In the future, should the network become congested, cells with CLP = 1 will be discarded first. Usually a switch will have

Table 7.18 QoS parameter summary

Attribute	Service class					
	CBR	rt-VBR	nrt-VBR	ABR	GFR	UBR
Traffic parameters						
PCR	Specified	Specified	Specified	Specified	Specified	Specified
SCR	n/a	Specified	Specified	n/a	n/a	n/a
MCR	n/a	n/a	n/a	Specified	Specified	n/a
CDVT	n/a	Specified	Specified	n/a	n/a	n/a
MBS	n/a	Specified	Specified	n/a	Specified	n/a
QoS Parameters						
CLR	Specified	Specified	Specified	Unspecified	Unspecified	Unspecified
CTD	Specified	Specified	Unspecified	Unspecified	Unspecified	Unspecified
CDV	Specified	Specified	Unspecified	Unspecified	Unspecified	Unspecified

a policing policy where minor breaches are punished with priority lowering, and major breaches with cell discard.

Note also from Table 7.18 that some of the service classes specify (or make optional) CLP = 1. This means that these classes, for example, ABR, will define two levels of QoS, one for CLP = 0 and one for CLP = 1. If this is the case, these are termed CLP 0 + 1 flows.

7.10 TRAFFIC SHAPING

Traffic shaping is the way that the network moulds data flows to the terms of the contract and meets the defined QoS parameters. The simplest method of introducing a shaping policy is using the *leaky bucket algorithm* (Figure 7.49). Consider a bucket with a small hole in the bottom. Water enters the bucket at the top, and the rate at which the water enters is variable. However, the entry rate is irrelevant, as water will always exit through the hole at a constant rate. If the bucket is full, any more water entering the bucket is lost over the sides.

In a communication system, a host is connected to the network through a leaky bucket. Thus it is implemented as a queue where there is a regular flow of cells exiting from the queue. If a cell arrives and the queue is full, it is discarded. To overcome this, bursts in cells are smoothed out and losses minimized.

7.10.1 Generic cell rate algorithm (GCRA)

The generic cell rate algorithm (GRCA) is the ATM implementation of the leaky bucket algorithm and is introduced in the switch to shape the traffic and hence meet the terms of the contract between user and network. The algorithm can be considered to operate at two points in the QoS process. At the input side, referring back to Figure 7.46, once cells arrive at the switch, it needs to police them to make sure that they are sticking to their

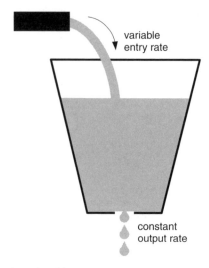

Figure 7.49 The leaky bucket algorithm

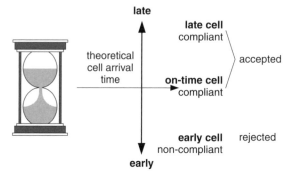

Figure 7.50 Virtual scheduling

defined QoS (policing function). At the output side, emerging cells are shaped to ensure that they meet the required QoS as they leave the switch.

GCRA is referred to as a virtual scheduling algorithm and, simplified, works on the following concept. A clock ticks at each node to set the pace of cell transmission. The clock ticks are the gaps between frames. A cell which arrives on the tick or after the tick is accepted and complies with the contract, but cells that arrive too early are rejected, i.e. fail to comply. The principle is shown in Figure 7.50.

Returning to the QoS parameters just defined, the two of concern to the GRCA are the PCR and the CVDT. As was seen, the gap between cells is $T = 1/PCR$. Figure 7.51 shows the different cases that may occur. In the diagram, E is the CDVT.

In the first case, the cell arrives exactly on time with a gap of T in between cells. The second case shows the cell arriving at a time greater than T, but this is also compliant. There are now two cases of early arrival to consider. The third case shows a cell arriving early, but the arrival time is still within the acceptable margin defined by E. However,

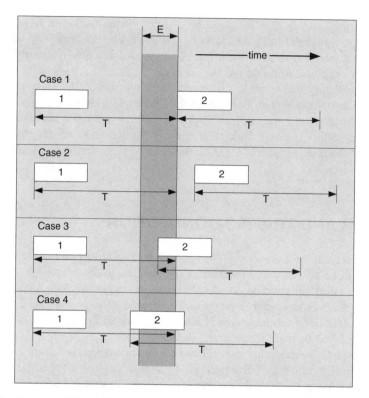

Figure 7.51   Generic cell rate algorithm

now the user is expected to transmit the next cell at a distance of twice the cell gap from the arrival time of the errant cell. This is to dissuade the user from continuing to transmit early, thus increasing the value of PCR. The last case is where the cell arrives early and beyond the error margin, E. The cell is therefore deemed to have breached the contract.

7.10.2   Usage parameter control

In the case of a contract breach, the actions taken by the network fall under the definition of usage parameter control (UPC). This is the policing function in the network used to actually enforce compliance with the traffic contract, as checked using the GCRA. The UPC function will enforce adherence to the speed limit, i.e. the PCR, through the following actions in the case of a breach:

- cell tagging: the CLP bit of the cell is marked to low priority (CLP = 1)
- cell discard.

Cell tagging is an option to offer a lower penalty for first-time offences or minor breaches, as defined within the network policy. Consider that cell discard for a breach when the network is not congested would be a rather harsh policy.

In terms of the leaky bucket analogy, the capacity of the bucket is T, with a gap left for more water, equivalent to E. As each cell arrives, it fills the bucket up to the level of T, and between then and the next cell, the bucket drains at a steady rate until empty. A cell that arrives late will find that the bucket is empty and fill it backup to T. A cell that arrives early will find that there is still some water remaining, but since there is still a gap left at the top of the bucket, E, it can fit into the bucket. However, once it arrives too early, the water left in the bucket will be greater than E, and the water would spill over the edge so the cell is rejected. Note, if it is expected that there will often be a burst of cells, then E should be set to a value such that $E >> T$, so that none of the cells in the burst will be lost.

7.11   ABR AND TRAFFIC CONGESTION

For the ABR service, how does the ATM network try to fulfil its end of the QoS contract when there will be so many users and so many switches to travel through? The network must attempt to deal with two broad classifications of congestion. Long-duration congestion is where there is more traffic entering the network than can be handled, causing a 'traffic jam'. Short-duration congestion is where there are traffic bursts. There are multiple techniques used and they are summarized in Figure 7.52.

At the long-duration congestion control end of the scale, admission control and resource reservation are the principle techniques. In admission control, congestion is prevented before it occurs using the QoS mechanisms, i.e. non-conforming traffic is not permitted to enter the network. Resource reservation is also a preventative measure, where for guaranteed services, enough bandwidth is allocated to handle the traffic. At the short-duration end, buffering is used. There have been many mechanisms proposed for congestion control based on providing feedback to the user for the ABR traffic, which inherently causes most of the congestion problems, because of bursts of traffic. One example is a rate-based scheme, which is an end-to-end approach and was implemented due to its simplicity. A switch need only monitor queues and if they become congested they set the middle bit

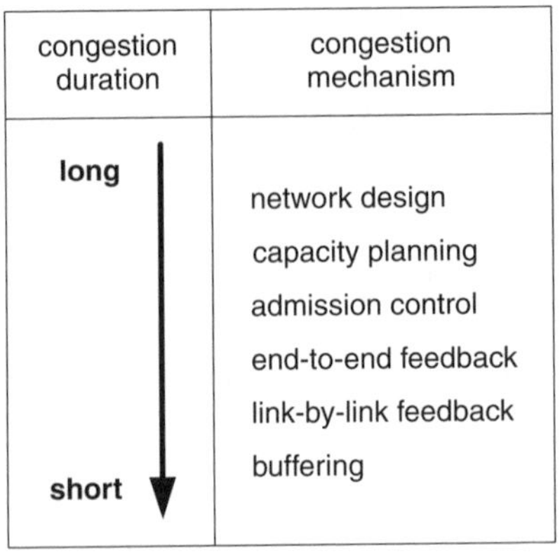

congestion duration	congestion mechanism
long ↓ short	network design capacity planning admission control end-to-end feedback link-by-link feedback buffering

Figure 7.52   Congestion control techniques

in the PTI field of the cell header to 1, referred to as the *explicit forward congestion indicator* (EFCI). At the destination, the congestion is measured for a given period and then a *resource management* (RM) cell is sent back to the sender if there is congestion. The sender, once it receives an RM, thus indicating congestion, will decrease the rate at which it is transmitting. There is a problem here, as what happens if RM cells are lost? The solution adopted was to send an RM cell when no congestion was experienced, informing the sender that the rate of traffic can be increased. The sender will keep decreasing the rate until it receives an RM cell, indicating no congestion. Again, there is a problem here, known as the 'beat down' problem, where virtual circuits whose paths are longer will have less chance of increasing their rate. A solution to this is to make the switch decide whether to set the EFCI bit by considering not only the congestion, but also the current rate of traffic on the virtual circuit. It should be noted that there are more complicated rate-based schemes which feedback more explicit control information in the RM cell.

7.12   NETWORK MANAGEMENT

For the management of an ATM network, a management architecture is needed. Referring again to the ATM reference model, the management plane covers all layers of the protocol architecture. The model for network management is defined by the ATM Forum and is illustrated in Figure 7.53. Here, each switch has its own network management system (NMS).

The model provides for a total network management solution. In the upper layers, management information is sent using the integrated local management interface (ILMI). This provides for configuration and alarm generation of the ATM interfaces, as well as establishing operations between two different ATM devices. Further down at the ATM layer, a layer management protocol is used to provide end-to-end circuit management and checking the circuit throughout the entire network. This management information is carried by operation administration and maintenance (OAM) cells. The model classifies management functions into five areas:

- *M.1*: management of ATM end devices
- *M.2*: management of private ATM networks

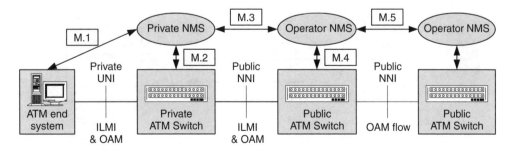

Figure 7.53   Network management model

- *M.3*: management of links between public and private networks
- *M.4*: management of public ATM networks
- *M.5*: management of links between two public networks

7.12.1   Integrated local management interface (ILMI)

ILMI is an implementation which uses the simple network management protocol (SNMP), the most popular management protocol in use today. It was primarily intended for use on TCP/IP networks but has been ported across to many other environments. SNMP allows the monitoring, evaluation and modification of networks from a single computer, called the SNMP manager. From here you can make inquiries to other SNMP devices on your network, such as routers and hosts. The devices questioned will return the requested information from their management information base (MIB). Figure 7.54 shows the model of the management structure.

7.12.2   Layer management

Layer management is associated with end-to-end connections and provides lower-level management functions such as fault detection. Operations and maintenance information

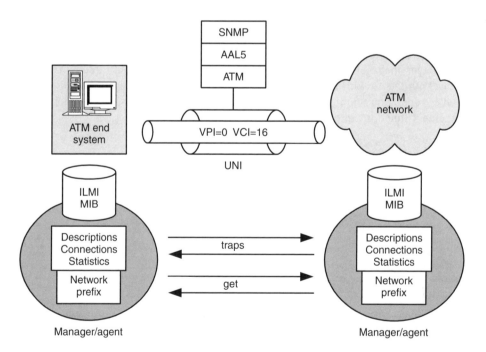

Figure 7.54   ILMI model

is transmitted between different network elements. At the physical layer, management information is carried in the SONET/SDH overhead, termed F1–F3, which cover management of the transmission path. At the ATM layer, management information is carried in OAM cells, and operates at two different levels:

- *Virtual path level*: F4
- *Virtual circuit level*: F5.

The concept is shown in Figure 7.55. Here it can be seen that the F4 operation terminates at the last switch whereas the F5 continues to the destination system.

To identify the OAM cells, for F4, the cells use the same virtual path as the user data, but travel on two reserved VCIs, as shown:

$$VCI = 3:\ \text{segment OAM F4 cells}$$

$$VCI = 4:\ \text{end-to-end OAM F4 cells}$$

For F5 cells, the PTI field in the cell header is used since the cells are travelling on the same VPI and VCI as the data:

$$PTI = 100_2:\ \text{segment OAM F5 cells}$$

$$PTI = 101_2:\ \text{end-to-end OAM F5 cells}$$

Segment OAM cells provide maintenance information between adjacent switches, while end-to-end OAM cells do the same between source and destination.

The structure of the OAM cell is as shown in Figure 7.56.

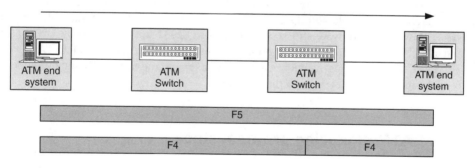

Figure 7.55 OAM flows

4 bits	4 bits	360 bits	6 bits	10 bits
OAM type	Function type	Function specific	Reserved	CRC

48 bytes

Figure 7.56 OAM cell structure

Table 7.19 OAM types

OAM type	Value	Fault type	Value
Fault management	0001	Alarm indication	0000
		Far end receive failure	0001
		OAM cell loopback	1000
		Continuity check	0100
Performance management	0010	Forward monitoring	0000
		Backward reporting	0001
		Monitoring and reporting	0010
Activation/deactivation	1000	Performance monitoring	0000
		Continuity check	0001

The OAM types and their respective function types (value) are outlined in Table 7.19.

7.13   ATM SIGNALLING

Signalling is the process by which ATM users and the network exchange control information, such as requesting network resources and establishing connections. The naming and scope of the standards is different in the definition from the ITU-T and the ATM Forum. For the ITU-T, the definitions lead on from those defined for narrowband ISDN. The standard for signalling in an N-ISDN network is Q.931. For B-ISDN, and hence ATM, the signalling standard is referred to as Q.2931 (formerly Q.93B). Q.2931 is a variant of signalling system 7 (SS7), the signalling system used in the international telephone network. The first signalling protocol devised by the ATM Forum was UNI3.0, which was unfortunately incompatible with Q.2931. Therefore the ATM Forum upgraded UNI3.0 to UNI3.1. However, there were still some aspects of Q.2931 not supported by UNI3.1, so eventually the two bodies worked together and the remaining features of Q.2931 were incorporated into the current version, UNI4.0. As its name suggests, UNI signalling deals only with signalling from a user to the network.

7.13.1   ATM signalling protocol stack

ATM provides the following protocol stack to support reliable transport of signalling protocols. Since the user plane of ATM does not provide for reliable end-to-end connections, additional layers are needed on top to provide this functionality. The signalling protocol stack sits on top of AAL5 and is known as the signalling AAL, or SAAL. The stack is shown in Figure 7.57.

The SAAL consists of the common part, which is the convergence sublayer of AAL5, and the service-specific part, as shown. It is defined in ITU-T recommendation Q.2100. AAL5 is used since it supports the transfer of large messages, up to 64 kbytes in size. However, if MTB3b is being transported, it restricts the message size to 4 kbytes.

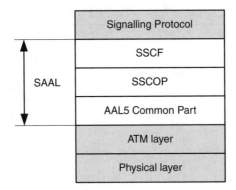

Figure 7.57 ATM signalling protocol stack

7.13.2 Service-specific connection-oriented protocol (SSCOP)

Signalling messages generally require assured delivery and this is provided by the SSCOP, as defined by the ITU-T in the Q.2110 specification. The rationale for separating the signalling AAL into two layers, the SSCOP and the Service-Specific Coordination Function (SSCF), is to allow the SSCOP to operate independently and offer the same service to a number of different layers above. The SSCF then coordinates the access between the SSCOP and these upper layers. Considering the UMTS applications alone, the SSCOP provides service to the transport layer AAL2 signalling and the UMTS application protocols such as NBAP.

The following key functions are performed by the SSCOP:

- *Transfer of user data in-sequence*: note that user data here refers to the user of the SSCOP and not end users. An example user is NBAP.

- *Error control*: detection of errors and correction through selective retransmission.

- *Flow control*: the receiver dictates the rate at which the transmitter can send signalling messages. This is implemented using a window mechanism, the size of which can be renegotiated once the connection is proceeding.

- *Keep alive*: a verification that the receiver and transmitter are still alive even if there has been little or no data transfer over a long period of time.

- Error and status reporting.

SSCOP has two principal modes of operation: assured and unacknowledged data transfer. In many ways, the SSCOP acknowledged mode can be considered analogous to the TCP protocol, where reliability is maintained using sequence numbers, flow control and timers. The unacknowledged mode is similar to UDP, where user data is not sequenced or acknowledged, and no delivery or integrity guarantees are made.

Table 7.20 SSCOP PDU types

Function	PDU	PDU field	Description
Establishment	BGN	0001	Begin connection
	BGAK	0010	Begin acknowledge
	BGREJ	0111	Begin reject
Release	END	0011	End connection
	ENDAK	0100	End acknowledge
Assured data transfer	SD	1000	Sequenced data
	POLL	1010	Information request
	STAT	1011	Information reply
	USTAT	1100	Unsolicited information
Unacknowledged data transfer	UD	1101	Unsequenced data
Resynchronization	RS	0101	Resynchronization
	RSAK	0110	Resynchronization acknowledge
Recovery	ER	1001	Recovery instruction
	ERAK	1111	Recovery acknowledge
Management data transfer	MD	1110	Management data

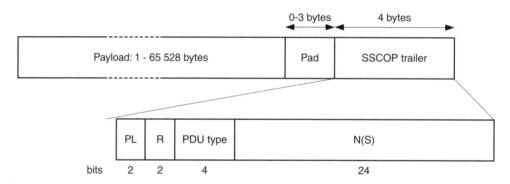

Figure 7.58 SSCOP SD PDU

Table 7.20 shows the SSCOP PDU types, and a brief description of each.

Most of the applications will request signalling messages to be carried in the Sequenced data (SD) PDU for assured data transfer. The SD appends a 4-octet trailer to the signalling message, as shown in Figure 7.58. The maximum permitted size of a message is 65 528 octets.

The fields are defined as follows:

- *Pad*: the pad field sizes the payload to a multiple of 4 octets
- *PL*: pad length, indicates the length of the pad (0–3)
- *R*: reserved for future use
- *PDU type*: the particular PDU type as defined in Table 7.20; 1000 for SD
- *N(S)*: sequence number of the SD PDU.

For assured data transfer, a signalling connection must be established. To initiate this, the transmitter sends a *begin connection* (BGN) message to the receiver, containing the initial sequence number, a window size and the first signalling payload. The receiver will respond with a *begin connection acknowledge* (BGAK) message, as illustrated in Figure 7.59.

The transmitter then proceeds to send signalling messages, incrementing the sequence number for each. Periodically the transmitter will send a POLL PDU to request the receiver's current state. The period is defined by an internal timer, and is based around either a time or a certain number of PDUs transmitted. The receiver will reply with a STAT PDU containing the next sequence number expected and a list of missing sequence numbers. The transmitter will then retransmit the missing PDUs. The STAT message also contains the window size, allowing it to be adjusted during a connection. This is referred to as a *selective ARQ* mechanism. An example of communication where PDUs are not received is illustrated in Figure 7.60. Here, the receiver has not received SD PDUs 3 and 5. At sequence 6, the transmitter sends a POLL PDU, the STAT reply to which indicates this to the sender. Once received, the sender retransmits 3 and 5.

There can be a problem with this mechanism when the time between POLLs is too large. Here the system will have a slow recovery time when PDUs are lost. Therefore,

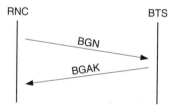

Figure 7.59   SSCOP signalling connection establishment

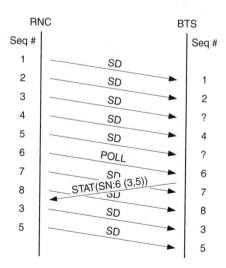

Figure 7.60   SD PDU transfer with PDU loss

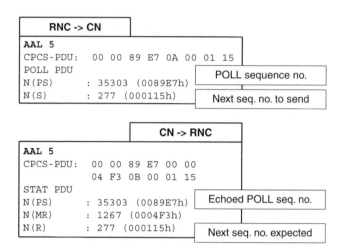

Figure 7.61 Iu interface POLL-STAT keep-alive procedure

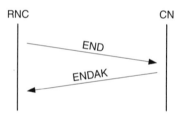

Figure 7.62 SSCOP signalling connection release

an additional PDU is defined, the unsolicited STAT (USTAT). This allows the receiver to send a status PDU (USTAT) once it detects a loss, even if it has not received a POLL PDU. The POLL and STAT function is also used as a keep-alive procedure when there is no network activity, with a single POLL sent and a STAT received. Figure 7.61, shows an example of this on the Iu interface.

Upon completion of the signalling connection, the transmitter will send an END PDU, which is acknowledged by the receiver using an ENDAK PDU (Figure 7.62).

7.13.3   Service-specific coordination function (SSCF)

The role of the SSCF is to interface the protocols accessing the SAAL to the SSCOP below; for example, mapping of Q.2931 signalling messages. The SSCF provides the following services to the upper layers:

- *Unacknowledged data transfer*: no guarantee of data integrity or delivery.
- *Assured data transfer*: guarantee of in-sequence delivery and loss/corruption protection.

Table 7.21   Reserved channels
for signalling

Signal	VPI	VCI
Meta	0	1
Broadcast	0	2
Point-to-point	0	5
ILMI	0	16
PNNI	0	18

- *Transparent transfer of data*: the contents of a payload is not restricted and is carried transparently.

- *Establishment and release of signalling connections*: for assured data transfer.

The SSCF can accept SDU sizes of up to a maximum of 4096 octets.

The ITU-T provides two recommendations for the SSCF: one for SSCF at the UNI (Q.2130) and one for SSCF at the NNI (Q.2140). In UMTS, the Node B or BTS is considered to be the host and therefore the Iub interface between the Node B and the first switch, the RNC, is a UNI interface. Therefore, SSCF-UNI is used on the Iub for both NBAP and AAL2 signalling (Q.2630.1). The other ATM interfaces, Iu and Iur, are considered NNI and use SSCF-NNI for RANAP, RNSAP and AAL2 signalling.

ATM signalling is sent over reserved channels. Simple point-to-point signalling uses VPI = 0, VCI = 5. The remaining reserved channels are outlined in Table 7.21.

Meta-signalling is used to establish signalling channels. It sends a one-cell signal, which can set up three different types of channel:

- Point-to-point
- General broadcast
- Selective broadcast.

It contains procedures to set up new channels, verify channels and release existing channels. For example, a packet sent on the meta-signalling channel, VPI 0/VCI 1, could indicate that point-to-point signalling will use VPI 0/VCI 25 instead of the default.

ILMI is the integrated local management interface, used for network management, and PNNI is the private network node interface, responsible for dynamic routing.

7.13.4   ATM addressing format

Within ATM, there are three different address formats currently specified. The formats are all based on the network service access point (NSAP) scheme outlined by the ISO. An example of an ATM address is:

47.246F.00.0E7C9C.0312.0001.000014362758.00

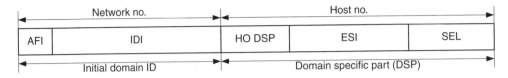

Figure 7.63 NSAP address format

Table 7.22 ATM addressing standards

AFI	Issuing authority	Format
39	ISO	Data country code (DCC)
45	E.164, B-ISDN	Public addresses
47	British Standards Institute	International code designator (ICD)
49	N/a – private addressing	Implementation specific

The address format of a NSAP is shown in Figure 7.63.

The initial domain part is globally unique, with the domain-specific part assigned locally. The AFI field is the authority and format indicator. It is a unique number allocated to each addressing authority. IDI, the initial domain identifier, is a unique identifier assigned by the addressing authority. HO-DSP is the high-order bits of the domain-specific part. It is similar to the network number in an IP address and can be used for subnetting. ESI, the end system identifier, specifies the end system, and corresponds to a host number in IP. SEL, the selector, selects a protocol within an end system. The three address formats used are allocated three unique AFIs for the organizations that define the address schemes, as shown in Table 7.22. Like IP, if the ATM addressing is completely private, then an organization need not apply for an address. This will be the case within the network for most mobile operators implementing an ATM transport layer for UMTS. However, the core network element, which bridges the divide between the ATM network of the RAN and the ISDN network of the circuit core, will most likely be given an E.164 format for compatibility with the ISDN network.

The address format for each is 20 bytes long, as shown in Figure 7.64. Note that the format for AFI = 49 is decided by the organization using this as an internal, private addressing structure.

- *DFI*: domain and format identifier
- *AA*: administrative authority
- *RD*: routing domain
- *AREA*: area identifier.

The ESI is typically derived from the 48-bit MAC address of the destination system. In the E.164 format, an extension is made to the standard N-ISDN E.164 number, which is up to 15 digits. Here, the 15-digit binary coded decimal (BCD) number is padded out with 0xFs to 8 bytes long.

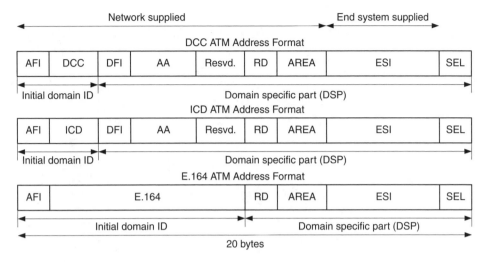

Figure 7.64 ATM address formats

Within UMTS, the ATM address formats E.164 and ICD are specified for use at both the ATM and AAL2 layers.

The ATM Forum has also defined an address naming system with resolution, similar to the domain name system (DNS) used with IP. The system is called the ATM Name Service (ANS). ATM automatic address registration is provided by the ILMI network management protocol.

An organization is free to implement its own addressing scheme either using format 0×49 or by applying for its own ICD. However, since ATM is generally only being used as a backbone, most commercial ATM equipment can automatically configure ATM addresses to interfaces connected to the switch. A typical example of how this is performed is Cisco ATM auto address configuration. Cisco has been assigned the ICD of 0×00.91. Therefore the ATM address of any devices manufactured by Cisco will begin 0×47.00.91. The first 13 bytes of an ATM address are the network identifier, and Cisco uses the MAC address to form a unique identifier, prefixing it with 0×81.00.00.00 to extend the MAC address beyond its normal 6 bytes.

Consider that the switch has a MAC address of 0×00.60.70.3E.F3.8C, and a device with a MAC address of 0×00.60.70.27.33.8A is connected to it, then the switch will automatically configure the following ATM addresses:

Switch:
0×47.00.91.81.00.00.00.00.60.70.3E.F3.8C.00.60.70.3E.F3.8C.00
Device:
0×47.00.91.81.00.00.00.00.60.70.3E.F3.8C.00.60.70.27.33.8A.00

7.13.5 UMTS signalling transport

Earlier the user planes for the UMTS network interfaces Iu and Iub were shown, illustrating the use of AAL2 and AAL5 for traffic transport. As has been seen, all signalling uses

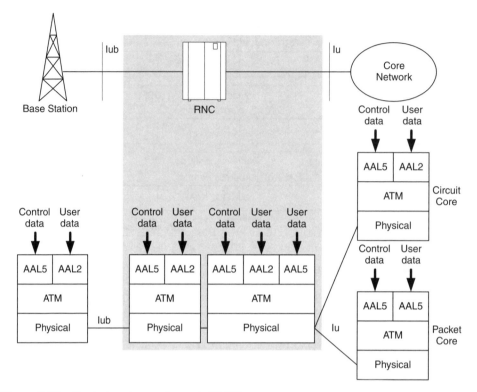

Figure 7.65   Control and user planes in UMTS

AAL5. The protocol stack can now be extended to show both user (traffic transfer) and control (signalling) planes for both interfaces. This is illustrated in Figure 7.65. Notice that there is no ATM used over the air interface and that signalling between the UE and the UTRAN or CN is considered higher layer and is all seen at the network interface as user data, and is thus transported over AAL2 on the Iub interface. It is only separated out into actual traffic (e.g. DTCH) and user signalling (e.g. DCCH/CCCH) at the logical layer above the UMTS MAC layer. This user signalling is all part of the radio resource control (RRC) protocol as discussed in Chapter 6.

7.13.6   UNI3.x signalling

UNI3.0 is the original signalling protocol defined by the ATM Forum, to provide basic functions for call setup, quality of service, etc. as required by a host connecting to an ATM network. It was upgraded to UNI3.1 to make it compatible with the ITU-T recommendations in Q.2931. The control plane protocol stack for signalling is shown in Figure 7.66.

Some of the main signalling functions defined in ITU-T recommendation Q.2931 are:

- Point to point connection setup and release
- VPI/VCI selection and assignment

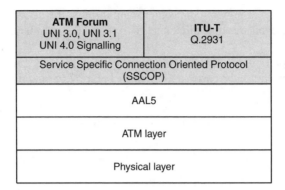

Figure 7.66   ATM control plane

- QoS class request
- Calling party identification
- Simple error handling.

7.13.7   Connection establishment

ATM describes two different types of virtual circuits, permanent (PVC) and switched (SVC). The permanent circuit is essentially like a leased line and entries are made in the network switches to indicate its route. PVCs have been enhanced by the introduction of soft PVCs where connection routes can be dynamically chosen. The SVC is established dynamically each time it is required in the same way that a circuit switched network creates a physical path for a phone call. To initiate an SVC, a connection setup phase is required. Referring to the ATM reference model shown earlier in Figure 7.6, the setup is handled by the control plane. The signalling protocol used is defined by the ITU-T and known as Q. 2931 (roughly equivalent to UNI3.1 as defined by the ATM Forum).

The procedure for establishing a connection has two phases. First, a number of cells are sent on virtual path 0, circuit 5, which is reserved for this purpose. The cells contain a request to open a new virtual circuit. The reason they are sent on this VCI = 5 is that since this circuit is only used for this purpose, it can be allocated minimum bandwidth, hence reducing overheads. Once a successful reply is received for the first phase, the new virtual circuit is established and now requests for connection setup are sent on this new circuit. The operation can be summarized as follows:

1. request a virtual channel to be established;
2. use this channel to send request for connection.

In some cases, a user may have virtual paths established on a permanent basis through an arrangement with the carrier. Then the user is free to allocate their own virtual circuits, since this operation would be transparent to the network switches which are routing based on the VPI. Alternatively, the virtual paths can be set up dynamically.

Figure 7.67, shows the flow of connection setup through the network, with each request and reply being relayed through the switches. Table 7.23 summarizes the various messages used.

To close the virtual circuit, the party that wishes to quit sends a release request, which is acknowledged by the other party, as shown in Figure 7.68 and Table 7.24.

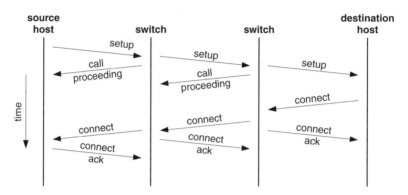

Figure 7.67 Connection setup

Table 7.23 Setup messages

Message	Interpretation	
	Sent by host	Sent by network
Setup	Establish this circuit	Call incoming
Call proceeding	Call seen	Attempting call request
Connect	Call accepted	Call request accepted
Connect ack	Accept acknowledged	Call acknowledged

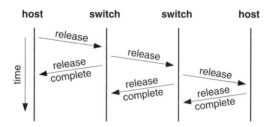

Figure 7.68 Connection release

Table 7.24 Connection release messages

Message	Interpretation	
	Sent by host	Sent by network
Release	Release this call	Host wants to release
Release complete	Release acknowledged	Release acknowledged

So far, all that has been considered is a point-to-point connection with only a sender and a receiver present. However, a facility to allow multicasting must also be provided, where one host is sending to multiple destinations. This is done by first establishing a point-to-point. Subsequently, an *add party* signal is sent which attaches other destinations to the current virtual circuit.

7.13.8 Signalling message structure

A signalling message is based around the Q.931 message format. It consists of a header and a variable number of information elements (IEs). The format is illustrated in Figure 7.69.

The format of the header is as shown in Figure 7.70.

Message header
IE
IE
...
IE

Figure 7.69 ATM signalling message structure

Bit 7	6	5	4	3	2	1	0	Byte
Protocol discriminator								1
0	0	0	0	Call reference value length				2
Flag	Call reference value							3
Call reference value (continued)								4
Call reference value (continued)								5
Message type								6
Message type (continued)								7
Message length								8
Message length (continued)								9
Infomation elements (variable length)								etc.

Figure 7.70 ATM signalling header

The protocol discriminator is used to distinguish UNI control messages from other messages. Two common values are:

$$08 - \text{Q.931 messages}$$

$$09 - \text{Q.2931 messages}$$

The call reference value is 3 bytes long and identifies the call to which this message is related, since a user may have many calls active at any given time. The first bit, the flag, is used to indicate whether the message is coming from or going to the originator of the call:

$$\text{Flag} = 0: \text{to call originator}$$

$$\text{Flag} = 1: \text{from call originator}$$

The next 2 bytes indicate the type of message being sent. A sample of some common message types are shown in Figure 7.71.

Value	Type
Call establishment messages	
00000010	Call proceeding
00000111	Connect
00001111	Connect ack
00000101	Setup
00001101	Setup ack
Call clearing messages	
01001101	Call release
01011010	Release complete
Information messages	
01110101	Status inquiry
01111101	Status

Figure 7.71   Message types

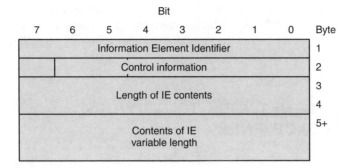

Figure 7.72 IE format

Table 7.25 Sample IE identifiers

Value	Information element	Description
01110000	Called party number	Destination of the call
01110001	Called party subaddress	Destination subaddress
01101100	Calling party number	Origin of the call
01101101	Calling party subaddress	Origin subaddress
01011000	AAL parameter	Requested AAL type plus other parameters
01011001	ATM traffic descriptor	Specifies traffic parameters
01011010	Connection identifier	Identifies connection and gives VPI/VCI values
01011100	QoS parameter	Indicates required QoS for the connection

Table 7.26 Call setup message

Message	Call setup
IEs	Setup message Called party address Calling party address Quality of service Traffic characteristics

Message length indicates the length of the contents of this message, since the information elements can be variable in length. Some IEs can appear only once in a message while others can appear more than once. The IEs included are dependent on the type of message; some elements are mandatory and some optional. The format of an IE is shown in Figure 7.72. Some sample information element identifiers are show in Table 7.25. As an example, Table 7.26 shows a call setup message containing a number of IEs.

7.13.9 UNI4.0

UNI4.0 is an enhanced system, adding on top of UNI3.1 to provide additional features recommended by the ITU-T. The new features include support for ABR services, enhanced

QoS, proxy signalling intended for use in video on demand services, and virtual UNIs. A significant change is the *leaf initiated join* function, which is an enhancement to multicasting, allowing for parties to be added without requiring signalling back to the call initiator.

7.14 PRIVATE NETWORK-TO-NETWORK INTERFACE (PNNI)

One key feature of ATM is its ability to be scalable, meaning it should be equally at home in a LAN environment and in a global backbone. A major stumbling block in achieving this goal has been the static routing mechanisms generally deployed for configuration of ATM switches in the backbone, where dynamic signalling of SVCs is not used. The method used requires manual entry of routing information for PVCs into the routing tables of switches, which had to be altered when the topology of the network changed, or each time a new switch was added to the network. The solution presented by the ATM forum is the PNNI standard. Figure 7.73 shows a conceptual view of PNNI implementation. While PNNI is not used in the first release of UMTS, it is set to play an important role as the network evolves to utilize packet switching exclusively. For example, the Release 5 all-IP network can be implemented on top of a large-scale ATM network.

PNNI is a hierarchical, dynamic link state routing protocol, which is designed to support establishment of soft PVCs across large-scale networks. With a soft PVC, the start and end points of the circuit are defined, and then PNNI is used to calculate the best route. However, PNNI provides more than a link state routing protocol and it can be described as a topology state protocol since it also exchange information relating to QoS parameters. Simply put, it dynamically updates switches about moves, additions and changes to the network topology, enhancing the ability of ATM to deliver QoS.

PNNI simplifies the configuration of large networks since it enables switches to learn about their neighbours and to dynamically distribute routing information. Switches can be arranged in a hierarchical structure, with each level representing one or more switches. A group of switches that are at the same level is referred to as a peer group. Topology information about the link states is passed around the members of the peer group. Note

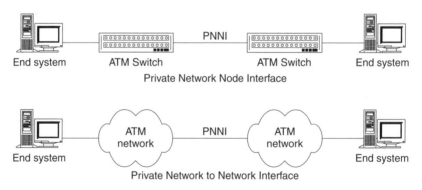

Figure 7.73   PNNI network

that if the network is of a smaller scale, it is possible to implement it as a single peer group for all the switches on the network.

7.14.1   Peer group

Within a peer group, each switch sends *hello* packets across the NNI between it and other switches. This is done to announce the existence of the switch so that each switch within the peer group is aware of all the others. Once it is established that two switches share the same peer group, they will exchange peer topology state packets (PTSPs). A PTSP contains one or more peer topology state elements (PTSEs) advertising the switch's resources available on each of its links. A PTSE will describe the attributes of a link, such as traffic type supported, maximum cell rate, cell delay variation, etc. Table 7.27 outlines some of the key parameters used, referred to as topology state parameters.

Based on this information, a switch can determine which is the best path to take to ensure that the QoS parameters are adequately met. Thus PNNI is more complicated than a standard routing protocol since a great deal more information is considered when deciding on which route to take. This is summarized in Figure 7.74.

Note that the result of the routing decision is to look for an acceptable, rather than an optimal path: PNNI only needs to fulfil the requirements of the caller. Once this initial exchange of topology information has occurred, the information will only be rebroadcast if there is a significant change in its topology information. For example, if the maximum cell rate of a link were to change, this would trigger a PTSE to be sent out from the attached switches, information neighbours of the change. How is a 'significant' change defined? Here, significance is defined for each parameter and in general means a specified percentage change in a given parameter. For example, a significant change in bandwidth could be defined as when the bandwidth changes by 15%. The change percentages are

Table 7.27   Topology state parameters

Acronym	Significance
MCTD	Maximum cell transfer delay
MCTV	Maximum cell transfer variation
MCLR	Maximum cell loss ratio
ACR	Available cell rate

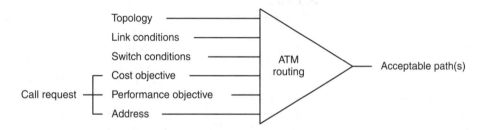

Figure 7.74   ATM routing

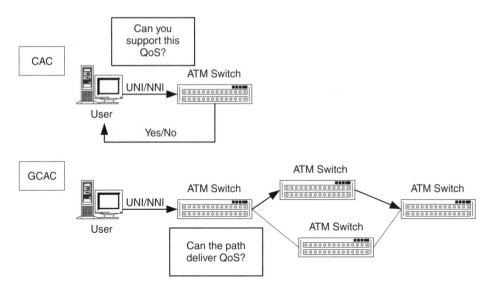

Figure 7.75   GCAC

up to the network administrator to decide. Since updates are triggered, this limits the amount of traffic generated by the network, and the overheads introduced to maintain up-to-date topology information. Therefore, a switch in a peer group has a detailed picture of the topology of the whole group, enabling each switch to perform the entire call setup procedure, thus speeding up the call setup time, i.e. call requests need not be handled at each hop. This is referred to as source routing. The QoS support is handled by the CAC of each switch. This means that the first switch making the routing decision does so based on the information it has received from the other peer group members. This decision is a best guess. The mechanism for deciding on a route is the generic call admission control (GCAC). Figure 7.75 illustrates the concept.

7.14.2   AAL2 signalling

Recently, the ITU-T has developed a protocol to allow the switching of individual AAL2 channels, based on their channel identification (CID), independently of the ATM layer. This has been possible due to the multiplexed nature of AAL2. This is particularly useful where a virtual circuit is externally owned, as a PVC can be set up through the external organization's infrastructure, and its use maximized by multiplexing many AAL2 connections through it. Note this may often be the case, for example, between a base station and RNC. This new protocol, Q.2630.1, was largely driven by requirements within mobile networks, particularly for UMTS, and involved the participation of many large cellular vendors, including Nokia, Ericsson, Lucent and Siemens.

A key feature of AAL2 signalling is that the AAL2 layer is independent of, and therefore transparent to, the ATM layer and the existing ATM switches. No changes are required to an underlying ATM network infrastructure, with the addition only of AAL2 end points.

If standard ATM switching was used, i.e. by the creation of an SVC to carry a voice call, this would require the maintenance of millions of SVCs, putting a great burden on the ATM switches. However, AAL2 switching allows for the switching of voice at the AAL2 layer. Its use in UTRAN is shown in Figure 7.76. The ATM layer is generally configured statically as PVCs in UMTS.

The protocol stack for AAL2 signalling is shown in Figure 7.77. It is used extensively in UMTS for the establishment of AAL2 connections over PVCs. For example, PVCs are configured between the base station and RNC, with dynamic AAL2 connections being created and torn down over them. Note that, like ATM signalling, AAL2 signalling messages are carried over AAL5.

The Q.2630.1 protocol defines the specification of AAL2 signalling, capability set 1. It covers the establishment, maintenance and release of AAL2 connections across ATM virtual circuits. The AAL2 signalling layer operates completes independently of the underlying ATM layers, and therefore allows a dynamic allocation of bearers over a static

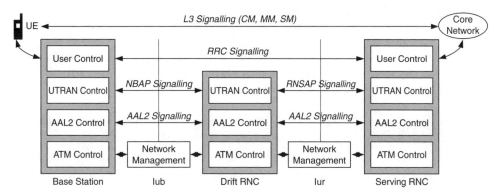

Figure 7.76   AAL2 use in UMTS

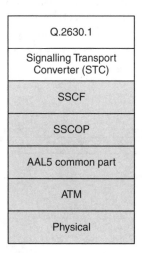

Figure 7.77   AAL2 signalling protocol stack

network of PVCs. AAL2 signalling uses the standard ATM signal AAL for transport, but is, however, independent of this transport, requiring only assured transport of messages. However, to adapt the underlying generic transport to the specifics of Q.2630.1, a signalling transport converter STC may be used and the ITU-T has defined two: Q.2150.1 and Q.2150.2. Q.2150.1 is used to convert to an MTP3/MTP3b transport layer, and Q.2150.2 to convert to an ATM SAAL transport layer. In the case of the Iub interface, Q.2150.2 is used. On the Iu and Iur interface, Q.2150.1 is used.

An AAL2 path is an ATM virtual circuit connection between two adjacent AAL2 nodes. For each AAL2 path, i.e. ATM virtual circuit designated to carry AAL2 traffic, an AAL2 path identifier is assigned which uniquely identifies the path. There are 248 channel IDs available on each path, from 8 to 255. An AAL2 node is identified by service end point address, which is the same as an ATM address, but supporting only NSAP and E.164 address formats. The addressing plan used at the AAL2 layer can follow the addresses used at the ATM layer. However an independent addressing plan may also be implemented. This essentially defines a separate AAL2 network, sitting on top of the ATM network. The actual definition of the addressing structure is implementation specific.

Consider the NBAP signalling message to request a radio link. Upon its return to the RNC, the BTS has enclosed its AAL2 address and a binding ID.

The RNC then uses these within an AAL2 establish request (ERQ), as shown in Figure 7.78. The BTS will respond with an establish confirm message (ECF).

The ERQ message will contain the following key parameters:

- *Connection Element ID*: this consists of a 4-byte path identifier, the coding of which is specific to the implementation, and a 1-byte channel ID.
- *Destination service end point address*: this is an E.164 or NSAP address. The E.164 address is used for the circuit core components (3G-MSC) since they already have an E.164 number.
- *Signalling association ID (SAID)*: OSAID/DSAID for originating and destination signalling association identifiers, respectively. This is a 4-byte field used as an ID for the signal connection. In an initial connection, the DSAID field will be null, and the OSAID will be a unique identifier of this signalling message.
- *Served user generated reference*: this is a field for the user to enter a value. In UMTS, it is used to transport the binding ID.

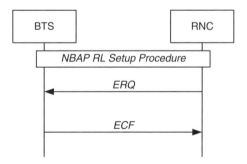

Figure 7.78 AAL2 signalling connection establishment

- *Link characteristics*: provides information on the size of the CPS SDU (1–45 bytes) and the bit rate (0–2048 kbps in steps of 64 kbps).

The establish confirm will then return to the RNC, containing only two parameters:

- *DSAID*: the DSAID will correspond to the OSAID contained in the ERQ message. This is used to reference signalling messages relating to one call.
- *OSAID*: an identifier for this ECF message.

To release the established connection, the release (REL) and release confirm (RLC) signalling messages are used.

The trace example shown in Figure 7.79 shows an AAL2 signalling exchange between an RNC and a BTS. In this example, the RNC is requesting an AAL2 channel, with a channel ID (CID) of 11 and a CPS-PDU size of 19 bytes.

In this example, the AAL2 address follows the private addressing 49 scheme. Here, the operator or vendor will designate the structure of the address. In this example, the

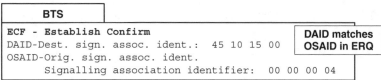

Figure 7.79 Trace example of AAL2 signalling connection establishment

format that has been adopted is to incorporate the identity of the base station and the cell into the address. The base station ID is 20 bits and the cell ID 12 bits. Therefore, from the address:

0x49.00.00.50.01.00.00.00.00.00.00.00.00.00.00.00.00.00.00.00

it can be seen that the BTS ID is 0x00005 and the ID is 0x001. This is also reflected in the path ID: 5001.

For UMTS implementation Release 5 onwards, the Q.2630.1 specification is superseded by Q.2630.2 (capability set 2), which expands to encompass link modifications in addition to all the procedures specified in Q.2630.1. Q.2630.2 uses the modify request (MOD) and modify acknowledged/modify reject (MOA/MOR) messages to perform link modification.

7.15 IP/ATM INTERNETWORKING

This section addresses the ATM methods used in UMTS to transport both real- and non-real-time traffic around the network, as well as describing the key functions of the major ATM components in a UMTS network: the radio network controller (RNC) and the media gateway/internetworking unit (MGW/IWU). The position of these components in an overall ATM network is shown in Figure 7.80.

In fact, despite the grand design of ATM and the endeavours of the ATM Forum, to date, the most popular and widely used application for ATM is as a transport mechanism, usually for IP traffic or to internetwork with a circuit switched network. The global Internet backbone consists of a considerable amount of ATM hardware, much of it not capitalizing on the benefits and facilities of ATM. This section addresses the mechanisms used in UMTS for transporting IP over ATM, and outlines several strategies for IP/ATM internetworking, particularly in the area of IP switching.

A typical deployment of ATM will establish a set of PVC connections between components for the transport of both user data and control information to both core network

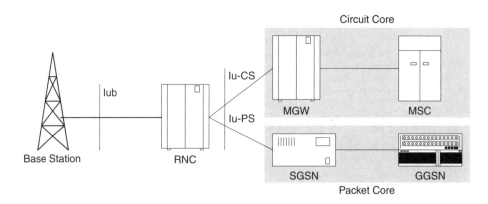

Figure 7.80 UMTS main components

domains. These connections are dimensioned in advance by transmission planning. This section also addresses the nature of the connections to both of these domains.

7.15.1   Packet core

To connect to the SGSN, the RNC transports IP packets over the Iu-PS interface. The RNC is responsible for segmenting and reassembling the TBs used for data transfer between the RNC and the UE. For the Iu-PS interface, the RNC has reassembled these TBs back to the IP packets from the originating application at the UE. It is now sending these to the SGSN using AAL5, which performs the segmentation and reassembly function to carry the traffic in ATM. However, to support this IP transport, ATM must also provide a mechanism to encapsulate the IP packets to allow the internetworking of the two protocols. It is essential that ATM provide mechanisms to transport legacy data traffic across its network infrastructure, the majority of this traffic being IP traffic. However, this interoperation of the two protocols is non-trivial.

It has been seen that ATM is a connection-oriented protocol where a connection is established between two parties prior to the transfer of data. In sharp contrast, IP uses a datagram approach with each packet treated individually and dealt with independently by each router.

Much of this IP traffic arrives at an ATM network from some other transport technology, such as Ethernet, as is used in the GPRS core. There are two approaches to internetworking with ATM:

- transfer the entire MAC frame (LAN emulation);
- extract the IP packet and transfer that (IP over ATM).

In addition, ATM networks are based on the provision of QoS to clients. IP uses a best-effort approach where each router tries its best to forward the data, but provides no guarantees of quality. To carry IP over ATM, the ATM protocol is viewed as a datalink layer, with IP running on top of it. The ATM network is operating with its own addressing scheme and so there needs to be a mapping between the IP addresses and the ATM addresses of the underlying network. Therefore there needs to be some address resolution protocol providing this mapping.

7.15.2   Data encapsulation

Two methods are defined for data encapsulation of IP over ATM (RFC1483). The first assigns a different virtual circuit (VC) for each protocol, whereas the second allows multiple protocols to be carried on a single VC. Both approaches use AAL5 to carry the IP protocol

- *VC-based multiplexing*: the VC is allocated to a single protocol, which is encapsulated in the PDU field of AAL5. This approach is advantageous if a large number of VCs can be established, and it is also quick and economical. However, it is not generally used in practice since it requires the maintenance of a large number of VCs.

- *LLC encapsulation*: this method allows a number of protocols to be carried over the one VC. The encapsulation system is designed for carrying both routed and bridged protocols. However, only routed protocols will be discussed.

The protocol is identified by prefixing the IEEE 802.2 LLC header. The format of the header is shown in Figure 7.81.

When the DSAP and SSAP are 0xAA and the control byte is 0x03, this indicates that a non-ISO protocol is being encapsulated and the extended subnetwork access point (SNAP) header is used, as shown in Figure 7.82. This header contains two extra fields, the OUI, which is a 3-byte organizationally unique identifier, identifying an organization, which assigns a 2-byte protocol identifier, PID.

The format used in the encapsulation is split between ISO protocols, such as IS-IS, and non-ISO protocols such as IP. The format of both is shown in Figure 7.83.

The ISO type is indicated by 0xFE FE 03. The 0x03 is for the control field and is common to both, indicating unnumbered information. A non-ISO type is indicated by 0xAA AA 03. The EtherType identifies the protocol that is encapsulated. The IP packet

DSAP	SSAP	Control

Figure 7.81 LLC header

DSAP 0xAA	SSAP 0xAA	Control 0x03	SNAP OUI	SNAP PID

Figure 7.82 Extended SNAP header

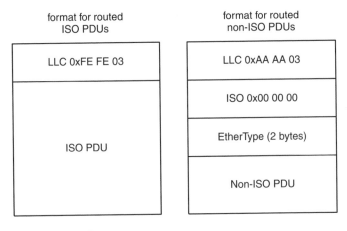

format for routed ISO PDUs

| LLC 0xFE FE 03 |
| ISO PDU |

format for routed non-ISO PDUs

| LLC 0xAA AA 03 |
| ISO 0x00 00 00 |
| EtherType (2 bytes) |
| Non-ISO PDU |

Figure 7.83 Routed protocols

format for IP

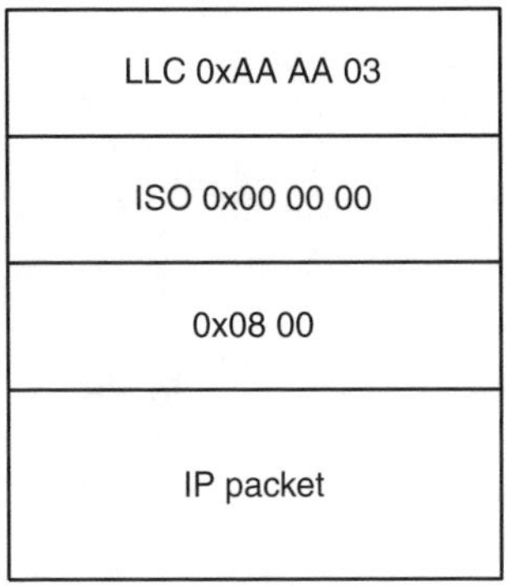

```
┌─────────────────────────────┐
│                             │
│      LLC 0xAA AA 03         │
│                             │
├─────────────────────────────┤
│                             │
│      ISO 0x00 00 00         │
│                             │
├─────────────────────────────┤
│                             │
│        0x08 00              │
│                             │
├─────────────────────────────┤
│                             │
│        IP packet            │
│                             │
│                             │
└─────────────────────────────┘
```

Figure 7.84   IP encapsulation over AAL5

is prefixed with the IEEE 802.2 logical link control (LLC) header before encapsulation in the PDU of AAL5. This is the scheme used for all IP over ATM systems. Figure 7.84 shows the format of the IP packet with the LLC prefix.

The LLC prefix is 8 bytes long and for IP the next 3 bytes indicate an EtherType protocol and contain zeros. The remaining two bytes, 0x08 00, identify the packet as being an IP packet. This is the encapsulation mechanism used to transport IP over AAL5 over ATM throughout the UMTS network.

There are two approaches to dealing with IP over ATM. The first sees the ATM network as a LAN and divides this LAN up into logical subnetworks based on the IP addresses of the hosts, similar to IP subnetting. This approach is known as classical IP over ATM. Here, hosts on the same subnet communicate with each other over ATM connections and use address resolution protocol (ARP) servers to resolve IP addresses to their corresponding ATM addresses. Each subnetwork is interconnected using a router, and traffic between subnets goes through the router. There is a problem here since even though hosts are on different subnets, they are on the same ATM network so the process of passing through one or more routers introduces a high degree of inefficiency into the system. This is because IP packet headers must be examined at each hop, requiring that the whole IP packet must be reconstructed by each router along the route.

The second approach expands on the classical IP over ATM model by providing a next hop resolution protocol (NHRP). This protocol allows hosts to resolve ATM addresses of hosts on other subnets and connect directly using an ATM end-to-end connection. The concept is similar to that of DNS address resolution. Both are now discussed in more detail.

7.15.3   Classical IP over ATM (CLIP)

As already mentioned, the ATM network is partitioned into logical IP subnetworks, referred to as an LIS, with each of these subnetworks interconnected by an IP router. An example of such a system is shown in Figure 7.85.

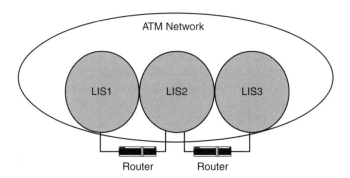

Figure 7.85 Classical IP over ATM

Each host on an LIS will share the same IP prefix and subnet mask, which is similar to a traditional IP subnetwork on a broadcast LAN. In contrast, though, the LISs are all on the same ATM network, hence the term 'logical' subnet.

There are two different types of communication that may occur.

1. Communication between hosts on the same LIS. This is done through an end-to-end ATM connection. However, prior to data transfer, a connection between the two hosts must first be established. Consider two hosts, A and B, where A wishes to send data to B. A knows the IP address of B, but requires its ATM address. As shown in Figure 7.86, each LIS must contain a server, called an ATMARP server, which provides the address resolution.
   A sends an ARP request to the ATMARP server containing the IP address of B. The server will look up B's IP address, extract the corresponding ATM address and return it to A. Actually, the ATMARP server sends an inverse ARP request to A, so that A will reply with its IP and ATM addresses, which the server can then store. This enables the ATMARP server to automatically build up a translation table. Once A receives the ATM address of B, it can establish a connection with B using ATM signalling.
2. Communication between hosts on different LISs. Each LIS will contain a router which will forward packets not destined for the local subnet. The router either forwards the packet to the destination host, or to another router, with the packet being routed on a per-hop basis. This scheme will introduce a considerable delay since each router is required to disassemble and reassemble the IP packet

CLIP and this form of address resolution is supported as an option in UMTS for use on the Iu-PS interface, and it is likely that, if implemented, the ARP server would be located in the SGSN. Originally, there was a requirement that CLIP be used, but this has subsequently been changed as certain parties argued that the use of PVC connections in UMTS was implemented for simplicity, which would then be negated by having to use CLIP. Thus a manufacturer is at liberty to implement an operations and maintenance (O&M) solution for address discovery.

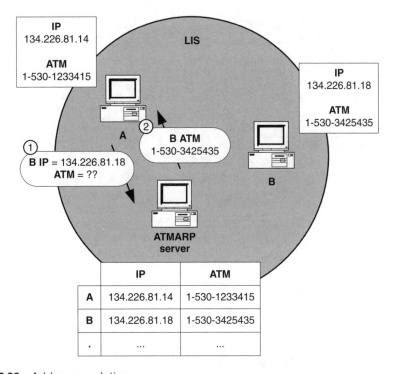

Figure 7.86 Address resolution

7.15.4 Next hop resolution protocol (NHRP)

It was seen that CLIP is limited since communication between hosts in different LISs is done through IP routers even though they are connected on the same ATM network. What is required to overcome this shortfall is a mechanism to resolve IP addresses to ATM addresses regardless of location in the network. This process is performed by NHRP.

Each LIS must contain an NHRP server. If a host needs an address resolution, it sends the request to the NHRP server, which contains a table of mappings between IP addresses and ATM addresses. The NHRP server performs the role of the ATMARP server for requests to resolve addresses within the LIS. However, if the request is for an address in another LIS, the NHRP server will forward the address resolution request to the correct NHRP server for the LIS on which the destination host resides. The network layout for the NHRP system, referred to as a non-broadcast multiple access (NBMA) subnet, is shown in Figure 7.87.

The NHRP servers are interconnected and use a routing protocol in which, like an IP router, an NHRP server can work out which NHRP server to forward the request to. Note that the request may pass through a number of servers before reaching the destination server. This NHRP server will then reply along the reverse path with the corresponding ATM address, thus allowing the source host to directly establish an end-to-end ATM connection with the destination. As the reply travels back to the source, the intermediate NHRP servers will cache the mapping to speed up subsequent requests. This approach is

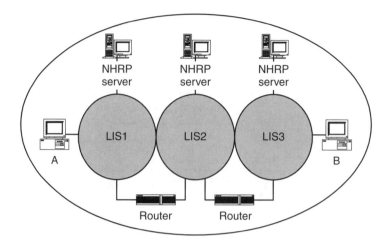

Figure 7.87   NHRP

known as cut-through routing. The system still employs IP routers, which may be used to transfer data while waiting for the direct connection to be established.

7.15.5   IP multicast over ATM

Both systems discussed so far are only for establishing a point-to-point or unicast connection. However, the IP protocol supports multicasting, where packet transfer can be point-to-multipoint or multipoint-to-multipoint. To service this, a mechanism is needed to resolve such a request into a list of ATM addresses. Once a set of destination hosts is established, a method is also needed to transfer data to all of these hosts.

To solve the problem, each LIS has a multicast address resolution server (MARS) which resolves the mappings. This resolution is performed in the same manner as for unicast address resolution by the ATMARP server.

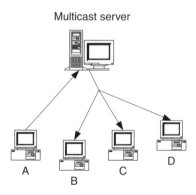

Figure 7.88   Multicast server

Once a set of destinations is known, a multicast connection needs to be established. This is done by one of two methods:

1. *VC mesh*: each host establishes a connection to each of the destinations.
2. *Multicast server*: here, each LIS contains a multicast server (MCS). A host will sent a request to the MARS, which will return the address of the MCS. Packets are then sent to the MCS, which will forward them to the multiple destinations. The MCS builds either a point-to-multipoint or multiple point-to-point connections with the destinations specified by the multicast packet. The system is shown is Figure 7.88. The MARS and the MCS are logical units and can reside in the same physical device.

7.16 SUMMARY

This chapter describes how ATM is used to provide the transport network for UMTS. A detailed explanation of ATM is presented, specifically with reference to its application to UMTS. The interaction between ATM and the physical medium is described, again with specific reference to the PDH and SDH infrastructures that are used for UMTS, and physical layer protocols such as inverse multiplexing for ATM and UTOPIA. In particular, the extensive use of AAL2 for transfer of data, both real-time and non-real-time, across the Iub/Iur interface is examined, as well as its use to transport real-time traffic between UTRAN and the core network on the Iu-CS interface. UMTS is the first major application for AAL2 switching, which allows AAL2 connections to be dynamically established and released. This allows the implementation to create a static ATM infrastructure based on PVCs and then dynamically run AAL2 over this. The AAL2 signalling protocol operation is examined, with real-world examples using network traces. Inherent to ATM is the provision of QoS, and the principles of operation of ATM QoS and traffic management, and its relationship to UMTS QoS, is explored in depth. Also, the mechanism for transport of IP over ATM, as used on the Iu-PS interface, is explained. Finally, some other aspects of ATM such as network management and PNNI are described. While not explicitly specified for use by 3GPP, they may be implemented to provide additional support and flexibility in the network.

FURTHER READING

3GPP TS25.414: UTRAN Iu interface data transport & transport signalling.
3GPP TS25.422: UTRAN Iur interface signalling transport.
3GPP TS25.424: UTRAN Iur interface data transport & transport signalling for CCH data streams.
3GPP TS25.426: UTRAN Iur and Iub interface data transport & transport signalling for DCH data streams.
3GPP TS25.432: UTRAN Iub interface: signalling transport.
3GPP TS25.434: UTRAN Iub interface data transport & transport signalling for CCH data streams.

RFC 1483: Multiprotocol Encapsulation over ATM Adaptation Layer 5. Juha Heinanen. July 1993.

ATM Forum: LAN Emulation over ATM 1.0, af-lane-0021.000.

ATM Forum: LANE v2.0 LUNI Interface, af-lane-0084.000.

ATM Forum: Multi-protocol Over ATM Specification, Version 1.1, af-mpoa-0114.000.

ATM Forum: MPOA v1.1 Addendum on VPN Support, af-mpoa-0129.000.

ATM Forum: Private Network-Network Interface Specification v.1.1, af-pnni-0055.001.

ATM Forum: ATM User Network Interface (UNI) Signalling Specification version 4.1, af-sig-0061.001.

ATM Forum: Utopia, af-phy-0017.000.

ATM Forum: Utopia Level 2, af-phy-0039.000.

ATM Forum: E-1 Physical Layer Interface Specification, af-phy-0064.000.

ATM Forum: Inverse ATM Mux Version 1.0, af-phy-0086.000.

ATM Forum: Inverse Multiplexing for ATM (IMA) Specification Version 1.1, af-phy-0086.001.

ATM Forum: ATM on Fractional E1/T1, af-phy-0130.000.

ATM Forum: Utopia 3 Physical Layer Interface, af-phy-0136.000.

ATM Forum: UTOPIA Level 4, af-phy-0144.001.

ATM Forum: Private Network-Network Interface Specification V. 1.0, af-pnni-0055.000.

ATM Forum: ATM Forum Addressing: User Guide Version 1.0, af-ra-0105.000.

ATM Forum: Circuit Emulation, af-saa-0032.000.

ATM Forum: ATM Names Service, af-saa-0069.000.

ATM Forum: UNI Signalling 4.0, af-sig.0061.000.

ATM Forum: Traffic Management 4.0, af-tm-0056.000.

ATM Forum: Traffic Management 4.1, af-tm-0121.000.

ATM Forum: ATM User-Network Interface Specification V3.1, af-uni-0010.002.

G.803: ITU-T: Architecture of transport networks based on the synchronous digital hierarchy (SDH).

G.804: ITU-T: ATM cell mapping into Plesiochronous Digital Hierarchy (PDH).

I.361: ITU-T: B-ISDN ATM layer specification.

I.363.1: ITU-T: B-ISDN ATM Adaptation Layer specification: Type 1 AAL.

I.363.2: ITU-T: B-ISDN ATM Adaptation Layer specification: Type 2 AAL.

I.363.3: ITU-T: Type 3/4 AAL.

I.363.5: ITU-T: Type 5 AAL.

I.366.1: ITU-T: Segmentation and Reassembly Service Specific Convergence Sublayer for the AAL type 2.

I.371: ITU-T: Traffic control and congestion control in B-ISDN.

I.371.1: ITU-T: Guaranteed frame rate ATM transfer capability.

I.432.1: ITU-T: General characteristics.

I.630: ITU-T: ATM protection switching.

Q.711: ITU-T: Functional description of the signalling connection control part.

Q.712: ITU-T: Definition and function of signalling connection control part messages.

Q.713: ITU-T: Signalling connection control part formats and codes.

Q.714: ITU-T: Signalling connection control part procedures.

Q.715: ITU-T: Signalling connection control part user guide.

Q.716: ITU-T: Signalling System No. 7–Signalling connection control part (SCCP) performance.

Q.2100: ITU-T: B-ISDN signalling ATM adaptation layer (SAAL)–Overview description.

Q.2130: ITU-T: B-ISDN signalling ATM adaptation layer – Service specific coordination function for support of signalling at the user–network interface (SSCF at UNI).

Q.2140: ITU-T: B-ISDN ATM adaptation layer – Service specific coordination function for signalling at the network node interface (SSCF AT NNI).

Q.2150.1: ITU-T: Signalling Transport Converter on MTP3 and MTP3b.

Q.2150.2: ITU-T: Signalling Transport Converter on SSCOP and SSCOPMCE.

Q.2210: ITU-T: Message transfer part level 3 functions and messages using the services of ITU-T Recommendation Q.2140.

Q.2630.1: ITU-T: AAL type 2 signalling protocol (Capability Set 1).

Q.2630.2: ITU-T: AAL type 2 signalling protocol – Capability Set 2.

A list of the current versions of the specifications can be found at http://www.3gpp.org/ specs/web-table_specs-with-titles-and-latest-versions.htm, and the 3GPP ftp site for the individual specification documents is http://www.3gpp.org/ftp/Specs/latest/

8

IP Telephony for UMTS Release 4

8.1 INTRODUCTION

With UMTS Release (R4), the architecture of the core network circuit switched domain is revised radically. The circuit traffic is delivered over an internal packet switched Internet protocol (IP) network with connections to external networks handled via media gateways (MGW). Figure 8.1 shows the architecture of an R4 network.

The architecture of the CS core is described by TS 23.205 'Bearer-independent circuit switched core network', termed bearer-independent since the core network can be asynchronous transfer mode (ATM) or IP, with many different Layer 2 options.

In this case traffic entering or exiting the circuit switch domain is controlled by the MGW. This is responsible for switching the traffic within the core network domain and performing data translation between the packet-based format used within the core network and the circuit switched data transmitted on the PSTN or ISDN external network. The MGW is controlled by the mobile switching centre (MSC) server, which sends the MGW control commands, for example to establish bearers to carry calls across the core network. The user data (i.e. voice traffic) within the CS-CN domain can be carried within ATM cells (ATM adaptation layer 2; AAL2) or IP packets. Note that from R99 to R4 the only part of the network to change is the CS domain; the PS domain and the radio access network (RAN) remain the same as before. For the user equipment (UE), circuit switched traffic will still be seen as a stream of bits, even though these bits are split into separate packets or cells as they are forwarded over the UMTS terrestrial RAN (UTRAN) and core network.

Convergence Technologies for 3G Networks: IP, UMTS, EGPRS and ATM   J. Bannister, P. Mather and S. Coope
© 2004 John Wiley & Sons, Ltd   ISBN: 0-470-86091-X

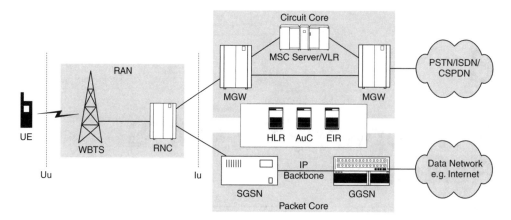

Figure 8.1   Release 4 UMTS architecture

8.2   R4 SOFTSWITCH ARCHITECTURE

In R4 the separation of the switching and call control functions within the core network is commonly referred to as a softswitch architecture. The call control component, i.e. the MSC server, is the softswitch in this case. This separation of functions makes it easier to scale the network as the traffic demand increases. If the network planners require more switching capacity they can add MGWs; if they require more call control capacity they then add more MSC servers. This is a clear distinction from the UMTS Release 99 and GSM networks, in which the call control and switching functions are all carried out within the MSC and gateway MSC (GMSC).

For R4 the GSM MSC functionality is split into two components, the MSC server and the MGW.

8.2.1   MSC server

This performs functions such as call control for mobile-originated and mobile-terminated calls, and mobility management in terms of maintenance of the registry of mobiles within its area of control. The MSC server integrates with the visitor location register (VLR) component, which holds location information as well as CAMEL (customized applications for mobile network enhanced logic) data for subscribers. Functions carried out by the MSC server include:

- controlling the registration of mobiles to provide mobility management;
- providing authentication functions;
- routing mobile-originated calls to their destination;
- routing mobile-terminated calls by using paging to individual mobiles.

The MSC server terminates signalling from the mobile network over the Iu interface to the radio network controller (RNC). It also controls the establishment of bearers across its core by the use of MGWs under its control.

8.2.2   Media gateway (MGW)

This translates media traffic between different types of network. Functionality carried out by the MGW includes:

- termination of bearer channels from the circuit switched and packet switched networks;
- echo cancellation for circuit switched circuits;
- translation of media from one CODEC form to another, e.g. G.711 to G.729;
- optional support of conferencing bridge function.

Each MGW is controlled by one or more MSC servers.

8.2.3   Gateway MSC server (GMSC server)

The GMSC server provides the same call control functionality that is provided by the GMSC in GSM. It works in conjunction with the home location register (HLR) to allow:

- calls from outside the operator's network to be routed to the appropriate MSC server.
- calls from within the operator's network to be forwarded on the PSTN.

The GMSC server uses the services of MGWs to control the establishment of bearer channels across the circuit core network and to terminate circuit switched bearers originating from the PSTN. Figure 8.2 shows the internal connections within the CS core network.

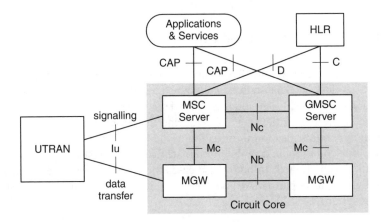

Figure 8.2   Core network CS domain release 4

Within the core network three separate interface are defined; Mc, Nc and Nb. The Mc interface carries control instructions and signals between the MSC servers (MSC and GMSC) and a MGW. Apart from control commands from the MSC server to the MGW this channel also allows the MGW to signal events back to the MSC server. For example, if the MGW receives dual tone multi-frequency (DTMF) signals from the phone connected in the PSTN domain these can be relayed back to the MSC server. The protocol that is used over this interface is called MEGACO (MEdia GAteway COntrol) and is defined in ITU recommendation H.248. Its functionality is described in Section 8.7.

The Nc interface carries signalling between two MSC servers. This allows the MSC server that is handling the incoming call MGW to signal its requirements to the other MSC server that controls the outgoing call MGW. The protocol that handles this signalling function is called bearer-independent call control (BICC). BICC is an ITU-T signalling protocol which provides the equivalent services to the ISDN call control protocol ISUP, regardless of both the underlying bearer used to transport the messages, and the media, hence the term bearer-*independent*. Its functionality is described in Section 8.8.

The Nb interface carries the user data between the MGWs. The protocol used to transport the data over the Nb interface is the user plane (UP) protocol as specified in 29.415 and 25.415. The interface offers both *transparent* and *support mode for predefined SDU sizes* and is identical to the Iu user plane protocol as described in Chapter 5. In support mode it allows strict control of the timing of the flow of media between MGWs and from MGW to the UTRAN (RNC). There are two transport protocol options available over this interface, IP and ATM. For the IP transport option user traffic will be carried in real-time transport protocol (RTP) packets, which are delivered over UDP/IP, and for the ATM option, traffic transport is AAL2 as it is in UTRAN.

Even though 3GPP does not require IP to be used for transport within the CN-CS domain there may be a number of benefits to be gained by doing so. This simplifies the call signalling for voice over IP (VoIP) calls. Also, the use of IP allows the operator to choose more freely from a range of network technologies at layer 2, including 10-gigabit Ethernet and IP over ATM. This allows the most appropriate choice to be made depending on bandwidth, quality of service (QoS) and budgetary requirements.

8.2.4   CS domain external interfaces

Apart from the interfaces to the RAN and the PSTN, the CS domain also needs to interface to a number of data sources including the HLR and other servers supporting user applications. Two traditional interfaces are defined between the CS core network and the HLR. The D interface is between the VLR and the HLR and allows the VLR to send updates about a mobile's location information to the HLR. The C interface is used by the GMSC when it needs to route a mobile-terminated call and must find out to which MSC server the call is to be routed. In both these cases the protocol used over these interfaces is the mobile application part (MAP), i.e. the mobility management application protocol for signalling system 7 (SS7).

8.2.5 CAMEL

CAMEL allows users to obtain the full range of services usually offered by their home network provider when roaming to a visited operator's network. Services such as pre-paid calling, caller ID and call forwarding on busy were traditionally provided within GSM networks using the intelligent network (IN). Unfortunately, the IN service could not be provided between operators since each IN service was proprietary and no standard protocol was developed so that operators could signal service capabilities for individuals roaming in a visited network. The CAMEL application part (CAP) provides a standardized protocol that can be used between operators. This protocol allows service descriptions and information such as pre-paid subscription details to be retrieved even when the user is roaming. The ability to provide pre-paid roaming is only a small part of the CAMEL service capability.

8.3 VOICE OVER IP (VoIP)

In UMTS from R4 onwards the use of Voice over IP (VoIP) becomes one of the dominant features. This provides a number of advantages over existing systems, including optimal use of bandwidth, integration of voice with data services and tandem free coding of calls made from mobile to mobile user. However, unlike its ATM counterpart the IP protocol suite was not originally designed to provide QoS. To ensure QoS a number of QoS enhancements for use with IP have already been developed such as differentiated services (DiffServ), resource reservation protocol (RSVP), RTP and multi-protocol label switching (MPLS).

8.3.1 VoIP call control

Within a conventional telephony network, SS7 protocols are used to handle call control by establishing a circuit for the data to follow. However, within an IP domain there is no such equivalent, since the IP address contained within the data packet is all that is required to route it to the correct destination. Two different general approaches have been proposed for handling call control within IP:

- Use the SS7 application protocols (and any other circuit switched call control protocols such as Q.931) unmodified, but transport them over the IP network transparently. For this approach a new transport protocol called stream control transmission protocol (SCTP), instead of transmission control protocol (TCP) or user datagram protocol (UDP), has been developed. This type of configuration is generally termed *sigtran* and has the advantage of being able to use the SS7 software already developed for the mobile network components with only slight modification.

- Replace the call control protocol completely with a new protocol optimized for the IP environment. There have been a number of different proposals for call control within IP networks, including H.323, session initiation protocol (SIP) and BICC. These new protocols have been optimized to work within the IP network.

Since some networks may be using SS7, some sigtran and some H.323, SIP or BICC, there is a need to be able to interwork between these protocols. Messages must be translated at the interfaces between the networks. For the R4 network show in Figure 8.2, the interworking function is performed by the MSC servers.

Also, as was discussed in Chapter 2, before voice data can be transmitted on a network it must be placed in packets. The original analogue signal must be sampled (or measured), converted to a digital form (quantized), coded, optionally compressed and encrypted and then put into packets.

8.4   REAL-TIME TRANSPORT PROTOCOL (RTP)

As mentioned, the data transport UP at the Nb interface is carried over either AAL2 (for ATM networks) or RTP/UDP (for IP networks). The RTP protocol (RFC 1889) provides end-to-end delivery of real-time audio and video over an IP network via the UDP. RTP provides payload-type identification, sequence numbering and delivery monitoring. Each payload that is sent using RTP is timestamped to ensure that it can be delivered to the CODEC at the correct rate. RTP is specified for use in R4 if IP transport is being used (29.414), to carry packets across the CS domain between the MGWs. It is also defined as the transport option for all real-time media for R5 in the IP multimedia subsystem (IMS) domain.

8.4.1   RTP at the Nb interface

Figure 8.3 shows the UP protocol stack for a voice call between two UEs. The UP protocol operating in *support mode* provides framing and timing functionality over both the Iu and Nb interfaces. With this means for communication within the R4 network, the delivery timing functionality of RTP at the Nb interface will be redundant. For the Nb interface, 3GPP states that the RTP timestamp can be ignored (29.414) and support for the real-time control protocol (RTCP) (see Section 8.4.5) is optional at the MGW. RTP in this case is only being used to encapsulate UP messages for IP transport.

For R5 networks, RTP is used to control the timing of media end-to-end, for example multimedia being transmitted between a pair of UEs. In this case the UP protocol will be operating in *transparent mode,* passing the data straight through, without the timing functionality.

The RTP header (Figure 8.4) starts with a version number, which is currently 2. The PTYPE defines the payload (for example G.711, G.723 audio or H.263 video).

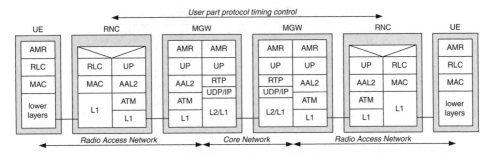

Figure 8.3   User plane for UMTS Release 4

0						31
VER	P	X	CC	M	PTYPE	Sequence Number
Timestamp						
Synchronisation source identifier						
Contributing source ID 1						
Contributing source ID 2						
...						
Contributing source ID N						

Figure 8.4   RTP header

Each payload type defines a profile, which not only defines the media type and CODEC but also specifies the granularity of the RTP timestamp and possible use of the M (marker bit) as described below. Table 8.1 shows how the payload types have been assigned. The payload has two types of payloads: static and dynamic. The static assignments are defined in a number of RFCs, in particular 1890. Table 8.2 describes some of the more common static payload types. The RTP clock frequency given in the table, defines how the RTP timestamp is to be incremented. So for an 8 kHz clock the timestamp will be incremented every 125 microseconds. Dynamic payloads are assigned when the session is created using some mechanism not covered by RTP (for example session description protocol (SDP) and SIP).

The comfort noise payload is for CODECs which do not provide a special format for comfort noise. RTP comfort noise packets contain a 7-bit value, specifying the required level of noise from 0 to -127 dBov (dBov $= 10\log_{10}(\text{S/Sov})$, where S $=$ signal strength and Sov $=$ signal strength that will overload system).

The RTP profile for adaptive multirate (AMR) is defined in RFC 3267. The RTP clock frequencies are assigned as 8 kHz and 16 kHz for AMR and AMR wideband, respectively. The payload type for AMR is assigned dynamically.

8.4.1.1   *RTP payload type in release 4*

For R4 networks the RTP payload is UP support mode packets; this UP/RTP configuration is referred to as the Iu framing protocol. In this case the payload type is dynamic and its

Table 8.1   RTP payload type assignment

Values	Assignment
0–34	Static
35–71	Unassigned
72–76	Reserved
77–95	Unassigned
96–127	Dynamic

Table 8.2   RTP static payloads

Payload type	Name	Type	RTP clock Frequency (Hz)	Tick time (s)
0	PCMA (G.711)	Voice	8000	125×10^{-6}
3	GSM	Voice	8000	125×10^{-6}
4	G.723.1	Voice	8000	125×10^{-6}
10	L16	Audio	44 100	22.6×10^{-6}
13	Comfort noise	Audio	8000	125×10^{-6}
31	H261	Video	90 000	11.11×10^{-6}

value negotiated when the bearer is established. The RTP clock frequency is 16 kHz, and the mime type is set to VND.3GPP.IUFP.

The sequence number, chosen at random at the start of each session, is used to keep the packets in order (remember RTP packets are carried using UDP with no guarantees that packets are delivered in the correct order).

The padding (p) bit indicates that the data contains padding. This is needed sometimes if the payload is encrypted. For padded RTP packets the last byte in the packet indicates the number of padded bytes added.

The X bit is set to 1 to indicate if the RTP header is followed by an extension header. The header extensions are application specific. The format of the header extension is a 16-bit field dependent on the application (extension type field) and a 16-bit length field giving the length of the extension in 32-bit words. RTP does not define any extensions itself.

The use of the marker (M) bit is dependent on the payload type. It can be used, for example, by a video stream to mark the beginning of each frame, when each frame is carried in multiple RTP frames.

The *TIMESTAMP* field tells the receiver when the media sample was taken. Two packets can contain the same timestamp if the data they contain was sampled at the same time. The meaning of the TIMESTAMP field is media dependent (i.e. dependent on the PTYPE field); for example, for G.711 and G.723.1 the RTP clock frequency is 8 kHz, therefore each new timestamp should be incremented by $125\,\mu s$. For video traffic, on the other hand, the sample time may require finer granularity (see Table 8.2). The initial timestamp value is chosen to be a random number at the start of the session and each subsequent value calculated relative to this initial value. To translate from packet timestamp to relative time from the beginning of the session the following formula is used:

$$Tr = (\text{Packet timestamp} - \text{initial\_timestamp}) \times \text{tick\_time}$$

8.4.2 Source identifiers

Each source of media in a session is defined by a unique 32-bit number called a *source identifier*. The identifier should be chosen randomly to reduce the chance of two stations picking the same identifier. Even though the chance of a collision (i.e. two stations using the same source ID) is small, each station is expected to provide some mechanism to detect its occurrence and remedy the situation. In particular, a station that detects another station transmitting with the same source identifier is expected to pick a new identifier randomly. It then will send an RTCP BYE message indicating that it is terminating the first media stream, and then use the new identifier in future transmissions. In the RTP header, two types of source identifier can be present, the *synchronization source identifier* and the *contributing source identifier*.

8.4.2.1 Synchronization source identifier

This is used to indicate the source host which sent the media packet. This could be the machine with the microphone which is sampling and coding the audio. Alternatively, if the audio is being mixed subsequently by a conferencing bridge, the identifier will be used to identify the bridge as the synchronization source.

8.4.2.2 Contributing source identifier and media mixing

RTP supports the mixing together of media streams, which is of particular use to applications such as voice and video conferencing. The CONTRIBUTING SOURCE ID fields define the source identities of the sources before mixing takes place. This field is variable length and can hold from 0 to 15 entries. The CC field contains a count of the number of contributing sources. An example of this is shown in Figure 8.5: four voice

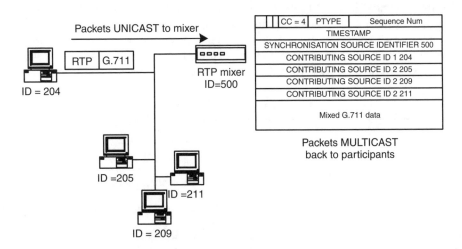

Figure 8.5 Conferencing with RTP

sources that have been mixed together are multicast to all the participants. Multicasting is desirable since it reduces the amount of traffic to be sent from the mixer by a factor of three. Use of the mixer also allows the participants to deal only with the well-known mixer address and not have to worry about directly communicating with other individual hosts.

8.4.3   Encryption with RTP

RFC 1889 states that RTP packets can be encrypted to provide security. The recommended technique is cipher block chaining using data encryption standard (DES). The correct decryption of the packet can be determined by certain header validity checks, for example checking that the version number is 2 and the payload type is known. There is no mechanism within RTP itself to allow for the exchange of encryption keys. This will usually be carried out at the call/session control stage of the call using protocols such as SIP and SDP.

8.4.4   Redundancy with RTP

The addition of redundancy with RTP allows some of the RTP packets to be lost in transit without loss of data. RTP redundancy format consists of a primary payload plus multiple secondary payloads. It is particularly useful for transmissions such as wireless where error rates may be high.

8.4.5   Real-time control protocol (RTCP)

RTCP (also covered in RFC 1889) provides control services for an RTP session. These include QoS feedback (e.g. round trip time, jitter, etc.), session identification and synchronization information. RTCP supports five different packet types, as shown in Table 8.3.

The RTCP packet types can be (and are expected to be) aggregated into compound RTCP packets to save on packet overhead. Since each RTCP packet type contains a fixed

Table 8.3   RTCP packet types

Type	Description
RR (receiver report)	Conveys reception statistics, sent by stations that only receive media
SR (sender report)	Conveys transmission and reception statistics, sent by active senders
SDES (source description)	Sent by a station that is sending media, provides text description of the sender
BYE	Sent by a station to indicate it is ending a session
APP	Used for application-specific extensions

header including a length field, demultiplexing the parts at the far end is not a problem. For compound RTCP packets, each packet should contain at least an SR or RR as well as a source description. Each of the five types is now described.

8.4.6   RTCP receiver report

Each RR packet contains zero or more report blocks. Each of these corresponds to a particular source of media and is prefixed with the 32-bit synchronization source identifier.

The format of the RR is shown in Figure 8.6. Each RTCP packet starts with the RTP *version number* (equal to 2) and the padding bit P as described previously. The next field, RC (report count), indicates how many receiver report blocks are contained in this report. The *packet type* is 201 to indicate receive report. The *length* indicates the total length of the packet and is represented in 32-bit units and actually stored as the length − 1. The final field in the common header is a 32-bit *synchronization identifier* indicating the sender of the report.

The receiver report also carries a number of report blocks. Each of these gives some indication of the quality of the transmission link, including jitter, packet loss statistics and timing delay between reports. The jitter is measured in the same units as the RTP timestamp. It is calculated as the smoothed average of the difference in spacing between the packets' actual sent time and the time they are received.

The deviation for a packet is defined as:

$$D_i = (AT_i - AT_{i-1}) - (TS_i - TS_{i-1})$$

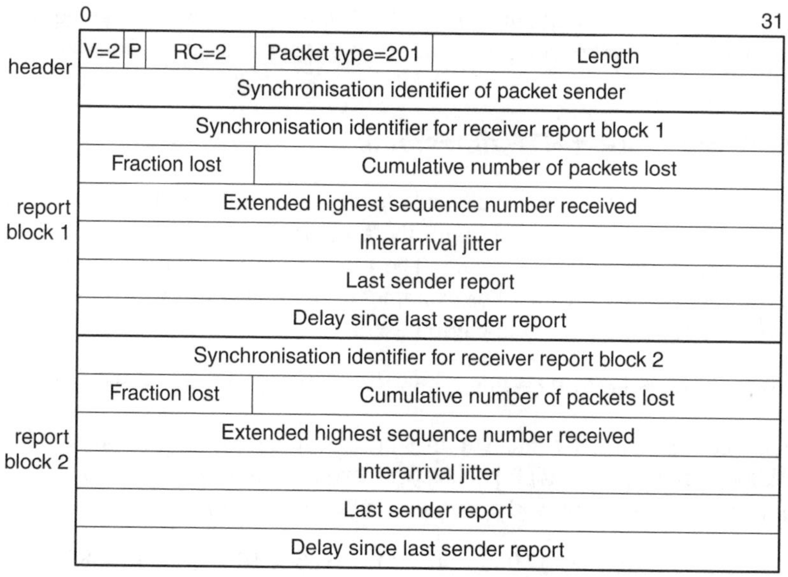

Figure 8.6   RTCP receiver report

where AT_i = arrival time of packet i and TS_i = timestamp of packet i. The jitter is calculated over the last 16 values using the following formula:

$$J_i = J_{i-1} + (|D_i| - J_{i-1})/16$$

The *last sender report* field holds the top 32 bits of the NTP timestamp associated with the last sender report packet received. The *delay since last sender report* field indicates the delay since receiving this sender report and sending the receiver report block. The round trip time can be calculated as follows:

RTT = Arrival time of receiver report block – Last sender report

– Delay since last sender report

8.4.7   RTCP sender report

An SR is transmitted by each active participant in a session. An SR also doubles up as a RR as it can also contain RR blocks. For senders with no statistics to report, the RC field is set to 0. The SR has a similar format to the RR but with some additional fields, namely the NTP timestamp, RTP timestamp, and packet and byte counts for the sender.

8.4.7.1   Network time protocol (NTP) timestamp

NTP is a protocol developed to allow a number of hosts to reach some agreement on absolute time (i.e. to synchronize internal clocks with each other.) NTP time is defined as the number of seconds elapsed since midnight, 1 Jan 1900. It is represented using 64-bit unsigned fixed-point storage composed of 32-bit integer and 32-bit fractional parts.

8.4.7.2   Sender report RTP timestamp

The RTP timestamp in the SR corresponds to the same time as the NTP absolute time. It is measured in the same units and with the same random offset of the RTP timestamp contained within the RTP media stream. The function of the RTP timestamp is to allow synchronization between streams. For example, to calculate the absolute NTP time for an RTP packet, the following formula can be used:

(RTP timestamp packet – RTP timestamp SR) $\times$ RTP tick time + NTP sender report

Consider that the last SR contained an RTP timestamp of 47 200 and an NTP time of 3 248 208 000 seconds. If an RTP packet is received with a timestamp of 47 900 using a G.711 payload (frequency of 8 kHz and tick time of 125 μs):

$$\text{Absolute time} = (47\,900 - 47\,200) \times 125 \times 10^{-6} + 3\,248\,208\,000$$

$$= 3\,248\,208\,000.0875 \text{ s}$$

Once the absolute time for various media streams can be established the streams can be synchronized by delaying the media packets which have the highest absolute time value. This is useful, for example, where video is sent in a separate stream from audio (lip sync), or multiple streams in the case of a conferencing application. The use of the timestamp to synchronize media from different hosts (each with a different NTP clock), is only possible if the hosts' NTP clocks are synchronized.

Possible solutions to this synchronization problem include the use of the NTP protocol or a standard external clock source such as one available from the global positioning system (GPS).

8.4.8 SDES source description

The SDES packet contains text descriptions about the sender of the media. Since the synchronization identifiers change, the SDES is needed to provide a binding between the synchronization identifiers and static descriptions of the participants in a session. Table 8.4 describes the different item types that can be used to identify the source.

8.4.9 BYE goodbye

This packet is used to tell the other end this stream of media is being terminated. Apart from the actual termination of a media stream, its use is also defined when handling synchronization number collisions (see Section 8.4.2). As well as a list of identifiers indicating which streams are being terminated, an optional string can be added giving the reason for the BYE message.

8.4.10 APP application defined

This message is defined to allow for experimental use to permit application developers to extended functionality without the RTP protocol needing modification.

Table 8.4 SDES item types

Name	Type	Description	Example
CNAME	1	Canonical end point identifier Provides a consistent name that the source can be identified from; usual format is user@host Expected to be unique for each source in session	user@pc1.orbitage.com
NAME	2	Real name for media source	Sebastian Coope
EMAIL	3	Email address of participant	sebcoope@orbitage.com
TEL	4	Telephone number of participant	+60356812
LOC	5	Location of participant	#14-15-06, Cambridge Business Park
APP	6	Name of application producing media stream	MS Netmeeting
NOT	7	Transient messages	'On-phone, can't talk'
PRIV	8	Private extensions	

8.4.11 RTP limitations

Even though RTP does have some loose session control via the use of SDES and BYE messages, it does not have the complex functionality required of a true call control protocol. For this reason RTP is invariably used with other session/call control protocols such as SIP, BICC or H.323. RTP also does not allow complex media control, such as the functionality provided by the real-time streaming protocol (RTSP), which allows media to be rewound, cued, forwarded etc.

8.5 SESSION DESCRIPTION PROTOCOL (SDP)

SDP (RFC 2327) is not really a protocol as such but a standard format used to represent session information. SDP descriptions include the type of media, including RTP payload type, timing of session, port numbers to send RTP data on etc. SDP is used by a wide range of other protocols to represent media options, including BICC, SIP, media gateway control protocol (MGCP) and MEGACO.

Table 8.5 shows some SDP fields and their meanings. Each field consists of a one-character token followed by the = sign and then a value.

The current version number is a major version number and is usually used to delimit multiple session descriptions within a block. The origin field defines who is announcing the session. This field starts with a username (sebcoope) followed by a session ID and a version ID. The session ID allows clients to distinguish between different sessions announced by the same media server. The version ID is used to distinguish between multiple announcements of the same session, which may have different media options. All this is followed by the origin address of the announcer of the session. The session name and description provide a short and long identifier for the session, the latter being used by clients to determine if they are interested in participating. The connection information defines the IP address of the source or destination of the RTP packets. This address is used by a receiver and tells a sender where to send RTP media or can be set by a multicast sender to indicate the address on which receivers should be listening.

Table 8.5 SDP field names

Token	Name	Example
v	Version	v = 0
o	Origin	o = sebcoope 345667 345668 IN IP4 128.5.6.7
c	Connection	c = IN IP4 128.5.6.7
s	Session name	s = SWDevel6
i	Session description	I = Development meeting for Orbitage msecure project
m	Media description	m = video 51372 RTP/AVP 31
a	Media attribute	a = rtpmap: 96 AMR a = ptime: 30
b	Bandwidth	b = 64
e	Email	sebcoope@orbitage.com

The media description field describes the type of media (video/audio), CODEC used and RTP port numbers. In the example given the media is video. This will be transported on port 51372 and is using CODEC type 31 (H.261). The attribute field allows the media description to be followed by a number of optional media attributes. One use of the attribute field is to allow the definition of RTP dynamic payload types. For example a = rtpmap 96 AMR defines dynamic payload type 96 to be AMR. The ptime attribute describes the packetization interval for the media CODEC in milliseconds.

The bandwidth field gives the bandwidth requirement for the session in kilobits per second, allowing the network to calculate QoS or admission requirements. In theory it is possible to work out the bandwidth requirement from the media description. This can be difficult, however, since it also depends on the overheads incurred when putting the data into packets. Finally, the email address of the originator of the session descriptor is given.

SDP is commonly used in an offer/response mode. In this case the first station may offer to communicate with various media options; the second station will respond with its preferred choice. The other possibility is to use SDP to announce sessions (such as a multi-casting transmission) to all possible participants, in which case no negotiation is possible.

8.6 MEDIA GATEWAY CONTROL

Media gateway control protocols were designed to allow the decomposition of gateway functions into separate control signalling and media translation functions. This arrangement can be seen in Figure 8.7.

The call signalling is separated from the media data and sent directly to the MSC server. The MSC server uses the H.248 messages to instruct the MGW. The media gateway provides a connection between the IP network and the external networks.

8.6.1 Evolution of media control protocols

The development of MGW control protocols commenced with simple gateway control protocol (SGCP). This was developed at Bell Research Centers (currently known as

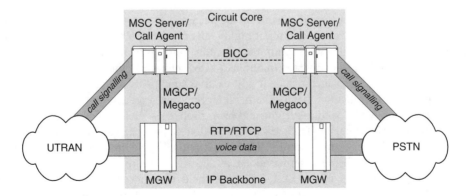

Figure 8.7 Media gateway control

Telcordia Technologies). It runs over UDP and was designed specifically to handle voice calls. Another similar protocol, IP device control protocol (IPDC) was developed by Level 3 Communication, Inc. and others.

In 1999 both IPDC and SGCP were merged into one new standard called MGCP. Without this merger the divergence of standards would limit the uptake of VoIP by causing interoperability problems between vendors. MGCP modifies and extends the functionality of both these earlier protocols. MGCP is a widely supported and somewhat mature protocol. It can be used to help scale fairly large VoIP networks and has wide support for most types of VoIP gateway. MGCP, however, has now been superseded by MEGACO or H.248, this is the protocol that is specified for UMTS R4 networks.

8.7 MEGACO

H.248, or MEGACO, as it is frequently known, has a number of improvements and enhancements in comparison with MGCP, this improves performance and provides a protocol which is better suited for multimedia and conferencing. MEGACO is actually a collaborative effort between the IETF and ITU-T. The specification for MGCP is now frozen and all future development work will be carried out on MEGACO instead. For MEGACO the softswitch component is referred to as the media gateway controller and in a 3G context this is integrated with the MSC server. To handle the 3G mechanisms, extensions are required to the base protocol and these are defined in TS29.232.

8.7.1 Terminations and contexts

Terminations and contexts exist at the MGW and are used as the basic abstraction to define connections across the gateway.

8.7.1.1 Terminations

A termination is the source and/or sink of media. Each termination can in fact sink or source multiple media streams, making it highly suitable for applications that need to, for example, synchronize video with audio. Each termination can relate to a physical resource such as a time slot in a time division multiplex (TDM) circuit. In this case the termination is referred to as semi-permanent since it will exist as long as there is provision made for it at the gateway. On the other hand ephemeral terminations created by the *add* command represent RTP or AAL2 media streams and have properties such as IP address and port number(s) or AAL2 channel ID(s). Each termination at the gateway is referred to by a unique name, called its termination ID. TS 29.232 defines a naming convention for TDM semi-permanent and ephemeral terminations, as shown in Figure 8.8.

Each termination ID consists of a 3-bit type field plus an extra 29-bit field which changes format depending on the type. Type values 000, 011–110 are reserved for future use. The root termination ID (type = 111) is used to refer to all the terminations within

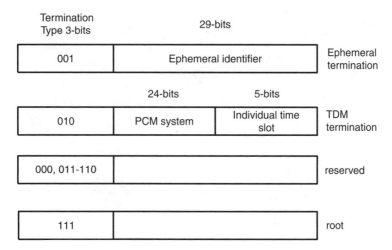

Figure 8.8 3GPP termination naming convention (ASN.1) (TS29.232)

the gateway in one go. In this case the remaining 24 bits are unused. For ephemeral termination (type $= 001$), the remaining 29 bits are unstructured and used to distinguish between the different terminations. For a TDM termination (type $= 010$) it is important that the MGC can distinguish between different PCM systems at the MGW, since these will represent different trunk connections. In this case the 29 bits carry a 24-bit PCM system identifier and a 5-bit time slot identifier (value $0-31$).

In augmented Backus–Naur form protocol syntax (ABNF) coding, the TDM termination is represented as **TDM_**PCM system/time_slot. For example TDM_5/20, would represent PCM trunk number 5, TDM slot 20. For ephemeral terminations the format is **Ephemeral_**ephemeral_id, for example Ephemeral_2050.

8.7.1.2 Contexts

Contexts define the connectivity of a termination. All terminations within a context will send and receive media between each other. If a termination is a media sink it will receive all media sent by other terminations in the same context. One can see that by moving terminations between contexts the connectivity between terminations can be controlled. There is a special context, called the null context. In this case the termination is not associated with any other termination.

Figure 8.9 shows how media can be connected using the context, termination model. In the top case two switched circuit network (SCN) channels are connected to a media stream on the IP network. This could be, for example, a conference call between three parties. The bottom example shows a simple one-to-one call. The middle example shows two terminations that are currently not associated. These terminations may have been moved into the null context. This would be the case, for example, if one of the parties had put the other on hold.

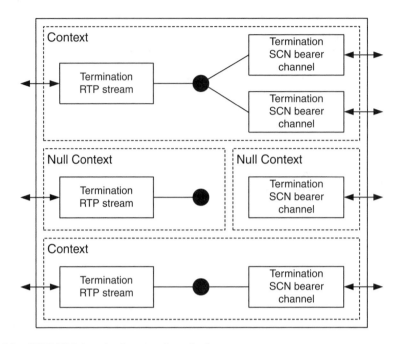

Figure 8.9 MEGACO terminations and contexts

8.7.2 Events and signals

Events are generated by a termination and detected by the MGW, which can then forward them to the MGW controller. An event, for example, could be a phone going offhook or the termination receiving some DTMF digits from a telephone. An MGW controller will request to be informed of certain events using the *modify* message. The MGW will inform the controller of the occurrence of an event using the *notify* message.

Signals are line conditions (such as ringing or busy tone) that can be applied to a termination within the gateway. The MGW controller can request the gateway to apply these signals to a termination using the *modify* command. Figure 8.10 shows the use of *modify* and *notify*. The call agent initially uses the *modify* command to request the gateway to listen for the offhook condition at termination T1. When the telephone receiver is picked up, this is detected at the gateway, which then sends the *notify*, containing the *offhook* event, back to the call agent. The call agent then sends another *modify*, instructing the gateway to apply the dial tone to the user's phone line.

All the events and signals are grouped into packages, some of which are shown in Table 8.6. A termination will support a particular package depending on its properties. For example, an analogue line will support the analogue line supervision, DTMF tone detection and call progress tone generator packages.

For each package the name of the events or signals within that package are prefixed by the package ID. For example, within the analogue supervision package the offhook event is named *al/of*, onhook *al/on* and the signal to apply ringing to the line *al/ri*. As

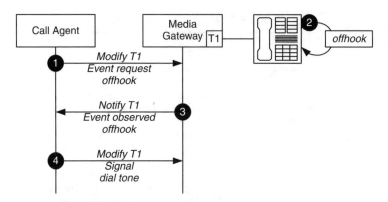

Figure 8.10 MEGACO modify and notify commands

Table 8.6 MEGACO packages

Package name	Package ID	Comment
Analogue line supervision	Al	Detects offhook, onhook, flashhook etc. Allows ringing to be applied to line
DTMF tone detection package	Dd	Detects DTMF digits dialed from a telephone
Call progress tone generator	Cg	Provides for the generation of call progression tones such as dial tone, busy tone, ringing tone etc.
Network package	Nt	Properties of network terminations, e.g. bytes received and sent
RTP package	Rtp	Provides statistics about RTP connection; for example, packets sent, received, average jitter and latency
TDM circuit	Tdmc	Controls functions such as echo cancellation and gain control

well as event and signals a termination's support for a given package defines properties and statistics.

8.7.2.1 Termination properties

The properties of a termination define its operational characteristics. Some properties are read-only and can be used by the gateway controller to determine what the termination is capable of. Other properties are read/write and can be modified to change how the termination will behave. One example of a property is the TDM circuit package echo cancellation property. This is a Boolean value with read/write permissions. It can be set to false to disable echo cancellation for the line.

8.7.2.2 Termination statistics

As well as properties, the termination may support a range of statistics (also defined by its package membership). For example RTP terminations supporting the RTP package, will record statistics such as jitter and delay (see Table 8.6).

Table 8.7 MEGACO commands

Command	From	Purpose
ServiceChange	Media gateway	Indicates gateway has come into or gone out of service (e.g. on reboot) Indicates termination has come into or gone out of service
ServiceChange	Media gateway controller	Instructs gateway to take termination in or out of service
AuditCapabilities	Media gateway controller	Gateway returns all possible values for a termination's properties (e.g. what CODECs it supports, data rate, etc.)
AuditValue	Media gateway controller	Determines current values of termination properties such as the current context, event values etc. for a given termination
Modify	Media gateway controller	Changes the properties of a termination, for example marking which events should be notified and setting signals on a line
Notify	Media gateway	Notifies the media gateway controller of an event occurring at the media gateway
Add	Media gateway controller	Adds a termination to a context
Subtract	Media gateway controller	Removes a termination from a context
Move	Media gateway controller	Moves a termination from one context to another

8.7.3 MEGACO commands and descriptors

MEGACO defines eight different commands, as shown in Table 8.7.

Parameters for MEGACO commands are called descriptors; some are optional while others are mandatory, depending on the specific command.

Table 8.8 shows some of the more commonly used descriptors. Some of the parameters support wild card options, which allows a command to be performed for multiple terminations or contexts at the gateway. The wild card symbol * is used to indicate all contexts or all terminations. It can be used as part of a termination's name to indicate a range of values; for example, the terminations aaln/1 aaln/2 aaln/3 could all be acted on together using the termination id aaln/*.

It is also possible to set a parameter to choose the value. This is indicated by the $ symbol. In this case the gateway is expected to choose a value and return it to the MGW controller. This is used when creating new terminations or contexts. For commands that have no context, the context ID is set to null, which is indicated by the − symbol. Finally, for a command to relate to all terminations at the gateway, the termination ID can also be set to a value of ROOT.

To save on overhead, MEGACO commands are grouped together. Each MEGACO message is called a transaction. A transaction consists of a series of actions and each action can consist of a series of commands, which all operate on a single context. This grouping is illustrated in Figure 8.11.

Table 8.8 MEGACO descriptors

Name	Purpose	Comment
ContextID	Identifies a context	Unsigned 32-bit
TerminationID	Identifies a termination	Arbitrary string, can have structure e.g. ds3/2/28 could be 28th slot of the DS3 card plugged into slot 2
LocalDescriptor	Local RTP parameters	SDP format for RTP parameters allocated by gateway when creating local IP terminations
RemoteDescriptor	Remote RTP parameters	SDP format for RTP parameters requested by gateway controller when connecting the gateway to a remote gateway
EventsDescriptor	Request for events to be detected at the gateway	Set by gateway controller
ObservedEvents Descriptor	Used by notify to indicate events detected at the gateway	Set by media gateway

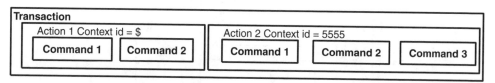

Figure 8.11 MEGACO transaction

8.7.4 Context and termination handling (bearer establishment)

For the MSC servers in the R4 network with a mobile originated call, the creation of the bearer to handle the call will be initiated by an CC SETUP message. Figure 8.12 shows how the MSC servers establish the bearer in the case of forward establishment (i.e. where the initiating gateway controls the network bearer creation). In this example the MSC server and MGW serving the RNC are referred to as local. The gateway MSC server and MGW connecting the core network to the external PSTN are referred to as remote. The steps are as follows:

1. The local MSC server receives CC SETUP from the originating UE.
2. The local MSC server determines which remote MSC server to forward call, then forwards the BICC IAM message.
3. The remote MSC server decides it needs to forward the call in PCM trunk 1. It instructs the remote MGW to create an association between the IP network (RTP media and TDM line on PSTN side) and forwards the IAM message onto the PSTN

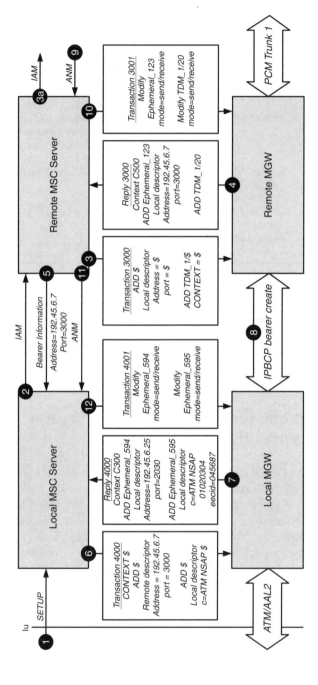

Figure 8.12 Forward bearer establishment

network (3a). Note that the MSC server uses the chosen format of the termination id (TDM_1/$).

4. The remote MGW creates an association and sends a reply back to the MSC server. This reply contains the RTP termination local media options (IP address and RTP port) as well as the chosen termination id for the TDM termination, TDM_1/20.

5. Media options from step 4 are sent from the remote to the local MSC server.

6. The local MSC server instructs the local MGW to create an association between the ATM/AAL2 and the IP network. To do this it adds two ephemeral terminations, IP and ATM. The ADD for the IP termination uses the RTP media options returned in step 4 to set the remote IP address for the IP bearer. The ADD for the ATM termination sets the connection address as c = ATM NSAP $, instructing the MGW to create an AAL2 connection.

7. The local MGW creates termination and context. It sends a reply back to the local MSC server containing termination and context IDs for the association.

8. The local MGW creates an IP bearer connecting to the remote MGW. This is done using the IP bearer control protocol (IPBCP; see Section 8.8.5).

9. The call is answered; ANM message is received at the remote MSC server.

10. The remote MSC server sets all the terminations for the call set to send/receive mode, i.e. full duplex, using the modify command.

11. ANM is forwarded to the local MSC server.

12. The local MSC server sets all the terminations for the call set to send/receive mode, i.e. full duplex, using the modify command.

Figure 8.13 shows the MEGACO commands for steps 3 and 4. The context ID is $, which means the media gateway is expected to create a new context and add in the terminations. The termination ID for the second add is $. This means that the termination is to be created by the gateway. The media requested for the RTP termination is an RTP payload type 96. Since this a dynamic payload, the media description is followed by the rtpmap attribute, which maps payload type 96 to 3GPP Iu framing protocol. Notice that the IP termination is set to send-only and the TDM termination to receive-only. This enables the media to flow across the remote MGW only from right to left (Figure 8.12). This is because the call has not been answered by the other party as yet, and the media stream is being used to pass call progression audio or early media (busy, ring tone etc.) back to the caller's phone. The gateway responds with the details of the media streams (payload type, IP address and port number) and the identity of the newly created or chosen terminations and context. The IP address and port number for the RTP termination are sent back in the bearer information message (step 5).

Figure 8.14 shows the MEGACO transaction to create the local association (step 6). The RTP termination is the first to be added. The MGW is passed the remote media options (IP address and RTP port) and will use this information to set up the bearer to the remote MGW using IPBCP (step 8).

The second termination represents the AAL2 connection to the RNC. Since no remote options are included in this request, the MGW will not set up the AAL2 connection at

```
Transaction = 3000 {
    Context = $  {
          Add = TDM_1/$  { Media {
              Stream = 1 {
                  LocalControl {Mode = ReceiveOnly} }},
          },
          Add  = $ {
              Media {
                      Stream = 1 {
                        LocalControl {
                        Mode = SendOnly,
                        },
                      Local {
                         v=0
                         c=IN IP4 $
                         m=audio $ RTP/AVP 96
                         a=rtpmap: 96 VND.3GPP.IUFP/16000
                      },
                  }
              }
          }
      }
}

Reply = 3000 {
    Context = C500 {
        Add = TDM_1/20,
        Add = Ephemeral_123 {
            Media {
                Stream = 1 {
                  Local {
                     v=0
                     c=IN IP4 192.45.6.7
                     m=audio $ RTP/AVP 96
                     a=rtpmap: 96 VND.3GPP.IUFP/16000
                  }
              }
          }
      }
}
```

Local descriptor
SDP format
v=SDP version number used as delimiter
c=connection information IP or ATM address
m=media descriptor, port number, CODEC
a=ptime: packtisation interval
a=rtpmap: dynamic payload definition

Figure 8.13   Remote association creation (transaction and reply) (steps 3 and 4)

this stage. Instead, the MGW prepares itself to receive an incoming AAL2 connection from the RNC and returns the ATM network service access point (NSAP) address in its reply back to the MSC server. This NSAP address is then passed to the RNC (in the RANAP RAB request message), which then uses it to establish the AAL2 connection with the MGW.

With the ATM termination the SDP local descriptor looks slightly different. The connection information defines an ATM address (c = ATM NSAP $). The media description defines a payload of AAL2/ITU with a payload type of 96. AAL2/ITU simply means the raw encapsulation of media within an AAL2 frame. The dynamic payload type for ATM is defined by the attribute a = atmap: 96 VND.3GPP.IUFP/16000.

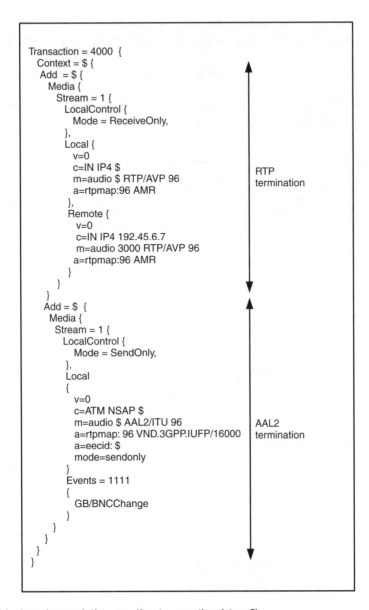

```
Transaction = 4000 {
  Context = $ {
   Add  = $ {
     Media {
       Stream = 1 {
         LocalControl {
           Mode = ReceiveOnly,
         },
         Local {
           v=0
           c=IN IP4 $
           m=audio $ RTP/AVP 96
           a=rtpmap:96 AMR
         },
         Remote {
           v=0
           c=IN IP4 192.45.6.7
           m=audio 3000 RTP/AVP 96
           a=rtpmap:96 AMR
         }
       }
     }
   }
   Add = $  {
     Media {
       Stream = 1 {
         LocalControl {
           Mode = SendOnly,
         },
         Local
         {
           v=0
           c=ATM NSAP $
           m=audio $ AAL2/ITU 96
           a=rtpmap: 96 VND.3GPP.IUFP/16000
           a=eecid: $
           mode=sendonly
         }
         Events = 1111
         {
           GB/BNCChange
         }
       }
     }
   }
  }
}
```

RTP termination

AAL2 termination

Figure 8.14 Local association creation transaction (step 6)

The eecid is the end-to-end call identifier. This is a unique value chosen by the MGW and passed back to the MSC server. This value (referred to as the binding ID) is then sent to the RNC along with the NSAP address and used in the AAL2 connection establishment request. Now the MGW can use the eecid value to match up the incoming AAL2 request from the RNC with its corresponding AAL2 termination. The BNCChange event occurs whenever there is a change in the status of the bearer network connection. It will be triggered when the RNC establishes the incoming connection to the MGW. This allows the MSC server to be informed as soon as the AAL2 bearer has been established from the RNC.

The reply to the local association creation transaction is shown in Figure 8.15. In this most of the fields marked $ (including the termination and context IDs) in the transaction have been filled in with values by the MGW. In particular, the values for the NSAP and the eecid have been allocated by the MGW.

```
Reply = 4000 {
  Context = C300 {
    Add  =  Ephemeral_594 {
      Media {
        Stream = 1 {
          LocalControl {
            Mode = ReceiveOnly,
          },
          Local {
            v=0
            c=IN IP4 192.45.6.25
            m=audio 2030 RTP/AVP 96
            a=rtpmap: 96 VND.3GPP.IUFP/16000
          },
          Remote {
            v=0
            c=IN IP4 192.45.6.7
            m=audio 3000 RTP/AVP 96
            a=rtpmap: 96 VND.3GPP.IUFP/16000
          }
        }
      }
    }
    Add =  Ephemeral_595  {
      Media {
        Stream = 1 {
          LocalControl {
            Mode = SendOnly,
          },
          Local
          {
            v=0
            c=ATM NSAP 47.0021.6732.0000.0060.3224.6701.9990.3e64.fd01.00
            m=audio $ AAL2/ITU 96
            a=atmmap: 96 VND.3GPP.IUFP/16000
            a=eecid: BB 56 90 12
            mode=sendonly
          }
          Events = 1111
          {
            GB/BNCChange
          }
        }
      }
    }
  }
}
```

Figure 8.15   Local association creation reply (step 7)

When the call is answered, indicated by the ISUP answer messages ANM, the remote and local MGWs are asked to modify their terminations to send/receive instead of just receive-only. This allows the speech to flow in both directions. The MEGACO transactions for this are illustrated in Figure 8.16.

8.7.5 Deleting contexts and bearers

When the call is finished the contexts and terminations have to be deleted at the MGWs to free up the resources. This is done using the subtract command. This transaction must be carried out at both the local and remote gateway to free up the resources there as well.

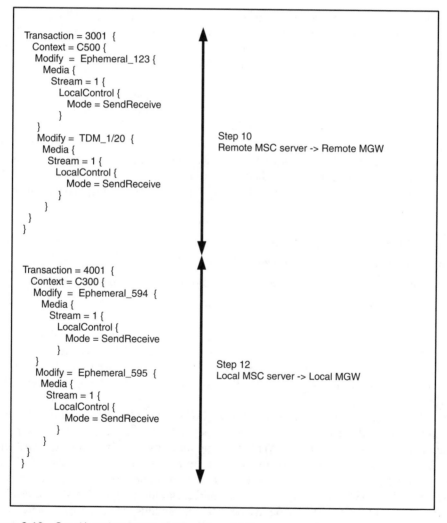

Figure 8.16 Send/receive enable (steps 10 and 12)

8.7.6   Summary

MEGACO is the enabling protocol for the separation of switching and call control functions within the network (softswitch architecture). It is recommended by 3GPP for UMTS networks since it allows for greater scalability and reliability. Call control capacity and switching capacity can be increased independently, allowing more flexibility of scaling. Multiple media gateways and MSC servers can be used within a redundant architecture to allow for carrier grade reliability.

8.8   BEARER-INDEPENDENT CALL CONTROL (BICC)

As mentioned previously, before a call can be transported across the IP core network, some type of call/session control has to be initiated. This is to allow a call which was initiated within the UTRAN to be terminated within the PSTN or vice versa. With current PSTN networks this call control is largely done using ISUP (ISDN user part). One possible approach to handle calls across the IP network is to use a new protocol such as SIP, and emulate all PSTN services using SIP call control. The problem with this, however, is that services are somewhat complex and emulation will not always produce the correct result. The ITU-T view on this was that only a protocol that was 100% compatible with ISUP would be suitable, and that all services would have to be emulated exactly. This would mean that two PSTN customers having their calls connected via an IP network would be unaware that their call had been handled any differently.

The protocol that was developed to achieve this is called bearer-independent call control (BICC), and has the following features:

- compatible with and based on the ISUP protocols;
- voice/data call capability (but no special multimedia features);
- independent of the underlying network technology (hence bearer-independent);
- support for tandem-free coding of mobile calls.

Since BICC is bearer-independent, it also requires a mechanism to establish bearers across the network it is using to transport its calls. The ITU-T has provided recommendations for running BICC over both ATM and IP. It is envisioned, however, that the long-term goal is to use BICC for IP networks with ATM being provided as an interim solution.

BICC development started with capability set 1 (CS1). For CS1 the only transport option available was ATM. In this case, a network operator can migrate their network from a traditional TDM solution to use of ATM, allowing for transport of real-time over AAL1 or AAL2. CS1 is essentially designed to allow packet-based networks to trunk ISDN calls. With BICC CS1, all supplementary service information is tunnelled transparently in BICC messages. BICC CS1 was released as an amendment to the original ISUP specifications.

BICC capability set 2 (CS2) builds on CS1 but in this case, a totally new set of specifications were released (Q.1902.x), and nodes are expected to support local exchange functionality. A BICC CS2 node therefore has to provide all the ISUP supplementary services (call forward on busy, etc.) as well as basic bearer establishment. CS2 allows the bearer options to include IP. To allow for bearers to be established across the IP network a new protocol, BICC IP bearer control protocol (Q.1970), was developed.

Figure 8.17 shows a simplified diagram of the BICC architecture. Notice that BICC, like MEGACO, splits the control of the network into call control and switching or bearer control. The call service function (CSF) sited at the MSC server performs the following functions:

- call control;
- interfaces to the ISUP network;
- interface to the bearer network to forward requests using BICC;
- interfaces to the bearer control function (BCF) to request for bearers to be established.

The bearer interworking function (BIWF, corresponding to the UMTS MGW) performs translation between the two different networks, for example between TDM and ATM or ATM and IP. Functions provided at the BIWF will include:

- interfacing to the user plane;
- CODEC translation;
- bearer establishment and tear down.

The BCF provides the signalling required to establish new bearers. The CSF controls the BCF, requesting the establishment of bearers and internetworking between them using the BICC bearer control protocol.

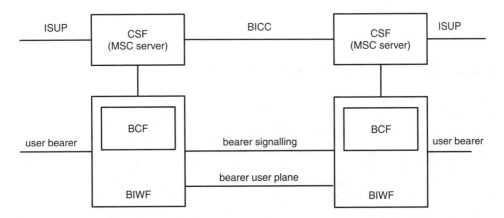

Figure 8.17   BICC architecture

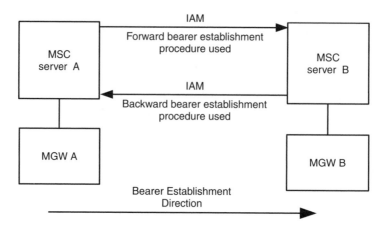

Figure 8.18   Use of forward/backward bearer establishment

8.8.1   Forward and backward bearer establishment

In forward bearer establishment the bearer setup control is initiated in the same direction as the IAM message which starts the call. It is possible, however, for the bearer to be established in the opposite direction to the IAM. This is referred to as a backward bearer establishment procedure. Allowing the transport bearer establishment to be performed in either direction (forward or backward) allows one MSC server to be in control of network resources, simplifying control. This is illustrated in Figure 8.18. MSC server A on the left is responsible for controlling admission by only allowing calls to be established if there is enough bandwidth available within the core network. For calls in which MSC server A originates the IAM message, forward bearer establishment is used; for calls forwarded from MSC server B, backward bearer establishment is used. In both case MSC server A is in control of bearer establishment.

Another example of forward/backward bearer establishment is with the AAL2 connection made from the RNC towards the core network, since the RNC always establishes the connection.

8.8.2   BICC messages and parameters

Table 8.9 shows some of the more commonly used BICC messages. The messages are used in very much the same way as their ISUP equivalent. Basic call setup progression for forward bearer establishment is shown in Figure 8.19. In this diagram CSF-N and BSF-N refer to CSF and BSF node, respectively.

The IAM message is used to initiate the call setup. This message contains the called party's number, and is forwarded to the CSF node serving the called party, or the CSF node connected to the PSTN which serves the called party. The APM message is a general purpose message used to forward non-BICC information between BICC users. In response to the IAM, an APM message is sent in the reverse direction. This message

Table 8.9   BICC messages

Message	Acronym	Meaning/purpose
Initial address	IAM	Initiates call setup
Application transport	APM	Carries non-BICC information
Address complete	ACM	Indicates all addressing information has been received
Answer	ANM	Called party has answered phone
Release	REL	Request release of bearer
Release confirm	RLC	Confirm release of bearer

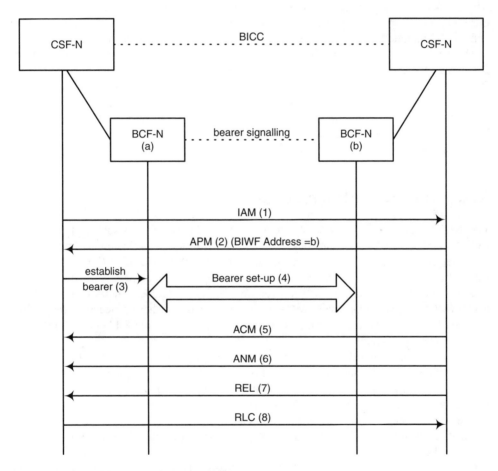

Figure 8.19   BICC forward bearer establishment

contains bearer information, including the remote gateway's local media options and its interworking address (i.e. IP address and RTP port number or ATM address). On receipt of the APM the local CSF-N requests its BSF-N to set up the bearer to the remote BSF-N using the bearer information contained in the APM message (this request is carried in an H.248/MEGACO transaction).

The bearer information is used to establish the bearer between the media gateways, which is performed using a non-BICC protocol. Once the call has been routed all the way to the destination, an ACM message is sent in the reverse direction. At this point, the destination phone starts ringing and a ringing tone is played to the caller. When the called party answers the phone, the ANM message is sent in the reverse direction, the ringing signals are removed from the lines and the media is connected in both directions. Finally, when one of the parties hangs up, release messages are used to clear the call.

Each BICC message can contain a number of parameters, a few of which are shown in Table 8.10. The call instance code (CIC) is used to relate a number of BICC messages which belong to a particular call. The called party number is used to help route the IAM message to the correct destination. The calling party number is used to identify the caller and can be included in the IAM message to allow for services such as caller ID.

The application transport parameter allows the standard BICC message to carry non-BICC data for another application. This could be, for example, charging information. In the examples given in this section this information is related to the bearer control.

The destination for the application transport message can be an application residing at the CSF-N, or if tunnelling is used the parameter contents will be forwarded to the BCF-N without being examined by the CSF. This mechanism is used by the IPBCP to allow transport of bearer establishment requests between peer BCF-Ns.

8.8.3   Bearer control function

The BICC bearer control protocol operates on the interface between the CSF-N and BCF-N. Defined in Q.1950 and based on H.248 with added packages, it defines a set of bearer control transactions and maps them onto H.248 transactions (i.e. add, modify, subtract etc.).

The 3GPP defines bearer control procedures for R4 in 29.232 (Media Gateway Controller (MGC) – Media Gateway (MGW) Interface). Most of these procedures map without modification to the transactions defined in Q.1950, and Table 8.11 lists some of the more common procedures.

As well as describing procedures for bearer establishment, TS29.232 also defines a number of new UMTS-specific packages. For example, the 3GPP 3gup/interface property is set, depending on which interface the termination resides on (possible values are RAN for the Iu interface and CN for the Nb interface).

The *prepare bearer* procedure is used to create terminations and contexts at the gateway but without bearer establishment taking place at the same time. This is used in a case

Table 8.10   BICC parameters

Name	Size (bits)	Use
Call instance code (CIC)	32	Relates multiple
Called party number	Variable	Destination of call
Calling party number	Variable	Source of call
Application transport	Variable	Application information, e.g. bearer control information

Table 8.11   Media gateway control procedures (29.232)

3GPP MGW control procedure	Q.1950 transaction	Possible H.248 commands
Prepare bearer	Prepare_BNC_Notify	Add, modify, move
Establish bearer	Establish_BNC_Notify	Add, modify, move
Change through-connection	Cut_Through	Add, modify, move
Active interworking function	N/A	Modify
Activate voice processing function	Echo canceller	Add, modify, move

where the bearer establishment is done by the remote media gateway or by the RNC. An example of the *prepare bearer* function is step 3 in the MEGACO call example shown in Section 8.7.4. A local termination is created and the internetworking address of the MGW is returned to the MSC server. This procedure is represented by H.248 add or modify commands. The modify command is used when an MSC server wishes to reuse a termination that already exists for a new call.

The *establish bearer* procedure requests that the media gateway creates the context and termination as well as establishing a bearer to the remote MGW. The MSC server provides the bearer address, binding reference and bearer characteristics requested for this connection. In this case the H.248 command also contains the remote address of the MGW to connect to. This procedure corresponds to step 6 in the MEGACO call example.

The *change through-connection* procedure alters the mode of a termination at the gateway to send/receive. It is commonly carried out at the same time as the *prepare bearer* or *establish bearer* procedures but can be executed on its own using the H.248 modify command.

The *activate interworking* function is entirely new (not from Q.1950) and defined wholly within TS29.232. This procedure activates the negotiation of any layer 2 protocols that may be required when making circuit switched data calls, for example when making a data call to a modem in the PSTN (see Figure 8.20). In this case, on receipt of the

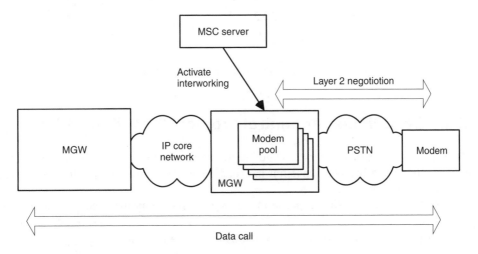

Figure 8.20   CS domain data call

Table 8.12 Circuit switched data package

Name	Type	Parameters	Description
Actprot	Signal	Local peer role Values orig or term	Activates layer 2 negotiation
Protres	Event	Result Values success, failure	Returns result of negotiation

activate interworking request the media gateway is allowed to allocate a modem out of its pool and start connection procedures to the remote modem. Once the connection is successful (or has failed) the MGW will notify the MSC server of the result. The *activate interworking* function is supported in H.248 by the addition of the circuit switched data package (29.232), some of which is described in Table 8.12.

The *activate interworking* procedure is executed after the call has been set up between the two end points. The procedure consists of an H.248 modify command being sent to the MGW applying the *actprot* signal and requesting notification of the result of the negotiation (notify on *protres*).

The *activate voice processing* function enables the echo cancellation function (and any other relevant voice processing functions) for terminations that support it. This procedure consists of an add, modify or move command which modifies the tdmc/ec (echo cancellation property).

8.8.4 Bearer control protocols

Since the CS core network can be based on ATM (AAL1 or AAL2) or IP, bearer establishment procedures have to be defined for both of these networks. The 3GPP specifies the following options for bearer control across the core CS domain (29.414):

- AAL2 connection signalling (Q.2630.2)
- BICC IP bearer control protocol (Q.1970).

For more detail on AAL2 connection signalling see Chapter 7.

8.8.5 BICC IP bearer control protocol (IPBCP, Q.1970)

IPBCP allows a pair of IP peers to negotiate IP addresses, RTP port numbers and media characteristics for a bearer connection between them. All media options for IPBCP are described using SDP.

The protocol uses an offer response mode, and the following messages are supported:

- *Request*: requests establishment or modification of media bearer
- *Accept*: accepts establishment or modification of media bearer

- *Confused*: response when peer does not understand request format
- *Rejected*: rejects request to establish or modify bearer

In general, an initiator will send a request message and wait for an accept or reject. If it does not receive a reply within a given time it will send the message again.

8.8.5.1   *IPBCP tunnelling*

The IPBCP messages, instead of being sent on a dedicated connection directly between MGWs (Nb interface), are tunnelled over the Nc and Mc interfaces. This is illustrated in Figure 8.21. IPBCP messages are encapsulated in the BICC tunnelling protocol (Q.1990). The BICC bearer control tunnelling messages are encapsulated in BICC APM messages or optionally piggybacked on other BICC messages (e.g. IAM) for transport over the Nc interface. The tunnelling information is passed over the Mc interface using events and signals defined in the tunnel package (Q.1950).

Tunnelling has two main advantages. First it reduces the time for bearer negotiation, since a new TCP connection setup is not required and the IPBCP messages can be piggybacked on the BICC call control messages. It also allows the IPBCP to be carried over the Nc signalling interface across the core, supporting the SCTP protocol, which is more suited for signalling transport.

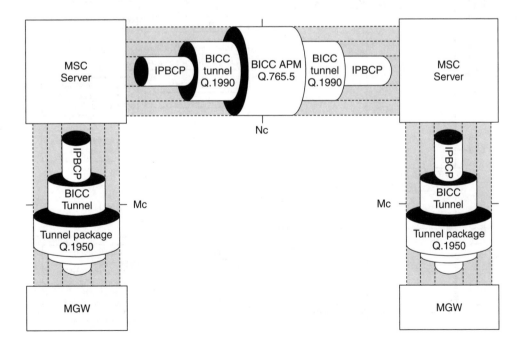

Figure 8.21   IPBCP tunnelling

8.8.6   BICC call flow examples for release 4

Figure 8.22 shows the call flow for BICC in a UMTS R4 network. The example given is for a mobile originated call with the bearer establishment in the forward direction.

1. The call starts with the mobile forwarding its call setup request to the MSC server. The MSC server replies to the mobile with a call proceeding message.
2. It then proceeds to prepare a bearer for the call on its core network side by issuing the *add* command to the local MGW. The local MGW responds with the identifiers for the new context and termination created. The local MGW will also pass up the description of its local media options in the IPBCP request message (in SDP format).
3. The MSC server then creates an IAM message and examines the called party number to determine the correct remote MSC server to which it should forward the call. This IAM message also contains the tunnelled IPBCP request.
4. The remote MSC server will prepare a bearer on its own MGW using the *add* command. This *add* command also contains the tunnelled IBCP message destined

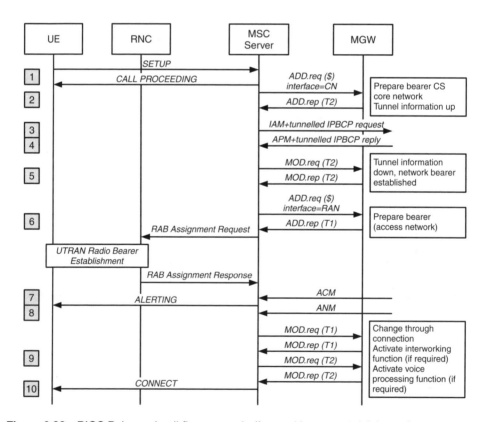

Figure 8.22   BICC Release 4 call flow example (forward bearer establishment)

for the remote MGW. Once the remote MGW has created the termination it will reply; its reply contains the tunnelled IPBCP response. This IPBCP response is tunnelled in an APM message which is sent back to the original MSC server.

5. The local MSC server passes the IPBCP response down to the local MGW using the *mod* request. At this point the bearer will have been established across the core network between the two MGWs.

6. The MSC server then prepares a bearer for the radio access side of the call. It uses the *ass* command to create a termination on the Iu interface. The reply to the *add* request at this point will return the IP address and RTP port number for the termination. The MSC server will then proceed to establish a radio access bearer over the RAN to the UE, by sending a *RAB assignment request* to the RNC. The RNC will then reply with the *RAB assignment response* to indicate that radio bearer has been established.

7. The ACM message being forwarded from the call destination initiates the MSC server to send the *altering* message to the UE. At this point, the destination will be ringing and the ringing tone will be being played to the caller.

8. The called party answers the phone and the ANM message is returned to the MSC server.

9. On call answer, the MSC server instructs the MGW to patch the call through. This will allow media to flow between the two parties. The *activate voice processing* function for voice calls and *activate interworking* function for data calls may also be requested at this point.

10. The connect message is sent back to the user equipment to signal connection success.

8.8.7   Tandem-free and transcoder-free operation

CODEC tandem operation is where media sent between two MGWs must be translated before transmission. For example, if one MGW is receiving a call on its external interface coded as G.723.1 and the other MGW is forwarded the media in GSM format, the speech will have to be transcoded before being forwarded. The usual way to do this is to translate the media into PCM before transmitting it between the two gateways. This process is illustrated in Figure 8.23. The reason for the use of PCM as a common CODEC is that every MGW must support it, therefore its use guarantees interoperability.

For the MGWs to operate tandem free, i.e. without transcoding to PCM, they would have to use common CODECs on their external interfaces. This is illustrated in Figure 8.24.

Figure 8.23   Tandem operation

Figure 8.24 Tandem-free operation

Figure 8.25 Transcoder-free operation

If the media for a call requires no transcoding between sender and receiver, this is referred to as transcoder-free operation (Figure 8.25). This will involve every MGW pair from source to destination to be operating tandem-free.

In general transcoding is undesirable since it can increase delay and distort the original signal. Also, since PCM has the highest bit rate of any telephony CODEC, there is also the issue of wasted bandwidth.

Figure 8.26 illustrates the transcoding issues for UMTS. For a call from the UTRAN to the PSTN, the AMR code will require transcoding between AMR and G.711, either at MGW-a or MGW-c. This is because the PSTN works on 64 kbps connections while the cellular network with its limited capacity, compresses the voice and works on less than 13 kbps.

To enable a transcoder-free operation from one UTRAN user to another UTRAN user, both AMR users will have to use the same CODEC.

For a call from the UTRAN to the GERAN (which is GSM-based), transcoding may or may not be required since the AMR CODEC supports GSM EFR as one of its options, therefore it may be possible to send packets from source to destination with no CODEC translation. It can be seen that within a single operator's realm, implementing transcoder-free operation should present a number of technical challenges. However, for the mechanisms to work between operators is a much more complex problem to solve.

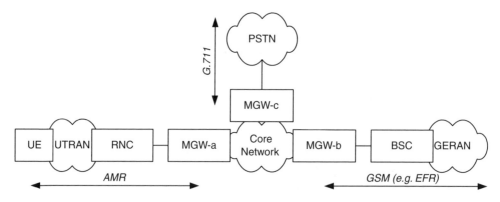

Figure 8.26 UMTS transcoding

To establish a common CODEC between end users, a process called CODEC negotiation is used. When a call is made by the UE in the UTRAN it can provide a list of CODECs it supports in the call setup messages (see 24.008). In the bearer setup, the initiating MSC server uses the list from the UE to generate a prioritized list of supported CODECs in the IAM message. For transit MSC servers each one can remove (puncture) CODECs from the list that are not supported before forwarding the IAM message. The MSC server terminating the IAM message then checks to see which CODECs are supported for the termination on its MGW, chooses a supported CODEC from the list and returns this choice back to the initiating MSC server in the APM message. The procedures for transcoder control are defined in detail in TS 23.153.

At the originating MSC, the CODEC choice is passed back to the UE using a NAS *synchronization indicator* information element. This is carried in RANAP messages to the RNC and from there in the RRC messages sent to the UE. Since the CC ALERTING message is sent via RANAP and RRC, it is convenient to piggyback the NAS synchronization identifier IE onto this message. This process is illustrated in Figure 8.27. Note that the NAS *synchronization indicator* is set to 2, i.e. the CODEC ID for GSM EFR.

The advantages of transcoder-free operation are:

- improved call quality (no distortion due to tandem coding);
- reduced end-to-end delay;
- freeing up of transcoding resources at the media gateways;
- possible saving of bandwidth use within the core network.

8.8.8 BICC summary

The use of BICC fits in well with the evolutionary model for UMTS development. It allows for easy internetworking between the TDM and packet switched technologies

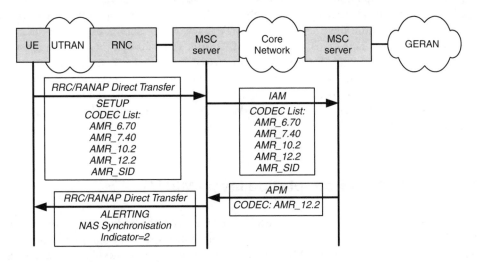

Figure 8.27 CODEC negotiation

without any loss of service. BICC was designed from the ground up to be fully compatible with ISUP, allowing for simpler migration from a GSM circuit switched core to an IP or ATM core network. 3GPP recommendations for the use of BICC are based on ITU-T recommendations but define a number of extensions. In particular, 3GPP defines its own set of UMTS bearer control procedures (TS29.232), with appropriate extensions for mobile data calls, control of user-part protocols and tandem-free operation.

8.9   SIGTRAN PROTOCOL

Signalling transfer (sigtran) is documented by the IETF under RFC 2719. It defines an architecture for transporting signalling packets over an IP network rather than a switched circuit network (SCN). It is designed to support current signalling messages, such as those used in SS7, transparently. In UMTS R4 networks with an IP core, sigtran can be used to transport BICC messages across the core network between MSC servers. In fact, for R4 three options for SS7 transport are specified as possible in 29.202: conventional MTP, AAL5/ATM and sigtran. In UMTS R5 and beyond as the network moves to an All-IP architecture, sigtran is used as transport for all SS7 messages.

SS7 over IP is an alternative to SS7 call control (as is SIP) whereby the SS7 messages are carried over IP rather than the traditional message transport. The transport for SS7 has traditionally been handled by a three-layer protocol stack called the message transfer part (MTP). MTP is composed of three layers: MTP-1, equivalent to the OSI physical layer, MTP-2 for datalink control and finally MTP-3, which performs the function of the OSI network layer. Since many data networks have already been built which support IP, the possibility of using these networks to carry voice traffic with SS7 signalling becomes attractive. To provide the transport mechanism, two extra layers are required, the MTP3 user adaptation layer (M3UA) and the stream control transmission protocol (SCTP; RFC 2960). These adaptation layers will offer services to support the SCCP which is currently used within the SS7 stack over MTP3. The SCCP should be unaware that the underlying transport network is IP and not MTP. It is assumed that these two layers will be carried over an existing IP network. As discussed in Section 3.7, telecommunications networks are designed to offer very reliable signalling and have a great deal of redundancy built in. To carry this signalling over an IP network reliably also requires that the IP network is a well-engineered network and has redundant links and redundant signalling points throughout.

8.9.1   MTP3 user adaptation layer (M3UA)

This protocol (RFC 3332) allows an SS7 application server residing on an IP network (e.g. MSC server, HLR, etc.) to communicate with entities residing on an SS7 network. As well as this, M3UA allows SS7 messages to be passed point-to-point between SS7 servers on the IP network. This use of M3UA is illustrated in Figure 8.28.

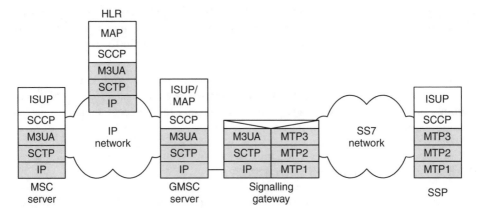

Figure 8.28   M3UA service

Version (8 bits) = 1	Reserved (8 bits)	Message class (8 bits)	Message type (8 bits)
Message length			

Figure 8.29   M3UA common header

Messages generated at signalling end points on the SS7 network (e.g. IAM for call setup) are forwarded via the signalling gateway using the M3UA service to SS7 application servers residing on the IP network (e.g. HLR, GMSC server, etc). The M3UA layer also allows the IP servers to send SS7 messages directly to each other.

Figure 8.29 shows the common header used for all M3UA packets. The current version number is 1. M3UA messages are split into six different classes (all other message class values are currently reserved). Within each class the 8-bit type field distinguishes between different messages. The length field indicates the length of the whole message in bytes including the common header.

The M3UA message classes are defined as follows:

- *Management messages (class 0)*: used to signal M3UA error conditions and other M3UA events between M3UA peers.

- *Transfer messages (class 1)*: used to move SS7 payload between M3UA peers. Only one message type defined, payload data.

- *SS7 signalling network management (SSNM) messages (class 2)*: used to indicate the status of the SS7 network. For example, the destination unavailable message indicates that a message could not be routed to an SS7 end point.

- *ASP state maintenance (ASPSM) messages (class 3)*: indicate the service status of application servers, for example the HLR or an MSC. Example messages for this class are ASP Up to indicate a server has come online, and ASP Down to indicate an ASP is currently out of service.

- *ASP traffic maintenance (ASPTM) messages (Class 4)*: indicates the routing status for various ASPs on the network. For example, the ASP active message indicates to a remote M3UA peer that it is ready to process signalling traffic for a particular application server.
- *Routing key management (RKM) messages (Class 9)*: allows routing key binding to be managed remotely. For example, an ASP which has been allocated a given destination point code (DPC) can then register this point code with the local signalling gateway using the registration request message

8.9.1.1 *Message routing with M3UA*

The addressing schemes used in IP and SS7 are quite different (SS7 uses a three-level address called a point code instead of IP addresses). The M3UA provides routing for SS7 messages across the IP network. Messages in M3UA are forwarded depending on a routing key, which consists of one or more of a number of parameters. Here are some examples:

- a DPC
- originating point code (OPC)
- SCCP subsystem number (SSN; defines which SS7 database to access), also used in UMTS to distinguish RANAP and RNSAP.

So, for example, a routing key could be defined based on DPC, DPC and OPC, or DPC and SSN. None of these examples is mandatory and the choice of routing key use is implementation dependent.

When the message is passed from the higher layers down to the M3UA layer, the routing key is examined against the entries in the message distribution table (similar to an IP routing table). If a routing key is found, the message will be forwarded to the associated IP address found in the table. It is possible for the destination of a routing key to resolve to a number of servers, which is used in the case of load sharing or backup. The M3UA layer in this case may forward the messages on a load-sharing basis (e.g. round robin scheduling), or may use one of the servers as the main server, with the second acting as backup.

If the routing key is not found, M3UA can be configured to forward the message to a default address. In the above example the servers on the IP network could be configured with a default route to the signalling gateway for all unknown routing keys.

The entries in the message distribution table can be configured statically by a management interface at the server or signalling gateway, or dynamically using M3UA routing key registration messages. The M3UA routing key registration procedure works as follows. The M3UA entity at the application server sends a routing key registration key request to the M3UA peer (usually residing at the signalling gateway) indicating which routing key to use for this server. The routing key must contain a DPC but can also contain other keying information as described before. The M3UA peer will reply with a registration response indicating success or failure for the registration.

In summary, M3UA will provides the same functionality as the SS7 MTP3 transparently over an IP network. It can be used to transport any SS7 protocols including BICC, ISUP and UMTS-specific protocols such as RANAP.

8.9.2 Streaming control transport protocol (SCTP)

As outlined in RFC 2960, the streaming control transport protocol has been designed to transport PSTN signalling over an IP network. It is a reliable transport protocol which resides on top of IP and is used by M3UA for transport across the IP network. Like TCP, SCTP requires a call establishment phase, referred to as handshaking, before transfer of data can commence. For many years the Internet has relied on TCP (RFC 793) to provide reliable transfer and UDP (RFC 768) to provide unreliable, unacknowledged transfer of data. Both of these protocols are lacking in functionality for many new applications such as transporting SS7 signalling and this is the reason for the introduction of this new transport protocol.

Although it is essentially designed for transporting PSTN signalling it is possible that SCTP may be used in many other applications. The word *stream* is used to refer to a sequence of messages that are required to be delivered in sequenced order to an application. Messages that are not in the same stream do not need to preserve this message order. While one stream of messages can be blocked, due to the loss of a packet, for example, another stream can continue to be delivered to the application. In contrast, a TCP stream is simply a sequence of bytes.

TCP is the standard connection-oriented transport protocol used on the Internet. It provides reliability, preservation of order, flow and congestion control for the data that it transfers. Data transferred is acknowledged as a byte stream; there is no inherent way for messages to be acknowledged individually or for the receiver to know where the message boundaries lie. This is a problem for signalling protocols such as SS7 where each message needs to be handled individually at the far end.

SCTP acknowledges *messages* rather than individual bytes, which is useful for passing messages that can be received out of order as it does not suffer from head of line blocking associated with TCP. If a single TCP byte is lost, the whole stream is held up while a retransmission is performed. By acknowledging messages, only the message that has the lost byte will be held up; other messages can be passed to the application without delay.

As discussed, SS7 networks are extremely reliable, and to maintain this robustness within an IP network SCTP introduces multi-homing. During the initialization phase, each of the end points exchanges a list of IP addresses through which the end point can be reached and also from which SCTP packets may be sent. A primary address is also chosen, which will be the source address used for normal sending of SCTP packets. This multi-homing only deals with hosts which have multiple IP addresses and/or multiple interfaces. These end points must be able to pass the received messages to the SCTP-user port. Multi-homing in SCTP does not include the use of multiple end points such as clustering or virtual router redundancy protocol (VRRP), where an end point can switch over to an alternative end point in the case of a failure of the original end point. As illustrated in Figure 8.30, a single port number is employed across the entire address list at an end point for a specific session.

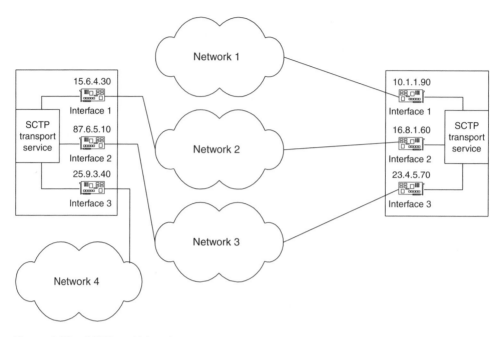

Figure 8.30 SCTP multi-homing

Like TCP, SCTP has a rate adaptive mechanism which will increase or decrease the data transfer across the network according to the prevailing load conditions. It is also designed to cooperate with TCP sessions using the same bandwidth.

8.9.2.1 SCTP security

It is well publicized that TCP is relatively vulnerable to denial of service attacks and also IP is an insecure transport protocol. SCTP introduces a verification tag which ensures the sender of a packet is authenticated. A cookie mechanism is also utilized, which helps in the prevention of denial of service attacks. If strong security and integrity protection are required it is suggested that IPSec be utilized.

8.10 SUMMARY

In Release 4 of UMTS, the network's core architecture evolves from the GSM TDM technology to a packet (or cell) switched core network using IP or ATM. R4 also introduces the use of the softswitch architecture where call and bearer control are separated. This not only provides for improved scalability and reliability but also allows for greater support for the control of multimedia streams through the use of protocols such as MEGACO and SDP. This evolution of UMTS reflects some of the trends that are currently prevalent in the fixed-line telecommunications industry, where network operators have for a while being looking at softswitches, VoIP and voice over ATM to reduce costs and improve scalability.

To bridge the gap between the traditional SS7-based call control and the new packet switched networks, a number of new protocols have been developed, namely BICC for call control over a range of different bearer options and M3UA and SCTP for the transport of SS7 signalling over the CS core. Release 4 is an important step in the evolution of UMTS to using IP for all transport (R5 and beyond).

FURTHER READING

RFC 1305: Network Time Protocol (Version 3) Specification, Implementation. D. Mills. March 1992.

RFC 1889: RTP: A Transport Protocol for Real-Time Applications. Audio-Video Transport Working Group, H. Schulzrinne, S. Casner, R. Frederick, V. Jacobson. January 1996.

RFC 1890: RTP Profile for Audio and Video Conferences with Minimal Control. Audio-Video Transport Working Group, H. Schulzrinne. January 1996.

RFC 2198: RTP Payload for Redundant Audio Data. C. Perkins, I. Kouvelas, O. Hodson, V. Hardman, M. Handley, J. C. Bolot, A. Vega-Garcia, S. Fosse-Parisis. September 1997.

RFC 2327: SDP: Session Description Protocol. M. Handley, V. Jacobson. April

RFC 2508: Compressing IP/UDP/RTP Headers for Low-Speed Serial Links. S. Casner, V. Jacobson. February 1999.

RFC 2719: Framework Architecture for Signaling Transport. L. Ong, I. Rytina, M. Garcia, H. Schwarzbauer, L. Coene, H. Lin, I. Juhasz, M. Holdrege, C. Sharp. October 1999.

RFC 2960: Stream Control Transmission Protocol. R. Stewart, Q. Xie, K. Morneault, C. Sharp, H. Schwarzbauer, T. Taylor, I. Rytina, M. Kalla, L. Zhang, V. Paxson. October 2000.

RFC 3015: Megaco Protocol Version 1.0. F. Cuervo, N. Greene, A. Rayhan, C. Huitema, B. Rosen, J. Segers. November 2000.

RFC 3267: Real-Time Transport Protocol (RTP) Payload Format and File Storage Format for the Adaptive Multi-Rate (AMR) and Adaptive Multi-Rate Wideband (AMR-WB) Audio Codecs. J. Sjoberg, M. Westerlund, A. Lakaniemi, Q. Xie. June 2002.

RFC 3332: Signaling System 7 (SS7) Message Transfer Part 3 (MTP3) - User Adaptation Layer (M3UA). G. Sidebottom, Ed., K. Morneault, Ed., J. Pastor-Balbas, Ed. September 2002.

RFC 3435: Media Gateway Control Protocol (MGCP) Version 1.0. F. Andreasen, B. Foster. January 2003.

3GPP TS23.153: Out of Band Transcoder Control; Stage 2.

3GPP TS23.205: Bearer-independent circuit-switched core network; Stage 2.

3GPP TS25.415: UTRAN Iu interface user plane protocols.

3GPP TS29.202: Signalling System No. 7 (SS7) signalling transport in core network; Stage 3.

3GPP TS29.205: Application of Q.1900 series to bearer-independent circuit-switched core network architecture; Stage 3.

3GPP TS29.232: Media Gateway Controller (MGC) – Media Gateway (MGW) interface; Stage 3.

3GPP TS29.414: Core network Nb data transport and transport signalling.

3GPP TS29.415: Core network Nb interface user plane protocols.

ITU-T Q.765.5: Application Transport Mechanism.

ITU-T Q.1902.1: Bearer Independent Call Control CS2 Functional Description

ITU-T Q.1902.2: Bearer Independent Call Control CS2 General Functions of Messages and Signals.

ITU-T Q.1902.3: Bearer Independent Call Control CS2 Formats and Codes.

ITU-T Q.1902.4: Bearer Independent Call Control CS2 Basic Call Procedures.

ITU-T Q.1902.5: Exceptions to the Application Transport Mechanism in the Context of Bearer Independent Call Control.

ITU-T Q.1902.5: Generic Signalling Procedures and Support of the ISDN User Part Supplementary Services with the Bearer Independent Call Control Protocol.

ITU-T Q.1950 Call Bearer Control Protocol.

ITU-T Q.2630.1-2: AAL type 2 signalling protocol.

ITU-T Q.1990 BICC tunnelling control protocol.

ITU-T Q.1970 IP Bearer Control protocol.

ITU-T Q.1912.1 ISUP-BICC Interworking.

ITU-T Q.1912.2 Interworking between selected Signalling System (PSTN Access DSS1, C5, R1, R2, TUP) AND THE Bearer Independent Call Control Protocol.

ITU-T Q.2150.0 Generic Signalling Transport Service.

ITU-T Q.2150.1 Signalling Transport Converter MTP and MTP3 B.

ITU-T Recommendation Q.2150.3 Signalling Transport Converter on SCTP.

ITU-T H.248: Media Gateway Control Protocol (06/00).

A list of the current versions of the specifications can be found at http://www.3gpp.org/specs/web-table_specs-with-titles-and-latest-versions.htm, and the 3GPP ftp site for the individual specification documents is http://www.3gpp.org/ftp/Specs/latest/

9

Release 5 and Beyond (All-IP)

9.1 INTRODUCTION

Release 5 (R5) builds on the partial implementation of Internet protocol (IP) packet switching within the core network, as discussed thus far, to move to an all-IP architecture. In this release, packets can be moved end-to-end using IP transport with an enhanced general packet radio service (GPRS) network connected to an IP multimedia subsystem (IMS). The GPRS backbone for R5 must be able to provide similar levels and classifications of quality of service (QoS) usually associated with asynchronous transfer mode (ATM) networks. This is to allow for the delivery of time-sensitive traffic such as voice and multimedia. As well as enhancements to the core network, the radio access network (RAN) also migrates from ATM to IP. Even though the vision for R5 is for a total IP solution, the operator may well still use ATM as a transport solution for some parts of the network. This is possible because all UMTS releases must provide backward compatibility with earlier releases. Figure 9.1 shows the architecture of the R5 network.

Notice that in the R5 network, the circuit switched (CS) domain can be dispensed with since the services associated with it, such as transfer of voice traffic, can be carried over the GPRS and IMS networks using IP QoS mechanisms. That given, many operators may still be using the R4 CS domain as well as the R5 IMS architecture. This allows for a gradual migration to an all-IP architecture with the minimal disruption to service. Some voice calls may be handled using the CS domain and some, for example video call services, via the IMS. The hybrid configuration is shown in Figure 9.2.

The major new components of the Release 5 architecture are now described.

9.2 IP MULTIMEDIA SUBSYSTEM (IMS)

R5 introduces a new network domain called the IP multimedia subsystem (IMS). This is an IP network domain designed to provide appropriate support for real-time multimedia services. Figure 9.3 shows the connections between the IMS and other networks.

Convergence Technologies for 3G Networks: IP, UMTS, EGPRS and ATM   J. Bannister, P. Mather and S. Coope
© 2004 John Wiley & Sons, Ltd   ISBN: 0-470-86091-X

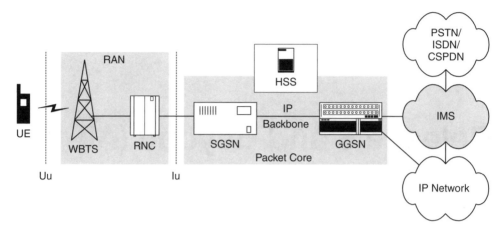

Figure 9.1 UMTS R5 network

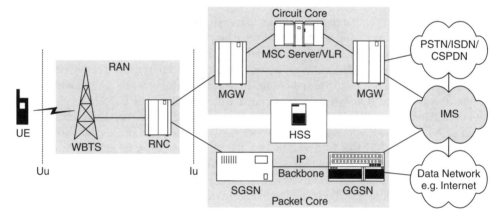

Figure 9.2 R4 to R5 migration

The user equipment (UE) communicates with the IMS using GPRS, with the IMS being directly connected to the GGSN. The IMS provides services to mobile users such as:

- real-time communication using voice, video or multimedia messaging (i.e. voice and video telephony);
- audioconferencing and videoconferencing;
- content delivery services such as video, audio or multimedia download;
- content streaming services such as video, audio or multimedia streaming (e.g. using video on demand server);
- multimedia messaging service (MMS).

Each operator's IMS can be connected to other operators' IMSs, allowing multimedia services between users on different networks. Connections to the public Internet allow

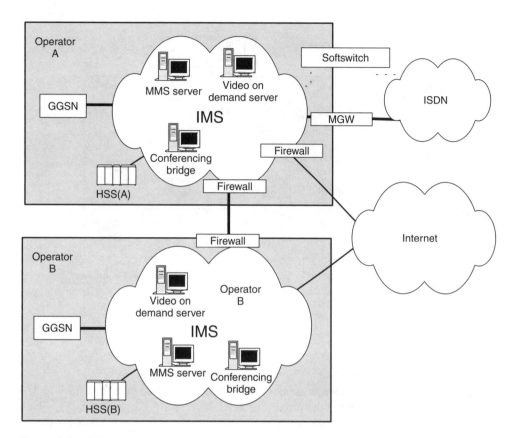

Figure 9.3   IMS connection

MMS messaging as well as voice over IP (VoIP) and video telephony between mobile and fixed-line users. Finally, the interface to the ISDN (or other circuit switched networks) allows VoIP calls to be connected through to conventional fixed-line and mobile users, e.g. global system for mobile communications (GSM). Connections between the IMS and other IP networks are controlled by firewalls to protect against hacking. The interface between the IMS and the CS network is controlled by the softswitch and media gateway (MGW) components. Within the operator's network the IMS is connected to the home subscriber server (HSS) to allow for subscriber authentication, authorization and mobility management. For R5 and beyond, the IMS can be used to provide transport for all of the operator's services, including conventional voice calls.

Figure 9.4 shows a functional diagram of the IMS. The dotted lines correspond to signalling paths and the solid lines to user data transport.

The IMS is made up of a number of component parts connected together using an IP backbone. This is implemented as a separate network from the IP backbone connecting the serving GPRS support node (SGSN) and the gateway GPRS support node (GGSN) for such purposes as security. Each mobile user must first obtain a GPRS connection (i.e. a packet data protocol (PDP) context) to the IMS prior to using its services, which

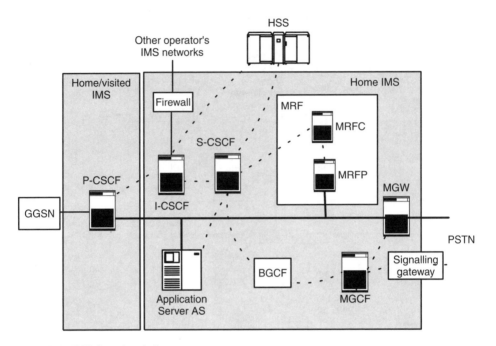

Figure 9.4 IMS functional diagram

means authentication/registration is done twice, first for GPRS and then for IMS. The UE always connects to the IMS via the proxy call session control function (P-CSCF), which may reside at the user's home network or in a roaming scenario within a visited network. On the right-hand side of the figure can be seen the connection to external CS networks via the MGW and media gateway control function (MGCF) (softswitch). The application server provides IMS value-added services. This could be, for example, a content server (video on demand) or an interactive voice/video mailbox server. The interrogating call session control function (I-CSCF) connects this network to other IMS networks. It acts as a point of entry for call signalling.

Each of the components shown in Figure 9.4 is a mandatory requirement with the exception of the application server and the media resource function (MRF) (which is only needed for conferencing applications). The components are defined in the following subsections.

9.2.1 Call session control function (CSCF)

Data transfer between users of the IMS is organized into sessions. The CSCF is responsible for session control and is the control point for the following functions:

- user authentication
- call routing

- establishing QoS over the IP network
- controlling the generation of call detail records (CDRs) for accounting purposes.

All call/session control signalling in the IMS is performed using the session initiation protocol (SIP). Three types of CSCF are defined: P-CSCF, S-CSCF and I-CSCF. Each network will typically provide multiple CSCFs of each type. This allows for load sharing and supports increased reliability through the use of backup servers.

Proxy CSCF (P-CSCF)

This acts as the first point of contact for call signalling coming from the UE. The P-CSCF forwards the call signalling to the serving CSCF (S-CSCF), which is the home network's point of control for the call. For a roaming subscriber, the P-CSCF will be located in the visited network, or more specifically the P-CSCF for a given user will be located in the same network as the GGSN from which they are receiving service. The P-CSCF is also responsible for controlling the generation of CDRs for mobile originated calls.

Serving CSCF (S-CSCF)

This carries out the call/session and accounting control for a given subscriber. The S-CSCF is always located within the subscriber's home network. This means that all mobile originated call signalling is routed via the user's home network. For example, a UK subscriber roaming in Malaysia who then phones Australia would have their call routed via the UK. The reason for this is that it allows a network operator to reconcile its call charging records with its overseas roaming partners for each subscriber. This non-optimal routing covers only signalling traffic; call traffic is still forwarded using standard IP routing between the Australian and Malaysian GGSNs.

Interrogating CSCF (I-CSCF)

The I-CSCF is located at the boundary of the IMS and acts as a point of entry for SIP signalling coming from outside the operator's network. This signalling could be:

- a SIP call setup request destined to a subscriber of the operator's network;
- a SIP call setup request destined to a *roaming* subscriber within the operator's network;
- a registration request.

For incoming registration requests the I-CSCF is responsible for assigning an S-CSCF to the subscriber. The choice of S-CSCF can be made dependent on the identity of the subscriber (SIP address or international mobile subscriber identify; IMSI), handled on a load sharing basis or using a main server/backup server arrangement.

9.2.2   Application server (AS)

This provides value-added services to a subscriber. This could be anything from receiving streaming video service (video on demand) to providing voice and video mail services.

9.2.3   Breakout gateway control function (BGCF)

This is used to select the appropriate gateway to forward calls destined for the CS domain (i.e. the CS breakout point). An S-CSCF will forward all call requests with CS destinations, which will then forward them to the appropriate MGCF.

9.2.4   Multimedia resource function (MRF)

The MRF is made up of two components, the MRF control and MRF processor, and is responsible for providing functions such as:

- mixing media for video/voice conferencing (conferencing bridge);
- providing multimedia announcements;
- processing media streams, e.g. audio transcoding.

The MRF functionality is split into a control (MRFC) and a media processing (MRFP) part, in much the same way as functionality is split between media gateway controller (mobile switching centre (MSC) Server) and media gateway. The interface between the two components is controlled using the H.248/MEGACO protocol. The MRFC receives call control signalling via the SIP protocol (e.g. to establish a Videoconference between a number of parties).

9.2.5   Media gateway control function and media gateway (MGCF and MGW)

The MGCF functionality and MGW provide a connection between the IMS and external CS networks such as ISDN or GSM. The MGCF controls the MGW and interfaces to the S-SCSF using the SIP protocol. Call signalling (e.g. SS7/ISUP) is forwarded from the CS network's signalling gateway to the MGCF using sigtran. The MGCF must translate messages between SIP and ISUP to provide interworking between the two protocols. These protocols are explained in the following sections.

9.3   HOME SUBSCRIBER SERVER (HSS)

The HSS contains a master database of all the subscribers on the network and contains the following information:

- identification information (user's telephone number, SIP addresses, IMSI);
- security information (secret authentication keys);

- location information (current serving GGSN, SRNC, IP address);
- user profile information (subscribed services).

It is also responsible for generating security information such as authentication challenges and integrity and ciphering keys. The HSS incorporates the home location regis-
ter/authentication centre (HLR/AuC) functionality defined in previous releases and provides service for three domains, as follows:

- authentication, service profile and location information for IMS (service for CSCF);
- HLR/AuC service for packet switched (PS) domain (service for SGSN and GGSN);
- HLR/AuC service for CS domain (service for R4 MSC server).

9.3.1 HSS Cx interface

The connection between the HSS and CSCF (I-CSCF and S-CSCF) is defined as the
Cx interface. The protocol for the Cx interface is based on DIAMETER and is specified in 3GPP 29.229. For more details of DIAMETER, please refer to Chapter 5. The
protocol uses a query/response approach where queries are sent to the HSS database,
which responds with information about a subscriber. The protocol can also be used by
the I-CSCF or S-CSCF to update values in the HSS database. The following types of
transaction are supported:

1. Registration authorization for subscriber (I-CSCF – HSS)
2. Query for authentication vectors for subscriber (S-CSCF – HSS)
3. Registration status notification (register or de-register) (S-CSCF – HSS)
4. Network-initiated de-registration (HSS – S-CSCF)
5. Location query for subscriber (I-CSCF – HSS)
6. Update of user profile (HSS – S-CSCF).

The first three transactions are used in the IMS registration process. The I-CSCF uses
the first transaction on incoming registration requests to determine if the subscriber can
register. The second and third transactions are used by the S-CSCF when authenticating
the user and update the registration status in the HSS. The details for these procedures
are given in Section 9.7.3.

The network-initiated de-registration allows the HSS to asynchronously de-register a
user. This might be used in the face of a network error due to the fact that the subscriber
has failed to pay their bill or perhaps fraud has been detected on the account.

The location query is used by the I-CSCF to determine how to route incoming calls
destined for subscribers belonging to this network. The HSS responds with the identity
of the S-CSCF with which the user is currently registered. A call flow example for this
is show in Section 9.7.5.

Table 9.1   Cx protocol to DIAMETER mapping

	Transaction	DIAMETER message header	Sent by
UAR	User authorization request	Type code = 1, R = 1	I-CSCF
UAA	User authorization answer	Type code = 1, R = 0	HSS
SAR	Server assignment request	Type code = 2, R = 1	S-CSCF
SAA	Server assignment answer	Type code = 2, R = 0	HSS
LIR	Location info request	Type code = 3, R = 1	I-CSCF
LIA	Location info answer	Type code = 3, R = 0	HSS
MAR	Multimedia authentication request	Type code = 4, R = 1	S-CSCF
MAA	Multimedia authentication answer	Type code = 4, R = 0	HSS
RTR	Registration termination request	Type code = 5, R = 1	HSS
RTA	Registration termination answer	Type code = 5, R = 0	S-CSCF
PPR	Push profile request	Type code = 6, R = 1	HSS
PPA	Push profile answer	Type code = 6, R = 0	S-CSCF

The last transaction (update of user profile) can be used by an HSS to update a subscriber's current service profile. For example, if the user has subscribed to a new video on demand service, this information would be stored in the service profile and delivered to the S-CSCF. The S-CSCF will then permit calls for the subscriber destined to the AS providing that service.

The Cx protocol maps to the DIAMETER base protocol as shown in Table 9.1. This usage of DIAMETER is defined strictly within UMTS (3GPP 29.229) and is not described by any IETF draft or RFC documents. All the information elements for the DIAMETER messages are sent as attribute value pairs (AVPs). Each message will always contain the user's public identity, which is used by the HSS to reference the subscription details. Other elements will include the user's current S-CSCF and subscription profile.

The UAR and UAA messages are used as part of the registration authorization transaction. The UAR (sent from the I-CSCF) contains the user's public identity as well as the information about the visited network. The HSS returns with details about the user's current registration information (for users already registered) or a decision about the permission for registration to proceed (reject or accept).

The SAR message is sent to the HSS to update the user's current S-CSCF. It contains the identity of the user's S-CSCF as well as a status value indicating if the user has recently registered or de-registered at this address. The HSS responds with an SAA message containing confirmation of the user's status as well as a user profile describing their current service profile.

The LIR is sent by the I-CSCF to the HSS to locate the current S-CSCF for a given subscriber. The HSS responds with an LIA message containing the name of the S-CSCF or a status value indicating that the user is not known at this HSS or is currently unregistered. The MAR is used to request authentication vectors for a particular subscriber from the HSS. The request contains both the user's public and private identity. The MAA message in reply contains one or more authentication vectors generated for the subscriber.

The network-initiated registration termination procedure is performed using the RTR and RTA messages. The HSS will send an RTR to the S-CSCF indicating which user it wants to de-register as well as a reason for the de-registration. The S-CSCF will send an RTA in reply confirming the de-registration. Finally, the PPR message is used by the HHS to update a user's profile at the S-CSCF. The S-CSCF confirms this request using the PPA message.

9.4   IP NETWORK DOMAIN SECURITY

Security within 2G networks has been identified as a significant weakness, one factor being that all signalling within the network is unprotected against snooping and active attacks such as injecting rogue packets. At the time this was not seen as a significant issue since the network was generally controlled by a number of large institutions which could provide suitable physical protection. With the introduction of 3G this may no longer be the case and much of the signalling will be carried over IP networks which afford little protection. Within the RAN, which is the most vulnerable, signalling integrity protection was introduced to R99, as described in Chapter 6.

TS33.210 specifies how the core network signalling carried using IP is to be protected. The standard security services of confidentiality, integrity, authentication and anti-replay protection are defined. The security model defined for UMTS is based around the concept of a security domain. Each security domain is administered and controlled by one organization. The architecture for UMTS IP domain security is shown in Figure 9.5.

The connection between two IP network entities on the same operator network or security domain is referred to as the Zb interface. Communication between security domains is controlled by a security gateway (SEG; i.e firewall); the interface between the two SEGs on different network domains is called Za. 3GPP mandates the use of integrity and origin authentication using IP security (IPSec) in encapsulating security payload (ESP) mode for all IP communication over the Za and Zb interfaces. It also states that IPSec ESP confidentiality must be used for all traffic over the Za interface but this is optional for traffic over the Zb interface.

The IPSec key exchange algorithm for UMTS is Internet key exchange (IKE). Within a security domain it would be possible to configure the security associations using manually pre-shared secrets. However, between domains this would be more problematic since the

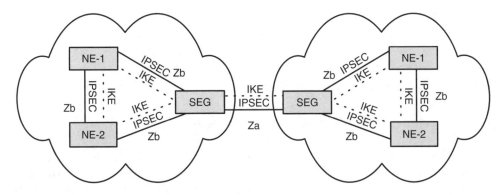

Figure 9.5   IMS IP domain security architecture

number of shared secrets increases as the square of the number of domains. In this case the use of public key cryptography and certificates is a more appropriate choice.

Within the definition of IPSec, support for the data encryption standard (DES) algorithm is mandatory. With 3GPP, however, due to the perceived weaknesses in the DES transform the advanced encryption standard (AES) must be used instead. AES is the latest US Federal Information Processing Standard (FIPS) cipher and is specified as a replacement for DES. It supports key lengths of 128, 192 or 256 bits.

9.5 SESSION INITIATION PROTOCOL (SIP)

SIP, as defined by RFC 3261, is designed to allow multimedia sessions to be established between users on an IP network. As well as call control, the baseline SIP RFC supports functions such as user mobility and call redirection. As well as RFC 3261, a number of extensions are defined in supplementary RFCs and IETF drafts covering topics such as SIP/PSTN interworking and SIP for instant messaging and presence detection. Currently the following types of services are supported (note this list is not exhaustive):

● multimedia call establishment
● user mobility
● conference call
● supplementary services (call hold, call redirect, etc.)
● authentication and accounting
● unified messaging
● instant messaging and user presence detection.

Although SIP can provide all these services, currently R5 does not define call scenarios for all of them. For example, multimedia conferencing will be made available as part of R6. This, however, does not preclude an operator or vendor introducing them as a value-added service.

The benefits of SIP when moving to an all-IP architecture are as follows:

● end-to-end SIP signalling between mobile and fixed-line IP users;
● Internet SIP servers can provide value-added services to mobile users;
● SIP is designed as an IP protocol, therefore it integrates well with other IP protocols and services;
● SIP is lightweight and (relatively) easy to implement.

The first point is of particular interest. As subscribers of the UMTS network start to use services based on an IP infrastructure they may wish to communicate to fixed-line Internet lines. Figure 9.6 shows an example of this. The mobile UE is participating in the voice portion of a video call with a user using a voice or video application on a desktop.

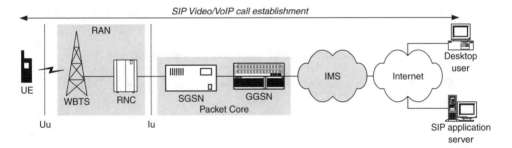

Figure 9.6   End-to-end SIP signalling

The SIP application server shown in the figure can be accessed by users on the Internet or connected via the mobile network.

The use of SIP does have one potential drawback. The emulation of existing PSTN supplementary telephony services using SIP may not provide an identical or complete set of services. This issue has been addressed by the ITU-T with the use of the bearer-independent call control (BICC) protocol for SS7/PS network interworking, as described in Chapter 8. This is not seen as a serious drawback, however, for two reasons. First, as SIP is maturing, so will SIP/PSTN interworking, with most of the service issues being resolved over time. Second, and probably more importantly, users of next generation networks are expecting a very different range of services (e.g. instant and multimedia messaging, email, unified message, videoconferencing) than was possible with the conventional telephone network. SIP is easily capable of delivering such a rich service set over IP. To provide such services using BICC would be difficult. This is because BICC was designed to emulate call control for existing voice and circuit services and has no facilities for multimedia and extra services such as instant messaging. Another added complication is that BICC addressing is not based on IP addressing but on E.164, whereas SIP supports both formats.

9.5.1   SIP addressing

SIP users are located via a SIP uniform resource indicator (URIs). Here are some examples of SIP URIs:

sip:icurtis@orbitage.com

sip: + 447789005240@wcom.com;user = phone

In the first case, the SIP address defines a user connected to the Internet; in the second, a telephone number is given and the user must be contacted via a PSTN gateway. For UMTS subscribers, the IP public identity of a subscriber (i.e. the SIP address) is contained within the subscriber's Internet multimedia system subscriber identity module (ISIM) application. This is an application on the universal integrated circuit card (UICC), as is the UMTS

SIM (USIM). For a UE without an ISIM application, the SIP address (or public identity) can be derived as follows:

sip:IMSI number @MNC.MCC.IMSI.3gppnetwork.org

where MNC is the mobile network code (the fourth and fifth digits of the IMSI) and MCC is the mobile country code (the first three digits of the IMSI). This hierarchical scheme makes it possible for DNS domains to be managed country by country and operator by operator, making it easier to locate users on the mobile network, given their IMSI. The following example shows an IMSI-derived SIP address:

sip:234150999999999@15.234.IMSI.3gppnetwork.org

This type of SIP address is only to be used within the IMS, i.e. between CSCF and HLR, and not used externally. In this case the anonymity of the IMSI appears to be somewhat compromised, since this temporary public identity must be used for all registrations between UE and the S-CSCF. However, the identity can be kept private within the operator's network since the GPRS link can provide encryption at the link layer and all other connections within the IMS are protected using IPSec.

9.5.2 SIP components

Figure 9.7 shows the basic component architecture for a SIP network. The SIP servers are responsible for assisting in call setup and mobility management but SIP also allows users to perform call setup directly with each other without a server being involved. Some SIP servers (proxy and redirect) can be stateless (i.e. do not need to store transaction state information). This is a desirable property since it makes it easier to scale the network to

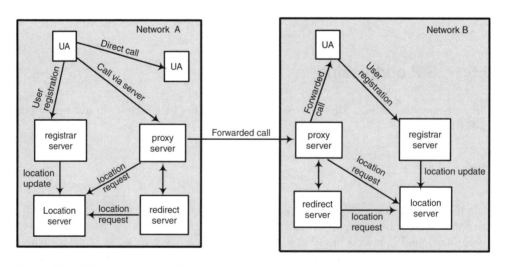

Figure 9.7 SIP component architecture

large numbers of nodes. Commonly, logical SIP servers are often co-located within the same physical device.

9.5.2.1 *User agent (UA)*

The user agent (UA) makes or accepts calls on behalf of a SIP user. The SIP user can be an actual person or an application on a server. The UA terminates the SIP call signalling and acts as the interface to the SIP network. The UA can act as either user agent server (UAS) or user agent client (UAC). The UAS processes incoming SIP requests (call setup) and generates appropriate responses (accept, reject, busy). The UAC generates requests on behalf of the SIP user. For a UMTS network each UE will contain a UA to handle user calls. UA functionality can also be located elsewhere in the network; for example, an application server providing a video on demand service would act as a UA, accepting incoming call requests from subscribers, and then streaming media to the calling parties. Another type of server that would act as a UA would be voice/video mail service. In this case it would accept calls that have been redirected in the case of the UE being switched off or the user busy.

9.5.2.2 *Proxy server*

The SIP proxy forwards requests on behalf of SIP agents. The SIP proxy may well rewrite a SIP message before forwarding it to its correct location. For example, a user may have moved from their work location where their proxy resides to a location at their home address. In this case the destination URI will have to be rewritten. For incoming calls the proxy will interrogate the location service to determine how to forward the call. The proxy can also be configured to provide authentication control and act as a point of control between the internal private network and an outside public network.

9.5.2.3 *Registrar server*

The registrar server allows SIP agents to register their current location. In 3G the location information includes the UE destination IP address assigned during PDP context establishment, the station's UTRAN cell ID and the identity of the P-CSCF. The registrar is responsible for keeping up-to-date information within the location service by sending updates.

9.5.2.4 *Redirect server*

The redirect server responds to a UA request with a redirection response indicating the current location of the called party. In this case the UA must establish a new call to the indicated location.

9.5.2.5 *Location service*

A location service is used by a redirect or proxy server to obtain information about a user's whereabouts. The location service can be co-located with other SIP servers. The interface between the location service and other servers is not defined by SIP.

9.5.3 SIP messages

Basic SIP performs all its functionality with only six basic message types called methods, as defined in RFC 3261, but supports a growing number of extensions covered by other RFCs or Internet drafts. The main methods are as shown in Table 9.2.

INVITE and ACK are used for basic SIP call setup. INVITE initiates the call and ACK is used to acknowledge the final response to the original INVITE. BYE is used to terminate a call when one of the parties hangs up. CANCEL is used to terminate a call that is currently being established. For example, if the called party's phone is ringing and the caller hangs up, CANCEL can be used to indicate that the INVITE is no longer requested.

REGISTER is used by a UA wishing to indicate their location and availability to the registrar server. Registration will occur on UA start-up as well as when the user is relocating to another network domain (for example when roaming in a visited network).

OPTIONS allows a UA to determine the capabilities (for example, which CODECs are supported) of either the UA being called or a proxy server.

The INFO request carries application signalling which does not affect the state of the SIP call. Examples of this signalling could be dual tone multi-frequency (DTMF) digits, account balance information or mid-call PSTN signalling between gateways.

PRACK (provisional acknowledgement) is an extra message allowing for the provisional response messages to be acknowledged. Usually 1xx messages (see Section 9.5.4) are carried in an unreliable manner. This message allows provisional responses to be delivered reliably and is used in UMTS to ensure the delivery of responses such as 180 RINGING (indicating that the called UA is alerting the user).

The UPDATE method is used to allow modification of media information about the session before it has been fully established. It is used principally to indicate QoS parameters that have been successfully negotiated for the session.

Table 9.2 SIP methods

Method	Function	RFC or draft
INVITE	Call setup	RFC 3261
ACK	Acknowledgement for response to INVITE	
BYE	Call termination	
CANCEL	Cancel call	
REGISTER	Register URL with SIP server	
OPTIONS	Check for capabilities	
INFO	Midcall signaling	RFC 2976
PRACK	Provisional acknowledgement	RFC 3262
UPDATE	Modify session information	RFC 3311
REFER	Transfer user to a URL	Draft-ietf-sip-refer-07.txt
SUBSCRIBE	Requests notification of an event	RFC 3265
NOTIFY	Event notification	RFC 3265
MESSAGE	Transports instant message	RFC 3428

The REFER message is used to instruct a UA to establish a session with a third party. It can be used to support various supplementary services such as call transfer or conferencing.

The SUBSCRIBE and NOTIFY methods support event notification. Events can be used, for example, to indicate new incoming email or a SIP UA becoming available after being busy. Event handling can also be used for presence detection for an instant messaging application. A SUBSCRIBE message is sent by a UA to another UA; within the request is a series of events that the sender wants to be notified of. The NOTIFY message is used to indicate the occurrence of the event to the listening UA.

Finally, the MESSAGE method is used to support instant messaging. It allows a short message to be sent from UA to UA without the need for session establishment. The messages are presented in multi-part MIME format (similar to email) and can contain multimedia as attachments. Instant messages in SIP can be delivered directly to their destination or via a message relay. In the latter case, again very much like email, the message is stored before being forwarded at a later time.

9.5.4   SIP responses

For each method, a SIP entity can respond to a request with a range of response codes organized into six classes, as shown in Table 9.3.

The 1xx provisional or informational responses are sent back when a request has been received and is being processed to indicate something about the progression of the call. For example, 180 RINGING tells the UA that the called party's phone is ringing; 100 TRYING indicates that the server is trying to contact the called party. All the other codes (2xx–6xx) are referred to as final response codes and indicate the end of the transaction. Each provisional response is expected at some later time to be followed up by a final response code.

The 2xx success responses indicate that the command has been successful. Currently there is only one 2xx response, 200 OK. 3xx responses are sent to indicate that the client should try to recontact the called party at a different location, as indicated in the response. For example, response 301 tells the calling party the new contact address is permanent,

Table 9.3   SIP response code classes

Class		Description	Examples
1xx	Provisional/ informational	Request received processing response	100 TRYING 180 RINGING 183 PROGRESS
2xx	Success	Request successful	200 OK
3xx	Redirection	Redirect client to try another location	301 Moved permanently 302 Moved temporarily
4xx	Client error	Bad syntax in message, client should rephrase request	407 Proxy unauthorized 404 Not found
5xx	Server error	Problem with server	503 Service unavailable
6xx	Global error	Error information about called party	603 Decline

302 is used for temporary redirection (when a user is roaming in a visited network). 4xx, 5xx and 6xx responses indicate certain error conditions. 4xx are sent back if there is a problem with the request from the client. For example, 407 proxy unauthorized indicates that authentication information is required with the request; 404 not found indicates that the user does not exist at the domain given. 5xx server responses indicate an error at the server. A server replies with 503 service unavailable to indicate that it cannot satisfy the request at this time. The client is expected to try another server or re-try the request at a later time. Finally, the 6xx global errors indicate error information about the called party. For example, 603 decline indicates that the called user was not willing to accept the call.

9.5.5   SIP transaction handling

Each SIP transaction consists of a request followed by one or more provisional responses (1xx) and then one or more final responses (2xx, 3xx, 4xx, 5xx or 6xx). Since SIP can use user datagram protocol (UDP) or transmission control protocol (TCP) for transport, a reliability mechanism must be included so that messages are not lost. For SIP messages over UDP the message is retransmitted if a response is not received within a given time interval. The time interval starts off set to T1 (default value 500 ms), and after each retransmission the timeout interval is doubled up to a maximum value of T2 (default 4 s). For all requests apart from INVITE, the retransmission will continue until the final response is received. This ensures that the final response is received by the initiator of the SIP transaction. For the INVITE request the retransmission stops as soon as the first provisional response is received; in this case the use of the extra ACK message ensures delivery of the final response. The special use of the ACK message is covered in more detail later.

In certain circumstances it is important to deliver provisional responses reliably. In a telecommunications network it is important that call progression messages such as 180 RINGING or 183 CALL PROGRESS are always delivered. To ensure this takes place the sender of the provisional response sets the *require* header of the response message as follows:

Require: 100rel

When the UA receives the response it is expected to acknowledge it. This is done using PRACK, which in turn will also require a final response, which, assuming there is no error condition, will be 200 OK.

9.5.6   SIP message transport

All SIP elements must support both TCP and UDP; for security transport layer security (TLS) can be used over TCP. TCP provides reliability but introduces an extra overhead in terms of connection establishment prior to sending the message. The reliability for TCP however is not needed since SIP provides its own retransmission mechanism. UDP transport has a lower overhead but does not provide congestion and flow control.

9.5.7 SIP server discovery

For the UA to gain service from the network, it must first establish the IP address and port number of the local proxy or registrar server to send the request to. The address of these can be configured manually or discovered dynamically. The dynamic host configuration protocol (DHCP) can be used to allocate the address of SIP servers on the attached network (RFC 3361). If the address returned by DHCP is a domain name then the domain name system (DNS) can be used to resolve this still further. The facilities of the local server (registrar, proxy, etc.) can be discovered by use of the SIP options request.

In dynamic server discovery (RFC 3263), the DNS naming authority pointer (NAPTR) mechanism is used to discover the URIs of SIP servers in a given domain. This allows messages to be forwarded from the outgoing proxy to the destination given in the request's URI. These URIs are then resolved into IP address and port number using the DNS SRV (server) resolution mechanism. If no port number is defined in the DNS resource record then the default values of 5060 for TCP or UDP and 5061 for TLS are used.

The source port number for the requests originating from the UA is assigned by the UA when the message is generated. The proxy server is expected to send responses back to the same port. The port number for requests terminated at the UA can be defined as part of the user's contact address URI. Again, the values 5060 and 5061 are used as default if no port number is specified.

Finally, an extra mechanism is available for registrars. These servers can be addressed by the multicast address sip.mcast.net (224.0.1.75); the port number will be 5060.

In summary, DHCP and DNS can be used to configure the local SIP server (proxy and registrar) addresses for a UA. DNS NAPTR and SRV records are then used by proxies to discover the address and port numbers of remote SIP servers where requests should be forwarded.

9.5.8 SIP headers

Table 9.4 shows a SIP request INVITE message. The format of the initial line is METHOD Request-URI SIP-Version (the format is similar for the response message but with the request method replaced by the response code). The current version of SIP is 2.0.

Each line of the request is followed by a carriage return (CR) and line feed (LF). After the initial request line comes a series of fields or headers which act as parameters to the method.

9.5.8.1 Via

Every time a message is received by a SIP server, it adds its own address to the beginning of the list in the Via header before forwarding the message. This information allows the response to be sent back along the same route as the original request. As the response is routed back each proxy along the route removes its own Via header from the top of the list; it then sends the response back to the address indicated by the new value at the beginning of the list.

Table 9.4 SIP INVITE example

Message line	Description
INVITE sip:seb@orbitage.com SIP/2.0	Request to start call to seb in domain orbitage.com
Via: SIP/2.0 TCP 128.7.8.9:2999 ;branch=z9hG4bK776asdhds	Who this request is directly from
To: seb@orbitage.com	Destination of INVITE
From: jeff@orbitage.com	Source of INVITE
Call-ID 12323121212:100.0.0.1	Random integer plus source host (globally unique)
Max-forwards	Integer giving maximum number of times the request can be forwarded by a server
Cseq: 314159 INVITE	Sequence number
Content-Length : 100	Length of SDP data
	Empty line
v = 0	SDP version number
O = sebcoope 345667 345668 IN IP4 128.5.6.7	o = originator of session
S = sales meeting	s = subject heading
I = Sales meeting about new VoIP product	i = information about session
U = www.orbitage.com/sales/conf1.html	u = url for session
T = 456789 458900	t = time limits for session
C = IN IP4 128.9.7.8	c = contact address
e = seb@orbitage.com	e = email contact address
m = video 45670 RTP/AVP 31	m = media descriptor
b = 20	b = bandwidth requirements
	Empty line

Each Via header also contains a branch parameter, which must be globally unique for each transaction. One common way of calculating the value is to perform a cryptographic hash on the values in the following headers: To, From, Call-ID, the Request-URI, the topmost Via header, Cseq, Proxy-Require and Proxy-Authorization. For RFC 3261 all branch values must start with z9hG4bK. This is to distinguish them from branch values generated by implementations compliant with the earlier SIP specification RFC 2543 (which does not require global uniqueness). The branch value is used as a transaction ID. For a UA or stateful proxy it allows responses to be matched with requests. It can also be used by SIP proxies in routing loop detection.

9.5.8.2 Record-route and route

The record-route header can be used by a proxy to request that future requests for a given dialogue will be sent via this proxy. This is different from the Via header in three ways: its use is optional, it is used to route requests (not responses) and its lifespan is over a given SIP session, not just one transaction. The route header is used to enforce the route of the message over a set of proxies. Here is an example of its use. First the call is established with an INVITE message from sebcoope to jsmith. The INVITE received is as follows:

> INVITE sip:jsmith@vodafone.net SIP/2.0
>
> Contact: sip:sebcoope@orbitage.com
>
> Record-Route: <sip:proxy.vodafone.net>
>
> Record-Route: <sip:proxy.maxis.net>
>
> Record-Route: <sip:proxy.orbitage.com>

to which jsmith responds with a 200 OK. Later, jsmith terminates the call using a BYE request as follows, which it sends to proxy.vodafone.com, the first in the list:

> BYE sip:sebcoope@orbitage.com SIP/2.0
>
> Route: <sip:proxy.vodafone.net>
>
> Route: <sip:proxy.maxis.net>
>
> Route: <sip:proxy.orbitage.com>

The proxy at vodafone.net will strip out its route header and forward the request to the next in the list (the proxy at maxis). The BYE will now look like this:

> BYE sip:sebcoope@orbitage.com SIP/2.0
>
> Route: <sip:proxy.maxis.net>
>
> Route: <sip:proxy.orbitage.com>

The use of route and record-route is important for proxies that need to be informed about both the start and end of a session. For example, a proxy accounting for usage in a visited network would need to be informed when the session was terminated so that the CDR can be updated correctly.

9.5.8.3 To and from

To and from are the final destination and source of the SIP request. Unlike the URI in the first line of the SIP message, these fields are not modified as the message is forwarded between the SIP servers.

9.5.8.4 Call-ID

This is set to a unique integer for each call so that the request and response codes can be matched for a particular call.

9.5.8.5 Max-forwards

This field is used to stop SIP messages looping the network indefinitely in case of a routing error. The field is decremented by each SIP server that receives the message. If the field reaches zero, the server will not forward the message and will return a *483 Too Many Hops* to the source. The default value for this field is 70.

9.5.8.6 CSeq

This is a traditional sequence number used to match requests with responses and is incremented for each new SIP request. The CSeq header contains not only the sequence number, but the method of the original request.

9.5.8.7 Content-length

This has the same meaning as its equivalent in the HTTP protocol, and indicates the length of the data following this request.

9.5.8.8 Content-type

The content type describes the format of the content as a MIME content type. For call establishment messages within UMTS the content is an SDP description of the session and the type is set to application/sdp. For other SIP messages the contents will vary depending on the message type.

9.5.8.9 Contents

The headers are followed by an empty line containing just the CR/LF sequence and then by a contents section. The contents section of a SIP INVITE request contains information about the session requested (i.e. type of media, voice or video, data rate requested, port number for RTP session). The parameters are as according to the session description protocol (SDP) protocol, as was defined in Chapter 8. The media being requested (or offered) by the caller in the example shown in Table 9.4 is a video session using UDP port number 45670 and RTP payload type 31, i.e. video encoded with the H.261 CODEC (m = video 45670 RTP/AVP 31).

9.5.9 SIP call establishment

A standard SIP call setup, as shown in Figure 9.8 consists of a three-way handshake. First the initiator sends an INVITE message to the recipient. The recipient can respond to the INVITE with a range of responses (provisional or final). In the example given, the phone starts ringing and responds with a RINGING information message. When the phone is picked up the call is accepted and 200 OK is sent back to the initiator. Finally, the calling UA sends an ACK back to the recipient to complete the call setup. Note, since the ACK is part of the original INVITE sequence it contains the same CSeq value.

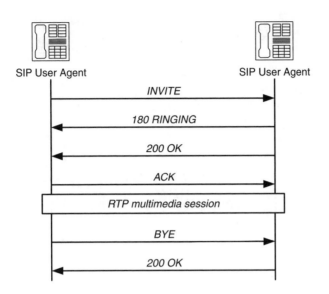

Figure 9.8   SIP call example

These SIP messages will include the SDP data, which will negotiate parameters for this particular call such as which CODEC to use.

Once the session is established, multimedia packets can be transferred between the two end points using the real-time transport protocol (RTP). When the call is finished the BYE message is sent which instructs the other SIP user that the call should be terminated. Finally, 200 OK is sent back to complete the call termination.

9.5.9.1 ACK request

For the INVITE request a large delay can occur between the user's phone ringing (180 RINGING) and the call being answered (200 OK). If the request was repeated until the final response a lot of unnecessary retransmissions of the INVITE would occur before the call was answered. To overcome this problem the sender of the INVITE stops retransmitting as soon as the first provisional (or final) response is received.

At this point in time the sender can be sure the original request has been delivered (or in the case of a stateful proxied request, accepted for delivery). Once the phone is answered the 200 OK is sent back to the originating UA. If the 200 OK message is lost in transmission, however, the calling UA would not know the call has been answered and continue to play the ringing tone, causing the call to fail. The use of the ACK message ensures the delivery of the 200 OK message. The called UA will therefore continue to retransmit the 200 OK until the ACK is received.

9.5.10   Cancel

CANCEL is used to terminate a SIP transaction before it is completed, i.e. before the final response has been received. This request should only be used for INVITE requests

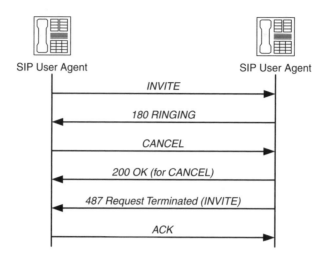

Figure 9.9 CANCEL example

because all other requests are expected to be completed without delay. Figure 9.9 shows a call cancelled before being answered because the caller has given up. The CANCEL is sent; the 200 OK acknowledges the CANCEL. The called UA then cancels the call and sends the *487 request terminated in response* to the original INVITE. The calling UA then sends the ACK in reply to the final 487 response.

9.5.11 Call establishment via proxy

Users within the IMS are expected to establish calls via a proxy server since it provides functions such as user authentication, accounting control and call routing (in particular mobility management). Figure 9.10 shows the call setup but this time with the messages being forwarded via a proxy. In this case, another response, 100 TRYING, is sent by the proxy to tell the UA that the request has been received and the call setup is in progress.

9.5.12 Stateless and stateful proxies

A stateless proxy only forwards the message it receives and does not take part in the client/server transaction. A stateful proxy, on the other hand, acts as a server for client requests it receives and a client for requests it forwards. The proxy shown in Figure 9.10 is stateful, since it accepts the SIP call on the UA's behalf and generates the 100 TRYING provisional response in reply. In this transaction the UAC of the caller is the client and the proxy acts as the server. The proxy then forwards the request acting as a SIP client in respect of the called UAS. The reliability of the message transfer is ensured hop-by-hop since the proxy will retransmit the INVITE message as part of its SIP transaction if the called party does not respond within the timeout period.

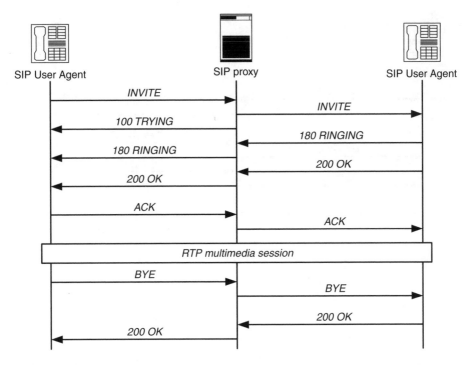

Figure 9.10   SIP call via proxy

Figure 9.11 shows how the same call would look for a stateless proxy. In this case the proxy will not generate the 100 TRYING message and the reliability mechanism is ensured end-to-end instead of hop-by-hop. This creates a problem if a stateless proxy receives a message using TCP in its first link and uses UDP on its second. The sending UA is using a reliable transport mechanism and will not retransmit its INVITE message to the proxy. If the proxy then sends the message using UDP and it is lost in transit, it will not retransmit, causing the call to fail. For this reason when a proxy server forwards a message using a different transport from the one which it is received on, some action is required by the proxy to ensure reliable delivery, i.e. the proxy must act statefully.

9.5.13   SIP offer/answer model

SIP uses an offer/answer model to negotiate media options between end points. This procedure is formally defined in RFC 3264. The protocol operates as follows:

- the initiator provides a series of media options in the initial SDP offer;
- the responder answers with its choice of media.

The original offer can be carried in the INVITE request, or, if omitted by the sender, the called party must include the offer in the 200 OK response. If the original offer is

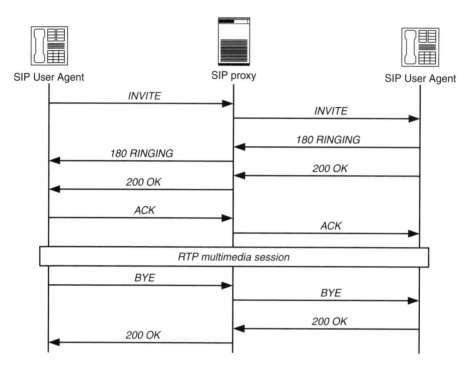

Figure 9.11 Call setup using stateless proxy

carried in the INVITE message then the answer must be carried in the 200 OK response. The answer can also optionally be included in the provisional 1xx responses. If, on the other hand, the original offer is made in the 200 OK response, the answer will be carried within the ACK message.

Figure 9.12 shows two offer/answer examples. In the first, Figure 9.12(a), the initiator is offering to communicate using both video and audio media streams. It is willing to offering the called party the following options:

Audio   PCM μlaw or GSM, RTP port 23099, RTP clock 8 kHz

Video   H.261 or H.263, RTP port 23101, RTP clock 90 kHz

The called UA replies choosing GSM and H.261. Note that static RTP payload types 0 (PCM μlaw), 3 (GSM) and 31 (H.261) are defined in the standard RTP audio visual profile (RFC 1890). In the second example, Figure 9.12(b), a UA is shown rejecting the offer of the video stream by setting the port number to 0.

After the initial offer/answer for any given call, a re-INVITE can be sent containing new SDP content negotiating the media streams options again. This would allow, for example, for two users participating in an audio call to add video. A simpler way to renegotiate media for a call is to use the UPDATE message. This is preferable to the re-INVITE since it only requires two messages (UPDATE and 200 OK) instead of the INVITE three-way handshake. One use of media renegotiation is to put a user on hold. To

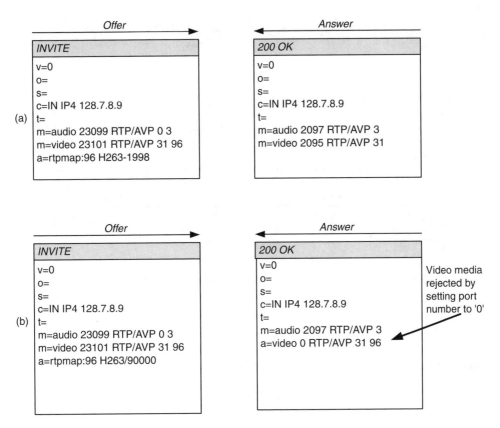

Figure 9.12   SIP offer/answer examples

do this the UA will send an SDP offer containing the SDP media attribute set to inactive. An example of this is shown in Figure 9.13. The first UPDATE puts the media on hold by sending a = inactive; the 200 OK confirms the hold status.

The media stream is resumed with the second UPDATE with the media attribute set as a = sendrecv.

9.5.14   SIP registration

SIP registration is used to allow users to indicate to the network their contact preferences and current location. Mobility is supported via the registration process since a user can indicate which visited network they currently reside in. To register, a REGISTER request is sent to the registrar server, which allows a user to submit one or more contact URIs where they can be reached. The contact header in the message contains the actual contact addresses. The contacts are usually presented as SIP URIs but any valid URI can be used, for example email addresses (mailto URL) and telephone numbers (tel URL). Figure 9.14 shows two registration requests. The first example shows the UA registering its local

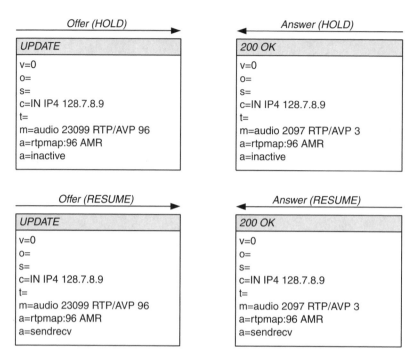

Offer (HOLD) →

```
UPDATE
v=0
o=
s=
c=IN IP4 128.7.8.9
t=
m=audio 23099 RTP/AVP 96
a=rtpmap:96 AMR
a=inactive
```

← *Answer (HOLD)*

```
200 OK
v=0
o=
s=
c=IN IP4 128.7.8.9
t=
m=audio 2097 RTP/AVP 3
a=rtpmap:96 AMR
a=inactive
```

Offer (RESUME) →

```
UPDATE
v=0
o=
s=
c=IN IP4 128.7.8.9
t=
m=audio 23099 RTP/AVP 96
a=rtpmap:96 AMR
a=sendrecv
```

← *Answer (RESUME)*

```
200 OK
v=0
o=
s=
c=IN IP4 128.7.8.9
t=
m=audio 2097 RTP/AVP 3
a=rtpmap:96 AMR
a=sendrecv
```

Figure 9.13 Use of UPDATE to provide call hold

```
REGISTER sip:vodafone.com SIP/2.0
Via: SIP/2.0/UDP 10.0.0.4:5060
To: seb <sip:seb@vodafone.com>
From: seb <sip:seb@vodafone.com>
Call-ID: 8435678684230@12345
Cseq: 2567 REGISTER
Contact: <sip:seb@10.0.0.4> ; q=0.5; expires 86400,
<mailto:seb@orbitage.com> ; q=0.3; expires 86400
```

```
REGISTER sip:vodafone.com SIP/2.0
Via: SIP/2.0/UDP 10.0.0.4:5060
To: seb <sip:seb@vodafone.com>
From: seb <sip:seb@vodafone.com>
Call-ID: 8435678684231@12345
Cseq: 2568 REGISTER
Contact: <mailto:seb@10.0.0.4> ; expires 0
<tel: +4478912345678> ; q=1.0; expires 3600
```

Figure 9.14 Registration examples

SIP IP contact and email address. The expires parameter controls the lifespan of the registration. In the example given, the expiry time is set to 86 400 seconds or 1 day. After this the registration will be deleted from the server. The q parameter gives the priority value. A UA receiving a list of contacts should try the contact with the highest q value first. This is a fractional value, with 1 being the highest priority and 0 the lowest. In the second example, the registration shows how the registration binding can be deleted by setting the expiry time to 0.

9.5.15   SIP call routing (direct, proxy and redirect)

Usually when an INVITE is sent out from the UA it is directed to the local outgoing proxy server. The local proxy will examine the URI in the SIP header. If the host part of the URI is an IP address, the request can be forwarded directly. More often, the host part of the URI is a domain name, in which case a DNS lookup will be performed to determine the associated IP address. For messages destined for a different domain, the message will be forwarded to that domain's proxy server. To do this, the local proxy must discover the address and port number of the remote domain's proxy server. This is done using the mechanisms described in Section 9.5.7.

When the SIP proxy serving the target domain receives the INVITE it will then interrogate its location service to determine the destination user's current contact details. If the SIP user's contact is local, the call can be forwarded directly to the UA. For non-local contacts, however, the INVITE must be proxied by the server or a redirect response sent back to the originating proxy or UA.

Figure 9.15 shows a call being proxied by the remote server. The call involves the following steps:

1. Originating UA sends INVITE, destination is seb@vodafone.com.
2. Local proxy forwards the INVITE to remote proxy at vodafone.com.
3. Proxy at vodafone.com interrogates location service for seb@vodafone.com.
4. Location service returns two SIP contacts: one at IP address 10.0.0.4, the other at seb@orbitage.com.
5. Proxy at vodafone.com picks seb@orbitage.com to forward the call to based on user preference settings and forwards INVITE to proxy at orbitage.com.
6. Proxy at orbitage.com interrogates location service for seb@orbitage.com.
7. Location service returns contact address at 192.10.4.5.
8. Call is forwarded to user's UA at 192.10.4.5.

Note that the SIP URI on the request line can be rewritten by the proxy servers, but the From and To headers remain unaltered as the call is forwarded across the network.

Figure 9.16 shows the call flow when the proxy sends a 3xx redirection response back to the originating UA. This has the benefit of putting the user in control, as at the point of receiving the response which contains the destination contact list, the UA can provide a list of contact options for the user to choose from. The stages for this call are:

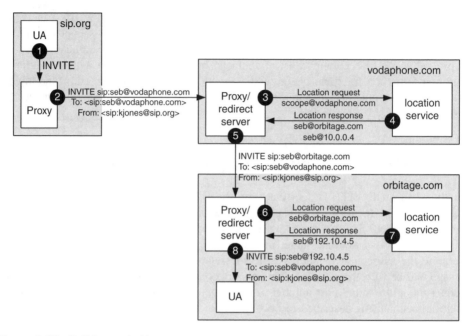

Figure 9.15   Call forwarded by proxy

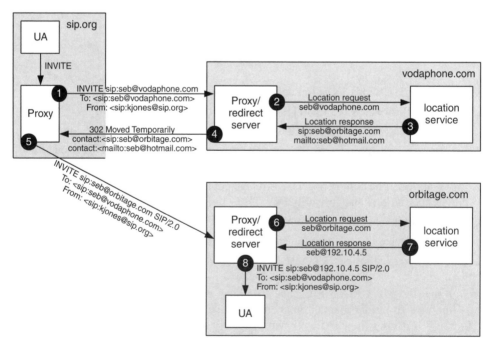

Figure 9.16   Call redirection using 3xx response

1. Originating UA sends INVITE, destination is seb@vodafone.com.
2. Local proxy forwards the INVITE to remote proxy at vodafone.com.
3. Proxy at vodafone.com interrogates location service for seb@vodafone.com.
4. Location service returns two contacts: one is an email address, seb@hotmail.com, the other a SIP contact at seb@orbitage.com.
5. Proxy at vodafone.com sends 302 moved temporarily redirect back to local proxy. This message contains the contact list for the destination.
6. 302 response sent back to originating UA.
7. Originating UA re-sends INVITE to seb@orbitage.com.
8. Proxy at orbitage.com interrogates location service for seb@orbitage.com.
9. Location service returns contact address at 192.10.4.5.
10. Call is forwarded to user's UA at 192.10.4.5.

9.5.15.1 Forking

It is possible for the proxy to forward the INVITE to a number of different contact addresses simultaneously. This process, called forking, allows the proxy to try to speed up the process of contacting the user. Forking can only be performed by a stateful proxy as the message reliability mechanism must be provided for each of the messages sent down a different branch. As more than one 200 OK message can be received, the originating UA needs to be able to distinguish between calls to different UAs belonging to the same user (i.e. registered with the same SIP URI).

To achieve this distinction the For and To headers carry a 32-bit random identifier called a *tag*. The tag on the To header is to distinguish between different UAs responding to the same INVITE. On the From header, the tag is used to separate different UAs using the same SIP URI as a source address. The format of the tag is as follows:

From: <sip:kjones@sip.org>; tag = 34231234
To: <sip:seb@vodafone.com>; tag = a5bfcc789

If a UA receives a provisional response (e.g. 180 RINGING) from more than one destination it can choose which user to talk to by dropping other calls using the CANCEL command. However, if a UA receives more than one 200 OK response from different destinations the UA must acknowledge each with the ACK message (otherwise the terminating UA will keep sending the 200 OK). For users that the UA does not want to talk to the call can be then terminated using the BYE message.

9.5.15.2 SIP routing with path header (RFC 3327)

The use of the path header allows calls routed from the home network to the UA to follow a fixed path. The use of the path header is very much like the record-route in that each proxy server that wants to be included on the path between the sender and receiver will

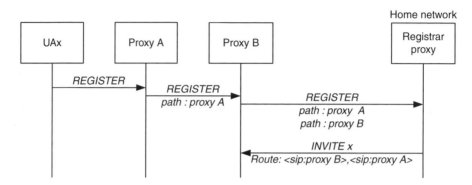

Figure 9.17   Use of path header

add a path header containing their own address. The difference is that the path header is only used with registration requests and is valid for all SIP messages being routed from the user's home network to the SIP user for the lifetime of that registration. This process is illustrated in Figure 9.17. User x sends a registration request to the home network. The proxies along the route both add path headers to the request before forwarding it. The home registrar then stores the contact for the SIP user as well as the associated path information. When an INVITE for the user arrives at the home proxy, it is forwarded via the intermediary proxies using the route header.

9.5.16   Provision of QoS with SIP

When SIP is used to set up multimedia streams across the network there is a requirement for QoS to be also provided. In fact, SIP itself does not provide QoS mechanisms but it does provide information via the SDP content about the service requirements. Each proxy server in the call path, by looking at the SDP content in the INVITE and 200 OK messages, can determine the required bandwidth and request the QoS in its part of the network. A number of extensions to SDP have been proposed to allow the UAs to signal their media streams' QoS requirements. These are outlined in RFC 3312.

However, there is a problem establishing QoS with the standard SIP INVITE transaction. The QoS requirements cannot be determined until the answer to the SDP offer in the INVITE has been returned in the response coming from the called party (e.g. 180 RINGING or 200 OK response). At this point the called party's UA will already be ringing. If the QoS request is rejected then the initiating UA will CANCEL the call; however, in the meantime, there may be 'ghost' ringing at the called party's UA. This problem is illustrated in Figure 9.18. The difficulty is that the destination UA should not proceed with alerting the end user until QoS has been established for the call. What is needed is a system to precondition the link before the remote UA starts to alert the user.

A mechanism for enabling QoS preconditioning with SIP is described in RFC 3312 'Integration of Resource Management and Session Initiation Protocol (SIP)'. The process operates in very much the same manner that media options are negotiated in a basic SIP

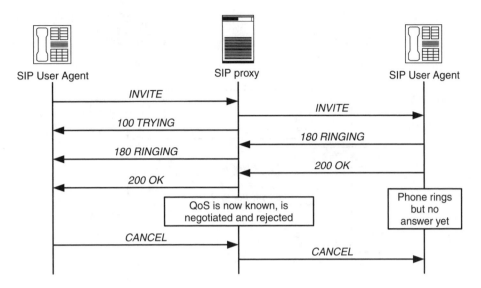

Figure 9.18   Ghost ringing problem

call, i.e. by using an offer/answer model, but in this case the handshake will usually be three-way:

1. UAa offers UAb a range of media streams; QoS is also requested.

2. UAb replies with answer confirming which media streams it would like to use as well as its QoS requirements.

3. UAa and UAb perform QoS reservation; UAa confirms QoS reservation to UAb, and UAb alerts user B.

The call flow for this example is shown in Figure 9.19. For step 1, the original offer is sent in the INVITE message; the SDP also contains the request for the QoS. The SDP for the answer in step 2 is sent using the 183 session progress response, SDP2. Since this message is crucial to the call, it must be sent reliably, and to achieve this the provisional acknowledgement message PRACK is used.

The 183 response will be retransmitted until a PRACK is received by UAb. Once the 183 message has been received by UAa, it can request the QoS from its own access network. When UAa receives local confirmation of its QoS request, it sends the UPDATE message containing SDP3. This contains the confirmed QoS provision. This UPDATE message (RFC 3311) is designed to allow details about the media stream to be updated without modifying the SIP call state. As well as QoS provision, UPDATE can be used to control early media streams (for example when sending call progression signalling in-band). At UAb, as soon as the local QoS request is confirmed and the UPDATE message received, the user can be alerted and the call proceeds as normal.

Table 9.5 shows the format of the SDP QoS parameters used in the call.

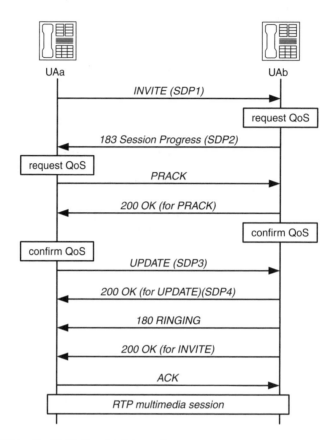

Figure 9.19 SIP call with QoS establishment

Table 9.5 SDP QoS parameters

Parameter	Description
Mode	This can be *send, receive, sendrecv* or *none*. This indicates the direction of the flow of media for the QoS. For none, this indicates that no QoS has been established for the stream to date.
Type	The type can be *local, remote* or *e2e* (or end-to-end). This parameter refers to the network in which the QoS is to be requested. For each UA, *local* means the UA's local access network and *remote* means its peer's local access network. *e2e* is used if the QoS can be established by either UA across the whole link as one request.
Strength	This indicates how critical the establishment of QoS is for the progress of the call. Possible values here are *none, optional* and *mandatory*. For a call to proceed, all *mandatory* QoS requests have to be satisfied. The *optional* strength allows the UA to request QoS establishment but the call may proceed if even if the request is denied.

For each SDP message two attributes are defined;

a = curr:qos the current status of the QoS.

a = des:qos the desired status of the QoS indicating the parties' QoS requirements

The QoS requirement is defined by three parameters – mode, strength and type – which are described in Table 9.5.

The parties negotiate the desired status (indicated by a = des:qos) in the initial exchange (i.e. INVITE/183 session progress). For each exchange, the UA is allowed to upgrade but not downgrade the QoS requirements of the call by modifying the a = des:qos field. If UAa sends a = des:qos remote none this indicates that it has no requirement for QoS in the remote network; UAb could reply with a = des:qos mandatory sendrecv, indicating that QoS is a mandatory requirement for both sending and receiving. Note that the use of local and remote is with respect to the message sender.

The current status of the QoS establishment is also agreed as the SDP is sent back and forth. Each end can upgrade the a = curr:qos field dependent on local information.

For the call to proceed, the current status of the QoS has to satisfy the desired QoS requirements negotiated. For *mandatory* strength requirements the QoS must be provided; if the desired strength is optional the QoS should be requested but the call may proceed even if it is denied. Use of the mode parameter is as follows: a send requirement is satisfied by a current QoS status with mode set to *send* or *sendrecv*, for a receive requirement the current status must be *receive* or *sendrecv*, and for a *sendrecv* requirement only *sendrecv* will suffice.

For example, if a UA sets a mandatory requirement that *send* QoS is achieved end-to-end, in the direction of send from this UA, the SDP would be a = des:qos mandatory e2e send. If it then receives a current QoS status update indicating end-to-end QoS has been established in both directions indicated by a = curr:qos e2e sendrecv, the QoS requirement will have been satisfied and the call can proceed.

Consider the example exchange shown in Figure 9.20. In the first message, the INVITE SDP, UAa is requesting a video stream. For this stream, it is stating that it requires a *sendrecv* QoS in its own access network (*des:qos local mandatory*). Currently, no QoS has been allocated (*curr:qos local none*). In the next SDP message (183 session progress), UAb signals that it requires QoS to be allocated within its own local access network (*des:qos local mandatory*). Notice that the meaning of local and remote are relative, i.e. local refers to the access network of the sender of a message.

In the UPDATE message, UAa informs UAb of the successful allocation of QoS in its access network by sending *curr:qos local sendrecv*. Finally, the 200 OK message contains a confirmation that UAb has also established QoS in its local network (*curr:qos local sendrecv*).

At each point in the offer/answer exchange of QoS parameters, the UAs are free to upgrade (but not downgrade) the QoS requirements and current status. For example, a desired QoS which is *send optional* could be upgraded to *sendrecv mandatory*. Downgrading QoS parameters is not allowed since this would allow either UA peer to deny QoS to the other.

If either party cannot provide QoS mandated by the other, either because of lack of resources, local policy or lack of support for QoS, the call should be dropped.

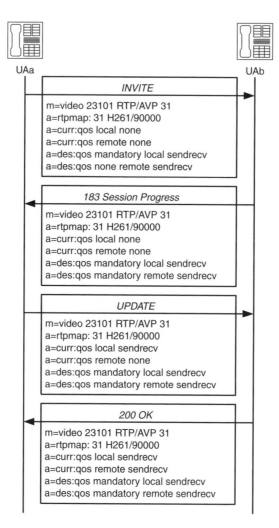

Figure 9.20   SIP QoS negotiation with SDP parameters

Figure 9.21(a) shows the call flow. When the called party's QoS fails, they respond to the INVITE with 580-precondition-failure final response rejecting the call. The calling UA releases the QoS requested locally and signals back the user that the call failed. Figure 9.21(b) shows the call flow when the caller's QoS request fails. In this case the CANCEL message is used to drop the call; the other party responds with 487 request terminated. Note that in both cases the called party's phone does not ring.

9.5.17   SIP security

Without some degree of security, attacking the services of a SIP network would not be difficult. Possible attack methods might be to:

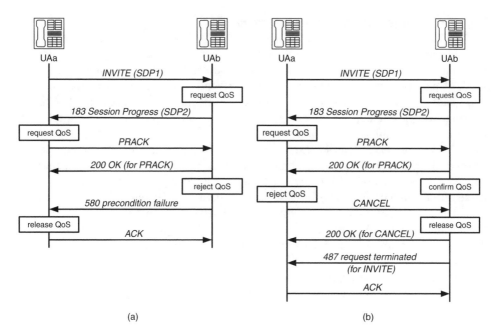

Figure 9.21   Call flow with failure to establish QoS

1. add a false registry binding, forwarding all sessions to the hacker's SIP phone or email;
2. introduce a spoof proxy server that would forward sessions to the hacker's SIP phone or email;
3. make calls on somebody else's account and at their cost, spoofing their SIP address;
4. perform a denial of service attack against the SIP network.

Each of these attacks is made possible by the fact that SIP messages are being carried on an IP network which may not be physically secure. An attacker could snoop the network using a network analyser, looking for SIP messages. This type of attack not only allows the user to invade the privacy of other users of the network, it also can be used as a platform to perform more complex hacking activities such as setting up false calls or interfering with SIP registrations. A denial of service attack could be carried out by possibly bombarding a SIP server with more call requests than it can cope with. To help protect against these attacks a number of security services are required: authentication and integrity protection, privacy and denial of service attack protection.

Three main approaches to providing security with SIP are commonly used: SIP-based, TLS or IPSec. SIP-based security allows for user authentication only. TLS security secures TCP connections for each SIP message transfer hop by hop. TLS is not used for secure IP communication in UMTS since it precludes the use of UDP. IPSec provides network level security for SIP communications regardless of the transport protocol, and for this reason, as well as the fact that IPSec is a mandatory requirement of IPv6, 3GPP recommends the use of IPSec ESP to protect all IP communications carrying core network signalling.

9.5.17.1 Authentication and integrity

Authentication services establish the identity of the origin of the message, while integrity checking ensures that the message has not been modified in transit. Messages sent between SIP UAs and SIP servers should be correctly authenticated and integrity protected. This is to guard against attacks based on user or server spoofing (attack scenarios 1, 2 and 3).

Using SIP messaging, authentication is the only service provided (no integrity protection). SIP authentication is defined as a mandatory requirement in RFC 3261 for all registration requests but can be used with any SIP message. This will protect against a hacker providing new false registrations but does not stop active attacks such as the contact field in the registration request being modified en route. For networks where active attacks are possible, TLS or IPSec should be used. SIP authentication is based on a challenge/response model. The initial message from the sender is challenged with a 401 unauthorized or 407 proxy authentication required (for proxy services) response. The challenge information is contained within WWW-Authentication and Proxy-Authenticate headers for 401 and 407 responses, respectively. The request must then be sent again, this time containing the appropriate authorization header or proxy authorization header containing the correct credentials for the user.

An authenticated registration request is shown in Figure 9.22. First, the registration is rejected by the proxy with a 407 response containing the challenge. The UA resubmits the request with a proxy authorization field containing the authentication response. The request is then forwarded to the registrar, which also requires authentication. It responds with a 401 containing its challenge, which is then forwarded by the proxy to the UA. The UA now resubmits the request a third time. This time it contains authentication credentials for both the registrar and the proxy. The request is forwarded to the registrar, which checks the credentials and the registration succeeds.

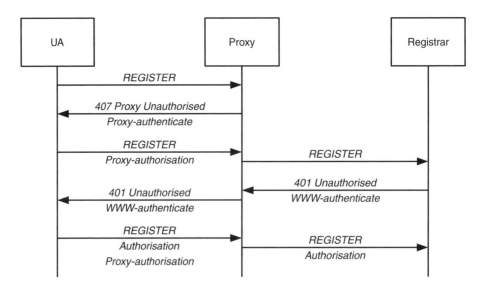

Figure 9.22   Authenticated registration

The exact method to produce the response can vary from application to application; 3GPP defines its own technique in 3GPP 33.203 for the authentication of registration requests.

3GPP 33.203 mandates the use of IPSec in ESP mode for the authentication and integrity protection of all other SIP requests/responses (i.e. not REGISTER) between the UA and P-CSCF. These techniques are described in Section 9.7.3.

Between all other core network elements using IP to communicate (e.g. SIP proxy or registrars) 3GPP 33.210 specifies the use of IPSec ESP headers to provide authentication/integrity protection.

9.5.17.2 Privacy

This is to guard against the snooping of SIP messages on the network. Without protection, a subscriber's confidential call data can be accessed. 3GPP 33.210 states that IPSec ESP confidentiality shall be used for all IP communications crossing the interface between different operators or security domains, but using the authentication/integrity only option is allowed. For all IP communication within an operator's network the use of IPSec ESP confidentiality is optional.

9.5.17.3 Denial of service attack

This is probably one of the most difficult attacks to protect the SIP network against. Techniques that may be deployed include:

- *Hot standby backups for SIP servers*: these can be configured as clusters and use the virtual router redundancy protocol (VRRP; see Chapter 5) or M3UA to communicate service state.
- *Stateless SIP proxy servers*: these are useful since they do not keep call state. A hacker attacking a *stateful* proxy could overwhelm the server by initiating many different INVITE requests. For each request the proxy needs to allocate memory; eventually the proxy may run out of memory and fail. All servers that provide authentication on a SIP network should perform the authentication in a stateless manner. This will protect against hackers who are currently not authenticated attacking the network. Attacks from nodes which are already authenticated can then be logged and the authenticated hacker identified.

9.5.18 SIP–PSTN interworking

Not all calls will use IP end-to-end. For example, many VoIP sessions from the IMS will need to be terminated within legacy PSTNs. At the interface between the two domains there will be a requirement for interworking. The SIP-T (SIP for Telephones, RFC 3372) defines an architectural framework in which legacy telephony signalling can be incorporated into SIP messages. Two types of service are defined:

- mapping between SIP and PSTN messages (e.g. ISUP)
- tunnelling of PSTN parameters not supported in SIP.

The second service is required since there may be no equivalent mapping for some PSTN signalling parameters within SIP.

In the examples given, the PSTN protocol shown is the ISDN User Part (ISUP), because of its wide-scale deployment in modern telecommunications networks. The mapping of ISUP to SIP is covered by RFC 3398. Some of the more important ISUP messages are shown in Table 9.6. Their operation in the context of GSM is outlined in Chapter 3.

Figure 9.23 shows how a call originating in the SIP network is terminated in the PSTN. To initiate the call, the SIP phone sends an INVITE to the gateway. This INVITE message is translated by the gateway into the IAM message and forwarded to the PABX for call routing to the final destination. If the address is valid and routable, the PABX will respond with an ACM. At this point the call is in progress and as soon as the remote phone starts to ring, an ALERTING signal will be sent back to the gateway from the PABX. This will be translated by the gateway to the 180 RINGING provisional response. In practice, calls over the PSTN cannot be guaranteed to be handled purely with ISDN end-to-end; for this reason call progression indicators (busy, ringing, number unobtainable, etc.) are therefore carried in-band as audio. This audio is generated at the terminating PSTN switch. This in-band technique also allows the PSTN operator to provide instructions to the caller, such as asking them to redial using a different prefix. To handle this situation, the SIP gateway uses the session progression indication (SIP provisional response 183), which indicates to the SIP phone that it should replay the audio stream or early media (containing the ringing or busy tone or other message) received from the remote PSTN switch.

When the call is answered, the ANM message is sent back, which is translated by the SIP gateway into the 200 OK response. The SIP phone sends an ACK to acknowledge and the two parties can now talk over the speech channel. When the SIP caller hangs up, the call is terminated by the SIP phone using the BYE message. This message is translated to the ISUP REL message. Finally, the PSTN confirms the call release using RLC and the gateway confirms the BYE message with a 200 OK.

Table 9.6   ISUP messages

ISUP message		Description
IAM	Initial Address Message	Sent from call originator to the remote node and is used to initiate call setup
ACM	Address complete message	Sent from remote node back to origin; confirms that call setup is in progress.
ANM	Answer message	Sent from remote node back to origin to indicate called party has answered the phone
CPG	Call progress	Sent from remote node back to origin to indicate how call is progressing
REL	Release	Sent from either end to request release of connection
RLC	Release complete	Sent from either end to confirm release of connection

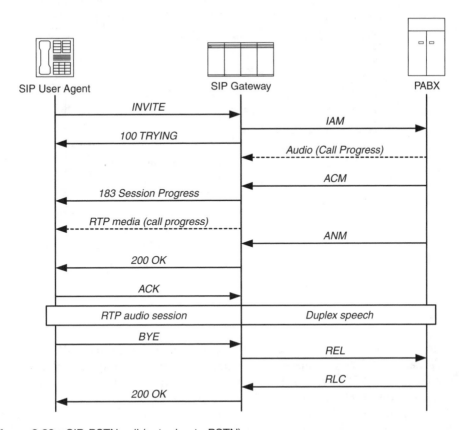

Figure 9.23 SIP-PSTN call (outgoing to PSTN)

Figure 9.24 shows the call sequence for an incoming call from the PSTN network into the SIP network. In this situation, the message translation is similar to the previous example except the direction is reversed. Note that in this case, the SIP gateway itself has to generate the in-band ring tone back to the calling PSTN phone.

For both call scenarios the gateway acts as a SIP UA, terminating and originating the SIP signalling.

9.5.19 SIP bridging

Interworking from PSTN to SIP and SIP to PSTN works well for the most part but will lead to loss of some of the signalling information. This is because a SIP message may not contain all the corresponding header fields in the ISUP message or vice versa. In fact, there are a lot more fields in ISUP than in SIP, leading to possibly considerable loss of service. This is not a problem in the interworking examples shown above since the ISUP information is not needed to route the call on the SIP network, but may cause problems when more complex tandem routing is involved.

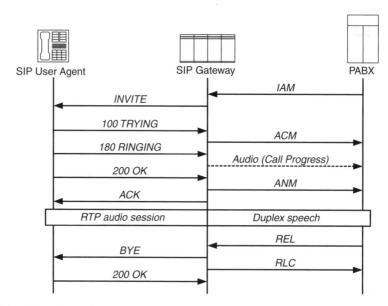

Figure 9.24   Incoming PSTN to SIP call

Figure 9.25 shows an example configuration, where an intermediate SIP network is used between two PSTN segments. This type of setup could be, for example, used to route long-distance calls over an IP backbone between two regional PSTN networks. In this case, the loss of information may generate significant problems since the ISUP fields in the originating network may need to be present to indicate SS7 service requirements in the destination network. To resolve this problem, a form of ISUP encapsulation called SIP bridging has been defined. In this, all the original data fields in the IAM message (or other ISUP messages) are attached to the SIP INVITE (and other SIP messages) when the translation is done at the ingress gateway. This encapsulated data is used to reconstruct the ISUP message at the egress gateway, with the SIP network providing transparent transport between the two PSTNs.

The SIP bridging solution is not ideal, and does have a number of drawbacks. First, the SS7 protocols come in different national variants, e.g. ITU ISUP based signalling

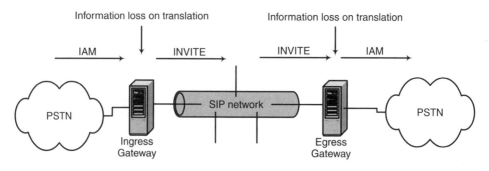

Figure 9.25   Tandem call routing

such as TUP (China), BTNUP (UK) and SSUTR2 (France), as well as ANSI ISUP type signalling used in North America. For the gateways to be able to handle each of these, complex ISUP to ISUP translation is required. Additionally, because of the sensitivity of the ISUP data (e.g. calling party number) it may be necessary to encrypt and authenticate the traffic.

9.5.20 Conferencing with SIP

Conferencing where many parties can participate in the same call is now a common feature of audio and multimedia communication systems. SIP supports three different modes of conference: meet-me, interactive broadcast and ad hoc. The meet-me conference involves all the media being mixed together using a central server commonly called a conferencing bridge. Each user is given a scheduled time to communicate with the conferencing bridge or multi-conference unit (MCU). Each user establishes a point-to-point call to the conferencing bridge which then mixes and forwards media on behalf of each client.

In an interactive broadcast conference, a conferencing bridge is also used but the mixed media is sent to a multicast address instead of being unicast to each participant in turn. This allows two types of conference participants, passive, who only receive media (i.e. listen and watch), and active, who send as well as receive. SIP signalling is only required for active participants. The interactive broadcast SIP signalling for active users is the same is the SIP signalling for the meet-me conference.

Finally, there is the ad hoc conference where a point-to-point conversation is expanded with a series of INVITEs.

9.5.21 SIP event notification

There are many circumstances when it is useful to be able to be informed of events as and when they happen. These events could be, for example:

- arrival of new email;
- arrival of a new instant message;
- a user becoming available after being busy;
- a user switching on their phone.

The event notification extensions to SIP are specified in RFC 3265, which defines two new messages: SUBSCRIBE and NOTIFY.

9.5.21.1 Subscribe

SUBSCRIBE is sent by a UA to indicate to another UA of which events it needs to be notified. SUBSCRIBE can be used to turn event notification either on or off. The header fields relevant to subscribe are:

- *Expires*: time for which the subscription is valid. A value of 0 indicates that the UA would like to unsubscribe
- *Event*: states the event package and optionally an integer ID.

Event packages

Events are grouped into packages. Each package defines the internal states of a resource, for example the state of an email inbox. If the internal state changes (e.g. a new email arrives), this event plus the current internal state (e.g. the total current number of unread emails) can be signalled to the subscriber.

9.5.21.2   Notify

NOTIFY is sent by a UA to another UA to signal the occurrence of an event. The event header in the NOTIFY message will contain the package name of the original SUBSCRIBE message plus the integer ID.

9.5.22   SIP and instant messaging services

UMTS specifies the use of SIP to provide instant messaging (IM) within the IMS. The use of SIP for instant messaging gives UMTS the simplest route to interoperating with other IM networks since it is IETF's preferred choice and the only IM proposed standard which is vendor independent.

The extension that provides instant messaging for SIP is defined in RFC 3428. This document defines a SIP method call MESSAGE. SIP instant messages are sent in what is called near real-time. This means that the message is expected to be delivered without delay directly to the receiver. Even though direct delivery is the case for most messages, SIP IM, like SMS, allows for instant messages to be stored (in case, for example, the user is offline).

Here is an example of a SIP instant message:

> MESSAGE sip: jeff@orbitage.com SIP/2.0
>
> Via: SIP/2.0/UDP proxy.orbitage.com
>
> To: sip:jeff@orbitage.com
>
> From: sip:seb@orbitage.com
>
> Call-ID: 123254caa121@pc1.orbitage.com
>
> Cseq: 1 MESSAGE
>
> Content-Type: text/plain
>
> Content-Length: 28
>
> Development meeting at 4 pm?

Usually the content in this case is text, but can be other formats. For example, the message format can support the delivery of multimedia content (e.g. for MMS service) by the use of the multi-part MIME format. This is essentially the same way that file attachments are handled by email.

If the destination for the MESSAGE is a UA and the message is accepted the response should be 200 OK. On the other hand if the destination is a message relay server, the response will be a 202 accepted.

9.6   E.164 NUMBERS (ENUM)

For calls handled purely within the IP network, SIP uses DNS and IP routing to forward the requests. For calls from IP to PSTN, the destination's telephone number can be represented as a SIP URI. For these calls, the gateway strips out the telephone number and uses it to initiate the call using ISUP signalling.

A problem arises when a caller on the PSTN network is attempting to call a user on the IP network. Since it is not possible to represent an IP address (or DNS name) in an ISUP message, when the call reaches the PSTN/SIP gateway there must be some mechanism to route the call to its final destination. The solution to this problem is to provide a mapping between E.164 telephone numbers and IP addresses using DNS. To do this the IETF has specified a domain for telephone numbers and a mechanism for generating DNS addresses from them. The process is defined in RFC 2916 and operates as follows.

As an example, consider the following E.164 telephone number, including country code (44):

441353789789

To perform the DNS conversion, the digits are reversed, dots put between them and *e164.arpa* concatenated to the end, so for the example the DNS address will be:

9.8.7.9.8.7.3.5.3.1.4.4.e164.arpa

This effectively splits up the DNS domain for telephone numbers based on country code then area code, since the DNS name defines domains from right to left in terms of coverage. The namespace of these numbers can be managed in the same way as any DNS namespace, so if an organization controls the telephone prefix 65780 (Singapore number) it can manage the set of DNS mappings that end with 0.8.7.5.6.e164.arpa.

To resolve this address, the DNS client makes its first query to the top-level domain, i.e. e164.arpa, to ascertain the name server for the particular country code, and will then contact this name server to look for the particular subscriber. In practice, it may be referred to another DNS server if the domain is large enough to be split into subdomains. It is also possible to translate this address using a lightweight directory access protocol (LDAP) directory server.

Using this scheme, each user on the IP network can be allocated a unique E.164 number which must also be entered into the DNS server. The DNS resource records either point to another name server (to indicate the ENUM service provider) or can use the naming authority pointer (NAPTR) resource records (RFC 2915) format.

9.6.1 NAPTR

NAPTR provides a DNS service which instead of resolving names directly to IP addresses, provides string translation. The result of this process is usually a URI. NAPTR is useful since it can translate between different types of protocol and service. For example, a user starting off with a URI as http://www.orbitage.com/~contacts might have it translated directly by DNS to mailto:mail1.orbitage.com, redirecting the user to send an email. This is particularly applicable for ENUM, which starts with telephone strings and needs to translate them to SIP URIs.

9.6.2 ENUM examples

Here are some examples of ENUM records.

$ORIGIN e164.arpa. 4.4 IN NS ns1.4.4.e164.arpa

This indicates the address of the name server for the UK ENUM domain (44). It is also possible to have a record indicating that the next server will be LDAP.

$ORIGIN 4.4.e164.arpa.

IN NAPTR 100 10 "u" "ldap+E2U" "ſ+44(.)$!ldap://ldap.uk/cn=01!".

The *in the record indicates that all addresses that start with 4.4.e164.arpa. will match the NAPTR record. For example, the number string +4467254907 would first be translated to the domain name:

7.0.9.4.5.2.7.6.4.4.e164.arpa

Then the original string is acted upon by the NAPTR record to produce the URI:

ldap://ldap.uk/cn =067254907

This request can then be sent by the client to the LDAP server.

The final example shows how a set of DNS entries for a particular user are represented in ENUM:

$ORIGIN 4.3.2.1.3.5.3.1.4.4.e164.arpa.

IN NAPTR 10 10 "u" "sip+E2U" "ſ.*$!sip:seb@orbitage.com".

IN NAPTR 10 10 "u" "mailto+E2U" ſ.*$!mailto:seb@orbitage.com".

IN NAPTR 10 10 "u" "tel+E2U" "ſ.*$!tel:+44 7789 123654".

This is for the telephone number +4413531234, and the possible resulting URIs would be:

sip:seb@orbitage.com

mailto:seb@orbitage.com

tel:+44 7789 123654

The client could then choose which way they would prefer to contact the caller. In the case that the telephone number is chosen then the ENUM translation process will start again.

9.7 UMTS IMS CALL SIGNALLING

The call signalling protocol for the IMS is based wholly on SIP and SDP and is described in 3GPP TS 24.229. The different CSCF components each act as SIP servers as well as containing extra functionality to perform functions such as accounting. The P-CSCF and I-CSCF act as proxy servers, the P-CSCF to provide service to the UE, and the I-CSCF as an incoming proxy for the home network.

The S-CSCF acts as a proxy and registrar server. It performs authentication of subscribers. It uses the registration contact information to route incoming calls for its own subscriber list to an appropriate P-CSCF, which then forwards them to the UE. For mobile originated calls it must decide if the call should be forwarded to a P-CSCF (for IMS to IMS calls), to a BGCF (for IMS to PSTN call) or to the Internet (for public SIP calls).

Before the subscriber can use the IMS to make a multimedia call, the following three operations must be carried out:

- perform a successful GPRS attach procedure to establish a bearer for the call;
- discover the address of their serving P-CSCF;
- register their current contact details and authenticate themselves with the S-CSCF.

Call flows for session initiation/termination use the same techniques as standard SIP except a number of new headers are added to the messages to assist in routing and accounting for network usage. Some examples of these are headers to provide information on radio access technology, network identity, cell identity, etc.

9.7.1 IMS security

Security within the IMS is provided as an overlay on top of the PS domain. There are five needs for security, as shown in Figure 9.26 (from TS 33.203):

1. Mutual authentication between the ISIM and the HSS. The HSS delegates the authentication function to the S-CSCF but is responsible for generating authentication challenges based on the subscriber's private key.
2. Mutual authentication and integrity checking for messages between the UA and the P-CSCF.
3. Mutual authentication and integrity checking for messages between the I-CSCF, S-CSCF and the HSS
4. Mutual authentication and integrity checking for messages between the P-CSCF and the S-CSCF/I-ISCF where the P-CSCF is sited within a visited network

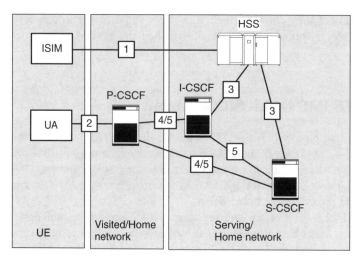

Figure 9.26 IMS security architecture

5. Mutual authentication and integrity checking for messages between the P-CSCF and the S-CSCF/I-ISCF where the P-CSCF is sited within the home network

Security for IMS service can be split into three areas, as follows.

- *IMS access security (1 and 2 in Figure 9.26)*: controls UE access to the network, and is used to protect against service theft and denial of service attacks. Also protects messages between the UE and P-CSCF. This is described in Sections 9.7.2 and 9.7.3.
- *IP network domain security (3,4, and 5 in Figure 9.26)*: securing signalling between IMS network elements. Security for these links is provided by the use of IPSec and IKE over the Za and Zb interfaces and is described in Section Section 9.4.
- *Application security*: providing end-to-end security for a user application, e.g. encryption of a multimedia call. This is negotiated end-to-end. Application security is not defined by 3GPP and developers are free to choose their own techniques. There are already a number of working and fully implemented security protocols for use by IP applications, namely IPSec for VPN service, TLS for secure web access, S/MIME for email authentication and encryption, and RTP encryption for multimedia calls

9.7.2 P-CSCF assignment

Before the user can register they must establish the address of the P-CSCF. There are two ways this can be done, using DHCP and DNS (see Section 9.5.7) or directly using the PDP context establishment procedure. Alternatively, the address of the P-CSCF can be contained within the protocol configuration options additional parameters list of the PDP context accept. This will be quicker than DHCP/DNS but requires a little more

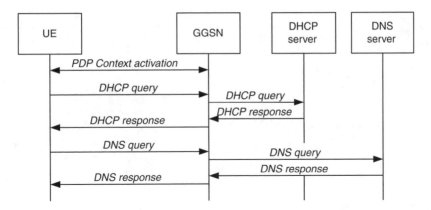

Figure 9.27   P-CSCF assignment using DHCP

complexity at the GGSN. For example, it may have to assign a P-CSCF address based on load sharing. The message flow for DHCP and DNS is shown in Figure 9.27.

9.7.3   IMS registration

Before the user can make use of the IMS, they must register with their home S-CSCF. This provides them with two services, an outgoing S-CSCF for mobile originated calls and a registration entry within the S-CSCF of their current contact address for mobile terminated calls. As the registration procedure is carried out, the user is also authenticated. The authentication procedure takes place between the subscriber's ISIM and the HSS. The technique uses the same algorithm as standard UMTS authentication (see Section 6.21); however, the subscriber's identity values and secret key are different. The following definitions are relevant to the discussion:

- *IM public identity (IMPU)*: essentially the user's SIP public address which is used to route calls to the subscriber.
- *IM private identity (IMPI)*: an identity used internally within the network for registration, admission and accounting purposes. It is not used for routing SIP calls.
- *IM services identity module (ISIM)*: an UICC application which performs security functions (e.g. authentication) for the IMS.
- *Private key*: encryption key shared between the HSS and ISIM, used to authentication both parties.

The IMPU, IMPI and private key are stored securely within the ISIM and cannot be modified. For a UE which does not have an ISIM application, the IMPU and IMPI can be derived from the IMSI as described in Section 9.5.1. The IMS registration/authentication procedure is shown in Figure 9.28.

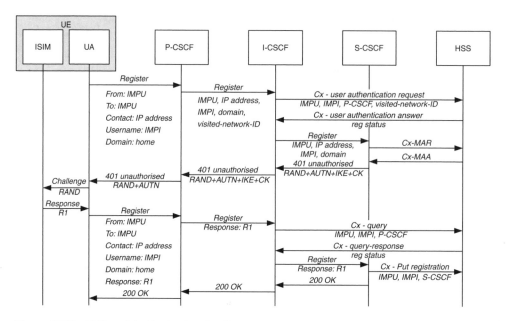

Figure 9.28   IMS registration/authentication

The UE initially builds a REGISTER request. The URI for this request points to the registrar at the user's home domain e.g. *sip:registrar.oribitage.com*. The request also contains the user's public and private identity (IMPU, IMPI), their assigned IP address in the visited network and home domain. The P-CSCF appends an identity for the visited network and then uses the URI to forward the request to the UA's home I-CSCF.

The I-CSCF is responsible for forwarding all registration requests to the S-CSCF. Before doing this it checks if the user is allowed access to the IMS by sending a Cx user authorization request (UAR) to the HSS. This message contains the user's public and private identity, as well as the visited network ID. The HSS responds with information about the subscriber's registration in the Cx-UAA response (see Table 9.7). For users who are already registered the response contains the name of the S-CSCF assigned; this is then used to forward the REGISTER request. For users who are not registered the I-CSCF must assigning an S-CSCF. If S-CSCF capabilities are present in the Cx-query response these are then used by the I-CSCF to determine which S-CSCF within its pool can be assigned. The format and use of the S-CSCF capabilities is not specified by 3GPP and their use is operator/vendor dependent. It is possible at this stage for the HSS to disallow the registration in two ways. Either the user is unknown or the registration is rejected (e.g. in the case of a visited network having no IMS roaming agreement with the home network).

If the I-CSCF determines the registration is permitted it must then assign an S-CSCF to the user (in the case of an unregistered user) and then forward the REGISTER request.

The S-CSCF is responsible for terminating the SIP registration requests (i.e. it is a SIP registrar) as well as performing the authentication checks for the subscribers. Before authentication can proceed the S-CSCF must obtain authentication vectors for the user

Table 9.7   Information elements in a Cx-UAA message

Name	Description
Registration status (mandatory)	Registration status of user; possible values are unknown, allowed, reject and registered already
S-CSCF name (only present if status = registered already)	Name of S-CSCF for users who are already registered
S-CSCF capabilities (Optional)	Capability requirements for S-CSCF

from the HSS. It does this by sending a Cx-MAR request to the HSS. This request contains the user's public and private identity. The HSS responds with the authentication data, i.e. AUTN, RAND, XRES, IK and CK.

The AUTN value authenticates the HSS to the UE. The RAND value is used as a challenge and the XRES is the expected response. CK and IK are the cipher key and integrity key, respectively. For more information about these parameters please refer to Section 6.21. The S-CSCF then sends a SIP 401 unauthorized response back, which contains the random challenge value (RAND) plus the authentication token AUTN as well as IK and CK values. The 401 response is routed back to the UA via the I-CSCF and P-CSCF. Note that the P-CSCF removes the IK and CK before forwarding the response. These values are used as shared secrets between the P-CSCF and the UE to provide integrity checking (and optionally privacy protection) for all future SIP signalling. The IK and CK values are generated by the UE using its private key and the RAND value.

The UA within the UE then uses the random challenge and requests the ISIM to produce the appropriate authentication response. It also sends the ISIM the AUTN token so that the appropriate checks on network authentication can be performed. Note that the private key is contained within the ISIM. This protects against illicit copying. The ISIM then sends the UA the response R1 to the challenge. The UA then produces a new registration request, this time containing the authentication data R1. The authenticated registration is then sent back to the S-CSCF. The S-CSCF checks that the value R1 = XRES and if so the user is authenticated and the 200 OK confirming the registration sent in reply. The S-CSCF also updates the registration status of the user at the HSS by sending the Cx-put message. This contains a notification of the registration event at the server. Possible values are registered, unregistered, unregistered due to timeout, and authentication failure.

Note that the authentication procedure is done twice in this case, once when establishing the GPRS context and secondly when connecting to the IMS. This separation of the authentication domains allows for more flexibility. For example, a user could send an authenticated registration request to the I-CSCF even if they are connected not via GPRS but using a fixed-line service. This is because the security is handled within SIP and is not dependent on the access technology (fixed line or wireless).

9.7.4   IMS mobile originated call

The mobile originated call is always routed via the S-CSCF. This is to ensure network accounting integrity (see Section 9.2.1). The mobile originated requests are routed from the UE to the P-CSCF then either directly to the S-CSCF or to the I-CSCF, which then forwards them to the S-CSCF. The use of the I-CSCF in call signalling is useful since it allows an operator to hide the internal structure of their network for security reasons. Note for all call flows shown in this section only message transfers within the IMS are shown.

Figure 9.29 shows the call flow for a mobile originated call routed without the aid of the I-CSCF. The initial INVITE message from the UE contains SDP containing a list of CODECs that the UE wants to support for the call as well as the QoS request. The P-CSCF then forwards the request to the S-CSCF. No QoS authorization is attempted at

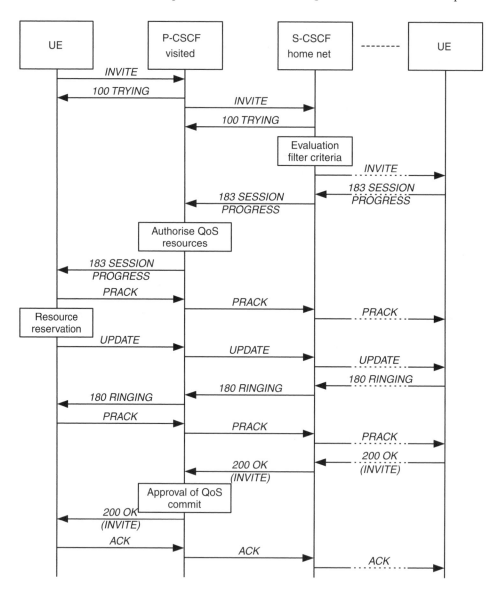

Figure 9.29 IMS Mobile originated call flow example

this stage since the CODEC for the session has not yet been established. The S-CSCF then checks the service profile for the subscriber, determines if the call is permitted by applying the user's filter criteria and then forwards the INVITE to the destination. The destination will then decides which CODEC(s) it wants to support then sends its reply in the SDP contained within a 183 session progress message.

When the P-CSCF receives the 183 message, it now knows the QoS requirements and can determine if it has enough resources within the IMS to support the call. If not it will send a 580 precondition failure to the UE and a CANCEL to the S-CSCF. For the example shown the QoS request is authorized and the 183 is forwarded to the UE. The UA then uses the PRACK message to acknowledge the 183 response and this in return gets a 200 OK from the remote user. The UE then proceeds to establish QoS suitable for the call over the GPRS network. This could involve the UE negotiating a bandwidth increase in the PDP context that it is already using for the SIP signalling. If the existing GPRS connection is sufficient for the call then nothing further needs to be done at this stage. Assuming the UE is successful in achieving QoS it will then send the UPDATE message confirming this to the far end. If it fails, it must cancel the call. Note that the 200 OK messages for the PRACK and UPDATE messages are omitted on the diagram for simplicity.

When the terminating UE receives the UPDATE message (and assuming all its local QoS requirements are established) it will apply ringing to the destination. It sends the 180 ringing provisional response, which is sent using the reliable mode in UMTS and is acknowledged with a PRACK, and 200 OK is sent in response to this message.

Finally, when the user at the far end answers the call, the 200 OK is sent for the INVITE. When the 200 OK is received at the P-CSCF this then confirms the QoS request and lets the call through by sending the 200 OK to the UE. At this point media can flow between the two parties and the ACK is sent to finally acknowledge the 200 OK.

9.7.5   IMS mobile terminated call

Figure 9.30 shows the call flow for a mobile terminated call. The call starts with the INVITE request entering the home network for the subscriber at the I-CSCF. The I-CSCF acting as an incoming proxy then interrogates the HSS, by sending a Cx location query to the HSS. This request contains the public ID, which is extracted from the URI in the INVITE message. The HSS now responds with a location response, which contains the address of the S-CSCF with which the destination user is registered. If the user does not exist then a 401 not found response will be returned. The I-CSCF then forwards the call to the S-CSCF.

The S-CSCF looks up the registration entry for the destination, which contains the contact details, including the UE IP address and serving P-CSCF address. It then forwards the INVITE to the P-CSCF, which in turn forwards it to the UE.

The called UE responds with a session progress message, which is acknowledged with PRACK. The called UE then proceeds to establish QoS for the GPRS connection. As soon as it receives the UPDATE from the far end confirming the remote QoS, it alerts the user and sends 180 RINGING. The user then answers the phone, generating a 200 OK

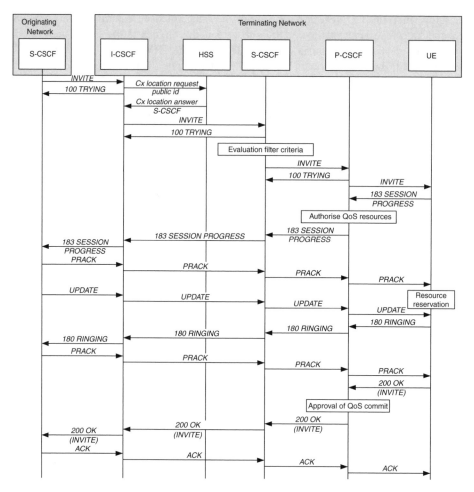

Figure 9.30 Mobile terminated call

message, which is also used to trigger the P-CSCF into enabling the QoS over the GPRS network. Once again, the 200 OK messages for the PRACK and UPDATE messages are omitted from the diagram for simplicity.

9.7.6 QoS reservation for IMS calls

The QoS reservation decisions for the network are based on two criteria, resources and policy. The resource criterion is simply stated as whether the network has enough capacity to accept this call. For the policy criterion the question is, does the network wish to allocate this subscriber this much bandwidth at this point in time? Resource decisions must be made locally at the point of control within the network, so the RNC controls resources in the RAN, the SGSN resources in the GPRS core, etc. Policy decision can be more centralized; in the case of the IMS it is carried out by an entity called the policy decision function (PDF). The PDF may be co-located with the P-CSCF and this is a logical choice since it will be the closest SIP point of contact for the UE within the IMS.

Three procedures for QoS control were shown in Figures 9.29 and 9.30; they are;

- authorize QoS resources at the P-CSCF
- resource reservation at the UE
- approval of QoS commit.

The first of these steps is always carried out near the beginning of a call as soon as the first 183 session progress message has been received at the P-CSCF. At this point in the call the CODEC and bandwidth requirements will have been agreed, and therefore the PDF can make a decision as to whether there are sufficient resources within the IMS network (not the GPRS or access network) to handle the call. As soon as the resources are reserved the session progress message can be forwarded to the UE.

At this stage the resource reservation at the UE can proceed. This is shown in Figure 9.31. First the UE uses the bandwidth requirements given in the SDP portion of the SIP response to generate an activate PDP context request. This is sent to the SGSN, and assuming the SGSN is able to fulfil this request, it then sends a create PDP context

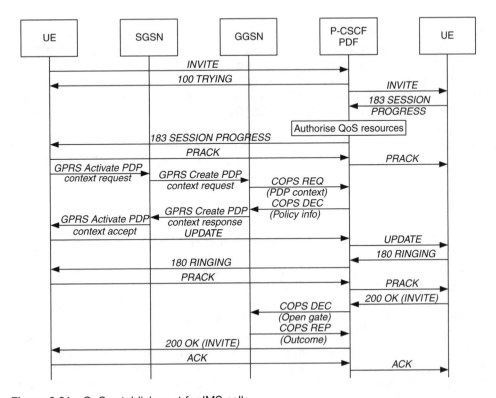

Figure 9.31   QoS establishment for IMS calls

message to the GGSN. The GGSN now needs policy information to decide on whether to allow the PDP context or not. It sends a request using the COPS REQ message to ask the PDF for a policy decision as regards the creation of the PDP context. The PDF knows which SIP call the QoS request relates to by the use of an authorization token. This is a unique value which the P-CSCF generates for each new call. It adds this value to the P-Media-Authorization header in the INVITE or 183 messages that are terminated at the UE. The policy information returns a decision accepting or rejecting the creation of the PDP context. Note at this stage, IP traffic for the given context may not be passed over the GGSN between the IMS and the GPRS core network. The GGSN then confirms the creation of the PDP context to the SGSN, which in turn confirms this to the UE. At the same time the GGSN sends a COPS REP message to the PDF confirming the creation of the PDP context. The call then proceeds in the standard manner until the 200 OK for the INVITE is received at the P-CSCF. At this point the PDF sends a COPS DEC message confirming the use of the resources at the GGSN. The GGSN will then proceed to forward traffic between the two networks for the newly created context. This request is called a gate open. The GGSN then reports back the status of the newly opened gate to the PDF. Note that throughout all these exchanges the authorization token is used to bind the SIP call to the PDP context.

9.7.7 IMS accounting

Accounting is an essential service within any commercial network operation. Network usage must be monitored and accounting records produced so to be able to produce the correct customer billing information. Accounting functions must be carried out at two different points: the home and visited network. The collating of accounting information within the IMS is performed by a network entity called the charging collection function (CCF). This is essentially a DIAMETER server which responds to accounting control messages.

The procedure is essentially as follows. Each P-CSCF and S-CSCF forwarding a 200 OK in response to an INVITE sends a request to the accounting server within their domain to open a CDR. At the end of the call when the BYE is sent the P-CSCF and S-CSCF send requests to update then close the CDR.

9.7.8 Common open policy service (COPS)

Whenever QoS is requested on a network two decisions have to be made. First, are there enough network resources to satisfy the request? This decision is decided locally by individual network elements, for example a GGSN providing GPRS connectivity into the IMS. The GGSN will determine if enough network resources are available to satisfy the request. The second is called the policy decision and this takes into account factors such as the user's service profile, time of day or even if they have enough credit on their account to decide whether to accept or reject the QoS request. In most networks it makes sense to take the policy decision at a single central point since subscriber data will be stored centrally in some database.

The COPS model for policy control (RFC 2748) uses this arrangement and requires two main components, PEP and PDP. The policy enforcement point (PEP) is the network element which receives the QoS request from the user and is responsible for enforcing network policy within the network element. The PEP receiving a request will first check to see if it has enough network resources and if not it will reject the request. If there are enough network resources it will then proceed to contact the policy decision point (PDP) by sending a COPS request message. This must contain enough information for the PDP to make an appropriate decision. In UMTS each request is bound to a SIP INVITE message via the use of the authorization token header. The PDP then determines if the QoS request is allowed by enforcing network policy. It then sends a COPS decision message back to the PEP. The PEP will then signal back to the user accepting or rejecting the request. In IMS the PEP is sited at the GGSN, which provides IMS connectivity; the PDP is sited within the user's P-CSCF.

9.8 IP IN THE RADIO ACCESS NETWORK (RAN)

For R5, all the UTRAN transport can be replaced with IP. In this case even though the datalink and physical layers could remain based on ATM, they need not be. This flexibility is provided to suit the needs of both new and existing operators. For new operators, it allows implementation of a network infrastructure based around any suitable L2 technology. However, for operators migrating to all-IP, it allows for an upgrade to use of IP, but without the need for a change in the existing ATM infrastructure. For example, a switched Ethernet network would be perfectly acceptable given that it could support the QoS requirements of UMTS.

In Release 99 of UMTS, transport within the UTRAN is based on an ATM transport, as was discussed in Chapter 6 and 7. Transport bearer services across the Iub, Iur and Iu-CS interfaces are based on AAL2, with the ALCAP protocols such as Q.2630.1 and Q.2630.2 used to dynamically establish channels within ATM circuits. The ALCAP protocols will be phased out as UMTS moves to an all-IP network but to aid interoperability IP-ALCAP (or ALCAP over IP) has been defined. This is used for networks supporting a mixture of old and new UTRAN elements (some using AAL2 and some IP) so that ALCAP messages can be forwarded by nodes that support IP only.

Within UTRAN, a clear distinction has been between the radio network layer in which the UTRAN functionality is carried out, and the transport layer, which is responsible for delivering this data across the access network. Because of the clear demarcation and logical independence of each of these layers, transport options within the RAN can be changed without affecting the radio network user or control protocols.

9.8.1 Support for IPv6

All IP nodes within a R5 RAN using IP must support IPv6. Support for IPv4 is optional for R5 but since network components require backward compatibility with earlier releases, it is clear that a dual stack configuration (v4 and v6) must be provided.

9.8.2   IP in the Iu interface

Options for supporting IP across the Iu interface for the user plane are defined in TS 25.414 (UTRAN Iu interface data transport and transport signalling). The R5 stack for IP transport of user plane (UP) traffic towards the Iu-CS domain interface is shown in Figure 9.32.

Each RTP payload contains a UP PDU. The functions of the RTP layer are somewhat redundant as timing control is performed using the UP protocol. Each radio access bearer (RAB) supported is sent and received using a separate transport bearer, defined by source and destination IP and port address. The IP address and port numbers are exchanged between the core network element and the RNC in the RANAP request/confirm messages. The RNC will send the IP address (in the transport address IE) and port number (in the binding ID IE) and associated RAB identifier to the MGW. The MGW will then use the address and port number received as the IP transport options for the IP transport bearer associated with that RAB. This process is illustrated in Figure 9.33. The MSC server indicates in the RAB assignment request that its MGW is prepared to receive user traffic on port number 5080 and IPv6 address 5555::123:456:789:900 for RAB ID = 1450. The RNC responds with its own incoming UDP port number, 8067, and IP address, 5555::456:123:089:912. Both parties now have transport information associated with the RAB ID and can forward traffic accordingly, the CN elements sending to port 8067, address 5555::456:123:089:912 and the RNC sending to 5080, address 5555::123:456:789:900 for the established RAB.

For the transport connection to the PS domain, the protocol stack is different (Figure 9.34(a)), since this time RTP is not used. The payload of the UDP protocol is GTP-U packets. The procedure is the same as was described in Chapter 6 with the exception of useing IPv6 instead of IPv4.

The protocol stack for the control plane in the Iu interface (both Iu-PS and Iu-CS) is shown in Figure 9.34(b). Since RANAP is an SS7 application part protocol and expects to run over MTP3, sigtran is used. M3UA and SCTP are covered in Chapter 8.

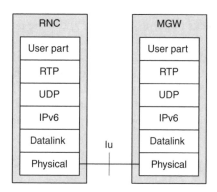

Figure 9.32   Iu-CS interface IP options for user plane traffic

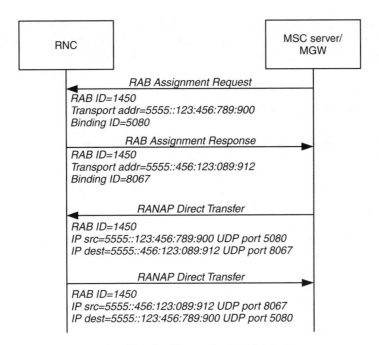

Figure 9.33   User transport signalling for IP over the Iu-CS interface

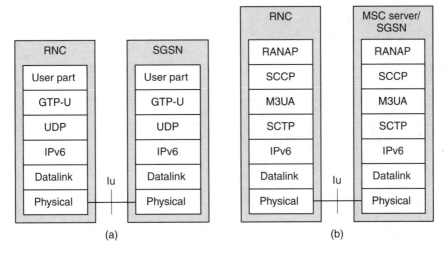

Figure 9.34   Iu-PS interface IP options for user and control plane traffic

9.8.2.1   Iu IP transport evolution

For R99 for the Iu interface IP is only used for the Iu-PS interface. For both signalling and user transport, IP over AAL5 is used (see description in Chapter 7). For the R4 network the Iu transport is the same as in R99, with IP being used only in the Iu-PS interface. For R5 IP can be used for both interfaces (Iu-CS and Iu-PS) and the datalink layer can be any suitable layer 2 technology and is not constrained to be AAL5.

9.8.3   IP in the Iur Interface

The Iur has to support RAN signalling as well as user data for functions such as soft handover. The transport options for the Iur (and Iub) interfaces for dedicated channels are as described in 3GPP 25.426 and for common data channels in 3GPP 25.424. Figure 9.35(a) shows the R5 IP protocol stack for the user plane traffic for the Iur interface. The various framing protocols for both dedicated and shared channels are carried over UDP. The source and destination IP addresses and UDP port numbers define the transport bearer channel carrying the framing protocols. This information is passed between the two RNCs using the RNSAP protocol. The IP address is carried in the transport address IE and UDP port in the binding ID IE.

Figure 9.35(b) shows the IP transport option protocol stack for the R5 Iur signalling plane; again, sigtran is used since RNSAP is an SS7 application protocol.

9.8.3.1   Iur IP transport evolution

In R99 and R4, IP over AAL5 can be used to transport both RAN signalling (i.e. RNSAP) and transport signalling in the form of ALCAP messages. The user plane transport for both R99 and R4 is AAL2. With the introduction of R5 IP can be used for user data and RAN signalling with any suitable layer 2 technology that can support the service requirements allowed.

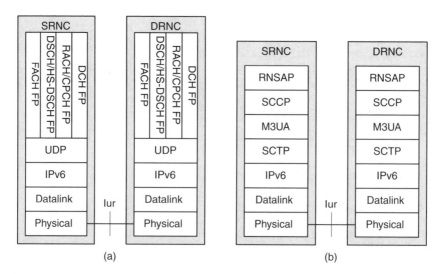

(a)   (b)

Figure 9.35   Iur IP transport option: user and control planes

9.8.4   IP in the Iub interface

The Iub interface connects the RNC to the BTS. its transport options are defined in TS 24.434 (dedicated channels) and TS 25.426 (common channels). Figure 9.36 shows the R5 protocol stack for a Iub interface which supports IP. Note that for user data the protocol stack is the same as is used for the Iur interface. Each transport bearer within the user plane of the Iub is distinguished by the use of UDP port number and IP address. The BTS control signalling in the form of NBAP messages is carried directly over the streaming control transmission protocol (SCTP). The IP transport bearers are defined by source and destination IP address and UDP port numbers, and these values are exchanged in Node B application part (NBAP) messages (again, using the transport address and binding ID information elements).

9.8.4.1   Iub IP transport evolution

Previous to R5 IP is not used at all on the Iub interface; all transport is based on AAL2 and AAL5. For R5 signalling is carried using SCTP over IP and user data using UDP over IP.

9.8.5   IP header compression in the RAN

All nodes that support low-speed links (e.g. E1/T1) are expected to support IP header compression using the scheme described in RFC 2507. The use of compression is established as part of the PPP NCP negotiation and is described in Chapter 5.

9.8.6   RAN IP datalink layer

When an RNC uses the IP transport option it must support the PPP protocol with HDLC framing at the datalink layer. However, this does not preclude support for other datalink

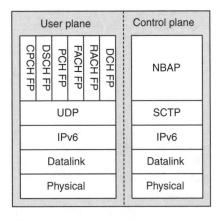

Figure 9.36   Iub IP protocol stack

protocols such as PPPMux, LIPE or CIP. These can be used to provide fragmentation and aggregation functions, which are important when providing efficient transport and QoS. These options are discussed in more detail in the following section.

9.8.7 IP QoS in RAN

Traffic moving within the RAN has certain QoS requirements. With the Release 99 version of UMTS these requirements are met by the use of ATM, in particular AAL2, to carry delay-sensitive traffic and AAL5 for signalling. With the use of IP within the RAN, 3GPP R5 specifies support for DiffServ marking for all RAN interfaces (Iu, Iur and Iub). For networks running IP over ATM, there are a number of approaches to QoS provision, including the mapping of codepoints to ATM SVC or PVC and the use of MPLS.

Another issue that needs to be addressed with regard to QoS for IP in the RAN is head of line blocking. This can happen when a high-priority frame is held up by the transmission of a long, lower priority frame. An example illustrates the point. What will be the delay that a high-priority frame will have to suffer given that it is at the head of its queue, assuming the link has already started to transmit a frame 1280 bytes in length? The number of bits to transmit will be $1280 \times 8 = 10\,240$.

$$\text{Delay for 128 kbps link: } 10\,240/128\,000 = 0.08 \text{ s} = 80 \text{ ms}$$

$$\text{Delay for 256 kbps link: } 10\,240/256\,000 = 0.04 \text{ s} = 40 \text{ ms}$$

$$\text{Delay for 1 mbps link: } 10\,240/1\,000\,000 = 0.001024 \text{ s} = 10.24 \text{ ms}$$

For ATM this problem is addressed with the use of AAL2 packet segmentation and reassembly (SSSAR; see Chapter 7). Each long packet is broken up into many small fragments before being transmitted in an ATM cell.

Different schemes have been proposed to allow for the fragmentation and aggregation of payloads within IP, including composite IP (CIP), lightweight IP encapsulation (LIPE) and multiplexed PPP. These protocols are currently classified as study areas for 3GPP (TS 25.933), meaning that more work is required before an actual standard can be produced. The operation of each proposed scheme is now explained.

9.8.8 Composite IP (CIP)

CIP provides both aggregation and fragmentation. Figure 9.37 shows the contents of how FP PDUs are packed into CIP containers. Each FP PDU (or fragment of FP PDU) is contained within a separate CIP payload. A long FP PDU will require multiple CIP payloads, each containing a fragment of the original. A header is then added to each CIP payload that allows for demultiplexing and reassembly at the far end.

A number of CIP packets are then assembled into a CIP container payload, a CIP container header is added and the whole container is then transmitted using UDP. Currently the specification defines a CIP container header of null.

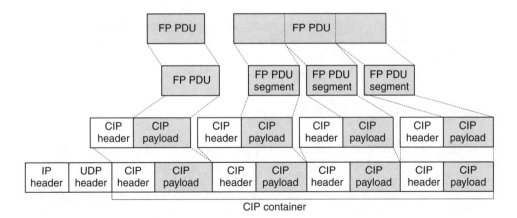

Figure 9.37   Transport using CIP packet format

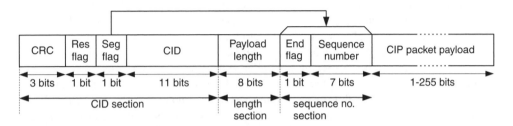

Figure 9.38   CIP packet header

The CIP packet header is shown in Figure 9.38. It is split into three sections. The first is the CID section, which identifies the stream to which the packet belongs, which is used for multiplexing. Each CIP packet belonging to a different FP PDU stream is assigned a different CID field value. It is equivalent to the AAL2 CID field. The segmentation flag is set to 1 if the payload following is a PDU segment and 0 if it contains a whole PDU. The 3-bit CRC protects the whole contents of the CID section, including the reserved flag (Res), which is available for further extensions to the protocol. The second section indicates the length of the payload, which is 8 bits and can thus indicate a payload length between 1 and 255 bytes. Finally, the optional sequence number section is used to aid segmentation and reassembly. This section is only present if the segmentation flag (Seg) is set to 1 in the CID section. The end flag is set to 1 for the last segment in a sequence, and the sequence number is used to order the fragments on reassembly. It is set to 0 at the start of each FP PDU and incremented by 1 for each subsequent segment.

9.8.9   Lightweight IP encapsulation protocol (LIPE)

Originally proposed in an Internet draft as a replacement for RTP, LIPE provides both fragmentation and aggregation. It also allows for extensions to packet headers so that

IP	UDP	MH1	MDP1	MH2	MDP2	MH3	MDP3

IP	TID	MH1	MDP1	MH2	MDP2	MH3	MDP3

Figure 9.39 LIPE encapsulation format

media dependent information can be sent to help decode the data stream. Figure 9.39 shows the format of a LIPE packet carried over UDP and also a format defined for IP (this has been dropped in the latest LIPE draft). The description given here relates to the LIPE draft titled draft-chuah-avt-lipe-02.txt. The LIPE PDU is comprised of multiplexed data (MD) payloads, each prefixed with a multiplexed header (MH) Each MD can contain whole parts or fragments of individual data streams.

The MH provides information on the channel ID for each of the MD packets plus fragmentation information. It is shown in Figure 9.40. The basic MH format is shown in Figure 9.40(a). The standard header length is 3 bytes if the E bit is set to 0. If E is 1 then an extended form of header of variable length up to a maximum of 19 bytes is defined. Figure 9.40(b) shows the basic header format without fragmentation, which is suitable for smaller packets. Only a sequence number and FlowID are provided. The sequence number allows out of order packets to be reordered (as in RTP). The FlowID allows different flows across the link to be demultiplexed. In Figure 9.40(c) the packet format which supports fragmentation is shown. In this case the EOF bit is used to indicate if this is the last frame in the fragment list. Also note that the FlowID for fragmented packets is

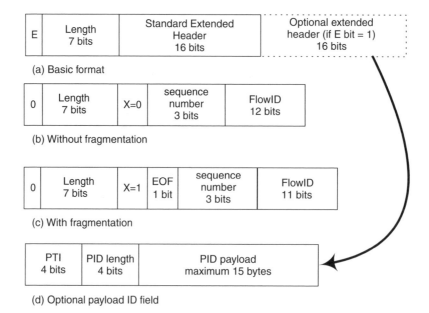

Figure 9.40 LIPE multiplexed header

reduced to 11 bits. If the E bit is set to 1 extra information is provided, called the payload ID field. This is shown in Figure 9.40(d). The payload type identifier (PTI) is used to indicate different types of wireless network (e.g. type 1 = IS95 network, type 2 = UMTS network). The PTI is followed by a length field and then some application-specific data called the PID payload. This could be timing information for the CODEC, long sequence numbers or voice quality indicators.

Since the FlowID must be interpreted at both ends of the link, LIPE also defines mechanisms to set up FlowIDs from one side to the other. Mechanisms are also provided to set up the UDP port number, which will be used to tunnel the LIPE packets. To enable this function LIPE provides a number of signalling messages, listed in Table 9.8.

Each end needs first to agree on a UDP port number to forward packets through. This process is carried out using the tunnel setup messages (request and response). Once a tunnel has been set up, the flow setup messages are used to set up LIP flows for the RABs. Each bearer is identified via the use of a 16 bit RAB ID. Finally, there is support for soft handover. This is carried out as follows. An RNC can send a software handoff request to another RNC (or Node B) asking for a LIPE flow to be set up for a given RAB ID. Given that the other UTRAN node can locate the appropriate radio bearer, the new flow will be allocated between the originator and the new node for the given RAB ID.

9.8.10 Multiplexed PPP

Multiplexed PPP allows multiple flows of data to be carried within one PPP frame. This reduces the amount of PPP overhead suffered when sending many small packets, a common problem with VoIP. Multiplexing is actually an option that can be established during the LCP negotiation phase of PPP. If the receiver offers to accept multiplexed frames the transmitter for each frame can choose to multiplex or not. Figure 9.41 shows the format of a PPP multiplexed frame using HDLC.

In this case each UDP payload is preceded by three fields. There is a protocol field flag, which is set to 1 if the protocol field is present. For subframes with the same protocol

Table 9.8 LIPE signalling message types

Message	Description
Tunnel setup	Sets up UDP tunnel for LIPE on a given UDP port
Tunnel tear down	Releases tunnel
Flow setup	Sets up flow for a given FlowId and RABID
Flow tear down	Release flow
Soft handover request	Sets up alternative flow for the same RABID

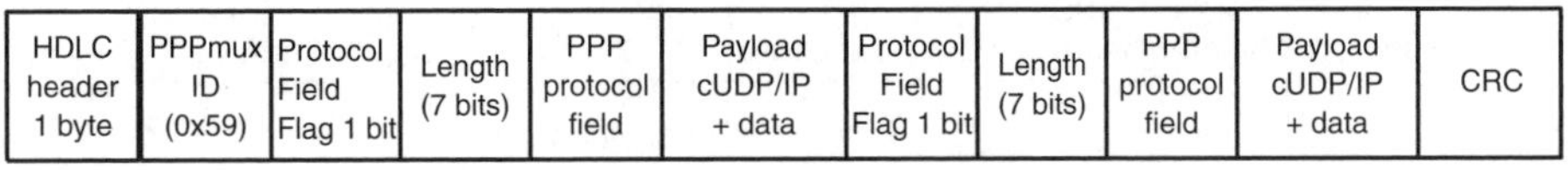

HDLC header 1 byte	PPPmux ID (0x59)	Protocol Field Flag 1 bit	Length (7 bits)	PPP protocol field	Payload cUDP/IP + data	Protocol Field Flag 1 bit	Length (7 bits)	PPP protocol field	Payload cUDP/IP + data	CRC

Figure 9.41 PPP multiplexed frame option

field as their predecessor this flag can be set to 0 and the protocol field omitted The length bits define the data length from 1–127 bytes. This is then followed by the protocol field, which is compressed using protocol field compression (PFC) for PPP. The payload for PPP multiplexed is carried within UDP/IP with the headers being compressed using the scheme defined in RFC 2508.

9.8.11   AAL2 over UDP

Another potential solution is to use AAL2 over UDP/IP. In this technique, only the SSSAR and CPS layers of AAL2 are needed. The packet format for this scheme is shown in Figure 9.42.

Note that with this scheme no start field is included. This is because the size of a UDP frame is variable in length, and therefore a CPS packet does not need to align to 48-byte ATM cell payloads, thus fragmentation of a CPS packet over two UDP packets is not required. For the very same reason padding is also omitted. AAL2 over UDP allows the delivery of AAL2 packets over an IP infrastructure. This is useful when internetworking between IP and ATM networks since it allows the AAL2 to be delivered transparently over both the ATM and IP infrastructure.

9.8.12   IP ATM interoperating

When migrating a network from R99 to an all-IP platform, a number of interoperability issues may arise; for example, mixing UTRAN nodes that support only ATM with R5 nodes supporting IP. Another example might be a core network upgraded to IP with the UTRAN still using ATM. Since ATM itself provides models for IP transport it is perfectly acceptable for IP messages to be moved transparently across an ATM link usually using AAL5. The added complication with UMTS is that protocols based on SS7 transport and ALCAP do not expect to be run over IP. To circumvent this problem for signalling transport, an SS7/SCTP/IP bearer has been proposed. This can carry both RAN signalling such as NBAP and RANAP as well as the transport control protocol ALCAP. For user data (carried in R99 over AAL2) AAL2 over UDP has been proposed.

There are two possibilities for internetworking between IP and ATM: dual stack or by using an internetworking unit to translate between the nodes. The dual stack configuration is shown in Figure 9.43 for the Iub interface. In this case the RNC will use ATM signalling with the R99 base station and IP with the R5 base station.

With the Iub interface the dual stack option is the simplest and most effective solution for interworking. This is because each BTS is expected to support only IP or ATM and

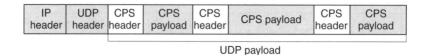

UDP payload

Figure 9.42   AAL2 over UDP

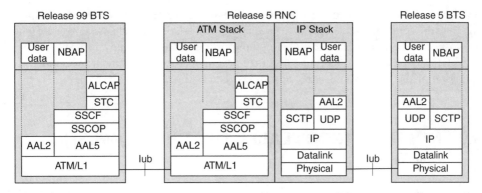

Figure 9.43   RNC dual stack IP/ATM (Iub)

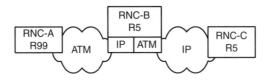

Figure 9.44   Interworking IP and ATM RAN domains

talk to only one RNC. For the Iur interface the dual stack option may not be the most suitable solution. Figure 9.44 shows why this is the case. The diagram shows two network domains: RNC-A supports only IP and RNC-C only ATM. RNC-B is connected to both networks and uses a dual stack to support connections to both networks. A can talk to B, C can talk to B but with the configuration shown A cannot talk to C. This is because the radio network layer signalling coming from A will be terminated at B; if it is not destined for B then it will be discarded. This is because B is not configured to forward messages between the two networks but only to provide connectivity from B to other RNCs directly connected. To provide full connectivity between the two domains, an interworking unit (IWU) is required.

An IWU (Figure 9.45) provides protocol translation between the IP and the ATM domains. The IWU sits between the two domains forwarding the packets from IP to

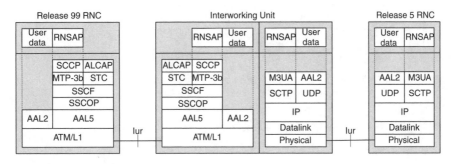

Figure 9.45   Interworking unit at Iur

ATM and vice versa. Since the transport addressing schemes used on the two domains are quite different, SS7 codepoints are used as a global addressing scheme across the whole network. The operation of the IWU is complex since it must not only forward messages between the two domains, but must also take part in bearer establishment procedures. The operation of the IWU is still work in progress and is discussed in TS 25.933.

9.9 MULTIPROTOCOL LABEL SWITCHING (MPLS) IN UMTS

MPLS combines the complexity and high levels of functionality of IP routing, QoS and security mechanisms with the speed and efficiency of layer 2 switching. The basic operation of MPLS is as follows. Before transmission a path, referred to as a label switched path (LSP), is established across the network between sender and receiver using a network layer or QoS protocol (e.g. OSPF or RSVP). The packets are then switched across this path using a fixed-length label. Since the switching is packet by packet at layer 2 using a fixed-length label, it can be done at great speed and implemented in hardware. Services that can be provided by MPLS include IP routing, QoS and virtual private networks using MPLS tunnels.

Although MPLS is focused on the delivery of IP services over an a ATM network, the specification is flexible and allows for:

- transfer of data over any combination of layer 2 technologies
- support of all layer 3 protocols.

Since UMTS is designed to deliver multimedia type services and will be using IP extensively, in particular as operators migrate to R5, the use of MPLS to provide QoS in the IMS is strongly indicated. 3GPP discusses the application of MPLS to provide QoS when using IP transport in the RAN (24.933). This is illustrated in Figure 9.46.

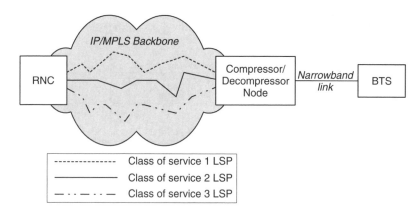

Figure 9.46 MPLS in the RAN

The RNC is connected to the BTS via a broadband routed IP/MPLS cloud, with the last mile being via a narrowband link, for example an E1 line. MPLS in this case is used to provide different classes of services for the UTRAN traffic. The compressor/decompressor node provides IP/UDP header compression to optimize the use of the bandwidth on the narrowband link.

9.9.1   MPLS terminology

Prior to a discussion of the specifics of MPLS operation, it is useful to define the key terms used in conjunction with MPLS:

- *Label*: a short fixed-length identifier indicating a path.
- *Forwarding equivalence class (FEC)*: a group of packets that should be forwarded the same way and therefore can be assigned the same label.
- *Label binding*: mapping between a label and a FEC.
- *Label information base (LIB)*: a database containing label bindings.
- *Label switched router (LSR)*: any router with MPLS functionality.
- *Label edge router (LER)*: LSR at the edge of an MPLS domain.
- *MPLS domain*: a set of nodes under one administrative domain capable of performing MPLS routing.

9.9.2   MPLS forwarding

With conventional routing, as a packet crosses the network, each router examines the header to extract all the relevant information to decide how to forward the packet. Usually, for IP, it is only the destination IP address that is relevant, but other components can also have relevance. Since this header inspection is done at each router, it limits the speed and efficiency of operation.

For MPLS, the LER that accepts the packet to the MPLS domain maps the layer 3 header information to a label using the label binding. This label is then used for subsequent forwarding decisions. Therefore, once the label is chosen, it is put at the front of the packet and the header need not be examined again throughout the MPLS domain. At each intermediary LSR the label is mapped to a FEC; at a minimum this will define a new label value and an outgoing interface at the LSR, but it may also control how the packet is queued and scheduled in the case of QoS provision. Within the RAN, the RNC and the compressor/decompressor node will act as LERs. The topology of an MPLS domain is shown in Figure 9.47.

Each LSR receiving a packet will look up the label binding in its LIB, add the new label then forward the packet on the given interface. This process is illustrated in Figure 9.48. The packet is received at the LER at the left of the diagram with an IP destination address of 130.24.45.6. The edge router looks up the IP destination address in its LIB, adds the

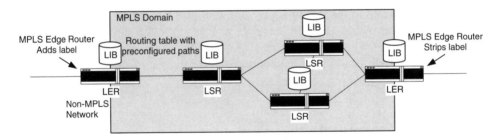

Figure 9.47   MPLS domain

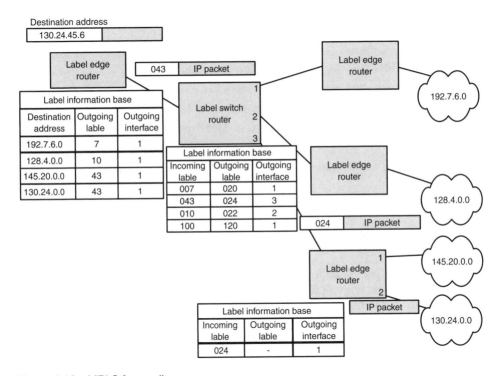

Figure 9.48   MPLS forwarding

label value (43) and then forwards the packet on interface 1 to the LSR in the centre. The LSR then looks up the value 43 in its own LIB and determines it must forward the packet on interface number 3 with a new label value of 24. The edge router at the lower right then has to forward the packet. It looks up the label binding and finds that the packet is to leave the MPLS domain (indicated by a null entry for the outgoing label). It then examines the IP header and determines that the packet should be forwarded on interface number 2. Note that it is only at the edge of the MPLS domain that the layer 3 header has to be examined.

9.9.3   Label switched paths (LSP)

The concatenation of a series of label bindings is called a label switched path (LSP). The LSP is created when the label bindings are distributed between the MPLS routers. LSP establishment is either done on demand, i.e. an MPLS edge node requests a binding when a new path is required, or is ongoing as part of standard routing table update procedures.

There are three different ways that an LSP can be established: hop-by-hop, explicit or constraint-based routing. With hop-by-hop routing each MPLS node determines the contents of the LIB by referring to its IP routing table. For each entry in its routing table it will advertise a binding (consisting of an IP network address and label) to its neighbours. Explicit routing, on the other hand, allows an MPLS edge router to establish an exact path across the network by specifying a router list. One example of explicit routing is when using MPLS with RSVP to provide QoS. The RSVP reservation message is carried along the route determined by the path message. For each hop a binding is carried which establishes the path. With constraint-based routing, the request to establish the path contains certain QoS requirements. Each MPLS node will determine on which interface to establish the path depending on local conditions. This is much like the way that an SPVC is established in ATM.

9.9.4   Label distribution

Label bindings can be distributed between MPLS nodes in two ways, either piggybacked on existing routing or QoS protocols or using the label distribution protocol (LOP). LDP provides for conventional hop-by-hop routing; an extension to LDP called constrained-based LDP (CR-LDP) provides for both explicit and constraint-based LSP establishment.

9.10   SUMMARY

This chapter has described how UMTS with Release 5 has moved to an all-IP architecture. Not only is IP used to transport signalling and user data, but SIP, an IP-based session control protocol, is used instead of SS7 call signalling. This has provided a number of advantages, including better interoperability with other users on the Internet as well as complex value-added services such as multimedia calls, conferencing and instant messaging. With R5 the RAN can use IP for all of its signalling. The options for IP in the RAN were described, as well as the issues involved when interworking between IP and ATM networks.

FURTHER READING

RFC 2507: IP Header Compression. M. Degermark, B. Nordgren, S. Pink. February 1999.

RFC 2508: Compressing IP/UDP/RTP Headers for Low-Speed Serial Links. S. Casner, V. Jacobson. February 1999.

RFC 2782: A DNS RR for specifying the location of services (DNS SRV). A. Gulbrandsen, P. Vixie, L. Esibov. February 2000.

RFC 2806: URLs for Telephone Calls. A. Vaha-Sipila. April 2000.

RFC 2915: The Naming Authority Pointer (NAPTR) DNS Resource Record. M. Mealling, R. Daniel. September 2000.

RFC 2916: E.164 number and DNS. P. Faltstrom. September 2000.

RFC 2976: The SIP INFO Method. S. Donovan. October 2000.

RFC 3031: Multiprotocol Label Switching Architecture. E. Rosen, A. Viswanathan, R. Callon. January 2001.

RFC 3036: LDP Specification. L. Andersson, P. Doolan, N. Feldman, A. Fredette, B. Thomas. January 2001.

RFC 3212: Constraint-Based LSP Setup using LDP. B. Jamoussi, Ed., L. Andersson, R. Callon, R. Dantu, L. Wu, P. Doolan, T. Worster, N. Feldman, A. Fredette, M. Girish, E. Gray, J. Heinanen, T. Kilty, A. Malis. January 2002.

RFC 3262: Reliability of Provisional Responses in Session Initiation Protocol (SIP). J. Rosenberg, H. Schulzrinne. June 2002.

RFC 3263: Session Initiation Protocol (SIP): Locating SIP Servers. J. Rosenberg, H. Schulzrinne. June 2002.

RFC 3264: An Offer/Answer Model with Session Description Protocol (SDP). J. Rosenberg, H. Schulzrinne. June 2002.

RFC 3265: Session Initiation Protocol (SIP)-Specific Event Notification. A. B. Roach. June 2002.

RFC 3268: Advanced Encryption Standard (AES) Ciphersuites for Transport Layer Security (TLS). P. Chown. June 2002.

RFC 3311: The Session Initiation Protocol (SIP) UPDATE Method. J. Rosenberg. October 2002.

RFC 3312: Integration of Resource Management and Session Initiation Protocol (SIP). G. Camarillo, Ed., W. Marshall, Ed., J. Rosenberg. October 2002.

RFC 3320: Signaling Compression (SigComp). R. Price, C. Bormann, J. Christoffersson, H. Hannu, Z. Liu, J. Rosenberg. January 2003.

RFC 3327: Session Initiation Protocol (SIP) Extension Header Field for Registering Non-Adjacent Contacts. D. Willis, B. Hoeneisen. December 2003.

RFC 3329: Security Mechanism Agreement for the Session Initiation Protocol (SIP). J. Arkko, V. Torvinen, G. Camarillo, A. Niemi, T. Haukka. January 2003.

RFC 3361: Dynamic Host Configuration Protocol (DHCP-for-IPv4) Option for Session Initiation Protocol (SIP) Servers. H. Schulzrinne. August 2002.

RFC 3372: Session Initiation Protocol for Telephones (SIP-T): (SIP-T): Context and Architectures. A. Vemuri, J. Peterson. September 2002.

RFC 3398: Integrated Services Digital Network (ISDN) User Part (ISUP) to Session Initiation Protocol (SIP) Mapping. G. Camarillo, A. B. Roach, J. Peterson, L. Ong. December 2002.

RFC 3428: Session Initiation Protocol (SIP) Extension for Instant Messaging. B. Campbell, Ed., J. Rosenberg, H. Schulzrinne, C. Huitema, D. Gurle. December 2002.

RFC 3485: The Session Initiation Protocol (SIP) and Session Description Protocol (SDP) Static Dictionary for Signaling Compression (SigComp). M. Garcia-Martin, C. Bormann, J. Ott, R. Price, A. B. Roach. February 2003.

3GPP TS23.002: Network Architecture.

3GPP TS23.003: Numbering, Addressing and Identification.

3GPP TS23.060: General Packet Radio Service (GPRS) Service description; Stage 2.

3GPP TS23.218: IP Multimedia (IM) session handling; IM call model; Stage 2.

3GPP TS23.228: IP Multimedia Subsystem (IMS); Stage 2.

3GPP TS23.815: Charging implications of IMS architecture.

3GPP TS24.228: Signalling flows for the IP multimedia call control based on SIP and SDP; Stage 3.

3GPP TS24.229: IP Multimedia Call Control Protocol based on SIP and SDP; Stage 3.

3GPP TS25.412: UTRAN Iu interface signalling transport.

3GPP TS25.414: UTRAN Iu interface data transport & transport signalling.

3GPP TS25.420: UTRAN Iur Interface: General Aspects and Principles.

3GPP TS25.422: UTRAN Iur interface signalling transport.

3GPP TS25.424: UTRAN Iur interface data transport & transport signalling for CCH data streams.

3GPP TS25.426: UTRAN Iur and Iub interface data transport & transport signalling for DCH data streams.

3GPP TS25.432: UTRAN Iub interface: signalling transport.

3GPP TS25.434: UTRAN Iub interface data transport & transport signalling for CCH data streams.

3GPP TS25.933: IP transport in UTRAN.

3GPP TS29.228: IP Multimedia (IM) Subsystem Cx and Dx Interfaces; Signalling flows and message contents.

3GPP TS29.229: Cx and Dx interfaces based on the Diameter protocol; Protocol details.

3GPP TS33.203: 3G security; Access security for IP-based services.

3GPP TS33.210: 3G security; Network Domain Security (NDS); IP network layer security.

draft-ietf-dhc-dhcpv6-23.txt: Dynamic Host Configuration Protocol for IPv6 (DHCPv6)", Ralph Droms, C Perkins, Jim Bound, Bernard Volz, M. Carney, Ted Lemon, 02/06/2002.

draft-ietf-enum-rfc2916bis-00.txt: The E.164 to URI DDDS Application", Michael Mealling, Patrik Faltstrom, 02/27/2002.

draft-ietf-sipping-dialog-package-01.txt: An INVITE Initiated Dialog Event Package for the Session Initiation Protocol (SIP), Jonathan Rosenberg, Henning Schulzrinne, 05-Mar-03.

A list of the current versions of the specifications can be found at http://www.3gpp.org/specs/web-table_specs-with-titles-and-latest-versions.htm, and the 3GPP ftp site for the individual specification documents is http://www.3gpp.org/ftp/Specs/latest/

Glossary of Terms

2G	2nd Generation
3G	3rd Generation
3GPP	Third Generation Partnership Project
8-PSK	8-state Phase Shift Keying
A3	Authentication algorithm A3
A38	A single algorithm performing the functions of A3 and A8
A5/1	Encryption algorithm A5/1
A5/2	Encryption algorithm A5/2
A8	Ciphering key generating algorithm A8
AAA	Authorisation, Authentication and Accounting
AAL	ATM Adaptation Layer
AAL1	ATM Adaptation Layer Type 1
AAL2	ATM Adaptation Layer type 2
AAL3/4	ATM Adaptation Layer Type 3/4
AAL5	ATM Adaptation Layer type 5
ABR	Available Bit Rate
AC	Authentication Centre
ACCH	Associated Control Channel
ACK	Acknowledgement
ACR	Allowed Cell Rate
ADPCM	Adaptive Differential Pulse Code Modulation
AESA	ATM End System Address
AFI	Authority and Format Identifier
AGCH	Access Grant CHannel

Convergence Technologies for 3G Networks: IP, UMTS, EGPRS and ATM   J. Bannister, P. Mather and S. Coope
© 2004 John Wiley & Sons, Ltd   ISBN: 0-470-86091-X

AH	Authentication Header
AI	Acquisition Indicator
AICH	Acquisition Indicator Channel
AK	Anonymity key
ALCAP	Access Link Control Application Protocol
AM	Acknowledged Mode
AMR	Adaptive Multi Rate
AMR-WB	Adaptive Multi Rate Wide Band
AN	Access Network
AP	Access preamble
APDU	Application Protocol Data Unit
API	Application Programming Interface
APN	Access Point Name
ARFCN	Absolute Radio Frequency Channel Number
ARP	Address Resolution Protocol
ARQ	Automatic Repeat Request
AS	Access Stratum
ASC	Access Service Class
ASN.1	Abstract Syntax Notation One
AT command	ATtention Command
ATM	Asynchronous Transfer Mode
AuC	Authentication Centre
AUTN	Authentication token
AWGN	Additive White Gaussian Noise
BA	BCCH Allocation
BCCH	Broadcast Control Channel
BCD	Binary Coded Decimal
BCH	Broadcast Channel
BER	Bit Error Ratio
BG	Border Gateway
BICC	Bearer Independent Call Control
B-ICI	B-ISDN Inter-Carrier Interface
B-ISDN	Broadband ISDN
BLER	Block Error Ratio
BMC	Broadcast/Multicast Control
BOM	Beginning of Message
BPSK	Binary Phase Shift Keying
BS	Base Station
BSC	Base Station Controller
BSS	Base Station Subsystem
BSSAP	Base Station Subsystem Application Part
BSSGP	Base Station Subsystem GPRS Protocol
BSSMAP	Base Station Subsystem Management Application Part
Btag	Beginning Tag
BTFD	Blind Transport Format Detection

BTS	Base Transceiver Station
BUS	Broadcast and Unknown Server
BVC	BSS GPRS Protocol Virtual Connection
BVCI	BSS GPRS Protocol Virtual Connection Identifier
BW	Bandwidth
C/R	Command/Response field bit
CA	Certification Authority
CAC	Connection Admission Control
CAMEL	Customised Application for Mobile network Enhanced Logic
CAP	CAMEL Application Part
CB	Cell Broadcast
CBC	Cell Broadcast Centre
CBCH	Cell Broadcast CHannel
CBR	Constant Bit Rate
CBS	Cell Broadcast Service
CC	Call Control
CCCH	Common Control Channel
CCH	Control Channel
CCITT	Comité Consultatif International Télégraphique et Téléphonique (The International Telegraph and Telephone Consultative Committee)
CCPCH	Common Control Physical Channel
CCTrCH	Coded Composite Transport Channel
CD	Collision Detection
CDMA	Code Division Multiple Access
CDR	Charging Data Record
CDV	Cell Delay Variation
CDVT	Cell Delay Variation Tolerance
CER	Cell Error Ratio
CES	Circuit Emulation Service
CFN	Connection Frame Number
CHAP	Challenge Handshake Authentication Protocol
CI	Cell Identity
CIP	Composite IP
CIR	Carrier to Interference Ratio
CIR	Committed Information Rate
CLK	Clock
CLP	Cell Loss Priority
CLR	Cell Loss Ratio
CM	Connection Management
CN	Core Network
COM	Continuation of Message
CONS	Connection-oriented network service
CPCH	Common Packet Channel
CPCS	Common Part Convergence Sublayer
CPCS-SDU	Common Part Convergence Sublayer-Service Data Unit

CPICH	Common Pilot Channel
CPS	Common Part Sublayer
CPU	Central Processing Unit
CRC	Cyclic Redundancy Check
CRNC	Controlling Radio Network Controller
C-RNTI	Cell Radio Network Temporary Identity
CS	Circuit Switched
CS	Convergence Sublayer
CSCF	Call Server Control Function
CSD	Circuit Switched Data
CSE	Camel Service Environment
CS-GW	Circuit Switched Gateway
CSPDN	Circuit Switched Public Data Network
CTCH	Common Traffic Channel
CTD	Cell Transfer Delay
CTS	Cordless Telephony System
DAC	Digital to Analog Converter
DCA	Dynamic Channel Allocation
DCC	Data Country Code
DCCH	Dedicated Control Channel
DCE	Data Circuit terminating Equipment
DCH	Dedicated Channel
DCN	Data Communication Network
DCS1800	Digital Cellular Network at 1800 MHz
DECT	Digital Enhanced Cordless Telecommunications
DHCP	Dynamic Host Configuration Protocol
DHCP	Dynamic Host Configuration Protocol
diff-serv	Differentiated services
DL	Downlink (Forward Link)
DLCI	Data Link Connection Identifier
Dm	Control channel (ISDN terminology applied to mobile service)
DMR	Digital Mobile Radio
DNS	Domain Name Service
DPCCH	Dedicated Physical Control Channel
DPCH	Dedicated Physical Channel
DPDCH	Dedicated Physical Data Channel
DRAC	Dynamic Resource Allocation Control
DRNC	Drift Radio Network Controller
DRNS	Drift RNS
DRX	Discontinuous Reception
DS-CDMA	Direct-Sequence Code Division Multiple Access
DSCH	Downlink Shared Channel
DTAP	Direct Transfer Application Part
DTCH	Dedicated Traffic Channel
DTE	Data Terminal Equipment

DTL	Designated Transit List
DTMF	Dual Tone Multiple Frequency
DTX	Discontinuous Transmission
DXI	Data Exchange Interface
EA	External Alarms
Ec/No	Ratio of energy per modulating bit to the noise spectral density
ECSD	Enhanced CSD
EDGE	Enhanced Data rates for Global/GSM Evolution
EFCI	Explicit Forward Congestion Indication
EFR	Enhanced Full Rate
E-GGSN	Enhanced GGSN
EGPRS	Enhanced GPRS
E-HLR	Enhanced HLR
EIC	Equipment Identity Centre
EIR	Equipment Identity Register
ELAN	Emulated Local Area Network
EMC	ElectroMagnetic Compatibility
EMI	Electromagnetic Interference
EOM	End of Message
ESP	Encapsulating Security Payload
ETR	ETSI Technical Report
ETS	European Telecommunication Standard
ETSI	European Telecommunications Standards Institute
FACCH	Fast Associated Control CHannel
FACCH/F	Fast Associated Control Channel/Full rate
FACCH/H	Fast Associated Control Channel/Half rate
FACH	Forward Access Channel
FAUSCH	Fast Uplink Signalling Channel
FAX	Facsimile
FB	Frequency correction Burst
FBI	Feedback Information
FC	Feedback Control
FCCH	Frequency Correction CHannel
FCS	Frame Check Sequence
FDD	Frequency Division Duplex
FDDI	Fiber Distributed Data Interface
FDM	Frequency Division Multiplex
FDMA	Frequency Division Multiple Access
FEC	Forward Error Correction
FEC	Forwarding Equivalence Class
FER	Frame Erasure Rate, Frame Error Rate
FFS	For Further Study
FH	Frequency Hopping
FM	Fault Management
FN	Frame Number

FP	Frame Protocol
FR	Full Rate
FR	Frame-Relay
FTP	File Transfer Protocol
GCAC	Generic Connection Admission Control
GCRA	Generic Cell Rate Algorithm
GERAN	GSM/EDGE Radio Access Network
GFC	Generic Flow Control
GGSN	Gateway GPRS Support Node
GMM	GPRS Mobility Management
GMSC	Gateway MSC
GMSK	Gaussian Minimum Shift Keying
GPRS	General Packet Radio Service
GRA	GERAN Registration Area
G-RNTI	GERAN Radio Network Temporary Identity
GSIM	GSM Service Identity Module
GSM	Global System for Mobile communications
GSN	GPRS Support Nodes
GT	Global Title
GTP	GPRS Tunneling Protocol
GTP-U	GPRS Tunnelling Protocol for User Plane
GTT	Global Text Telephony
HCS	Hierarchical Cell Structure
H-CSCF	Home CSCF
HDLC	High Level Data Link Control
HEC	Header Error Control
HFN	HyperFrame Number
HHO	Hard Handover
HLR	Home Location Register
HN	Home Network
HO	Handover
HPLMN	Home Public Land Mobile Network
HPS	Handover Path Switching
HR	Half Rate
HSCSD	High Speed Circuit Switched Data
HSS	Home Subscriber Server
HTTP	Hyper Text Transfer Protocol
HTTPS	Hyper Text Transfer Protocol Secure
I/O	Input/Output
IAM	Initial Address Message
ICC	Integrated Circuit Card
ICD	International Code Designator
ICGW	Incoming Call Gateway
ICM	In-Call Modification
ICMP	Internet Control Message Protocol

ID	Identifier
IDN	Integrated Digital Network
IE	Information Element
IEC	International Electrotechnical Commission
IEEE	Institute of Electrical and Electronics Engineers
IEI	Information Element Identifier
IETF	Internet Engineering Task Force
IGRP	Interior Gateway Routing Protocol
IK	Integrity key
ILMI	Interim Link Management
IMA	Inverse Multiplexing on ATM
IMEI	International Mobile Equipment Identity
IMS	IP Multimedia Subsystem
IMSI	International Mobile Subscriber Identity
IMT-2000	International Mobile Telecommunications 2000
IN	Intelligent Network
INAP	Intelligent Network Application Part
IP	Internet Protocol
IPBCP	IP Bearer Control Protocol
IP-M	IP Multicast
IPv4	Internet Protocol Version 4
IPv6	Internet Protocol Version 6
IPX	Novell Internetwork Packet Exchange
IR	Infrared
ISDN	Integrated Services Digital Network
ISO	International Organisation for Standardisation
ISP	Internet Service Provider
ISUP	ISDN User Part
ITU	International Telecommunication Union
ITU-T	International Telecommunications Union Telecommunications
IWF	InterWorking Function
IWMSC	InterWorking MSC
IWU	Inter Working Unit
JAR file	Java Archive File
JPEG	Joint Photographic Experts Group
k	Windows size
K	Constraint length of the convolutional code
kbps	kilo-bits per second
Kc	Ciphering key
Ki	Individual subscriber authentication key
ksps	kilo-symbols per second
L1	Layer 1 (physical layer)
L2	Layer 2 (data link layer)
L3	Layer 3 (network layer)
LA	Location Area

LAC	Location Area Code
LAI	Location Area Identity
LAN	Local Area Network
LANE	LAN Emulation
LAPB	Link Access Protocol Balanced
LAPD	Link Access Procedure D
LAPDm	Link Access Protocol on the Dm channel
LAU	Location Area Update
LB	Leaky Bucket
LCP	Link Control Protocol
LCS	Location Services
LE	Local Exchange
LE	LAN Emulation
LEC	LAN Emulation Client
LECS	LAN Emulation Configuration Server
LEN	LENgth
LES	LAN Emulation Server
LI	Length Indicator
LIB	Label Information Base
LIPE	Lightweight IP Encapsulation
LLC	Logical Link Control
LNS	L2TP Network Server
LPLMN	Local PLMN
LR	Location Register
LSB	Least Significant Bit
LSP	Link State Protocol
LSR	Label Switched Router
LU	Location Update
MA	Multiple Access
MAC	Medium Access Control
MAC	Message Authentication Code
MAP	Mobile Application Part
MBS	Maximum Burst Size
MCC	Mobile Country Code
MCDV	Maximum Cell Delay Variance
MCLR	Maximum Cell Loss Ratio
Mcps	Mega-chips per second
MCR	Minimum Cell Rate
MCS	Modulation and Coding Scheme
MCTD	Maximum Cell Transfer Delay
MCU	Media Control Unit
ME	Mobile Equipment
MEGACO	Media Gateway Control Protocol
MEHO	Mobile evaluated handover
MExE	Mobile Execution Environment

MF	MultiFrame
MGCF	Media Gateway Control Function
MGCP	Media Gateway Control Part
MGCP	Media Gateway Control Protocol
MGW	Media GateWay
MIB	Management Information Base
MIP	Mobile IP
MM	Mobility Management
MMI	Man Machine Interface
MNC	Mobile Network Code
MNP	Mobile Number Portability
MO	Mobile Originated
MOHO	Mobile Originated Handover
MOS	Mean Opinion Score
MPEG	Moving Pictures Experts Group
MPLS	Multiprotocol Label Switching
MPOA	Multiprotocol over ATM
MRF	Media Resource Function
MS	Mobile Station
MSB	Most Significant Bit
MSC	Mobile Switching Centre
MSCS	MSC Server
MSID	Mobile Station Identifier
MSIN	Mobile Station Identification Number
MSISDN	Mobile Subscriber ISDN Number
MSRN	Mobile Station Roaming Number
MT	Mobile Terminated
MTP	Message Transfer Part
MTP3-B	Message Transfer Part level 3
MTU	Maximum Transfer Unit
MVNO	Mobile Virtual Network Operator
NAS	Non-Access Stratum
NBAP	Node B Application Part
NE	Network Element
NEHO	Network evaluated handover
NM	Network Manager
NMS	Network Management Subsystem
NNI	Network-Node Interface
NRT	Non-Real Time
NSAP	Network Service Access Point
NSAPI	Network Service Access Point Identifier
NSS	Network Sub System
NT	Network Termination
O&M	Operations & Maintenance
OAM	Operations Administration and Maintenance

OSI	Open System Interconnection
OSPF	Open Shortest Path First
OUI	Organizationally Unique Identifier
OVSF	Orthogonal Variable Spreading Factor
PABX	Private Automatic Branch eXchange
PACCH	Packet Associated Control Channel
PAD	Packet Assembler/Disassembler
PAD	Packet Assembler and Disassembler
PAGCH	Packet Access Grant Channel
PAP	Password Authentication Protocol
PBCCH	Packet Broadcast Control Channel
PBX	Private Branch eXchange
PC	Power Control
PCCCH	Packet Common Control Channel
PCCH	Paging Control Channel
P-CCPCH	Primary Common Control Physical Channel
PCH	Paging Channel
PCM	Pulse Code Modulation
PCPCH	Physical Common Packet Channel
P-CPICH	Primary Common Pilot Channel
PCR	Peak Cell Rate
PCS	Personal Communication System
PCU	Packet Control Unit
PD	Protocol Discriminator
PDCH	Packet Data Channel
PDCP	Packet Data Convergence Protocol
PDH	Plesiochronous Digital Hierarchy
PDN	Packet Data Network
PDP	Packet Data Protocol
PDSCH	Physical Downlink Shared Channel
PDTCH	Packet Data Traffic Channel
PDU	Protocol Data Unit
PHS	Personal Handyphone System
PHY	Physical layer
PhyCH	Physical Channel
PI	Page Indicator
PICH	Page Indicator Channel
PIN	Personal Identification Number
PLMN	Public Land Mobile Network
PMD	Physical Media Dependent
PN	Pseudo Noise
PNNI	Private Network-Network Interface
POTS	Plain Old Telephony Service
PP	Point-to-Point
PPCH	Packet Paging Channel

PPP	Point-to-Point Protocol
PRACH	Physical Random Access Channel
PRI	Primary Rate Interface
PS	Packet Switched
PSC	Primary Synchronization Code
PSCH	Physical Shared Channel
PSPDN	Packet Switched Public Data Network
PSTN	Public Switched Telephone Network
PTCCH	Packet Timing advance Control Channel
PTI	Payload Type Indicator
PTM	Point-to-Multipoint
P-TMSI	Packet TMSI
PTP	Point to point
PU	Payload Unit
PUSCH	Physical Uplink Shared Channel
PVC	Permanent Virtual Circuit
PVCC	Permanent Virtual Channel Connection
QoS	Quality of Service
QPSK	Quadrature (Quaternary) Phase Shift Keying
R4	Release 4
R5	Release 5
R99	Release 1999
RA	Routing Area
RAB	Radio Access Bearer
RAC	Routing Area Code
RACH	Random Access Channel
RADIUS	Remote Authentication Dial In User Service
RADIUS	Remote Authentication Dial-In User Service
RAI	Routing Area Identity
RAN	Radio Access Network
RANAP	Radio Access Network Application Part
RAND	RANDom number (used for authentication)
RAT	Radio Access Technology
RAU	Routing Area Update
RB	Radio Bearer
RBER	Residual Bit Error Ratio
RDF	Resource Description Format
RED	Random Early Detect
REJ	REJect(ion)
REL	RELease
Rel-4	Release 4
Rel-5	Release 5
REQ	REQuest
RF	Radio Frequency
RFC	Request For Comments

RFCH	Radio Frequency CHannel
RFN	Reduced TDMA Frame Number
RIP	Routing Information Protocol
RL	Radio Link
RLC	Radio Link Control
RLCP	Radio Link Control Protocol
RLP	Radio Link Protocol
RM	Resource Management
RNC	Radio Network Controller
RNS	Radio Network Subsystem
RNSAP	Radio Network Subsystem Application Part
RNTI	Radio Network Temporary Identity
RPLMN	Registered Public Land Mobile Network
RR	Radio Resources
RRC	Radio Resource Control
RRM	Radio Resource Management
RSCP	Received Signal Code Power
R-SGW	Roaming Signalling Gateway
RSSI	Received Signal Strength Indicator
RST	Reset
RSVP	Resource Reservation Protocol
RSVP	Resource ReserVation Protocol
RT	Real Time
RTCP	Real Time Control Protocol
RTE	Remote Terminal Emulator
RTP	Real Time Protocol
RU	Resource Unit
RX	Receive
RXLEV	Received signal level
RXQUAL	Received Signal Quality
SAAL	Signalling ATM Adaptation Layer
SABM	Set Asynchronous Balanced Mode
SACCH	Slow Associated Control Channel
SAP	Service Access Point
SAPI	Service Access Point Identifier
SAR	Segmentation and Reassembly
SAT	SIM Application Toolkit
SB	Synchronization Burst
SBSC	Serving Base Station Controller
SBSS	Serving Base Station Subsystem
SC	Service Centre (used for SMS)
SCCH	Synchronization Control Channel
SCCP	Signalling Connection Control Part
S-CCPCH	Secondary Common Control Physical Channel
SCH	Synchronization Channel

SCN	Sub-Channel Number
SCP	Service Control Point
S-CPICH	Secondary Common Pilot Channel
SCR	Sustainable Cell Rate
S-CSCF	Serving CSCF
SCTP	Streaming Control Transport Protocol
SDCCH	Stand-Alone Dedicated Control Channel
SDES	Source Descriptor
SDH	Synchronous Digital Hierarchy
SDT	Structured Data Transfer
SDU	Service Data Unit
SEAL	Simple and Efficient Adaptation Layer
SF	Spreading Factor
SFH	Slow Frequency Hopping
SFN	System Frame Number
SGSN	Serving GPRS Support Node
SHCCH	Shared Channel Control Channel
SID	SIlence Descriptor
SIM	GSM Subscriber Identity Module
SIP	Session Initiation Protocol
SIR	Signal-to-Interference Ratio
SLA	Service Level Agreement
SLIP	Serial Line Interface Protocol
SM	Session Management
SMDS	Switched Multimegabit Data Service
SMS	Short Message Service
SMS/PP	Short Message Service/Point-to-Point
SMS-CB	SMS Cell Broadcast
SMS-SC	Short Message Service - Service Centre
SN	Serving Network
SN	Sequence Number
SNDCP	Sub-Network Dependent Convergence Protocol
SNMP	Simple Network Management Protocol
SONET	Synchronous Optical Network
SPC	Signalling Point Code
SQN	Sequence number
SRB	Signalling Radio Bearer
SRES	Signed RESponse (authentication)
SRNC	Serving Radio Network Controller
SRNS	Serving RNS
S-RNTI	SRNC Radio Network Temporary Identity
SS7	Signalling System No. 7
SSCF	Service Specific Co-ordination Function
SSCF-NNI	Service Specific Coordination Function – Network Node Interface
SSCOP	Service Specific Connection Oriented Protocol

SSCS	Service Specific Convergence Sublayer
SSDT	Site Selection Diversity Transmission
SSN	Sub-System Number
SSSAR	Service Specific Segmentation and Re-assembly sublayer
STC	Signalling Transport Converter
STP	Signalling Transfer Point
STTD	Space Time Transmit Diversity
SVC	Switched Virtual Circuit
SVCC	Switched Virtual Circuit Connection
T	Transparent
TA	Timing Advance
TBF	Temporary Block Flow
TBR	Technical Basis for Regulation
TC	TransCoder
TC	Transmission Convergence
TCH	Traffic Channel
TCP	Transmission Control Protocol
TDD	Time Division Duplex
TDM	Time Division Multiplexing
TDMA	Time Division Multiple Access
TD-SCDMA	Time Division-Synchronous Code Division Multiple Access
TE	Terminal Equipment
TEID	Tunnel End Point Identifier
TF	Transport Format
TFC	Transport Format Combination
TFCI	Transport Format Combination Indicator
TFCS	Transport Format Combination Set
TFI	Transport Format Indicator
TFI	Temporary Flow Identity
TFS	Transport Format Set
TFT	Traffic Flow Template
TLLI	Temporary Logical Link Identity
TLS	Transport Layer Security
TLV	Tag Length Value
TM	Telecom Management
TM	Traffic Management
TMSI	Temporary Mobile Subscriber Identity
TOA	Time of Arrival
TPC	Transmit Power Control
TPDU	Transfer Protocol Data Unit
TR	Technical Report
TRAU	Transcoder and Rate Adapter Unit
TrCH	Transport Channel
TRX	Transceiver
TS	Technical Specification

TS	Time Slot
TSC	Training Sequence Code
TSG	Technical Specification Group
TTI	Transmission Timing Interval
TUP	Telephone User Part (SS7)
TX	Transmit
UARFCN	UTRA Absolute Radio Frequency Channel Number
UARFN	UTRA Absolute Radio Frequency Number
UART	Universal Asynchronous Receiver and Transmitter
UBR	Unspecified Bit Rate
UDP	User Datagram Protocol
UE	User Equipment
UI	User Interface
UIC	Union Internationale des Chemins de Fer
UICC	Universal Integrated Circuit Card
UL	Uplink (Reverse Link)
UM	Unacknowledged Mode
UMTS	Universal Mobile Telecommunications System
UNI	User-Network Interface
UP	User Plane
URA	UTRAN Registration Area
URAN	UMTS Radio Access Network
URB	User Radio Bearer
URI	Uniform Resource Identifier
URL	Uniform Resource Locator
U-RNTI	UTRAN Radio Network Temporary Identity
USB	Universal Serial Bus
USC	UE Service Capabilities
USCH	Uplink Shared Channel
USF	Uplink State Flag
USIM	Universal Subscriber Identity Module
USSD	Unstructured Supplementary Service Data
UT	Universal Time
UTOPIA	Universal Test & Operations PHY interface for ATM
UTP	Unshielded Twisted Pair
UTRA	Universal Terrestrial Radio Access
UTRAN	Universal Terrestrial Radio Access Network
UUI	User-to-User Information
VA	Voice Activity factor
VAD	Voice Activity Detection
VASP	Value Added Service Provider
VBR	Variable Bit Rate
VC	Virtual Circuit
VC	Virtual Circuit/Channel
VCI	Virtual Channel Identifier

VHE	Virtual Home Environment
VLAN	Virtual Local Area Network
VLR	Visitor Location Register
VMSC	Visited MSC
VoIP	Voice Over IP
VP	Virtual Path
VPI	Virtual Path Identifier
VPLMN	Visited Public Land Mobile Network
VPN	Virtual Private Network
VRRP	Virtual Router Redundancy Protocol
WAE	Wireless Application Environment
WAN	Wide Area Network
WAP	Wireless Application Protocol
WCDMA	Wideband Code Division Multiple Access
WDP	Wireless Datagram Protocol
WG	Working Group
WRED	Weighted Random Early Detect
WSP	Wireless Session Protocol
WTLS	Wireless Transport Layer Security
WWW	World Wide Web
XRES	EXpected user RESponse

Index

Convergence Technologies for 3G Networks: IP, UMTS, EGPRS and ATM   J. Bannister, P. Mather and S. Coope
© 2004 John Wiley & Sons, Ltd   ISBN: 0-470-86091-X